BIODIVERSITY OF PANTEPUI

BIODIVERSITY OF PANTEPUI

The Pristine "Lost World" of the Neotropical Guiana Highlands

Edited by

Valentí Rull
Institute of Earth Sciences Jaume Almera (ICTJA-CSIC), Barcelona, Spain

Teresa Vegas-Vilarrúbia
Department of Evolutionary Biology, Ecology and Environmental Sciences, University of Barcelona, Barcelona, Spain

Otto Huber
Botanical Institute of Venezuela Foundation, Caracas, Venezuela

Celsa Señaris
Center of Ecology, Venezuelan Institute for Scientific Research (IVIC), Caracas, Venezuela

ELSEVIER

ACADEMIC PRESS
An imprint of Elsevier

Academic Press is an imprint of Elsevier
125 London Wall, London EC2Y 5AS, United Kingdom
525 B Street, Suite 1650, San Diego, CA 92101, United States
50 Hampshire Street, 5th Floor, Cambridge, MA 02139, United States
The Boulevard, Langford Lane, Kidlington, Oxford OX5 1GB, United Kingdom

British Library Cataloguing-in-Publication Data
A catalogue record for this book is available from the British Library

Library of Congress Cataloging-in-Publication Data
A catalog record for this book is available from the Library of Congress

ISBN: 978-0-12-815591-2

For Information on all Academic Press publications
visit our website at https://www.elsevier.com/books-and-journals

Publisher: Charlotte Cockle
Acquisition Editor: Anna Valutkevich
Editorial Project Manager: Ruby Smith
Production Project Manager: Kiruthika Govindaraju
Cover Designer: Limbert Matthew

Typeset by MPS Limited, Chennai, India

Contents

12. Amphibians and reptiles

CELSA SEÑARIS AND FERNANDO J.M. ROJAS-RUNJAIC

13. Birds

JORGE L. PÉREZ-EMÁN, MIGUEL LENTINO AND ELISA BONACCORSO

14. Mammals

DANIEL LEW AND BURTON K. LIM

15. Vertebrate parasites

RICARDO GUERRERO

Part IV

BIODIVERSITY CONSERVATION

16. Conservation of Pantepui: between complex emergency and climate change

MARIAPIA BEVILACQUA, CELSA SEÑARIS AND OTTO HUBER

17. Pantepui and global warming

VALENTÍ RULL, SANDRA NOGUÉ, ELISABET SAFONT AND TERESA VEGAS-VILARRÚBIA

List of contributors

Paul E. Berry University of Michigan Herbarium, Department of Ecology and Evolutionary Biology, Ann Arbor, MI, United States

Mariapia Bevilacqua Venezuelan Association for the Conservation of Natural Areas - ACOANA, Caracas, Venezuela

Elisa Bonaccorso Institute BIOSFERA and Laboratory of Evolutionary Biology, University San Francisco de Quito, Quito, Ecuador

Abraham S.H. Breure Naturalis Biodiversity Center, RA Leiden, The Netherlands; Royal Belgian Institute of Natural Sciences, Brussels, Belgium

Charles Brewer Carías Venezuelan Society for Natural Sciences, Caracas, Venezuela

Mauro Costa Museum of the Institute of Agricultural Zoology (MIZA), Faculty of Agronomy, Central University of Venezuela, Maracay, Venezuela

Tomáš Derka Department of Ecology, Faculty of Natural Sciences, Comenius University, Bratislava, Slovakia

Karolina Fučíková Assumption College, Worcester, MA, United States

Ricardo Guerrero Institute of Tropical Zoology, Faculty of Sciences, Central University of Venezuela, Caracas, Venezuela

Otto Huber Trauttmansdorff Botanical Gardens, Merano, Italy; Botanical Institute of Venezuela Foundation, Caracas, Venezuela

Burton K. Lim Department of Natural History, Royal Ontario Museum (ROM), Toronto, ON, Canada

Jan Kaštovský University of South Bohemia, České Budějovice, Czech Republic

Miguel Lentino Phelps Ornithological Collection, Caracas, Venezuela

Daniel Lew Biodiversity Unit, Venezuelan Institute of Scientific Research (IVIC), Caracas, Venezuela

Fabián A. Michelangeli The New York Botanical Garden, Bronx, NY, United States

Encarni Montoya Institute of Earth Sciences Jaume Almera (ICTJA-CSIC), Barcelona, Spain

Sandra Nogué Department of Geography and Environment, University of Southampton, Southampton, United Kingdom

José A. Ochoa Faculty of Sciences, Department of Biology, National University of San Antonio Abad del Cusco, Cusco, Peru

Jorge L. Pérez-Emán Institute of Tropical Zoology and Ecology, Central University of Venezuela, Caracas, Venezuela; Phelps Ornithological Collection, Caracas, Venezuela

Ricarda Riina Royal Botanical Garden, RJB-CSIC, Madrid, Spain

Fernando J.M. Rojas-Runjaic La Salle Natural History Museum, La Salle Natural Sciences Foundation, Caracas, Venezuela; Laboratory of Vertebrate Systematics, Zoology Postgraduate Program, School of Sciences, Pontifical Catholic University of Rio Grande do Sul (PUCRS), Porto Alegre, Brazil

Valentí Rull Institute of Earth Sciences Jaume Almera (ICTJA-CSIC), Barcelona, Spain

Elisabet Safont Department of Evolutionary Biology, Ecology and Environmental Sciences, Faculty of Biology, University of Barcelona, Barcelona, Spain

Celsa Señaris Laboratory of Population Genetics, Center of Ecology, Venezuelan Institute for Scientific Research (IVIC), Caracas, Venezuela

José Manuel Tierno de Figueroa Department of Zoology, Faculty of Sciences, University of Granada, Granada, Spain

Teresa Vegas-Vilarrúbia Department of Evolutionary Biology, Ecology and Environmental Sciences, Faculty of Biology, University of Barcelona, Barcelona, Spain

Jana Veselá University of South Bohemia, České Budějovice, Czech Republic

Ángel L. Viloria Center of Ecology, Venezuelan Institute for Scientific Research (IVIC), Altos de Pipe, Venezuela

Carmen Zamora-Muñoz Department of Zoology, Faculty of Sciences, University of Granada, Granada, Spain

Foreword

Eastern tepui chain. *Photo: V. Rull.*

Having visited tepuis in Venezuela, Guyana, and Brazil, I am fully aware of the biological importance of the Pantepui region and of the relevance of this book. Each mountain there is an isolated island where plants and animals are also relatively isolated from their congeners and so may evolve into separate species. The region is famous for the endemicity of many of the species of the organisms that occur there. It has been the joy and privilege of many naturalists to discover new species as they explore the tops of these mountains. On two expeditions to the isolated Brazilian tepui of Aracá in 1984 and 1985 we discovered 25 new species and one new genus of plants on that mountain, and this has been the experience of many other explorers of Pantepui.

The early biological work that stimulated interest in Pantepui was mainly from birds and plants. Here we see that now a lot of other disciplines have been involved in biological research in the region. I am glad that this volume brings together data from a large number of different fields and about such a wide range of organisms from algae to mammals and higher plants. The published information on the various organisms that inhabit Pantepui is very scattered and often hard to access. It is so good summarize all this information together in a single volume. This book will be a reference volume that unites a lot of scattered information ranging from paleoecology to the most modern phylogenetic and evolutionary concepts. There is a lot of information here, but it is still a fact that the majority of the tepuis remain virtually unexplored. I hope that this book will challenge future researchers to find out many more of the secrets of this fascinating region. It certainly serves to show that Pantepui is a dynamic biogeographical concept.

Thankfully the Pantepui region is still relatively pristine, and this book shows its biological and evolutionary importance. It will certainly help to support the efforts to preserve the region and keep it pristine for future generations, and so I am most happy to see a final section on the important topic of biodiversity conservation and climate change.

Sir Ghillean T. Prance

Acknowledgments

Tirepón-tepui (Chimantá massif). *Photo: V. Rull.*

The editors are very grateful to all the authors who contributed to this book for their dedication and hard work leading to high-quality chapters. We also wish to thank the reviewers for their time and expertise. They are (in alphabetic order) Alexandre Antonelli, Gerardo Aymard, Paul Berry, Elisa Bonaccorso, Israel Cañizares, Jürgen Demarmels, Carl Dick, Mark Engstrom, Ángel Fernández, José María Fernández-Palacios, Suzette Flantua, Sherilyn Fritz, Stefan Gorzula, Marinus Hoogmoed, Bruce Host, Carlos Lasso, Daniel Lew, Burton Lim, Célio Magalhães, Fabián Michelangeli, Shinichi Nakahara, Jorge Pérez-Emán, Lorenzo Prendini, Gustavo Romero, Donald Taphorn, Rodrigo Salvador, Luiz Ricardo Lopes de Simone, Nathan Stansell, Fred Stauffer, and Zicheng Yu. We are also very grateful to people who provided photographs for free, especially Javier Mesa, and also Stefan Gorzula. We greatly acknowledge the invitation of Elsevier/Academic Press for writing/editing the book and the managing editors Kiruthika Govindaraju, Ruby Smith and Anna Valutkevicth for their constant support in editorial and production tasks.

Introduction

Valentí Rull[1], *Teresa Vegas-Vilarrúbia*[2], *Otto Huber*[3] *and Celsa Señaris*[4]

[1]Institute of Earth Sciences Jaume Almera (ICTJA-CSIC), Barcelona, Spain [2]Department of Evolutionary Biology, Ecology and Environmental Sciences, University of Barcelona, Barcelona, Spain [3]Botanical Institute of Venezuela Foundation, Caracas, Venezuela [4]Center of Ecology, Venezuelan Institute for Scientific Research (IVIC), Caracas, Venezuela

To our knowledge, this is the first attempt to summarize the existing knowledge on the Pantepui biota, including most plant and animal groups studied to date. The available information on the biodiversity of Pantepui is dispersed in hundreds of books and articles, including gray literature, and a synthesis of the state of the art in this matter seemed pertinent. Knowing the extant biodiversity patterns and the potential environmental drivers and evolutionary mechanisms involved in their generation and maintenance is essential, not only for the progress of fundamental science but also for biodiversity conservation (Rull, 2019). Unfortunately, such knowledge is still very limited for Pantepui, and only a preliminary assessment can be offered due to incomplete exploration. However, it is hoped that this book will provide a basis to advance the ecological and evolutionary study of the unique Pantepui biota and will help set the basis for its conservation.

Aims and scope

The main aims of this book are (1) to summarize the present state of knowledge on the extant biodiversity of the Pantepui biogeographical province of the Guiana region, (2) to provide insights on relevant biogeographical and endemism patterns, (3) to initiate discussions on the potential origin and evolution of the unique Pantepui biota as a whole, and (4) to evaluate the main current and potential threats to Pantepui biodiversity and to envisage some potential conservation actions. The peculiar physiographic features of these spectacular, remote, and nearly pristine Guiana Highlands and their geological origins have already been highlighted in scientific and popular literature, movies, and websites. Here the emphasis is on the biotic component.

The terms "Pantepui" and "Guiana Highlands" (or Guayana Highlands, to refer specifically to those in the Venezuelan territory) are often used as synonyms in the literature. Although the first is a biogeographical term and the second a physiographical one, both allude to the summits of the characteristic table mountains (tepuis) above ~1500 elevation, which is the lower boundary defined for the highlands of the Guiana region (Huber, 1995). The first chapter of this book is more explicit in this respect. For a more extended account on geological, geographical, physiographical, and climatic aspects, we recommend the first volume of the *Flora of the Venezuelan Guayana*, edited by Berry et al. (1995). The long-term geomorphological evolution of the Pantepui area leading to its present topographical configuration was summarized by Briceño and Schubert (1990) and Huber and García (2011). The history of scientific exploration of Pantepui is a fascinating subject (Huber, 1995; McDiarmid and Donnelly, 2005) that requires a monography by itself and is beyond the scope of this book.

The lack of autoecological and synecological studies developed on Pantepui organisms and ecosystems prevents us from providing more detailed ecological information on the unique biological communities of this biogeographical province. Therefore for the time being, a book on Pantepui biota should necessarily be focused on taxonomic and biogeographic aspects and the eventual evolutionary insights derived from them. It is hoped that this will change in the near future, and a book like this could provide the basis for further studies focused on specific ecological and evolutionary topics. The book has been organized into four main parts, namely (1) general aspects, (2) the photosynthetic world, (3) animal biodiversity, and (4) biodiversity conservation.

General aspects

The first part of the book presents the general aspects of Pantepui biodiversity and is composed of four chapters. The first chapter (Rull et al.) defines explicitly the Pantepui biogeographical province, mainly on the basis of vascular plant biogeography, and includes a list of all the tepuis and other highlands falling within the definition of Pantepui, as well as a location map for all of them. To avoid repetitions and eventual geographical inconsistencies, this map is the reference for all the remaining chapters. The second chapter (Rull et al.) is a summary of paleoecological works developed to date on Pantepui in an attempt to understand the ecological state and composition of modern Pantepui ecosystems in light of the environmental shifts experienced since the Last Glacial Maximum, which has been recognized worldwide as the resetting point for modern living communities. Chapter 3 (Rull & Vegas-Vilarrúbia) attempts to show that Pantepui is not a static biogeographical concept as it has changed through time, especially during the Quaternary, due to the significant ecological changes caused by the glacial–interglacial recurrence. The final chapter of this section (Rull) is a an attempt to unravel how the Pantepui biota has originated and evolved since the Mesozoic until its present conditions, as well as the geological and environmental drivers that have influenced this process.

The photosynthetic world

The second part is focused on algae and plant diversity and biogeography and consists of three chapters. The first chapter (Kaštovský et al.) is about Cyanobacteria and Eukaryotic algae inhabiting several types of extreme habitats with relatively uniform, yet very specific, algal floras across all investigated summits. The role of these organisms as founders and promoters of more complex pioneer communities on rocky substrates is highlighted. The second chapter (Riina et al.) updates the main patterns of diversity and endemism of the Pantepui flora, including mosses and vascular plants, and discusses the existing knowledge about the phylogenetics of some plant lineages to provide insights in the evolution and historical biogeography of the unique Pantepui flora. The third chapter (Huber and Rull) provides an overview of the vegetation cover of Pantepui in terms of plant communities. Forty vegetation types are described and classified into four formations: forests, shrublands, herbaceous, and pioneer communities. The occurrence of some types of vegetation endemic to Pantepui is also highlighted.

Animal diversity

The third part of the book deals with animal biodiversity and includes the major taxonomic groups present in Pantepui that have been studied to date. Other groups may occur, but they are rare or studies are inexistent or still embryonic. The case of fishes is remarkable, as no highland species have been found and lowland/upland species rarely cross the lower elevational boundary of Pantepui, with very few exceptions (Lasso et al., 1989).

This section has been subdivided into eight chapters organized taxonomically. The first chapter (Derka et al.) summarizes the available information on the aquatic insect fauna, mainly mayflies, stoneflies, and caddisflies, which are confined to the nutrient-poor waters of the Guiana Shield. The second chapter (Viloria and Costa) is concerned with the Pantepui-endemic butterflies relevant for the definition of the Pantepui biogeographic province within the Neotropical Amazonian subregion. The next chapter (Ochoa and Rojas-Runjaic) summarizes the scorpion fauna of Pantepui and peripheral areas of the Guiana region, which is composed almost exclusively of endemic species, with species richness decreasing with elevation. Noteworthy is the occurrence of four endemic genera. The fourth chapter (Breure) updates knowledge about land snail fauna, mostly from the eastern Pantepui area, which shows a high level of endemism. Some preliminary ecological and biogeographical insights are provided. It follows a chapter (Señaris and Rojas-Runjaic) on the Pantepui herpetofauna (amphibians and reptiles), a remarkable group with many endemisms of very restricted distribution. This chapter updates the richness and endemism patterns of this group and summarizes the existing molecular phylogenetic studies, which provide insights on the origin and evolution of the Pantepui herpetofauna.

The next chapter (Pérez-Emán et al.) is on birds, one of the better-known groups of organisms that had provided the basis for the first definition of Pantepui. A review of the new information on bird distribution furnishes the basis for understanding their current biogeographical patterns and for inferring the potential processes that have generated their diversity and endemism on Pantepui. The chapter on mammals (Lew and Lim) summarizes the richness of this group, which in Pantepui is represented mainly by bats and rodents. In contrast with other zoological groups, endemism is low, with only one rodent species restricted to a tepuian summit (although some opossums and rodents are exclusively from tepuian slopes). The final chapter (Guerrero) of the zoological part is a brief preliminary account of the ectoparasites (mainly insects and acari) and endoparasites (chiefly worms) found on Pantepui vertebrates, especially mammals. The scarce information available prevents sound inferences on biogegraphy and endemism, but it is believed that parasites are largely dependent on the geographical patterns of their respective hosts.

Biodiversity conservation

Finally, the fourth part of the book is about the direct and indirect threats to Pantepui biodiversity and the more suitable conservation actions to avoid or palliate eventual extinction. The first chapter (Bevilacqua et al.) explores why the performance and commitments of regional governments towards the practice of conservation in the highlands of the Guiana Shield have been insufficient and discusses the complex emergency of climate change as new threats to Pantepui biodiversity. The final chapter of the book (Rull et al.) reports the GIS-based estimates of potential extinction by habitat loss of the Pantepui vascular flora, given the influence of the global warming predicted for the end of this century. Further ecological and genetic studies of the most endangered species needed to refine these estimates are suggested. This would require national actions and international collaboration, especially among Guiana Shield countries, including a thorough revision of the process for obtaining official field-work permits, which is now the most important handicap in terms of scientific study.

References

Berry, P., Holst, B.K., Yaskievych, K., 1995. Flora of the Venezuelan Guayana, Introduction, vol. 1. Missouri Botanical Garden Press, St. Louis.

Briceño, H.O., Schubert, C., 1990. Geomorphology of the Gran Sabana, Guayana Shield, southeastern Venezuela. Geomorphology 3, 125–141.

Huber, O., 1995. Geographical and physical features. In: Berry, P., Holst, B.K., Yaskievych, K. (Eds.), Flora of the Venezuelan Guayana. Introduction, vol. 1. Missouri Botanical Garden Press, St. Louis, pp. 1–61.

Huber, O., García, P., 2011. The Venezuelan Guayana region and the study areas: geo-ecological characteristics. In: Zinck, J.A., Huber, O. (Eds.), Peatlands of the Western Guayana Highlands, Venezuela, Ecological Studies 217. Springer-Verlag, Berlin Heidelberg, pp. 29–89.

Lasso, C., Machado-Allison, C., Pérez, R., 1989. Consideraciones zoogeográficas del los peces de la Gran Sabana (Alto Caroní) Venezuela, y sus relaciones con las cuencas vecinas. Mem. Soc. Cien. Nat. 49, 109–129.

McDiarmid, R.W., Donnelly, M.A., 2005. The herpetofauna of the Guayana Highlands: amphibiands and reptiles of the Lost World. In: Donnelly, M.A., Crother, B.I., Guyer, C., Marvalee, H., White, M.E. (Eds.), Ecology and Evolution in the Tropics. University Chicago Press, Chicago, pp. 461–560.

Rull, V., 2019. Neotropical diversification: historical overview and conceptual insightsIn: Rull, V., Carnaval, A. (Eds.), Neotropical Diversification. Springer, Berlin. Peer J Preprints 6: e27294v1. Available from: http://doi.org/10.7287/peerj.preprints.27294v1.

PART I

General Aspects

CHAPTER

1

Definition and characterization of the Pantepui biogeographical province

Valentí Rull[1], Otto Huber[2], Teresa Vegas-Vilarrúbia[3] and Celsa Señaris[4]

[1]Institute of Earth Sciences Jaume Almera (ICTJA-CSIC), Barcelona, Spain [2]Botanical Institute of Venezuela Foundation, Caracas, Venezuela [3]Department of Evolutionary Biology, Ecology and Environmental Sciences, Faculty of Biology, University of Barcelona, Barcelona, Spain [4]Laboratory of Population Genetics, Center of Ecology, Venezuelan Institute for Scientific Research (IVIC), Caracas, Venezuela

Introduction

Pantepui has been defined as a discontinuous biogeographical province of the Guiana region north of the Neotropical realm. The Guiana region is surrounded by the Caribbean and the Amazon regions and is subdivided into four provinces: the Eastern Guiana province, the Western Guiana province, the Central Guiana province, and the Pantepui province (Fig. 1.1). The Guiana region lies on the Guiana Shield (Fig. 1.2A), one of the oldest geological areas of South America, which forms an igneous-metamorphic basement of Archaeo-Proterozoic granites and gneisses. On this basement, there is a cover of Precambrian sandstones and quartzites—the Roraima Group—on which the typical tabular Guiana mountains have been modeled. These tabular mountains, locally known as *tepuis*, are the remnants of a long-standing erosional process starting probably in the Mesozoic (Briceño and Schubert, 1990) (see Fig. 4.4 of Chapter 4: Origin and evolution of the Pantepui biota). The basic elements of a typical tepui are a basal slope emerging from the surrounding lowlands/uplands, a vertical cliff, and a largely flat summit in the highlands (Fig. 1.3). Different combinations and compositions of these elements, as well as clusters of this type of mountains (tepuian massifs), also exist (Plates 1.1–1.12).

Pantepui is the assemblage of the tepui summits, which form a highly fragmented surface. This assemblage has been recognized as an archipelago of sky islands, that is, high-elevation habitats that are geographically subdivided and isolated among different

Biodiversity of Pantepui
DOI: https://doi.org/10.1016/B978-0-12-815591-2.00001-X

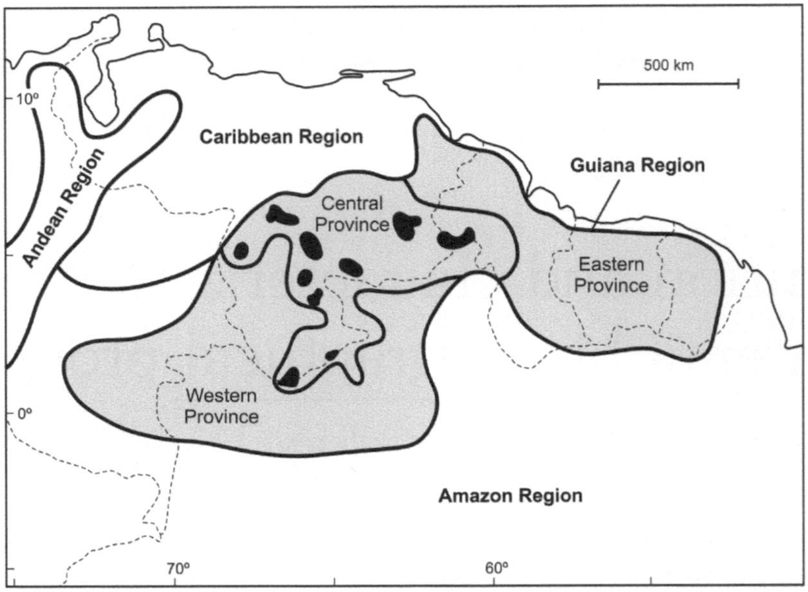

FIGURE 1.1 Map of northern South America showing the Guiana region (*gray area*) and its four biogeographical provinces. The Pantepui province is represented as a discontinuous surface of black spots representing the major tepuis and tepuian massifs. *Source: Redrawn from Huber, O., 1994. Recent advances in the phyotgeography of the Guayana region, South America. Mém. Soc. Biogéogr. (3ème série) IV, 53–63.*

mountain ranges and that are topographically separated by the surrounding uplands and lowlands (McCormack et al., 2009). Due to their remoteness, difficult access, and lack of natural resources to exploit, the tepuian summits remain virtually pristine, although tourist activities are increasing (Huber, 1995a) and might be a threat to biodiversity in the near future (Rull et al., 2016). The expression "Lost World" comes from the famous fantastic novel of the same title published by Sir Arthur Conan Doyle in 1912, which was inspired by the Pantepui landscape, probably the Roraima-tepui (Plate 1.1). Some examples of the unique and spectacular nature of this landscape can be seen in Plates 1.1–1.12. The tepuis and tepuian massifs, as well as other nontabular mountains with summits that fall within the Pantepui province, are depicted in Fig. 1.2B, which is the geographical reference for the whole book. The main features of these mountains are available in Table 1.1, where it can be seen that most of the tepuis are situated in Venezuela, with few representatives in Guyana and Brazil.

Life on Pantepui is also exceptional. The biota and ecosystems of the tepuian summits exhibit very special, often unique, features and a high level of endemism, which has been the basis for the definition of Pantepui as a separate biogeographical province within the Guiana region. To document the variety of life and its geographical patterns of the Pantepui province and their possible evolutionary origin is the main aim of this book. In this chapter, we will focus on describing the main physiographic, environmental, and biotic traits used to define Pantepui as a separate biogeographical unit. This chapter is neither a thorough review nor a formal assessment on Pantepui as a biogeographical province, but a summary description to set the stage for the reader to realize the framework of the book. Further chapters will be more specific with regard to their respective taxonomic

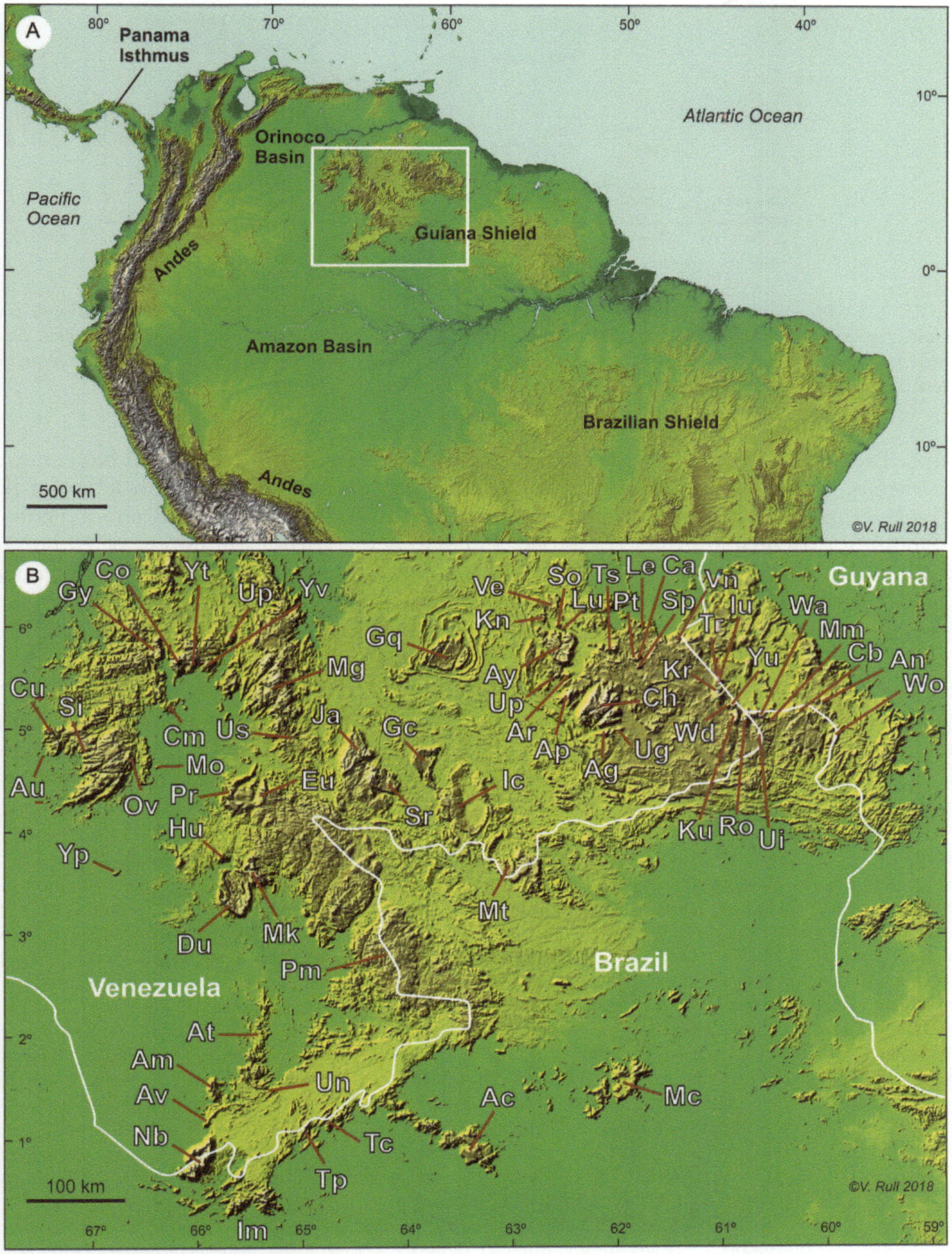
A
80°
70°
60°
50°
40°
Panama Isthmus
Orinoco Basin
Atlantic Ocean
10°
Pacific Ocean
Andes
Guiana Shield
0°
Amazon Basin
Brazilian Shield
10°
Andes
500 km
©V. Rull 2018
B
Gy
Co
Yt
Up
Yv
Ve
So
Ts
Le
Ca
Vn
Guyana
Lu
Pt
Sp
Iu
Kn
Tr
Wa
Mm
Gq
Mg
Ay
Kr
Yu
Cb
An
Wo
Cu
Si
Up
Ch
Ja
Gc
Cm
Us
Ar
Wd
Ap
Ug
Ag
Mo
Eu
Ic
Au
Pr
Ov
Sr
Hu
Ku
Ro
Ui
Yp
Mt
Du
Mk
Pm
Brazil
Venezuela
At
Am
Un
Ac
Mc
Av
Nb
Tc
Tp
100 km
©V. Rull 2018
Im
67°
66°
65°
64°
63°
62°
61°
60°
59°
6°
5°
4°
3°
2°
1°

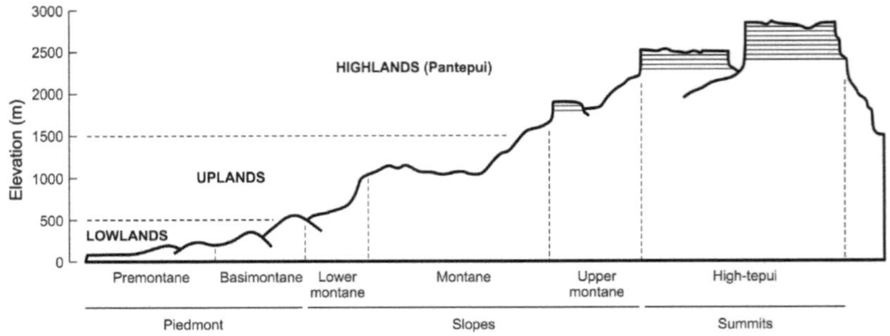

FIGURE 1.3 Physiographic sketch of the Guiana region around Pantepui indicating the elevational stages (lowlands, uplands, highlands) and the altitudinal sequence of life zones (piedmont, slopes, summits). *Source: Redrawn from Huber, O., 1995c. Vegetation. In: Berry, P., Holst, B.K., Yaskievych, K. (Eds.), Flora of the Venezuelan Guayana, Introduction. vol. 1. Missouri Botanical Garden Press, St. Louis, pp. 97–160.*

groups. This chapter begins with a brief account of the origin of the Pantepui biogeographical concept. Then, the main physical, environmental, and biotic (mainly in terms of flora and vegetation) features of Pantepui are described. The biogeographical subdivisions of the Pantepui province are then presented. Finally, as the current definition of Pantepui is based mainly on phytogeographical criteria, some comments on the different zoogeographical views are summarized.

Some terminological notes seem pertinent to facilitate reading. The term "Guayana" (Guayana Shield, Guayana region) is often used instead of "Guiana" mostly in the Spanish but also in the English literature, especially in reference to the Venezuelan part of this physiographic region. These terms are equivalent and can be used interchangeably (Berry et al., 1995a). In this book, the term "Guiana" is adopted to avoid confusion. The term "Guyana" refers to the country situated between Venezuela and Suriname. The terms

◀ FIGURE 1.2 Topographic maps of the studied region. (A) Northern South America, with the main geographical and physiographic features. The Guiana Highlands are indicated by a white box. (B) The Guiana Highlands, with the location of its tepuis and tepuian massifs (Table 1.1). *Ac*, Serra do Acará; *Ag*, Angasima-tepui; *Am*, Cerro Arakamuni; *An*, Mount Ayanganna; *Ap*, Aprada-tepui; *Ar*, Araopán-tepui; *At*, Cerro Aratitiyope; *Au*, Cerro Autana; *Av*, Cerro Avispa; *Ay*, Auyán-tepui; *Ca*, Carrao-tepui; *Cb*, Monte Caburaí; *Ch*, Chimantá massif (summits of undifferentiated tepuis: Abakapá, Akopán, Agparamán, Amurí, Apakará, Chimantá, Churí, Eruoda/Murey, Tirepón, Toronó); *Cm*, Cerro Camani; *Co*, Coro Coro; *Cu*, Cerro Cuao; *Du*, Cerro Duida; *Eu*, Cerro Euaja; *Gc*, Cerro Guanacoco; *Gq*, Cerro Guaiquinima; *Gy*, Cerro Guanay; *Hu*, Cerro Huachamakari; *Ic*, Cerro Ichún; *Im*, Sierra Imerí; *Iu*, Ilú-tepui; *Ja*, Cerro Jaua; *Kn*, Kurún-tepui; *Kr*, Karaurín-tepui; *Ku*, Kukenán (Matauí)-tepui; *Le*, Sierra de Lema; *Lu*, Cerro La Luna; *Mc*, Serra da Mocidade; *Mg*, Sierra de Maigualida; *Mk*, Cerro Marahuaka; *Mm*, Mount Maringma; *Mo*, Cerro Moriche; *Mt*, Sierra Marutaní (Pia-Zoi); *Nb*, Sierra de la Neblina; *Ov*, Cerro Ovaña (Ouana); *Pm*, Sierra de Parima; *Pr*, Cerro Parú (A'roko); *Pt*, Ptari-tepui; *Ro*, Roraima-tepui; *Si*, Cerro Sipapo; *So*, Cerro El Sol; *Sp*, Sororopán-tepui; *Sr*, Cerro Sarisariñama; *Tc*, Cerro Tamacuari; *Tp*, Serranía Tapirapecó; *Tr*, Tramen-tepui; *Ts*, Los Testigos massif (Aparamán-tepui, Murisipán-tepui, Tereké-yurén-tepui, Kamarkawarai-tepui); *Ua*, Uaipán-tepui; *Ug*, Upuigma-tepui; *Ui*, Uei-tepui; *Un*, Sierra Unturán; *Up*, Cerro Ualipano; *Us*, Serranía de Uasadi; *Ve*, Cerro Venado; *Vn*, Cerro Venamo; *Wa*, Wei-Assipu-tepui; *Wd*, Wadakapiapué-tepui; *Wo*, Mount Wokomong; *Yp*, Cerro Yapacana; *Yt*, Serranía Yutajé; *Yu*, Yuruaní-tepui; *Yv*, Cerro Yaví. The location of these summits follows Huber and Berry (1995). Names are according to Table 1.1. *Source: Radar image courtesy NASA/JPL/SRTM, February 2000 (freely available at https://photojournal.jpl.nasa.gov/catalog/pia03388).*

PLATE 1.1 Eastern tepui chain. (A) Roraima-tepui. (B) Kukenán-tepui. *Source: Photos: Javier Mesa.*

PLATE 1.2 Eastern tepui chain. (A) Roraima-tepui (front) and Kukenán-tepui (back). (B) General view showing the tepuis of the Eastern chain (from front to back): Roraima, Kukenán, Yuruaní, Wadakapiapué, Karaurín, Ilú, and Tramen. *Source: Photos: Javier Mesa.*

PLATE 1.3 Eastern tepui chain. (A) Tramen-tepui seen from the summit of the Ilú-tepui. (B) Tramen-tepui. *Source: Photos: Javier Mesa.*

PLATE 1.4 Auyán massif. (A) Angel Falls emerging from the summit of the Auyán-tepui. (B) Auyán-tepui flank from the Guayaraca savannas. *Source: Photos: Javier Mesa (A) and Celsa Señaris (B).*

PLATE 1.5 Auyán massif. (A) Cliffs and slopes of the Auyán-tepui. (B) Cerros El Sol and La Luna. *Source: Photos: Javier Mesa.*

PLATE 1.6 Chimantá massif. (A) Amurí-tepui (front) and Abakapá-tepui (back). (B) Apakará-tepui. *Source: Photos: Javier Mesa.*

PLATE 1.7 Surroundings of the Chimantá massif. (A) Angasima-tepui from the Amurí-tepui summit. (B) Apaurai-tepui. *Source: Photos: Javier Mesa.*

PLATE 1.8 Isolated tepuis from the Eastern district. (A) Upuigma-tepui. (B) Ptari-tepui. *Source: Photos: Javier Mesa.*

PLATE 1.9 Los Testigos massif. (A) Tereké-yurén-tepui, Murisipán-tepui, and Aparamán-tepui from the summit of Kamarkawarai-tepui. (B) Aparamám-tepui (*left*), Murisipán-tepui (*center*), and Kamarkawarai-tepui (*right*). *Source: Photos: Javier Mesa.*

PLATE 1.10 Duida-Marahuaca massif. (A) General view of Cerro Huachamakari. (B) Cerro Huachamakari from the Cunucunuma River. *Source: Photos: Javier Mesa.*

PLATE 1.11 Jaua-Sarisariñama massif. (A) Summit of Cerro Sarisariñama with the characteristic sinkholes. (B) Summit of Cerro Jaua. *Source: Photos: Javier Mesa.*

PLATE 1.12 Cuao-Sipapo massif. (A) Aerial view of Cerro Autana. (B) General view of Cerro Autana from the Sipapo River. *Source: Photos: Javier Mesa.*

TABLE 1.1 High mountains and tepuis of Pantepui (see Fig. 1.2 for location). Countries: *BR*, Brazil; *GU*, Guyana; *SU*, Suriname; *VE*, Venezuela; summit/slope connection: 0, no connection; 1, deep rocky canyons; 2, valleys; 3, slopes without major interruptions; N.A., not ascertained/not applicable. Based on Huber (1995b) and personal observations by Otto Huber.

Mountain complex	Tepui/mountain unit	Country	Maximum elevation (m)	Summit area (km^2)	Summit–slope connection	Slope area (km^2)
Eastern tepui chain	Uei-tepui	GU	2150	2.5	3	20
	Roraima-tepui	VE-GU-BR	2723	34.38	0	300
	Kukenán (Matauí)-tepui	VE	2650	20.63	0	
	Yuruaní-tepui	VE	2400	4.38	0	
	Wadakapiapué-tepui	VE	2000	<0.01	0	
	Karaurín-tepui	VE	2500	1.88	3	
	Ilú-tepui	VE	2700	5.63	0	
	Tramen-tepui	VE				
	Wei-Assipu-tepui	GU-BR	c. 2260	3	N.A.	N.A.
	Mount Ayanganna	GU	2080	N.A.	0	N.A.
	Mount Wokomong	GU	1680	300	3	N.A.
	Mount Maringma	GU	2134	1.7	N.A.	N.A.
	Monte Caburaí	GU-BR	1465	N.A.	N.A.	N.A.
Gran Sabana uplands	Cerro Venamo (Waukauyengtipu)	VE-GU	1600	N.A.	N.A.	N.A.
	Sierra de Lema	VE	800–1650	N.A.	3	N.A.
Ptari massif	Ptari-tepui	VE	2400	1.25	0	28
	Carrao-tepui	VE	2200	1.25	N.A.	N.A.
	Sororopán-tepui	VE	2050	N.A.	N.A.	30
Los Testigos massif	Kamarkawarai-tepui	VE	2400	5	0	N.A.
	Tereke-yurén-tepui	VE	1900	0.63	0	N.A.
	Murisipán-tepui	VE	2350	5	1	88
	Aparamán-tepui	VE	2100	1.25	0	28
Auyán massif	Auyán-tepui	VE	2450	666.9	2	715
	Cerro La Luna	VE	1650	0.2	1	N.A.
	Cerro El Sol	VE	1750	0.6	1	N.A.
	Uaipán-tepui	VE	1950	2.5	3	60
Canaima uplands	Cerro Venado	VE	1320	0.6	1	17
	Kurún-tepui	VE	1100	0.6	1	N.A.

(*Continued*)

TABLE 1.1 (Continued)

Mountain complex	Tepui/mountain unit	Country	Maximum elevation (m)	Summit area (km^2)	Summit–slope connection	Slope area (km^2)
Aprada massif	Aprada-tepui	VE	2500	4.37	1	210
	Araopán-tepui	VE	2450	1.25	1	
Chimantá massif	Sumits of 10 undifferentiated tepuis: Abapaká, Akopán, Agparamán, Amurí, Apakará, Eruoda (Murey), Chimantá, Churí, Tirepón and Toronó.	VE	2200–2650	615	2–3	915
	Angasima-tepui	VE	2250	2	0	32
	Upuigma-tepui	VE	2100	0.63	0	13
Paragua uplands	Sierra Marutaní (Pia-Zoi)	VE	1500	740	3	N.A.
	Cerro Ichún	VE	1400	2460	1	798
	Cerro Guanacoco	VE	1500	526.25	2	400
	Cerro Guaiquinima	VE	700–1650	1096.26	2	410
Jaua massif	Cerro Sarisariñama	VE	2350	546.88	2	286
	Cerro Jaua	VE	2250	625.62	2	482
Maigualida massif	Sierra de Maigualida	VE	2400	440	3	N.A.
	Serranía de Uasadi	VE	1300–1800	N.A	3	N.A
Yaví massif	Cerro Ualipano	VE	1800	N.A	N.A.	N.A
	Cerro Yaví	VE	2300	5.62	1	70
Yutajé massif	Serranía Yutajé	VE	2140	95.63	1	N.A.
	Coro Coro	VE	2400	179.38	2	143
	Cerro Guanay	VE	2080	165	1	113
	Cerro Camani	VE	1800	1.88	3	N.A.
Sipapo uplands	Cerro Ovaña (Ouana)	VE	1800	N.A.	N.A.	N.A.
	Cerro Moriche	VE	1250	0.2	3	31
Parú massif	Cerro Parú (Asisa)	VE	2200	724.38	2	580
	Cerro Euaja	VE	2000	205.62	2	N.A.
Cuao-Sipapo massif	Cerro Autana	VE	1300	1.88	0	N.A.
	Cerro Cuao	VE	2000	80	2	282
	Cerro Sipapo	VE	1800	56	3	N.A.

(Continued)

TABLE 1.1 (Continued)

Mountain complex	Tepui/mountain unit	Country	Maximum elevation (m)	Summit area (km^2)	Summit–slope connection	Slope area (km^2)
Yapacana upland	Cerro Yapacana	VE	1300	10.5	1	38
Duida-Marahuaka massif	Cerro Huachamakari	VE	1900	8.75	1	60
	Cerro Marahuaka	VE	2800	121	0	325
	Cerro Duida	VE	2358	1089	2	715
Parima uplands	Sierra de Parima	VE	800–1600	N.A.	3	N.A.
Unturán uplands	Cerro Aratitiyope	VE	1700	<0.01	0	N.A.
	Sierra Unturán	VE	1600	N.A.	3	N.A.
Tapirapecó massif	Cerro Tamacuari	VE	2340	<0.01	N.A.	N.A.
	Serranía Tapirapecó	VE-BR	2000	N.A.	1?	N.A.
Aracamuni-Avispa uplands	Cerro Arakamuni	VE	1600	N.A.	2	N.A.
	Cerro Avispa	VE	1600	238	1	658
Imeri massif	Sierra de la Neblina	VE-BR	2994	235	2	857
Mocidade massif	Serra da Mocidade	BR	1900	N.A.	N.A.	N.A.
Aracá massif	Serra do Aracá	BR	1700	N.A.	N.A.	N.A.

"Pantepui" and "Guiana Highlands" are also frequently used as synonyms because of their spatial coincidence, but it should be noted that the first is a biogeographical concept, whereas the second is a physiographic setting (Fig. 1.3).

History of the Pantepui biogeographical concept

The term "Pantepui" was coined by the ornithologists Mayr and Phelps (1967) to refer to "the sandstone tabletop mountains in the Venezuelan Territorio Amazonas and Estado Bolivar and in the adjacent border regions of Brazil and Guyana." According to these authors, the "subtropical" bird fauna of Pantepui was highly endemic and strikingly different from the "tropical" fauna of the surrounding lowlands. Although Mayr and Phelps (1967) did not provide any explicit assessment on elevational boundaries for Pantepui, they mentioned that the Pantepui avifauna was characteristic of the summits and the upper slopes of the tepuis. These researchers also noted that within the large area where the tepuis are present (c. 80,000 km^2), only a small fraction was suitable for the characteristic Pantepui "subtropical" avifauna and emphasized that the total area above 1500 m—including most tepuian summits—does not exceed 10,000 km^2. In spite of its evident biogeographical meaning, Mayr and Phelps (1967) considered Pantepui an artificial unit and did not assign any specific biogeographical category (region, subregion, province) to it. Interestingly, Mayr and Phelps (1967) highlighted that, in spite of their differences from the lowlands, "the bird fauna of Pantepui is far less distinct than the flora."

This first definition of Pantepui was applied, with some modifications, to other animal groups (Müller, 1973; Brown, 1975; Hoogmoed, 1979); a detailed account is available in Huber (1987). However, other authors defined Pantepui in a very different manner. For example, Steyermark (1979) considered Pantepui a plant refugial complex, including not only the tepuian summits and the upper slopes, that is, the Pantepui of Mayr and Phelps (1967), but also most areas south of the Orinoco River, including all of the Central Guiana province and part of the Western and the Eastern Guiana provinces (Fig. 1.1). According to Steyermark (1979), the Pantepui refugial complex included six refuge areas (Venamo, Gran Sabana, Tepui, Amazonas Savannas, Rio Negro, Atures) where rainforests and other lowland tropical plant communities would have persisted through time, even during the assumedly arid glacial phases of the Pleistocene. Such refugia would have provided not only survival but also speciation centers during the Pleistocene glacial-interglacial recurrence (see Chapter 2: Climatic and ecological history of Pantepui and surrounding areas, for more details on the refuge hypothesis).

A first attempt to unify the concept of Pantepui as a biogeographical unit was undertaken by Huber (1987), who noted that latitudinal terms, such as "tropical," "subtropical," and "temperate," were erroneously used as altitudinal subdivisions by Mayr and Phelps (1967) and other zoologists. In the Guiana region, it seemed more appropriate to refer to three altitudinal levels defined climatically: macrothermic (0–800 m elevation; average annual temperature c. 24°C), mesothermic (800–2000 m; 12°C–24°C), and submicrothermic (>2000 m; 8°C–12°C) (Huber, 1987). According to Huber (1987), the significant faunal change used by Mayr and Phelps (1967) to define their "subtropical" fauna, and therefore their Pantepui, occurs within the mesothermic zone at approximately 1200–1500 m. Above these elevations, there is a conspicuous shift in geological, geomorphological, geochemical, and biological features, including flora, fauna, and vegetation. Therefore, Huber (1987) proposed that 1200/1500 m would be the lower altitudinal boundary of Pantepui as a biogeographical unit.

Several years later, Huber (1994) formally described the Pantepui phytogeographical province within the Guiana region as depicted in Fig. 1.1. The Guiana region was defined on the basis of geographical, floristic, and ecological criteria (Table 1.2). Within this region, the four provinces recognized (Eastern Guiana, Central Guiana, Western Guiana,

TABLE 1.2 Criteria used to define the phytogeographical Guiana region (Huber, 1994; Berry et al., 1995b).

Geography	Floristics	Ecology
Location: Guiana Shield of northern South America	*Flora:* very rich (c. 15,000 species of vascular plants)	*Characteristic habitats:* nutrient-poor substrates
Geology: igneous-metamorphic (granite, gneiss) and sedimentary (quartzite, sandstone)	*Endemism:* very high (four families and 140 genera of vascular plants)	*Plant characteristics:* high diversity of growth forms in numerous plant families
Physiography: montane uplands and highlands, peripheral lowlands	*Phytosociology:* many highly evolved plant communities	*Vegetation:* high diversity of vegetation types (forests, shrublands, meadows)

TABLE 1.3 Criteria used to define the four phytogeographical provinces of the Guiana region (Huber, 1994).

	Eastern province	Central province	Pantepui province	Western province
Location	NE: Venezuela, Guyana, Suriname, French Guiana	Guyana, Venezuela, Brazil, Suriname	Guyana, Venezuela, Brazil	SW-Venezuela, SE-Colombia, NW-Brazil
Physiography	Lowlands	Lowlands, uplands	Highlands	Lowlands, uplands
Relief	Plains, low hillands	Hillands, mountain slopes, uplands	Mountains, high plains (tepuis)	Plains, low hillands
Geology	Granites	Granites, sandstones, diabases	Sandslopes, granites, diabases	Sandslopes, granites
Soils	Oxisols, ultisols, entisols, podsols	Ultisols, entisols, alfisols, histosols	Histosols, entisols	Oxisols, ultisols, entisols, podsols
Temperature	Macrothermic	Submesothermic, mesothermic	Mesothermic, submicrothermic	Macrothermic, submesothermic
Vegetation	Lowland forests, scrub, savannas	Montane forests, scrub, savannas	Tepui scrub, meadows	Forests, scrub, savannas
Endemism	Moderate	High	Very high	High
Key families	Leguminosae, Lauraceae, Lecythidaceae, Chrysobalanaceae	Tepuianthaceae, Monotaceae, Combretaceae, Ochnaceae, Rubiaceae	Saccifoliaceae, Theaceae, Rapateaceae, Sarraceniaceae, Ericaceae	Leguminosae, Euphorbiaceae, Rapateaceae, Theaceae, Tepuianthaceae

Pantepui) were differentiated using criteria such as geography, geology, physiography, geomorphology, edaphology, climate, vegetation, plant taxonomy, and the degree of endemism (Table 1.3). In this formal definition, the Pantepui province coincided with the Guiana Highlands, defined by Huber (1995b) as the altitudinal interval between 1500 and 3000 m elevation (Fig. 1.3) situated above the uplands (500–1500 m) and lowlands (<500 m), which belong to the Central Guiana province (Fig. 1.1). Recently, Huber et al. (2018) situated the lower boundary of Pantepui at 1300/1500 m elevation, thus including the upper part of the uplands. Whereas the upper boundary of Pantepui is defined by the highest tepui summit, corresponding to the Sierra de la Neblina (2994 m elevation), the lower boundary seems less clear and may fluctuate between 1200 and 1500 m.

In a recent biogeographical classification of the Neotropical region based on terrestrial plants and animals, Morrone (2014) situated the Pantepui province within the Boreal Brazilian dominion of the Brazilian subregion (note that in this classification, the Neotropical area is not considered a realm but a region). In this framework, Pantepui was considered a synonym of the Pantepui area of Mayr and Phelps (1967), the Pantepui center of Müller (1973), and the Pantepui province of Huber (1994) and was characterized mainly by its arthropod and vertebrate endemisms without any specific reference to elevational limits or any other geological, physiographic, and environmental features. Although it is not explicitly stated, this classification does not differentiate between the Pantepui province (highlands) and the Central Guiana province (uplands and lowlands) *sensu* Huber (1994) (Fig. 1.1). The geographical setting was described as "Northwestern South America, in the Guianan Shield, between Venezuela, Colombia, Guyana, Suriname and northern Brazil, where there are sandy plateaus or tepuis higher than 2000 m altitude" (Morrone, 2014).

Phytogeographical Pantepui

The only formal thorough and specific characterization available for the Pantepui province in terms of geography, geology, physiography, environment, biota, and ecosystems is the phytogeographical Pantepui (Huber, 1994; Berry et al., 1995b), which also includes further subdivisions into districts and subdistricts with peculiar biotic characteristics. This section briefly explains the main traits that characterize this biogeographical province with a focus on three main aspects: the physical setting, the climatic conditions, and the biotic component.

Physical setting

According to the current phytogeographical definition, the Pantepui province ranges from 1200/1500 m to 2994 m elevation, which includes all of the Guiana Highlands and the uppermost part of the uplands (Fig. 1.3). Thus tepui summits below 1500 m fitting the Pantepui environmental and/or biotic criteria—for example, the Cerros Autana (Plate 1.12), Venado, or Yapacana, ranging from 1200 to 1300 m (Table 1.1 and Fig. 1.2B)—can be included. As noted in the introduction, the tepuis have been modeled by erosion of the Precambrian sandstones/quartzites of the Roraima Group probably since the Mesozoic (Briceño and Schubert, 1990). This long erosional process, which is still active, has led to the formation of several erosion surfaces, of which the Auyán-tepui surface is the highest (2000–2900 m elevation) and forms the summits of the highest tepuis. The lower summits are part of the Wonkén surface (up to 1200 m elevation) (see Fig. 4.4 of Chapter 4). The erosion has also exposed several igneous rocks, such as local diabase intrusions into the Roraima Group. Some granitic mountains and massifs are also present in the region, and although they do not have the typical tepuian morphology, the elevation, climate, and biotic features of their summits fall within the Pantepui phytogeographical province. Examples are the Sierra de Maigualida (2400 m elevation) and the Cerro Sipapo (1800 m) (Fig. 1.2 and Table 1.1).

The substrate for life in the tepui summits is a combination of bare rock, extensive peat accumulations of several meters' depth, and diabase outcrops. Rocky and peaty substrates are poor in nutrients, and peats are highly acidic, which constrains the type of communities growing on them (Zinck and Huber, 2011). Diabase outcrops are more nutrient rich and form more fertile soils for vegetation development, as well as the establishment of different ecosystems (Huber, 1995c). The rocky cliffs between the basal slopes and the tepui summits may reach up to ~1000 m and are usually marked by high waterfalls fed by summit rivers and subsurface waters running between the base of the permeable peats and the top of the impermeable Roraima sandstones/quartzites. The most popular example is Angel Falls with a height of c. 980 m (Plate 1.4A). These highly organic and acidic blackwaters (Briceño et al., 1990) are part of the headwaters of important tributaries of the Orinoco and Amazon river networks (Fig. 1.1). Not all of the tepui summits are totally isolated from the surrounding lowlands/uplands by vertical cliffs, as many of them are topographically connected in some parts by extensive valleys, ridges, and rocky canyons (Huber, 1988) (Table 1.1). The surface of the tepui summits is variable, ranging from <0.01 km^2 (e.g., Wadakapiapué-tepui) to >1000 km^2 (e.g., Cerro Duida) (Table 1.1). The total surface of Pantepui has been estimated to be c. 5000 km^2 (Huber, 1995b).

Climatic conditions

The available meteorological data from Pantepui is still insufficient for a sound climatic assessment, and most inferences have been based on the altitudinal range of the tepui summits. According to Huber (1995b), the lower summits, situated between 1500 and 2400 m elevation, would have mesothermic ombrophilous climates with annual average temperatures (AAT) of 12°C–18°C and total annual precipitation (TAP) between 2000 and 3500 mm, with fewer than one dry month (<50 mm/month). The higher summits (2400–3000 m elevation) would have a submicrothermic ombrophilous climate with AAT between 8°C and 12°C and precipitation regimes similar to the mesothermic ombrophilous climate.

The only data available to corroborate these inferences are from three tepui summits situated at different elevations: Auyantepui (Plates 1.4 and 1.5A), Guaiquinima and Kukenán (Plates 1.1B and 1.2), where some basic meteorological parameters were measured during roughly a decade (1997–2009). These data showed that AAT decreased with elevation from 16.5°C (1750 m) to 11.4°C (2600 m), which is consistent with a transition from mesothermic to submicrothermic climates. This finding also fits with the general adiabatic lapse rate of the Guiana region, which is of −0.6°C/100 m elevation and largely determines the altitudinal distribution of the life zones (Fig. 1.3). TAP, on the contrary, increased with elevation from 2800 to 5300 mm at the same elevations (Huber and García, 2011), which is also consistent with ombrophilous climates. Therefore previous climatic inferences from altitudinal patterns in Pantepui are supported by the existing data, at least in terms of general climatic characterization. In the absence of more detailed meteorological surveys, the mesothermic/submicrothermic ombrophilous climatic type could be considered the best approach for the whole range of the tepui summits.

The biotic component

In this section, we will consider the phytogeographical Pantepui concept because it is the only formal biogeographical unit fully described and characterized to date. The more recent definition of the biotic component of the Pantepui province is by Huber et al. (2018), who highlighted that the typical ecosystems of Pantepui are dominated by broadleaved meadows and sclerophyllous woodlands growing on peat, pioneering saxicolous plant communities and small patches of low forests growing in depressions and crevasses. The most characteristic plant families of Pantepui are the herbaceous Rapateaceae, the woody Bonnetiaceae, and the pitcher plant family Sarraceniaceae with their endemic pantepuian genus *Heliamphora*. In general, montane shrublands, meadows, and open rock communities dominate the landscape. Forests play a subordinate role, and they are typically low, relatively species-poor, and restricted to depressions and along water courses. Shrublands are highly varied, ranging from dense high-tepui scrub to paramoid and ericoid scrub. Meadows are well differentiated into broad-leaved, tubiform, and rosette meadows with several areas of grasslands (Berry et al., 1995b). More details on the vegetation types are given in Chapter 7: Plant communities.

Richness and endemism are important parameters to define the Pantepui phytogeographical province. A general compilation by Berry et al. (1995b) reported a total of 158 vascular plant families (1 endemic), 630 genera (23 endemic), and 2322 species (766 endemic). A decade later, Berry and Riina (2005) updated this information to 156 families (none endemic), 626 genera (23 endemic), and 2447 species (1034 endemic). The latest estimates are in similar magnitudes and are available in Chapter 6: Vascular plants and bryophytes. This implies a vascular plant richness of 4500–5000 species per 10,000 km^2 (considering an approximate extension of 5000 km^2 for Pantepui, see earlier), which falls within the most biodiverse areas worldwide (Mutke and Bathlott, 2005). The specific endemism is approximately 35%–40%, which is also very high and comparable to many oceanic archipelagos (Rull, 2009).

Pantepui floristic subdivisions

The tepui summits are not regularly distributed in space, but rather occur in discrete blocks with relevant phytogeographical differences. Based on floristic and geographical criteria, four phytogeographical districts can be distinguished, namely, the Eastern district, the Jaua-Duida district (also known as Central district), the Western district, and the Southern district (Fig. 1.4). The following descriptions are based on Berry et al. (1995b).

Eastern district

This district includes tepui summits between 1600 and 2750 m elevation, such as the Roraima-tepui (Plates 1.1Aand 1.2A), the Auyán-tepui (Plates 1.4 and 1.5A), and the Chimantá massif (Plate 1.6). This is one of the more floristically and ecologically diverse districts of the Pantepui province. Key taxa that define the district are *Bonnetia roraimae* (Theaceae) and species of the endemic genera *Quelchia* (Asteraceae), *Conellia* (Bromeliaceae), and *Tepuia* (Ericaceae). Other relevant endemic genera are *Chimantaea* (Asteraceae), *Mallophyton* (Melastomataceae), and *Coryphothamnus* (Rubiaceae). This district has been subdivided into the Roraima and the Chimantá subdistricts.

1. The *Roraima subdistrict* includes Mount Ayanganna, Mount Wokomong, and the Eastern Tepui Chain (Plate 1.2B). The dominant vegetation consists of small islands of meadows and scrubs within bare rock.
2. The *Chimantá subdistrict* includes the Auyán-tepui and the Chimantá massif.

The transition between the Roraima and the Chimantá subdistricts includes the Aparamán-tepui and Los Testigos massif (Plate 1.9).

Jaua-Duida district (Central district)

Characteristics of this district are large tepuis with elevations between 1500 and 2800 m. Despite the distance between Cerro Jaua and Cerro Duida, they have relevant affinities,

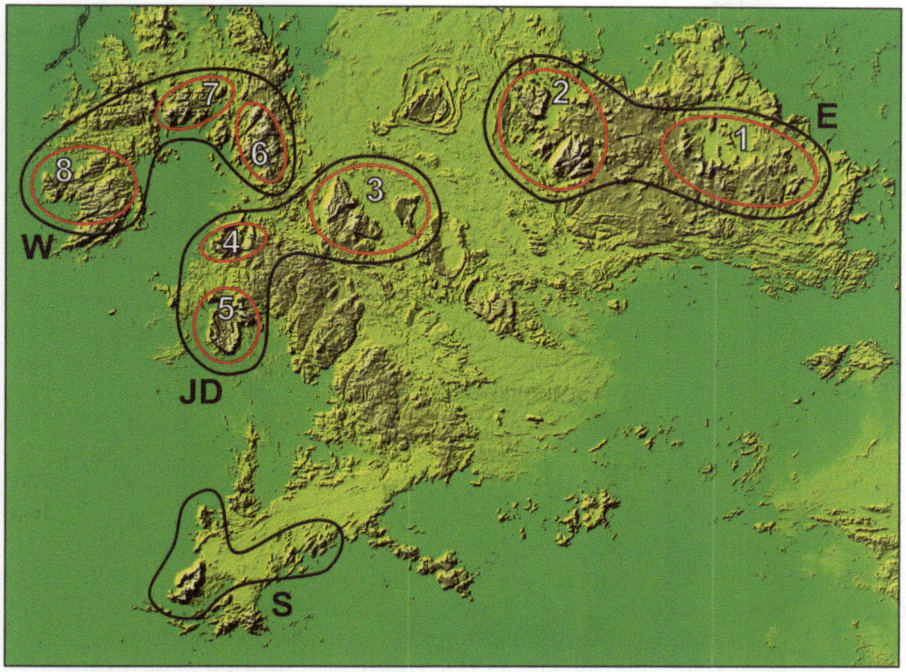

FIGURE 1.4 Pantepui phytogeographical districts and subdistricts, according to Berry et al. (1995b). Districts (*black lines*): *E*, Eastern district; *JD*, Jaua-Duida district; *W*, Western district; *S*, Southern district. Subdistricts (*red lines*): 1, Roraima subdistrict; 2, Chimantá subdistrict; 3, Jaua-Sarisariñama subdistrict; 4, Asisa subdistrict; 5, Duida-Marahuaka subdistrict; 6, Maigualida subdistrict; 7, Yutajé subdistrict; 8, Cuao-Sipapo subdistrict. Base map as in Fig. 1.2.

especially in terms of vegetation types, and share ecologically important genera, such as *Tyleria* (Ochnaceae), *Neotatea* (Clusiaceae), and *Tepuianthus* (Tepuianthaceae), as well as the dominant species of the tepui meadows: *Stegolepis grandis* (Rapateaceae). Three subdistricts have been recognized: the Jaua-Sarisariñama subdistrict, the Asisa subdistrict, and the Duida-Marahuaka subdistrict.

3. The *Jaua-Sarisariñama subdistrict* includes the summits of Cerro Guanacoco, Cerro Sarisariñama (Plate 1.11A), Cerro Jaua (Plate 1.11B), and possibly the highest parts of the Serranía Marutaní, which is still poorly explored. The vegetation is dominated by shrublands and tubiform meadows. *Tyleria* is a characteristic element, and several *Stegolepis* species (*Stegolepis breweri*, *Stegolepis jauaensis*, *Stegolepis microcephala*) are endemic of this subdistrict.
4. The *Asisa subdistrict* comprises the summits of Cerro Asisa (the highest part of Cerro Parú) and the Cerro Euaja. *Phelpsiella* (Rapateceae) is endemic, and common genera are *Coccochondra* (Rubiaceae), *Tepuianthus*, and the Euphorbiaceae *Celianella* and *Mabea*.
5. The *Duia-Marahuaka subdistrict* is characterized by the summits of Cerros Duida, Marahuaka, and Huachamakari (Plate 1.10). Typical Pantepui-endemic genera of this subdistrict are *Salpinctes* (Apocynaceae), *Duidaea* and *Tyleropappus* (Asteraceae), *Duidania* (Rubiaceae), and *Marahuacea* (Rapateaceae).

Western district

This district is the most extensive and topographically irregular district of Pantepui, with summit elevations ranging from 1300 to 2350 m. The most characteristic species is *Kunhardtia rhodantha* (Rapateaceae), together with species of *Graffenrieda* and *Meriania* (Melastomataceae), as well as a large number of endemic species of *Phyllanthus* (Euphorbiaceae). This district comprises three subdistricts, namely (from east to west), the Maigualida subdistrict, the Yutajé subdistrict, and the Cuao-Sipapo subdistrict.

6. The *Maigualida subdistrict* is represented by the summits of the granitic Sierra de Maigualida, which is not a table mountain but bears typical high-tepui ecosystems on its top, ranging from 1800 to 2350 m elevation. This subdistrict is characterized by the easternmost boundary of *Kunhardtia*, the occurrence of the endemic genus *Huberopappus* (Asteraceae), and the noticeable absence of *Bonnetia*.
7. The *Yutajé subdistrict* includes tepui summits between 1800 and 2300 m elevation (Cerro Yaví, Serranía Yutajé-Coro Coro, Cerro Guanay) and is characterized by a high variety of shrublands and high levels of endemism. A feature of some tepuis of this subdistrict (Yaví, Yutajé) is the absence of peats characteristic of most Pantepui summits. The occurrence on Cerro Yaví of the rare species *Pleurostima celiae* (Velloziaceae) is also noteworthy.
8. The *Cuao-Sipapo subdistrict* includes the summit of the small Cerro Cuao (c. 2000 m elevation) and the top of the granitic Sipapo complex (1400–1800 m elevation). Characteristic vegetation types are high-tepui scrub, meadows, and open rock communities. Well-diversified genera in this subdistrict include *Graffenrieda* (Melastomataceae), *Phyllanthus* (Euphorbiaceae), *Schefflera* (Araliaceae), and *Diacidia* (Malpighiaceae).

Southern district

This district includes the upper sections of the Neblina massif (1600–2994 m elevation) and possibly the botanically unexplored summit of the granitic Cerro Tamacuari (2350 m elevation). The characteristic vegetation types of this district on the Neblina massif are extensive shrublands, including *Bonnetia maguireorum* scrub, and broad-leaved or tubiform high-tepui meadows. Low forests of other *Bonnetia* species are also relatively frequent in depressions and small valleys of the upper summit plains. Floristically, this is the section of Pantepui—and of the entire Guiana region—with the highest number of endemic taxa, with one endemic family (Saccifoliaceae) and the endemic genera *Glossarion*, *Imeria*, and *Neblinaea* (Asteraceae); *Neblinantha* (Gentianaceae); *Pyrrorhiza* (Haemodoraceae); *Comoliopsis* and *Neblinanthera* (Melastomataceae); *Adenarake* (Ochnaceae); *Cephalodendron* and *Neblinathamnus* (Rubiaceae); *Rutaneblina* (Rutaceae); and *Achlyphila* (Xyridaceae).

Other tepuis

Some tepuis/mountains from Table 1.1/Fig. 1.2 do not fall into any of these four districts, mainly because of the lack of knowledge of their flora and vegetation. These include the Serra do Acará, Cerro Aratitiyope, Sierra Imerí, Cerro Ichún, Serra da Mocidade, Sierra

Marutaní, Sierra de Parima, and Sierra Unturán. One exception is the peculiar Cerro Guaiquinima, the vascular flora of which is relatively well known but cannot be assigned to any of the defined districts. The largely forested Guaiquinima summit is an inclined surface ranging from 700 m to c. 1700 m elevation, and its flora is a mixture of tepuian elements and lowland/upland elements (Steyermark and Dunsterville, 1980) that do not fit any of the Pantepui districts. It should also be noted that the four Pantepui districts were defined for the Venezuelan part of Pantepui (Berry et al., 1995b), and only two mountains outside this country—Mount Ayanganna and the Wokomong massif from Guyana (Fig. 1.2B and Table 1.1)—were included in this classification.

Zoogeographical insights

From a zoogeographical point of view, there is not a unified formal definition of Pantepui, probably because different taxonomic groups exhibit contrasting biogeographical patterns and/or because the respective specialists have diverse ideas on what Pantepui is. A number of zoologists follow the phytogeographical definition. For example, McDiarmid and Donnelly (2005), studying the Pantepui herpetofauna, explicitly used the definition and classification into districts and subdistricts based on botanical criteria. However, these authors defined some new subdistricts, as is the case with the Los Testigos subdistrict within the Eastern district or the Guaiquinima subdistrict within the Central (Jaua-Duida) district. They also distinguished the Sierra Tapirapecó and the Neblina-Aracamuni massif within the Southern district, although without considering them subdistricts. Other zoologists consider areas of lower elevation than the phytogeographical Pantepui in their studies. This is the case of the ornithologists Borges et al. (2018), who included mountains and hills below 1500 m elevation from Venezuela and Colombia, arguing that they are geologically and phytogeographically similar to Pantepui. However, these authors do not redefine the Pantepui biogeographical province.

Others recover the idea of Steyermark (1979) and include the lowlands and uplands in Pantepui. For example, Kok (2013), working on amphibians and reptiles, considered Pantepui *sensu lato* to be subdivided into a lower Pantepui zone, between 200 and 1200 m elevation, and an upper Pantepui zone, or Pantepui *sensu stricto*, above 1200 m, although "...only the upper Pantepui zone should probably be seen as a 'natural' biogeographical unit" (Kok, 2013). Thus defined, the Pantepui *sensu stricto* zone coincides with the phytogeographical Pantepui province, as defined earlier, and Pantepui as a whole is considered by the author a "bioregion" rather than a specific biogeographical unit. No formal biogeographical description or characterization of Pantepui was provided by this author, but the idea of including lowlands, uplands, and highlands in the definition was derived from the fact that single-tepui endemism among amphibians and reptiles was significantly lower than expected, likely due to the historical biotic interchange among summits by vertical migrations through the uplands and lowlands (Kok et al., 2012). Details on vertical migrations in relation to the origin of Pantepui biota, are provided in Chapter 4.

The already referred to biogeographical regionalization of Morrone (2014) follows similar criteria and is eminently zoogeographical. Indeed, the diagnostic endemic taxa used by this author to characterize the Pantepui province included nearly 100 arthropod and vertebrate species and only one plant, although the known flora is notably richer and more endemic than all faunal groups studied. Some zoologists have proposed that the current biogeographical concept of Pantepui should be reconsidered. For example, Costa et al. (2013), working on butterflies, suggested that Pantepui should be redefined as a discontinuous endemism area and the current category of biogeographical province of the Neotropical realm should be reconsidered.

The general impression is that the lower Pantepui boundary defined on the basis of botanical criteria is more diffuse and variable from a zoological point of view. Explicitly or not, zoologists commonly use the phytogeographical concept of Pantepui, as they can clearly observe significant differences in the Guiana Highlands' fauna, but they have disparate opinions on the lower Pantepui limit. This is not a minor subject, as any change in this limit directly affects the definition of the other provinces of the Guiana region, especially the Central Guiana province. A detailed and comprehensive biogeographical study, similar to that already existing for plants, is needed to advance the zoogeographical characterization of Pantepui.

Conclusions and final remarks

Currently, the formal description, characterization, and subdivision of the Pantepui biogeographical province is based on phytogeographical criteria. This approach is not uncommon in biogeography, as vegetation is not only the primary food source but also a major landscape element that often determines the main structural traits of ecosystems and largely contributes to niche compartmentalization, thus potentially affecting all components of the biotic community. However, the geographical distribution of animals may be different due to their mobility and their capacity for exploiting resources from different biomes and ecosystems. Several zoologists have followed the phytogeographical definition of Pantepui, as the distribution of the fauna they study shows similar patterns. Others have suggested different definitions, including the surrounding uplands and lowlands. It has also been realized that levels of richness and endemism–two key parameters in the definition of biogeographical units—may vary across the different taxonomic groups according to their particular biological and ecological features.

As already suggested by earlier biogeographers, the richness and endemism patterns of the vascular plants of Pantepui are unparalleled by other taxonomic groups, which could facilitate biogeographical analysis and classification. Another constraint to comparing different taxonomic groups from a biogeographical point of view is the degree of knowledge of each of them. In Pantepui, vascular plants are the better-known organisms, and this could create a bias in the progress of phytogeography with respect to zoogeography. A book such as this is a good opportunity to compare the geographical patterns of plants and animal groups studied to date in the tepui summits in relation to the biogeographical concept of Pantepui. For this reason, the authors of the zoological chapters have been free to adopt their own concept of Pantepui according to the particular biogeographical traits of their respective taxonomic groups. In light of the available knowledge, it seems too soon for a sound synthesis, but this may be a first step towards a definition of Pantepui as a comprehensive biogeographical entity. This does not mean that the phytogeographical definition should be revised in depth or abandoned, as its definition is based on strong botanical and ecological evidence, but it could possibly be refined and enriched with equally firm evidence from zoological fields.

Acknowledgments

The authors are grateful to Javier Mesa for providing color pictures (Plates 1.1–1.12) and to Daniel Lew and Jorge Pérez-Emán for their critical reviews that contributed to the improvement of the manuscript.

References

Berry, P.E., Riina, R., 2005. Insights into the diversity of the Pantepui flora and the biogeographical complexity of the Guayana Shield. Biol. Skr. 55, 145–167.

Berry, P.E., Holst, B.K., Yatskievych, K., 1995a. Introduction. In: Berry, P.E., Holst, B.K., Yatskievych, K. (Eds.), Flora of the Venezuelan Guayana. Introduction, vol. 1. Missouri Botanical Garden Press, St. Louis, pp. xv–xx.

Berry, P.E., Huber, O., Holst, B., 1995b. Floristic analysis and phytogeography. In: Berry, P.E., Holst, B.K., Yatskievych, K. (Eds.), Flora of the Venezuelan Guayana. Introduction., vol. 1. Missouri Botanical Garden Press, St. Louis, pp. 161–191.

Borges, S.H., Santos, M.P.D., Moreira, M., Baccaro, F., Capurucho, J.M.G., Ribas, C., 2018. Dissecting bird diversity in the Pantepui area of endemism, northern South America. J. Ornithol. 159, 1073–1086.

Briceño, H.O., Schubert, C., 1990. Geomorphology of the Gran Sabana, Guayana Shield, southeastern Venezuela. Geomorphology 3, 125–141.

Briceño, H., Schubert, C., Paolini, J., 1990. Table-mountain geology and surficial geochemistry: Chimantá Massif, Venezuelan Guayana Shield. J. South Am. Earth Sci. 3, 79–94.

Brown, K., 1975. Geographical patterns of neotropical Lepidoptera. Systematics and derivation of known and new Heliconiini (Nymphalidae: Nymphalinae). J. Entomol. 44, 201–242.

Costa, M., Viloria, A.L., Huber, O., Attal, S., Orellana, A., 2013. Lepidoptera del Pantepui. Pate I: endemismo y caracterización biogeográfica. Entomotropica 28, 193–217.

Hoogmoed, M.S., 1979. The herpetofauna of the Guianan region. In: Duellman, W.E. (Ed.), The South American Herpetofauna: Its Origin, Evolution and Dispersal. Monogr. Mus. Nat. Hist. 7, Lawrence, pp. 241–279.

Huber, O., 1987. Consideraciones sobre el concepto de Pantepui. Pantepui 1, 2–10.

Huber, O., 1988. Guayana highlands versus Guayana lowlands, a reappraisal. Taxon 37, 595–614.

Huber, O., 1994. Recent advances in the phyotgeography of the Guayana region, South America. Mém. Soc. Biogéogr. (3ème série) IV, 53–63.

Huber, O., 1995a. Conservation. In: Berry, P., Holst, B.K., Yaskievych, K. (Eds.), Flora of the Venezuelan Guayana, Introduction, vol. 1. Missouri Botanical Garden Press, St. Louis, pp. 193–218.

Huber, O., 1995b. Geographical and physical features. In: Berry, P., Holst, B.K., Yaskievych, K. (Eds.), Flora of the Venezuelan Guayana, Introduction., vol. 1. Missouri Botanical Garden Press, St. Louis, pp. 1–61.

Huber, O., 1995c. Vegetation. In: Berry, P., Holst, B.K., Yaskievych, K. (Eds.), Flora of the Venezuelan Guayana, Introduction, vol. 1. Missouri Botanical Garden Press, St. Louis, pp. 97–160.

Huber, O., Berry, P.E., 1995. Venezuelan Guayana. Topographic map (1:2,000,000). In: Berry, P., Holst, B.K., Yaskievych, K. (Eds.), Flora of the Venezuelan Guayana, Introduction, vol. 1. Missouri Botanical Garden Press, St. Louis.

Huber, O., García, P., 2011. The Venezuelan Guayana region and the study areas: geo-ecological characteristics. In: Zinck, A., Huber, O. (Eds.), Peatlands of the Western Guayana Highlands, Venezuela. Springer, Berlin, pp. 29–89.

Huber, O., Prance, G.T., Kroonenberg, S., Antonelli, A., 2018. The tepuis of the Guiana Highlands. In: Hoorn, C., Perrigo, A., Antonelli, A. (Eds.), Mountains, Climate and Biodiversity. Wiley-Blackwell, Hoboken, pp. 339–353.

Kok, P.J.R., 2013. Islands in the Sky: Species Diversity, Evolutionary History, and Patterns of Endemism of the Pantepui Herpetofauna (Ph.D. dissertation). Univ. Leiden, The Netherlands.
Kok, P.J.R., MacCulloch, R.D., Means, D.B., Roelants, K., Van Bocxlaer, I., Bossuyt, F., 2012. Low genetic diversity in tepui summit vertebrates. Curr. Biol. 22, R589–R590.
Mayr, E., Phelps, W.H., 1967. The origin of the bird fauna of the South Venezuelan Highlands. Bull. Am. Mus. Nat. Hist. 136, 269–328.
McCormack, J.E., Huang, H., Knowles, L.L., 2009. Sky islands. In: Gillespie, R., Clague, D. (Eds.), Encyclopedia of Islands. Univ. of California Press, Berkeley, pp. 839–843.
McDiarmid, R.W., Donnelly, M.A., 2005. The herpetofauna of the Guayana Highlands: amphibiands and reptiles of the Lost World. In: Donnelly, M.A., Crother, B.I., Guyer, C., Marvalee, H., White, M.E. (Eds.), Ecology and Evolution in the Tropics. Univ. Chicago Press, Chicago, pp. 461–560.
Morrone, J.J., 2014. Biogeographical regionalisation of the Neotropical region. Zootaxa 3782, 1–110.
Müller, P., 1973. The dispersal centres of the terrestrial vertebrates in the Neotropical Realm. Biogeographicala 2, 1–244.
Mutke, J., Bathlott, W., 2005. Patterns of vascular plant diversity at continental to global scales. Biol. Skr. 55, 521–531.
Rull, V., 2009. Pantepui. In: Gillespie, R., Clague, D. (Eds.), Encyclopedia of Islands. Univ. of California Press, Berkeley, pp. 717–720.
Rull, V., Vegas-Vilarrúbia, T., Safont, E., 2016. The lost World's pristinity at risk. Divers. Distrib. 22, 995–999.
Steyermark, J.A., 1979. Plant refuge and dispersal centres in Venezuela: their relict and endemic element. In: Larsen, K., Holm-Nielsen, L. (Eds.), Tropical Botany. Academic Press, New York, pp. 185–221.
Steyermark, J.A., Dunsterville, G.C.K., 1980. The lowland floral element of the summit of Cerro Guaiquinima and other cerros of the Guayana Highlands of Venezuela. J. Biogeogr. 7, 285–303.
Zinck, J.A., Huber, O., 2011. Peatlands of the Western Guayana Highlands, Venezuela. Springer, Berlin.

CHAPTER

2

Climatic and ecological history of Pantepui and surrounding areas

Valentí Rull[1], Encarni Montoya[1], Sandra Nogué[2], Elisabet Safont[3] and Teresa Vegas-Vilarrúbia[3]

[1]Institute of Earth Sciences Jaume Almera (ICTJA-CSIC), Barcelona, Spain [2]Department of Geography and Environment, University of Southampton, Southampton, United Kingdom [3]Department of Evolutionary Biology, Ecology and Environmental Sciences, Faculty of Biology, University of Barcelona, Barcelona, Spain

Introduction

Living communities can be viewed as aggregates of transient species changing over time due to the ongoing action of shifting biotic and abiotic ecological drivers (Jacobson and Grimm, 1986). Thus, present-day biodiversity and biogeographical features of a given region or ecosystem are merely a snapshot of a long-term ecological continuum running from the past to the future (Rull, 2010a, 2012). Therefore, current community composition and dynamics should be understood in light of both present-day environmental constraints, notably climate, and past historical developments. The configuration of extant communities is a relatively recent phenomenon that has been influenced by Quaternary environmental changes, especially the Pleistocene glacial–interglacial cycles during the last 2.6 million years. These changes have determined the recurrent disruption and reassembly of living communities, with corresponding shifts in diversity and composition (Wright, 1977; West, 1980). In such a dynamic context, the communities we observe today were assembled after the Last Glacial Maximum (LGM)—which occurred between 26 and 20 calendar kiloyears before present (cal kyr BP), depending on the region considered—under the action of continued range shifts of the species involved, each responding individually to the Late Glacial climatic oscillations and the ensuing Holocene warming (Davis, 1981) (Fig. 2.1).

Biodiversity of Pantepui
DOI: https://doi.org/10.1016/B978-0-12-815591-2.00002-1

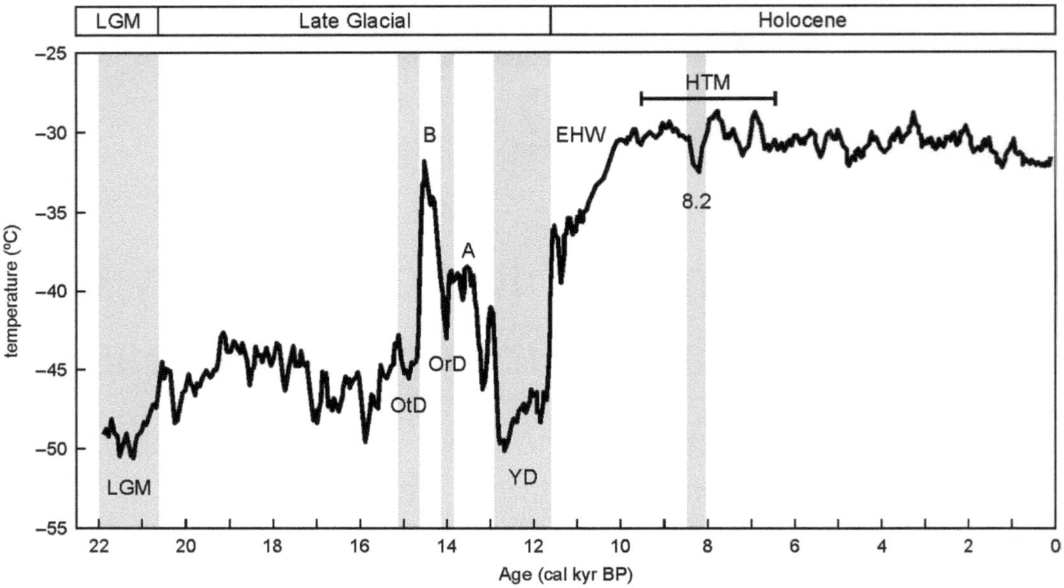

FIGURE 2.1 Temperature trends recorded in the Greenland GISP2 ice core as a reference for global changes since the Last Glacial Maximum (LGM). Absolute temperature values correspond to central Greenland, where the ice core was retrieved, but trends are considered of global validity. The most significant cold events are highlighted by gray bands. A, Allerød; B, Bølling; EHW, Early Holocene Warming; HTM, Holocene thermal Maximum; OrD, Older Dryas; OtD, Oldest Dryas; YD, Younger Dryas (Stuiver et al., 1995). *Source: Raw data of Alley, R.B., 2000. The Younger Dryas cold interval as viewed from central Greenland. Quat. Sci. Rev. 19, 213–226 (Alley, 2000) downloaded from the NOAA World Data Center for Paleoclimatology at: https://www.ncdc.noaa.gov/data-access/paleoclimatology-data/datasets.*

Community turnover during the Quaternary has been dramatic in temperate latitudes, where the effects of glaciations and their environmental consequences were devastating. For example, polar ice sheets and cold steppes covered most of Europe and North America during the LGM, and the species of present-day temperate forests were refugiated into small areas of the southernmost part of these continents. With the Holocene warming, these forests initiated a northward colonizing trend that has led to the present distribution and composition of temperate forests (Davis, 1981; Huntley and Birks, 1983; Hewitt 1999; Sommer and Zachos, 2009). Until the mid-20th century, tropical areas, especially lowlands, were considered to have been little affected by glacial–interglacial cyclicity, and this assumed environmental stability was proposed as a major cause for the comparatively higher tropical biodiversity (Fischer, 1960; Pianka, 1966; Slobodkin and Sanders, 1969). However, widespread evidence of Pleistocene climatic shifts in tropical regions has accumulated, and today we know that the tropics were also deeply affected by the glacial–interglacial cycles (Flenley, 1979; Morley, 1999; Bush et al., 2011).

In the American tropics, thereafter the Neotropics, abundant evidence of glacial climates has been gathered since the late 20th century. For example, it was shown that during the LGM the Andean highlands were subject to environmental shifts similar to those in the temperate zones, that is, downward glacier expansion and downward migration of ecological belts (van der Hammen, 1974). Concerning the lowland tropical areas, a lively debate persists between the defenders and the detractors of the refuge hypothesis. The defenders contend that LGM Neotropical climates were arid and that most of the current rainforest areas were occupied by savannas or deserts, whereas forests were fragmented into small refugia restricted to sites with favorable climatic conditions (Prance, 1982; Whitmore and Prance, 1987). Detractors of the refugium hypothesis propose that during the LGM Neotropical lowlands were not arid, but rather cooler and lower in atmospheric CO_2 concentration, a situation found today at higher elevations. This would have caused downward range shifts of sensitive montane species and the ensuing compositional changes in a continuous lowland forest cover, rather than its fragmentation (Bush, 1994; Colinvaux et al., 2000). This has been called the disturbance-vicariance hypothesis.

The Guiana Highlands, where the Pantepui biogeographical province is centered (Chapter 1: Definition and characterization of the Pantepui biogeographical province), are not high enough to have been glaciated during the LGM, as it occurred on the Andean summits of similar latitudes, which are significantly higher (>5000 m). Indeed, no geological evidence of glaciation has been found in the tepui summits. On the other hand, LGM glaciers in the northern tropical Andes, situated at the same latitude as the Guiana Highlands, descended down to elevations of ~4000 m (Rull et al., 2010), which is still ~1000 m above the highest Pantepui summit. Therefore the disturbance-vicariance hypothesis seems more plausible, a priori, for these summits. If this were the case, Late Glacial climatic variability would have determined minor elevational biotic migrations of the sensitive Guiana Highland species, leading to changes in both diversity and composition of their communities. Finally, these communities would have stabilized after the Early Holocene Warming (EHW), which ended with the Holocene Thermal Maximum (HTM). In order to test these hypotheses this chapter reviews the more significant environmental and ecological changes that occurred in the Guiana Highlands and the surrounding lower terrains since the LGM to the present. As most of these studies are based on pollen analysis, emphasis has been placed on the reconstruction of vegetation shifts as representative of ecological dynamics in terrestrial ecosystems.

Some knowledge gaps complicate this task, however. First, the main bulk of palynological analyses have been conducted in the eastern part of the Pantepui province, which precludes attaining a truly regional picture. Some studies on peat properties and formation and characteristics have been carried out in the western sector (Zink and Huber, 2011), but paleoclimatic and paleoecological information is still lacking. Second, sediments corresponding to the LGM are still unknown from the tepui summits, where only Late Glacial and Holocene peats have been found so far (Zink et al., 2011; Rull et al., 2010). Third, the rarity of lakes and their corresponding sediments atop the tepuis has constrained the development of multiproxy studies able to provide independent paleoecological and paleoclimatic evidence. Lakes are especially well-suited archives of past environmental and ecological dynamics due to the abundance and diversity of well-preserved biotic and abiotic paleoecological indicators (proxies) they hold (Smol et al., 2001–2012). In an attempt to address these knowledge gaps, this review includes lake records from the surrounding Amazon lowlands (200–300 m elevation) and the Gran Sabana uplands (800–1200 m), which is the basal topographic level from which the tepuis emerge (Fig. 2.2).

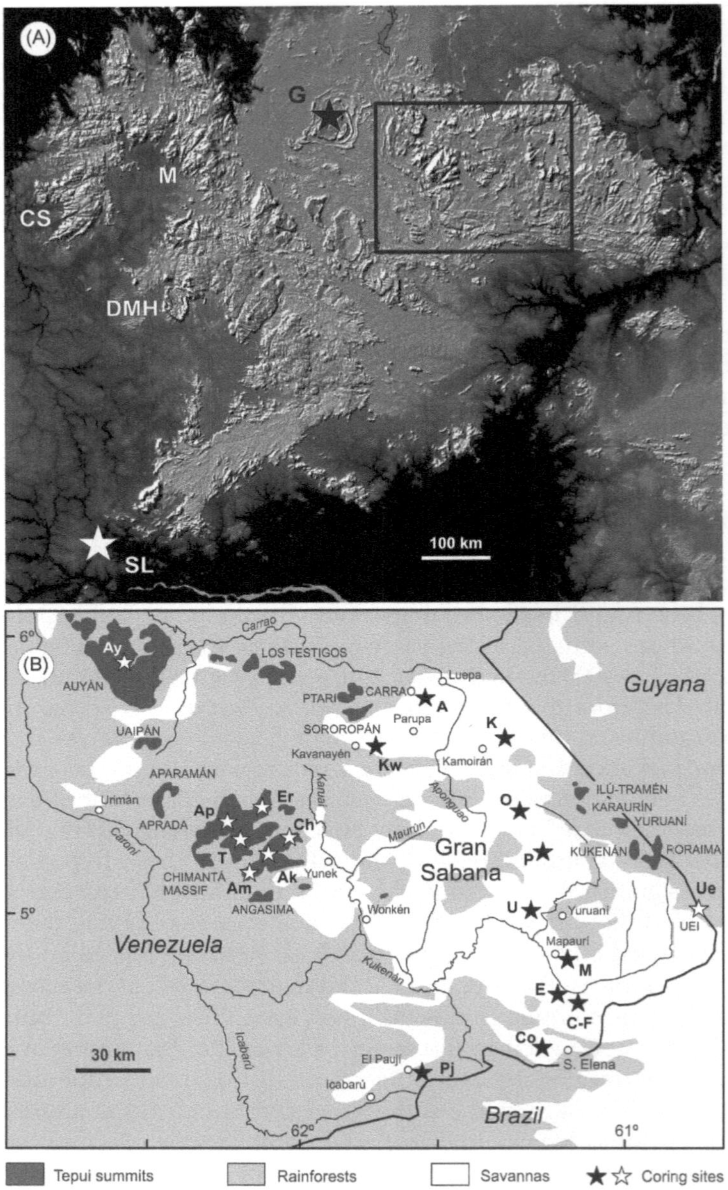

FIGURE 2.2 Location maps. (A) Radar image (NASA/JPL/NIMA) of the Guiana Highlands (GH) showing the eastern Pantepui region where most paleoecological studies have been carried out *(black box)* and two other coring localities corresponding to the Guaiquinima massif (G) and the Hill of the Six Lakes (SL). Other sites from the western sector with studies on peat formation and characteristics (Zink and Huber, 2011) are also indicated: CS, Cuao–Sipapo massif; DMH, Duida–Marahuaka–Huachamakari massif; M, Maigualida massif. (B) Enlargement of the black box of eastern Pantepui with details on the location of coring sites available to date. A, Ariwe; Ak, Akopán; Am, Amurí; Ap, Apakará; Ay, Auyán; C, Chonita; Ch, Churí; Co, Colonia; E, Encantada; Er, Eruoda; F, Fidencio; K, Kamarata; Kw, Kowana; M, Mapaurí; O, Oso; P, Pacheco; Pj, Paují; T, Toronó; U, Urué; Ue, Uei. *Source: Modified from Rull, V., Montoya, E., Nogué, S., Vegas-Vilarrúbia, T., Safont, E., 2013. Ecological palaeoecology in the neotropical Gran Sabana region: long-term records of vegetation dynamics as a basis for ecological hypothesis testing. Persp. Plant Ecol. Evol. Syst. 15, 338–359 and Rull, V., Montoya, E., 2014. Mauritia flexuosa palm swamp communities: natural or human-made? A palynological study of the Gran Sabana region (northern South America) within a neotropical context. Quat. Sci. Rev. 99, 17–33.*

Due to the absence of present and historical human disturbance, most tepui summits are virtually pristine and have been considered natural laboratories to study the evolutionary origin of their species and the biotic responses to natural environmental changes (Rull, 2010b). In contrast, the Gran Sabana uplands are, and have been historically, under significant human pressure (particularly fire), which could be useful to disentangle natural from anthropogenic drivers of ecological change (Vegas-Vilarrúbia et al., 2011; Rull et al., 2013). Hopefully the information obtained about past climatic changes, human disturbance, and the corresponding biotic responses at individual and community levels will be useful to improve conservation strategies in the face of future global change.

Regional paleoclimatic trends

An appropriate knowledge of the main past and present regional climatic features and their controlling factors is essential for paleoecological interpretation. In the study area this regional overview is even more important due to the scarcity of weather stations and local climatic data. The basic climatic parameters associated with the main vegetation types and ecosystems of the Guiana region are average annual temperature (T), average annual precipitation (P), and seasonality (S), expressed as the number of dry months (i.e., the months with less than 50 mm of total rainfall). Temperature is controlled mainly by the near-equatorial position of the Guiana region and topography. In general, there is a standard lapse rate temperature decrease of 0.6°C per 100 m increase in elevation, and the climate of the Guiana Highlands is (sub)microthermic ($T < 12°C$) and ombrophilous (P between ~2000 and 4000 mm/year) (Huber, 1995a; Huber and García, 2011). In the Guiana region, precipitation regimes are controlled mainly by the seasonal oscillation of the Intertropical Convergence Zone (ITCZ), where northeast and southeast tropical trade winds converge in response to seasonal changes in insolation and energy balance asymmetries between the Northern and Southern Hemispheres (Poveda et al., 2006; Schneider et al., 2014). Precipitation maxima occur between April and November, when the ITCZ is situated in its northernmost position, near the southern Caribbean coasts, and minima during the rest of the year, when the ITCZ lies over southern Brazil (McGregor and Nieuwolt, 1998). This seasonality is much more pronounced in the Guiana lowlands and uplands, with 2–5 dry months, than in the highlands, with less than 1 dry month per annum (Huber, 1995a). Interannual variability is primarily controlled by the quasi-periodic El Niño Southern Oscillation (ENSO), which has an average period of 4 years and a range of 2–10 years. In tropical South America, the ENSO variability is characterized by the recurrence of anomalously warmer and drier "El Niño" years and anomalously colder and wetter "La Niña" years (McGregor and Nieuwolt, 1998; Poveda et al., 2006).

Past regional changes in these climatic parameters have been recorded at high resolution in marine cores from the Cariaco Basin, situated just 400 km north of the study area along the eastern coast of Venezuela. Recently some additional paleoprecipitation records have been obtained in speleothems from Amazonian caves (Chen et al., 2013; Novello et al., 2017), but they are too far from Pantepui to represent its past paleoclimatic trends. Temperature trends obtained from Cariaco records parallel the Greenland paleotemperature trends (Fig. 2.3), which highlight the global character of these changes. The same

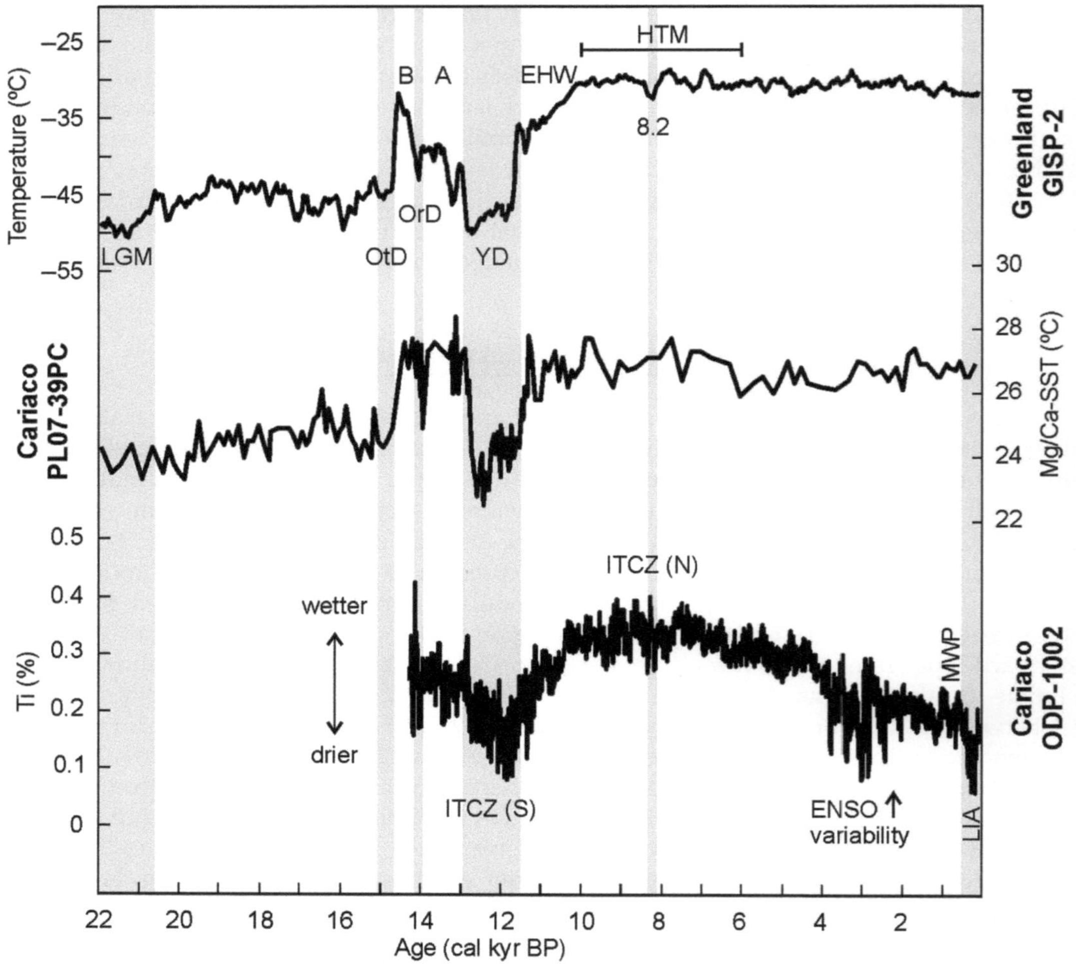

FIGURE 2.3 Regional paleotemperature and paleoprecipitation reconstructions for northern South America based on marine cores from Cariaco Basin, using the Greenland standard (GISP2). In the absence of atmospheric temperature reconstructions, sea surface temperature (SST) is used as the best approach. Precipitation trends are expressed in terms of titanium (Ti) content of marine sediments. This element is absent in the sea, and its abundance in marine sediments depends on the terrigenous input, a straightforward manifestation of river discharge and, hence, of continental precipitation regimes. ITCZ, Intertropical Convergence Zone (N and S mean northern and southern migration, respectively); ENSO, El Niño Southern Oscillation; LIA, Little Ice Age; MWP, Medieval Warm Period (Haug et al., 2001). *Source: Modified from Rull, V., Abbott, M.B., Vegas-Vilarrúbia, T., Bezada, M., Montoya, E., Nogué, S., et al., 2010. Paleoenvironmental trends in Venezuela during the last glacial cycle. In: Sánchez-Villagra, M.R., Aguilera, O.A., Carlini, A.A. (Eds.), Urumaco and Venezuelan Palaeontology. The Fossil Record of the Northern Neotropics. Indiana University Press, Bloomington, pp. 55–83.*

LGM and Late Glacial stadial/interstadial pattern and the Holocene stabilization after the EHW are evident in both curves. Paleoprecipitation trends have a more regional character, and the Cariaco records, which are available for the Late Glacial and the Holocene (Fig. 2.3), have been considered representative of most areas of northern South America, including the Guiana region (Haug et al., 2001). A decreasing precipitation trend initiated at 14 cal kyr BP or before culminated in the Younger Dryas (YD) cooling, which was also a dry event (Fig. 2.3). This was interpreted as the consequence of a southern migration of the mean position of the ITCZ and, therefore, of the higher precipitation tropical belt (Haug et al., 2001; Schneider et al., 2014). An ensuing northward shift of the ITCZ would have been the cause of the EHW precipitation increase, whose maximum was attained during the HTM, spiked by a small reversal around the 8.2 cal kyr BP cooling. A trend toward drier conditions started at about 6 cal kyr BP and extended until the present, probably due to a farther southward ITCZ shift. However, this decreasing trend was disrupted by two conspicuous events. The first, between about 4 and 2 cal kyr BP, consisted of an increase in the amplitude of precipitation fluctuations that was attributed to an increase in the variability of the ENSO activity and the predominance of El Niño warmer/drier conditions (Haug et al., 2001). The second event began around 1 cal kyr BP and consisted of a precipitation increase that coincided with the Medieval Warm Period (MWP). This event was then followed by a significantly drier phase starting around 0.7 cal kyr BP, coeval with the Little Ice Age (LIA) cooling. The causes for these two recent climatic reversals in the Cariaco record remain unknown (Black et al., 2007).

Paleoecology of Pantepui and adjacent areas

This section synthesizes the development of vegetation on Pantepui and the surrounding lowlands and uplands since the LGM to the present in light of the regional climatic changes reviewed earlier. Vegetation types and their taxonomical composition follow the standards of Chapter 7 (Plant communities). Emphasis is placed on Pantepui, while lower elevation records are considered only when they are necessary to understand the dynamics of Pantepui plant communities. More detailed paleoecological reviews of the Gran Sabana region (Fig. 2.2), the better-known uplands of the Guiana Highland region, are available elsewhere (Montoya and Rull, 2011; Rull et al., 2013, 2015a; Rull and Montoya, 2014). This paleoecological account is subdivided into the LGM, the Late Glacial, and the Holocene. The last millennium is analyzed separately because of its special features, including both climatic forcing and human disturbance.

The Last Glacial Maximum

Based on phytogeographical criteria and following the refuge hypothesis, Steyermark (1979) defined the Pantepui refuge, which would have remained climatically stable and fully vegetated not only in the LGM but also during the whole Pleistocene. This refuge would have included the tepui summits, the Gran Sabana, and some Amazon lowlands. However, the first paleoecological studies of Pantepui and the Gran Sabana that reported the absence of LGM peats seemed to dismiss this hypothesis (Schubert and Fritz, 1985). In the tropics, peat formation requires humid to perhumid climates to favor waterlogging and anoxia, which allows organic matter to accumulate by minimizing decomposition (Zinck, 2011). The absence of LGM peats atop the tepuis, a widespread phenomenon across Pantepui (Zink et al., 2011), was taken as indicative of arid climates and vegetation scantiness, which was incompatible with the occurrence of a Pantepui refuge (Schubert et al., 1994). In this situation the continuity of vegetation atop the tepuis during the Pleistocene glaciations was explained assuming the eventual occurrence of hypothetical microrefugia on very localized summit sites with suitable microclimatic conditions (Rull et al., 1988). Another possibility is that pre-Holocene peat layers have been washed out by subsurface water currents flowing between the Holocene peats and the bedrock (Rull et al., 1988; Rull, 1991). This is a common phenomenon in the tepui summits that can ultimately lead to peat detachment and sliding under heavy rainfall, which restarts the accumulation of younger peats (Zink et al., 2011). Therefore the lack of peat records of a given age could be due either to unfavorable dry climates preventing peat formation or to peat removal under wet climates (Rull et al., 2013). Finally, the purported absence of LGM peats could be a sampling bias due to the low number of peat sequences analyzed and their limited geographic distribution (Zink et al., 2011) (Fig. 2.2). Much more coring effort is needed before claiming that the tepui summits are devoid of LGM peats.

The Hill of Six Lakes, located in a 300-m high inselberg from the Amazon lowlands, at a short distance south of the southernmost tepuis, has been a key paleoecological site in the debate between the defenders and detractors of the refuge hypothesis in the Neotropics (Fig. 2.2). Despite being a lowland site, the paleoecological information obtained in its lake sediments is relevant for the reconstruction of Pantepui vegetation during the LGM. The first pollen analyses from this site indicated that rainforests were continuously present during the last 50 cal kyr BP (Bush et al., 2004), which contradicted the hypothesis of extended LGM glacial aridity, according to which Neotropical lowlands would have been covered by deserts and savannas. However, these forests were not constant in species composition. Indeed, during full glacial conditions, between 35 and 22 cal kyr BP, they were formed by a combination of lowland and montane taxa, which was interpreted in terms of downward migration of cold-loving montane forest taxa such as *Podocarpus*, *Hedyosmum*, or *Weinmannia* to the lowlands. The descent was estimated to be on the order of 800–900 m and was attributed to a cooling of 4°C–5°C below present-day temperatures, a scenario that is more consistent with the disturbance-vicariance hypothesis. The most likely source for temperature-sensitive montane forest elements, especially *Podocarpus*, was considered to be Sierra de la Neblina—the southernmost tepui in the Guiana Highlands (Bush et al., 2004). Further analyses of the same lake sediments were used to argue that full glacial climates were drier than previously considered, emphasizing that there were discontinuities in the pollen record and the forests were not continuous through the LGM (D'Apolito et al., 2013). However, these reanalyses did confirm the presence of normally montane forest elements in the lowlands, and therefore the effect of glacial cooling was supported.

The case of the Hill of Six Lakes is more consistent with the disturbance-vicariance hypothesis and sets the Guiana Highlands within the framework of the tropical Andes and most mountain ranges of the world, where downward biotic migrations have been the norm during the LGM. However, this does not clarify the nature of the vegetation on the tepui summits. If we consider that about half of the tepui summits are above 2200 m elevation (Huber, 1995a) and that during the LGM a downward shift of ~900 m occurred, the glacial biota of these tepuis should have corresponded to present-day elevations above 3100 m, which are nonexistent in the Guiana Highlands. Therefore the glacial Pantepui biota would have been composed of unknown extinct species, the so-called "glacial ghosts" (Rull, 2005a), and/or species with large ranges of temperature tolerances and wide elevational distributions. The more likely modern analogues for these unknown LGM Pantepui plant communities might be the Andean páramos, situated at the same latitude above 3000 m elevation. The páramos are open plant communities characterized by the presence of conspicuous columnar rosettes of several species of the genus *Espeletia* (Asteraceae). A physiognomically analogous and floristically similar vegetation type, the paramoid shrublands, has been described for the summits of the Chimantá massif, where *Espeletia*-like columnar rosettes correspond to several species of the endemic genus *Chimantaea* in the same family (Huber, 1992). It is possible that, during the LGM, Pantepui summits were inhabited by páramo-like communities of which the *Chimantaea* paramoid shrublands would be the remnants. In this case, the "glacial ghosts" would have become extinct during warmer interglacials by habitat loss due to the impossibility of migrating upwards (Rull, 2005a).

The Late Glacial

The only Late Glacial record available to date for Pantepui is from the summit of Eruoda-tepui, in the Chimantá massif (Fig. 2.2). It dates back to 12.7 cal kyr BP, corresponding to the YD chron (Nogué et al., 2009) (Fig. 2.1). However, this evidence is restricted to a single sample at the basal layer of the peat sequence, which allows inferring the vegetation of this particular time slice but precludes interpreting its ecological dynamics. The pollen assemblage was co-dominated by herbs (mainly Poaceae, *Xyris*, and *Brocchinia*) and shrubs or small trees, notably Ericaceae, Melastomataceae, Ochnaceae, and *Cyrilla*. The abundance of algae in this basal peat sample suggests permanent flooding. In general there are no significant differences with present-day vegetation (Plate 2.1), as shown in modern pollen assemblages representing local peat vegetation. Therefore in spite of the significantly cooler climate of the YD, at a regional scale, the vegetation of the summit of Eruoda-tepui was likely similar to today, with some minor compositional differences. This could be due to the dominance of local hydrological factors over regional climatic features. According to Zink et al. (2011), local peat-forming vegetation is relatively insensitive to regional climate changes, as "long-term water saturation sustained by abundant rainfall causes inertia, buffering against temperature changes" (p. 223). In the Gran Sabana uplands two sediment cores from lakes Chonita and Fidencio (Fig. 2.2) that also correspond to the YD chron showed a conspicuous savannization trend from either shrublands (Chonita) or forests (Fidencio) to open treeless savannas under the synergistic action of fire and climatic aridity (Montoya et al., 2011a; Rull et al., 2015a). These contrasting ecological and hydrological conditions between uplands and highlands suggest that during the YD, the Gran Sabana ecosystems were more heavily influenced by regional aridity, whereas tepuian ecosystems, especially peat-forming vegetation, were more constant through time due to more stable microclimatic conditions. In the Eruoda sequence there is a conspicuous time gap between the YD basal layer and the overlying peats, which are early Holocene in age. According to Zink et al. (2011), gaps of similar ages are common in many tepuian peats and may be due to a climate change, rather than peat sliding.

PLATE 2.1 Relevant aspects of selected coring sites mentioned in the text (Eruoda-tepui and Apakará-tepui). (A) Eruoda summit. Peat bog dominated by *Brocchinia hechtioides* (Bromeliaceae) and several Xyridaceae and Cyperaceae species surrounded by tree stands of *Bonnetia roraimae* (Bonnetiaceae). (B) Eruoda summit. Closer view of *B. hechtioides*. (C) Eruoda-tepui summit. Inundated rocky environment with small stands of pioneer vegetation. (D) Eruoda summit. First stages of peat formation on rocky substrate. (E) Apakará summit. Paramoid shrubland of *Chimantaea mirabilis* (Asteraceae) with the bambusoid *Myriocladus steyermarkii* (Poaceae) in the herbaceous layer. (F) Closer view of *C. mirabilis*. *Source: Photos: V. Rull.*

The Holocene

Besides the YD peat layer found on the summit of Eruoda-tepui, all other Pantepui peat sequences are Holocene, ranging from 8.6 cal kyr BP to the present (Zink et al., 2011; Rull and Montoya, 2017). Palynological studies on these peats have shown two contrasting situations regarding vegetation dynamics. In one group of tepui summits (Akopán, Amurí, Eruoda, Guaiquinima, and Toronó), the vegetation has remained nearly constant in species richness and composition during the last 6 cal kyr BP, whereas in others (Apakará and Churí), significant changes have been recorded since the early Holocene. Such a disparity may seem surprising, since all these summits except for Cerro Guaiquinima belong to the Chimantá massif (Fig. 2.2).

In Akopán, Amurí, and Toronó tepuis the Holocene vegetation was largely dominated by *Stegolepis* meadows with the presence of *Bonnetia* gallery forests and shrublands with *Cyrilla*, Ochnaceae, Myrsinaceae, and Ericaceae, among others (Rull, 2005b) (Fig. 2.4). On Cerro Guaiquinima similar trends have been recorded, with the addition of a different type of gallery forests dominated by *Archytaea* (Rull, 2005c). The summit of Eruoda-tepui (2650 m) lies above the upper elevational limit of *Stegolepis* meadows (2200–2300 m), and the Holocene vegetation of the coring site was characterized by local peat-forming communities (Nogué et al., 2009), as described in the "Late Glacial" section. In all these summits minor variations in pollen percentages of taxa from these communities indicated long ecological constancy over time with no species turnovers that might be indicative of climatic changes (Rull, 1996, 2005b). It is also noteworthy that known regional climate changes outlined earlier seem not to have affected Holocene vegetation of these tepuis (Fig. 2.4).

The longest sequence showing significant vegetation shifts is located at 2170 m elevation, on the summit of Apakará-tepui, and consists of a continuous palynological record since the early Holocene to the present (Fig. 2.5). In this record two main vegetation shifts are apparent, the first from a *Myrica* forest surrounding a shallow lake to a *Stegolepis* meadow (Plate 2.2) on nonflooded soils, and the second a replacement of these meadows by a *Chimantaea* shrubland (Plate 2.1), with vegetation currently growing at the site (Rull and Montoya, 2017). *Myrica* forests dominated during the HTM and reached their maximum importance at the end of this phase. A sudden replacement of these forests by tepui meadows dominated by *Stegolepis* took place after the HTM, just at the beginning of the regional cooling and drying trend initiated at ~6 cal kyr BP. *Myrica* forests never returned to the site. The only species of this genus living today in the Guiana region, *Myrica rotundata*, is endemic to Pantepui and occurs on the slope forests of the Apakará-tepui, with an upper distribution limit near the coring site (Miller, 2001). It has been suggested that during the HTM, warmer and wetter climates would have favored upslope migration of *Myrica* forests to higher elevations, which could explain their dominance in the Apakará summit. The subsequent post-HTM cooling would have returned *Myrica* to lower elevations favoring the local expansion of meadows. HTM climatic conditions never recovered during the rest of the Holocene, and *Myrica* remained at lower elevations until today (Rull and Montoya, 2017).

Stegolepis meadows dominated until the middle of the phase of increased ENSO and precipitation variability, when they started to decrease, reaching a minimum at the end of this phase, when *Chimantaea* shrublands underwent an abrupt increase peaking at

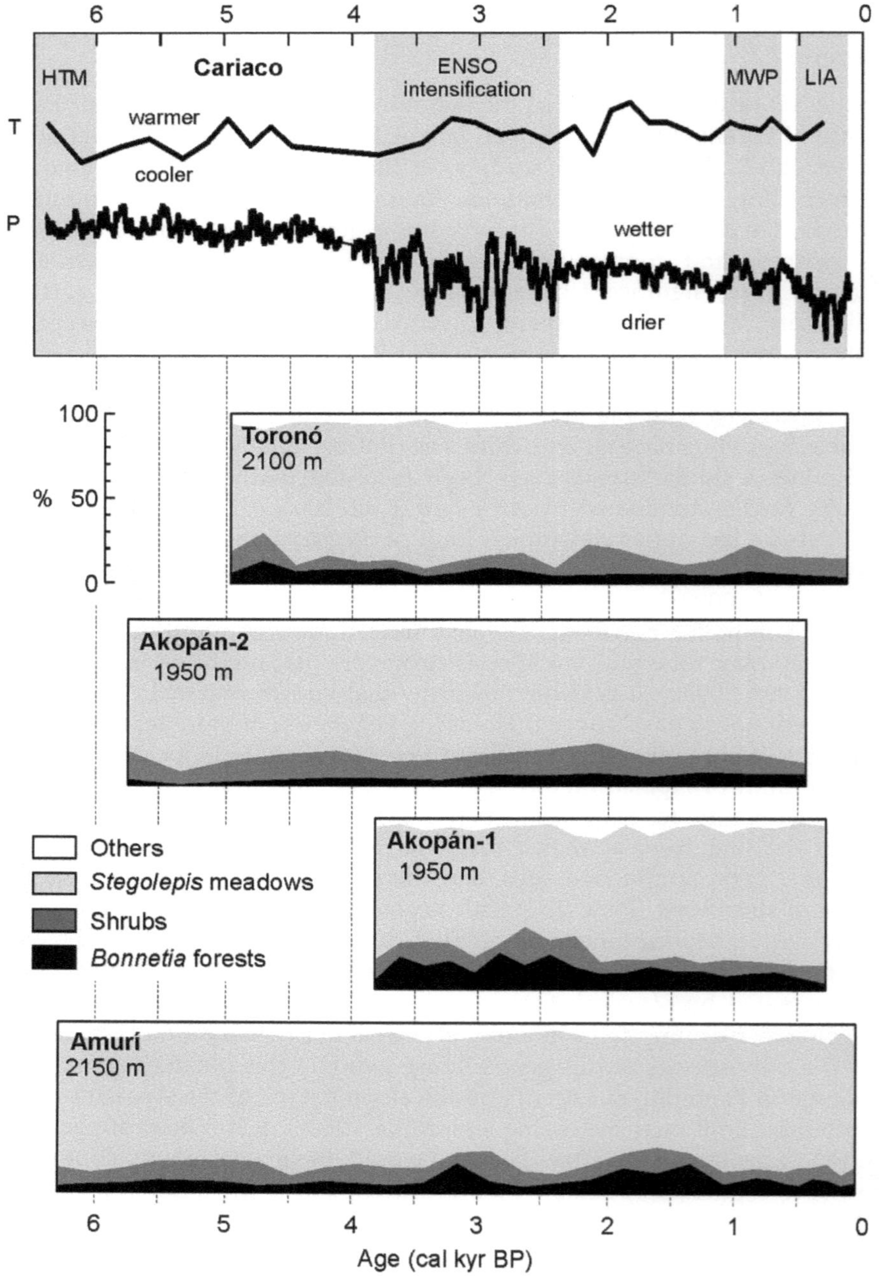

FIGURE 2.4 Summary pollen diagrams of several tepui summits from the Chimantá massif (see Fig. 2.2 for locations) showing the vegetation constancy of the vegetation on coring sites. The landscape was dominated by *Stegolepis* meadows, with some minor asynchronous expansions and contractions in gallery forests attributed to local processes such as river meandering (Rull, 2005b). The uppermost curves are regional temperature (T) and precipitation (P) reconstructions from the Cariaco Basin cores, indicating the main events discussed in the paleoclimatic section, according to Haug et al. (2001) (Fig. 2.3). The Toronó-tepui percentage scale is the same as for the other three graphs. *Source: Redrawn from Rull, V., Montoya, E., Nogué, S., Vegas-Vilarrúbia, T., Safont, E., 2013. Ecological palaeoecology in the neotropical Gran Sabana region: long-term records of vegetation dynamics as a basis for ecological hypothesis testing. Persp. Plant Ecol. Evol. Syst. 15, 338–359.*

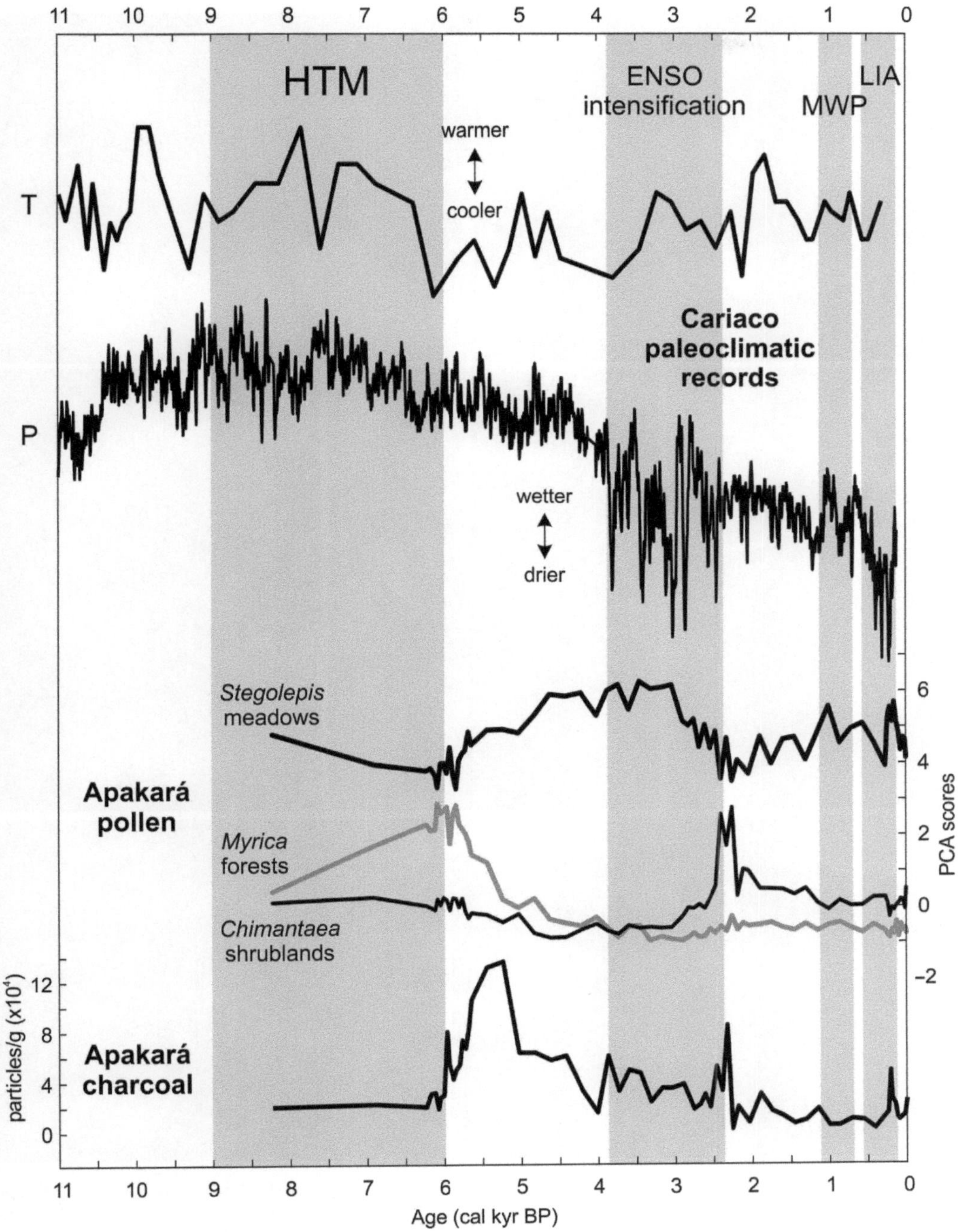

FIGURE 2.5 Holocene vegetation and fire records from the summit core site on Apakará-tepui at 2170 m (Chimantá massif; Fig. 2.2) compared with the Cariaco paleoclimatic records shown in Fig. 2.3 (Haug et al., 2001). Vegetation types are expressed as Principal Component Analysis (PCA) scores from pollen records, and fire trends are inferred from the concentration of charcoal particles in the sediment. See Figs. 2.1 and 2.2 for explanations of the acronyms. *Source: Redrawn from Rull, V., Montoya, E., 2017. Holocene vegetation dynamics on the Apakará summit of the neotropical Guayana Highlands and potential environmental drivers. Rev. Palaeobot. Palynol. 240, 22–32.*

PLATE 2.2 Relevant aspects of selected coring sites mentioned in the text (Apakará-tepui and Uei-tepui). (A) Apakará summit. Plant of *Stegolepis ligulata* (Rapateaceae). (B) Apakará summit. Inflorescences of *S. ligulata*. (C) Uei summit. General view of the meadows dominated by *Stegolepis guianensis* (Rapateaceae), with *Xyris* sp. (Xyridaceae) and several species of Eriocaulaceae. Note the conspicuous presence of yellow *Brocchinia tatei* (Bromeliaceae) rosettes. (D) Uei summit. Closer view of *B. tatei*. (E) Charred trunk of *Bonnetia* cf. *tepuiensis* (Bonnetiaceae) in growth position, whose burnt bark was used for radiocarbon-dating the fire event. (F) Uei summit. *S. guianensis* meadows growing among recently burnt *Bonnetia* cf. *tepuiensis* trunks. Note a brown-colored *Bonnetia tepuiensis* forest at the background. (G) Uei summit. *Cyrilla racemiflora* (Cyrillaceae) colonizing burnt areas. *Source: Photos: V. Rull.*

2–2.5 cal kyr BP. The lack of autoecological information on the species involved in this vegetation shift prevents relating it with changes in the regional precipitation regime, but it seems very likely that the precipitation anomalies recorded between ~4 and ~3 cal kyr BP affected the Apakará plant communities in some way. The decline of *Stegolepis* pollen was gradual, but its replacement by *Chimantaea* communities was abrupt (Fig. 2.5), which suggests the potential existence of a tipping point beyond which ecological changes accelerated. Late Holocene climatic shifts such as the MWP and the LIA seem not to have affected the Apakará communities, but this observation could also be due to the lack of resolution to record short events.

Besides climate, fire was likely a significant driver of vegetation change. In this case three main charcoal peaks were recorded about 6–5 cal kyr BP, 2.5 cal kyr BP, and during the last centuries (Fig. 2.5). Due to the small size of charcoal particles, which favors wind dispersion, and the fact that fires are rare on tepui summits possibly due to their perhumid climates (Huber, 1995b), it was argued that these particles probably originated in the surrounding Gran Sabana uplands, where Holocene fires were frequent (Rull, 2007; Montoya and Rull, 2011; Montoya et al., 2011a,b,c). It is also noteworthy that the first charcoal peak occurred at the beginning of the post-HTM decreasing precipitation trend and the other two coincided with precipitation minima. It is known that savanna wildfires are usually associated with climatic dryness, which drives positive feedbacks and leads to burning (Jacobs et al., 1999; Beerling and Osborne, 2006). As discussed before, the Gran Sabana uplands were significantly influenced by regional climatic trends, especially droughts, which would have led to increased fire occurrence during the dry phases in which the Apakará charcoal peaks were found (Rull and Montoya, 2017). With the available evidence it is not possible to know whether these fires were natural or anthropogenic in origin.

On the summit of Churí-tepui in the Chimantá massif (Fig. 2.2), a conspicuous shift from *Chimantaea* shrublands to *Stegolepis* meadows was recorded between 6.5 and 1.5 cal kyr BP, followed by a second change to the present-day vegetation at the coring site (Rull, 2004a,b). The *Chimantaea* to *Stegolepis* shift was tentatively dated to 2.5 cal kyr BP by linear interpolation, but this date is now questioned by the possibility of a sedimentary hiatus between 6.5 and 1.5 cal kyr BP. What is evident is that *Stegolepis* was absent from the coring site before the vegetation shift. Since the current uppermost elevational limit of *Stegolepis* meadows (2200–2300 m) is around the coring site (2250 m), it was suggested that before the time of community replacement, this boundary would have been lowered due to climates that are cooler and drier than today. The subsequent increase of *Stegolepis* meadows could have been due to a shift to climates that are warmer and wetter than today. The present situation would have been reached after a recent cooling and drying trend initiated after 1.5 cal kyr BP, which was not intense enough to remove *Stegolepis* communities from around the site. The dating uncertainty of this *Chimantaea* to *Stegolepis* shift prevents any specific correlation with known regional climatic changes.

The contrast between tepui summits showing ecological shifts associated with regional climatic changes and those exhibiting long-term ecological constancy throughout the Holocene can be explained in terms of site sensitivity and the intensity of environmental shifts. Rull (2005b) proposed that vegetation changes are best detected on sites close to an elevational ecotone, where vertical displacement promoted by climatic shifts could be better detected by pollen analysis. On the other hand, sites situated close to the mean elevation of a given vegetation type and, therefore, far from its corresponding lower and upper ecotones, would be less suitable to capture the palynological signal of eventual elevational shifts. This, combined with the capacity of the tepuian perhumid climates to buffer regional climatic changes (Zink et al., 2011), has been used to explain why some tepui sites do not show Holocene climatic and ecological shifts. In these summits, the recorded vegetation constancy over millennia could be interpreted as supporting the concept of potential natural vegetation, which once a hypothetical equilibrium with bioclimatic conditions is reached, remains unchanged (Chiarucci et al., 2010). However, the case of the summits whose vegetation changed in agreement with regional climatic shifts does not support such an idea (Rull, 2015).

The last millennium

Most tepuian paleoecological records are too low in temporal resolution to account for the events corresponding to the last millennium, but a peat sequence from the summit of Uei-tepui (Fig. 2.2) is suitable for this purpose. Fire is a key element in the interpretation of this paleoecological record in terms of vegetation dynamics (Safont et al., 2016). Two main microcharcoal peaks as proxies for fire events are conspicuous during the last millennium (Fig. 2.6). The first occurred during the 12th century and coincided with a gentle decrease of the *Bonnetia* gallery forests in an open landscape of herbaceous meadows dominated by *Stegolepis*, *Brocchinia*, *Xyris*, Poaceae, and Cyperaceae (Plate 2.2). This occurred during a drying trend within the MWP, and it was suggested that such aridity would have facilitated the occurrence and expansion of fire (Safont et al., 2016). At the beginning of the LIA, after a significant increase in regional precipitation and the decrease of fire incidence, forests began to recover and continued their expansion in spite of drier climates, suggesting that fire was more important than climate for forest dynamics.

The second fire event started at ~1700 CE (Common Era, formerly AD for *Anno Domini*) and increased rapidly after ~1800 CE, peaking shortly after 1900 CE (Fig. 2.6). At the beginning (~1700 CE), fire increase did not interrupt forest recovery, probably due to the low burning intensity and the occurrence of internal wet microclimatic conditions of *Bonnetia* gallery forests (Huber, 1995c; Chen et al., 1999), which prevented the penetration of fire into these forests. However, the acceleration of fires starting ~1800 CE was more destructive and caused the decline of forests and meadows in favor of shrublands dominated by *Cyrilla*, whose only species living on Uei-tepui, *Cyrilla racemiflora* (Plate 2.2) is a well-known early colonizer of burned-over areas (Ewel, 1990). This second peak of fire coincided with a conspicuous increase in regional precipitation, which is usually not favorable for wildfire occurrence, suggesting that humans may have been involved. However, there is no evidence of human occupation on the summit of Uei-tepui, which is a rather unique case in that it lacks the typical tabular morphology of most tepuis. Instead it shows a more pyramidal structure with gradually sloping sides, which could easily have facilitated the spread of fires from the surrounding Gran Sabana all the way up to the summit (Safont et al., 2016).

According to the available historical records, the Pemón people who currently inhabit the Gran Sabana arrived 300–500 years ago (Thomas, 1982; Colson, 1985; Huber, 1995a),

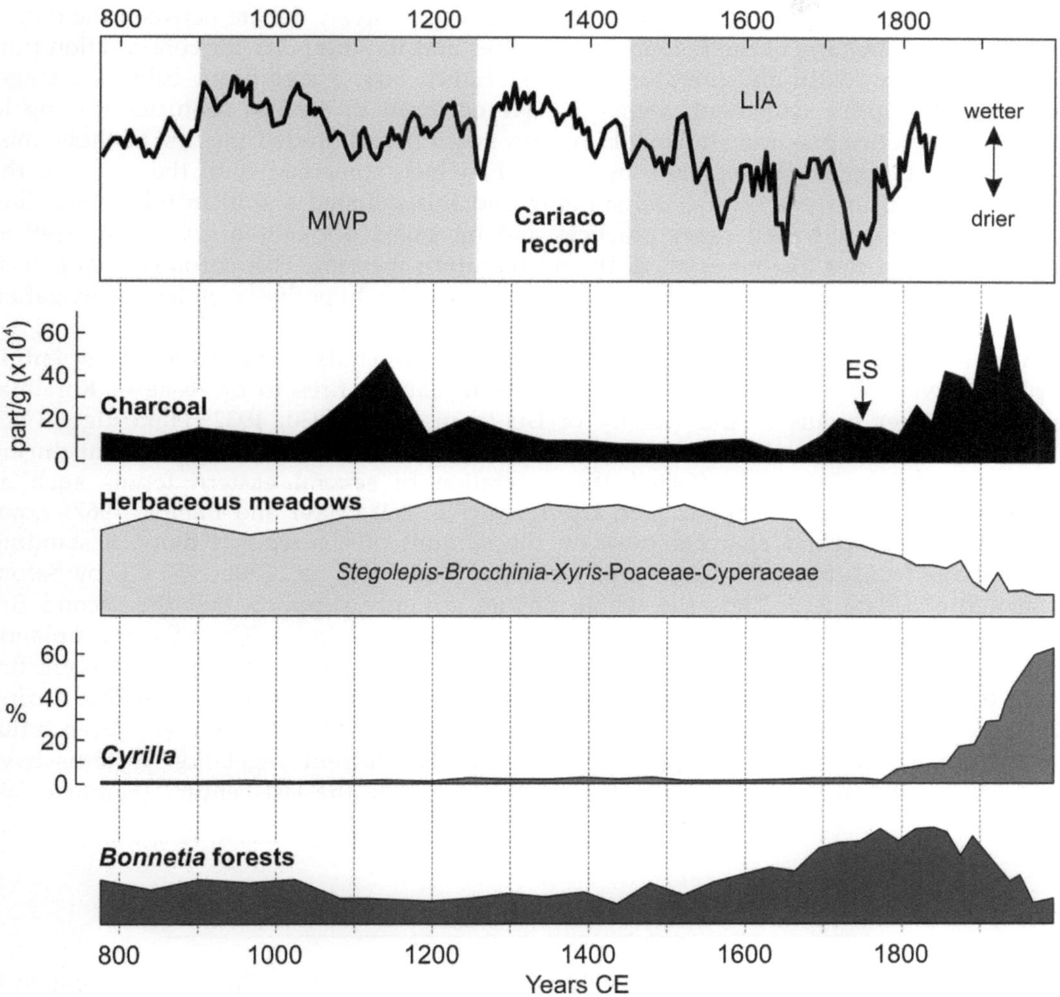

FIGURE 2.6 Main pollen types from the Uei-tepui record for the last millennium compared with regional humidity trends from the Cariaco record (Fig. 2.3). MWP, Medieval Warm Period; LIA, Little Ice Age (Haug et al., 2001). The temperature curve is not represented because of its very low resolution. Charcoal is expressed in particle concentration. *ES*, European settlement. *Source: Redrawn from Safont, E., Rull, V., Vegas-Vilarrúbia, T., Montoya, E., Huber, O., Holst, B.K., 2016. Late-Holocene vegetation and fire dynamics on the summits of the Guayana Highlands: the Uei-tepui palynological record. Palaeogeogr. Palaeoclimatol. Palaeoecol. 455, 33–43.*

although some paleoecological records suggest that they might have been present in the Gran Sabana for the last 2000 years (Montoya and Rull, 2011; Montoya et al., 2011c). Fire is a key element in the culture of the Pemón people. It is estimated that 5000–10,000 fires are lit by them every year, and most Gran Sabana areas are burned every 1–3 years (Huber, 1995a; Hernández and Fölster, 1994). This has contributed to the historical expansion of savannas to the detriment of forests, which has caused a lively debate between the defenders and the detractors of the Pemón fire practices and its suitability for conservation purposes (Rull et al., 2016 and literature therein). Either way, a significant cultural change, driven by European settlement, seems to have occurred in the last centuries leading to present Pemón fire practices. Indeed, European missionaries funded the first Catholic missions in the Gran Sabana uplands by 1770 CE, which coincided with the onset of the second charcoal increase on the Uei summit, and this initiated a shift in the Pemón lifestyle from nomadic to sedentary practices and increased population growth, as well as increased use of fire (Safont et al., 2016 and literature therein). This chronological coincidence would be considered an indirect support for the hypothesis of fire propagation from the Gran Sabana uplands to the summit of Uei-tepui.

A more direct support for fire propagation from the uplands comes from early scientific explorers, who observed the propagation of Gran Sabana fires to the base of Roraima-tepui and other tepuis in the vicinity of Uei-tepui (Tate, 1930, 1932; Holdridge, 1933; Fig. 2.2). It is noteworthy that a historical report exists on an extensive fire event during the 1920s that profoundly affected the vegetation of several eastern tepuis such as Roraima, Uei, Auyán, Sororopán, and Upuigma (Fig. 2.2) (Mayr and Phelps, 1967), coinciding with the second charcoal peak on the summit of Uei-tepui. Finally, a standing charred tree trunk from the summit of Uei-tepui was dated to 1730–1950 CE by Safont et al. (2016) (Plate 2.2). Thus the whole picture strongly supports that the second fire occurrence recorded in the summit of Uei-tepui originated in the Gran Sabana uplands after the European settlement. The charcoal peak would likely correspond to a recent fire event that occurred during the 1920s. In light of these results, it is quite clear that during the last millennium the vegetation dynamics on the summit of Uei-tepui have been influenced primarily by fire, rather than climate, and that current vegetation patterns have been shaped by anthropogenic fires that occurred during the last century (Safont et al., 2016).

Final remarks

After three decades of research, paleoecological knowledge of the Guiana Highlands still has significant gaps. A major challenge is to find sediments, most likely peats, corresponding to the LGM, provided they exist. Several hypotheses—including glacial aridity, microrefugia, downward migration, and "glacial ghosts," among others—can be readily tested with such sediments, and other hypotheses may emerge with new findings. Holocene trends are better known, but additional pollen cores, preferably from elevational ecotones, are needed for a more complete picture. A single record from Uei-tepui is available for the last millennium with enough resolution to reconstruct paleoecological trends. This record is probably a special case of upland anthropogenic fires propagating to a tepui summit and shaping its current vegetation patterns, but this should be confirmed with more sequences from other tepuis. In general, one of the most relevant results is that temperature-sensitive species have experienced elevational migrations likely linked to regional climatic changes, which is important not only to understand the biodiversity and composition of present communities but also to test classical biogeographical and evolutionary hypotheses on the origin of biodiversity and endemism patterns, as, for example, the relative importance of vicariance and dispersal (*sensu lato*) or the potential role of Quaternary climate changes.

Important handicaps for this type of study are the remoteness and difficult accessibility of the tepui summits and, especially, the difficulty-to-obtain permits for field work (Rull and Vegas-Vilarrúbia, 2008; Rull et al., 2009). It is also important to note that, to our knowledge, all palynological studies published to date have been carried out by the same research group, which is probably due, at least in part, to the difficulties mentioned earlier. However, paleoecological and paleoclimatic research has not stopped in places far more difficult to reach and with harder working conditions, for example, Greenland and Antarctica. The eventual involvement of more scientific teams with different backgrounds, approaches, methodologies, and perspectives in Pantepui paleoecological research would be very desirable and beneficial.

Acknowledgments

Paleoecological results summarized in this chapter were obtained in research projects funded by Spanish institutions, including the Ministry of Education and Science (project BIOSABANA; CGL2006-00974/BOS to V.R.), the Ministry of Science and Innovation (project ECOPAST; CGL2009-07069/BOS to V.R. and grant BES-2007-16308 to E.M.), the BBVA Foundation (projects PANTEPUI; BIOCON 2004-90/05 and PANTEPUI-II; BIOCON 2008-031 to V.R.), the Higher Council for Scientific Research or CSIC (projects PATAM; 200830I258 and PALAB; 201130E075 to V.R.), the Catalan Agency for Management of University and Research Grants (grant GREB; 2009-SGR439 to V. R., T.V., and E.M.), and the University of Barcelona (grant APIF to E.S.). Critical reviews by Paul Berry, Otto Huber, Celsa Señaris, Nathan Stansell, and Zicheng Yu contributed to the improvement of the manuscript.

References

Alley, R.B., 2000. The Younger Dryas cold interval as viewed from central Greenland. Quat. Sci. Rev. 19, 213–226.

Beerling, D.J., Osborne, C.P., 2006. The origin of the savanna biome. Glob. Change Biol. 12, 2023–2031.

Black, D.E., Abahazi, M.A., Thunell, R.C., Kaplan, A., Tappa, E.J., Peterson, L.C., 2007. A 8-century tropical Atlantic SST record from the Cariaco Basin: baseline variability, twentieth-century warming, and Atlantic hurricane frequency. Paleoceanography 22, PA4204. Available from: https://doi.org/10.1029/2007PA001427.

Bush, M.B., 1994. Amazonian speciation: a necessarily complex model. J. Biogeogr. 21, 5–17.

Bush, M.B., De Oliveira, P.E., Colinvaux, P.A., Miller, M.C., Moreno, J.E., 2004. Amazonian paleoecological histories: one hill, three watersheds. Palaeogeogr. Palaeoclimatol. Palaeoecol. 214, 359–393.

Bush, M., Flenley, J., Gosling, W. (Eds.), 2011. Tropical Rainforest Responses to Climate Change. Springer, Berlin.

Chen, J., Saunders, S.C., Crow, T.R., Naiman, R.J., Brosofske, K.D., Mroz, G.D., et al., 1999. Microclimate in forest ecosystem and landscape ecology. Bioscience 49, 288–297.

Chen, H., Sinha, A., Cruz, F.W., Wang, X., Edwards, R.L., d'Horta, F.M., et al., 2013. Climate change patterns in Amazonia and biodiversity. Nat. Commun. 4, 1411. Available from: https://doi.org/10.1038/ncomms2415.

Chiarucci, A., Araújo, M.B., Decocq, G., Beierkuhnlein, C., Fernández-Palacios, J.M., 2010. The concept of potential natural vegetation: an epitaph? J. Veg. Sci. 21, 1172–1178.

Colinvaux, P.A., De Oliveira, P.E., Bush, M.B., 2000. Amazonian and neotropical plant communities on glacial time-scales: the failure of the aridity and refuge hypothesis. Quat. Sci. Rev. 19, 141–169.

Colson, A.B., 1985. Routes of knowledge, an aspect of regional integration in the circum-Roraima area of the Guayana highlands. Antropológica 63/64, 103–149.

Davis, M.B., 1981. Quaternary history and the stability of forest communities. In: West, D.C., Shugart, D.B., Botkin, D.B. (Eds.), Forest Succession, Concepts and Applications. Springer, New York, pp. 152–153.

D'Apolito, C., Absy, M.L., Latrubesse, E.M., 2013. The Hill of Six Lakes revisited: new data and re-evaluation of a key Pleistocene Amazon site. Quat. Sci. Rev. 76, 140–155.

Ewel, K.C., 1990. Swamps. In: Myers, R.L., Ewel, J.J. (Eds.), Ecosystems of Florida. University of Central Florida Press, Orlando, pp. 281–323.

Fischer, A.G., 1960. Latitudinal variation in organic diversity. Evolution 14, 64–81.

Flenley, J.R., 1979. The Equatorial Rain Forest: A Geological History. Butterworth, London.

Haug, G., Hughen, K.A., Sigman, D.M., Peterson, L.C., Röhl, U., 2001. Southward migration of the Intertropical Convergence Zone through the Holocene. Science 293, 1304–1308.

Hernández, L., Fölster, H., 1994. Vegetación en transición. Scientia Guayanae 4, 118–144.

Hewitt, G.M., 1999. Post-glacial re-colonization of European biota. Biol. J. Linn. Soc. 68, 87–112.

Holdridge, D., 1933. Pindorama, Jungle—To You!. Minton Balch, New York.

Huber, O., 1992. El macizo del Chimantá, un ensayo ecológico tepuyano. Oscar Todtmann Editores, Caracas.

Huber, O., 1995a. Geographical and physical features. In: Berry, P., Holst, B.K., Yaskievych, K. (Eds.), Flora of the Venezuelan Guayana. Introduction, vol. 1. Missouri Botanical Garden Press, St. Louis, pp. 1–61.

Huber, O., 1995b. Conservation of the Venezuelan Guayana. In: Berry, P.E., Holst, B.K., Yatskievych, K. (Eds.), Flora of the Venezuelan Guayana., vol. 1. Missouri Botanical Garden Press, St. Louis, pp. 193–218. Introduction.

Huber, O., 1995c. Vegetation. In: Berry, P.E., Holst, B.K., Yatskievych, K. (Eds.), Flora of the Venezuelan Guayana. Introduction, vol. 1. Missouri Botanical Garden Press, St. Louis, pp. 97–160.

Huber, O., García, P., 2011. The Venezuelan Guayana region and the study areas: geo-ecological characteristics. In: Zink, J.A., Huber, O. (Eds.), Peatlands of the Western Guayana Highlands, Venezuela. Springer, Heidelberg, pp. 29–89.

Huntley, B., Birks, H.J.B., 1983. An Atlas of Past and Present Pollen Maps for Europe, 0–13000 Years Ago. Cambridge University Press, Cambridge.

Jacobs, B.F., Kingston, J.D., Jacobs, L.L., 1999. The origin of grass-dominated ecosystems. Ann. Missouri Bot. Gard. 86, 590–643.

Jacobson, G.L., Grimm, E.C., 1986. A numerical analysis of Holocene forest and prairie vegetation in central Minnesota. Ecology 67, 958–966.

Mayr, E., Phelps, W.H., 1967. The origin of the bird fauna of the South Venezuelan Highlands. Bull. Am. Mus. Nat. Hist. 136, 269–328.

McGregor, G.R., Nieuwolt, S., 1998. Tropical Climatology. Wiley, Chichester.

Miller, J.S., 2001. Myricaceae. In: Berry, P.E., Yaskiewych, K., Holst, B.K. (Eds.), Flora of the Venezuelan Guayana. Liliaceae–Myristicaceae, vol. 6. Missouri Botanical Garden Press, St. Louis, pp. 733–734.

Montoya, E., Rull, V., 2011. Gran Sabana fires (SE Venezuela): a paleoecological perspective. Quat. Sci. Rev 30, 3430–3444.

Montoya, E., Rull, V., Stansell, N.D., Bird, B.W., Nogué, S., Vegas-Vilarrúbia, T., et al., 2011a. Vegetation changes in the neotropical Gran Sabana (Venezuela) around the Younger Dryas chron. J. Quat. Sci. 26, 207–218.

Montoya, E., Rull, V., Nogué, S., 2011b. Early human occupation and land use changes near the boundary of the Orinoco and the Amazon basins (SE Venezuela): palynological evidence from El Paují record. Palaeogeogr. Palaeoclimatol. Palaeoecol. 310, 413–426.

Montoya, E., Rull, V., Stansell, N.D., Abbott, M.B., Nogué, S., Bird, B.W., et al., 2011c. Forest-savanna-*morichal* dynamics in relation to fire and human occupation in the southern Gran Sabana (SE Venezuela) during the last millennia. Quat. Res. 76, 335–344.

Morley, R.J., 1999. Origin and Evolution of Tropical Rain Forests. Wiley, Chichester.

Nogué, S., Rull, V., Montoya, E., Huber, O., Vegas-Vilarrúbia, T., 2009. Paleoecology of the Guayana Highlands (northern South America): Holocene pollen record from the Eruoda-tepui, Chimantá massif. Palaeogeogr. Palaeoclimatol. Palaeoecol. 281, 165–173.

Novello, V.F., Cruz, F.W., Vuille, M., Strikis, N.M., Edwards, R.L., Cheng, H., et al., 2017. A high-resolution history of the South American Monsoon from the Last Glacial Maximum to the Holocene. Sci. Rep. 7, 44267. Available from: https://doi.org/10.1038/srep44267.

Pianka, E.R., 1966. Latitudinal gradients in species diversity: a review of concepts. Am. Nat. 100, 33–46.

Poveda, G., Waylen, P.R., Pulwarty, R.S., 2006. Annual and inter-annual variability of the present climate in northern South America and southern Mesoamerica. Palaeogeogr. Palaeoclimatol. Palaeoecol. 234, 3–27.

Prance, G.T., 1982. Biological Diversification in the Tropics. Columbia University Press, New York.

Rull, V., 1991. Contribución a la paleoecología de Pantepui y la Gran Sabana (Guayana Venezolana): clima, biogeografía y ecología. Scientia Guayanae 2, 1–133.
Rull, V., 1996. Holocene vegetational succession in the Chimantá and Guaiquinima massifs (SE Venezuela). Interciencia 21, 7–20.
Rull, V., 2004a. An evaluation of the lost world and the vertical displacement hypotheses in the Chimantá massif, Venezuelan Guayana. Glob. Ecol. Biogeogr. 13, 141–148.
Rull, V., 2004b. Is the 'Lost World' really lost? Palaeoecological insights into the origin of the peculiar flora of the Guayana Highlands. Naturwissenschaften 91, 139–142.
Rull, V., 2005a. Biotic diversification in the Guayana Highlands: a proposal. J. Biogeogr. 32, 921–927.
Rull, V., 2005b. Vegetation and environmental constancy in the Neotropical Guayana Highlands during the last 6000 years? Rev. Palaeobot. Palynol. 135, 205–222.
Rull, V., 2005c. Palaeovegetational and palaeoenvironmental trends in the summit of the Guaiquinima massif (Venezuelan Guayana) during the Holocene. J. Quat. Sci. 20, 135–145.
Rull, V., 2007. Holocene global warming and the origin of the neotropical Gran Sabana in the Venezuelan Guayana. J. Biogeogr. 34, 279–288.
Rull, V., 2010a. Ecology and palaeoecology; two approaches, one objective. Open Ecol. J. 3, 1–5.
Rull, V., 2010b. The Guayana Highlands: a natural laboratory for the biogeographical and evolutionary study of the neotropical flora. In: Sámchez-Villagra, M.R., Aguilera, O.A., Carlini, A.A. (Eds.), Urumaco and Venezuelan Paleontology. The Fossil Record of the Northern Neotropics. Indiana University Press, Bloomington, pp. 84–102.
Rull, V., 2012. Community ecology: diversity and dynamics over time. Commun. Ecol. 13, 102–116.
Rull, V., 2015. Long-term vegetation stability and the concept of potential natural vegetation in the Neotropics. J. Veg. Sci. 26, 603–607.
Rull, V., Montoya, E., 2014. *Mauritia flexuosa* palm swamp communities: natural or human-made? A palynological study of the Gran Sabana region (northern South America) within a neotropical context. Quat. Sci. Rev. 99, 17–33.
Rull, V., Montoya, E., 2017. Holocene vegetation dynamics on the Apakará summit of the neotropical Guayana Highlands and potential environmental drivers. Rev. Palaeobot. Palynol. 240, 22–32.
Rull, V., Vegas-Vilarrúbia, T., 2008. Biopiracy rules hinder conservation efforts. Nature 453, 26.
Rull, V., Schubert, C., Aravena, R., 1988. Palynological studies in the Venezuelan Guayana Shield: preliminary results. Curr. Res. Pleistocene 5, 54–56.
Rull, V., Vegas-Vilarrúbia, T., Nogué, S., Huber, O., 2009. Conservation of the neotropical vascular flora from the neotropical Guayana Highlands in the face global warming. Cons. Biol. 23, 1323–1327.
Rull, V., Abbott, M.B., Vegas-Vilarrúbia, T., Bezada, M., Montoya, E., Nogué, S., et al., 2010. Paleoenvironmental trends in Venezuela during the last glacial cycle. In: Sánchez-Villagra, M.R., Aguilera, O.A., Carlini, A.A. (Eds.), Urumaco and Venezuelan Palaeontology. The Fossil Record of the Northern Neotropics. Indiana University Press, Bloomington, pp. 55–83.
Rull, V., Montoya, E., Nogué, S., Vegas-Vilarrúbia, T., Safont, E., 2013. Ecological palaeoecology in the neotropical Gran Sabana region: long-term records of vegetation dynamics as a basis for ecological hypothesis testing. Persp. Plant Ecol. Evol. Syst. 15, 338–359.
Rull, V., Montoya, E., Vegas-Vilarrúbia, T., Ballesteros, T., 2015a. New insights on paleofires and savannisation in northern South America. Quat. Sci. Rev. 122, 158–165.
Rull, V., Vegas-Vilarrúbia, T., Montoya, E., 2016. The neotropical Gan Sabana region: palaeoecology and conservation. Holocene 26, 1162–1167.
Safont, E., Rull, V., Vegas-Vilarrúbia, T., Montoya, E., Huber, O., Holst, B.K., 2016. Late-Holocene vegetation and fire dynamics on the summits of the Guayana Highlands: the Uei-tepui palynological record. Palaeogeogr. Palaeoclimatol. Palaeoecol. 455, 33–43.
Schneider, T., Bischoff, T., Haug, G.H., 2014. Migrations and dynamics of the intertropical convergence zone. Nature 513, 45–53.
Schubert, C., Fritz, P., 1985. Radiocarbon ages of peat, Guayana Highlands (Venezuela). Naturwissenschaften 72, 427–429.
Schubert, C., Fritz, P., Aravena, R., 1994. Late quaternary paleoenvironmental studies in the Gran Sabana (Venezuelan Guayana Shield). Quat. Int. 21, 81–90.

Slobodkin, L.B., Sanders, H.L., 1969. On the contribution of environmental predictability to species diversity. In: Woodwell, G.M., Smith, H.H. (Eds.), Ecological Systems. Brookhaven Symposia in Biology, vol. 22. pp. 82–93.

Smol, J.P., Last, W.M., Birks, H.J.B., Lotter, A.F., Juggins, S. (Eds.), 2001–2012. Tracking Environmental Change Using Lake Sediments, 5 vols. Kluwer, Dordrecht.

Sommer, R.S., Zachos, F.E., 2009. Fossil evidence and phylogeography of temperate species, 'glacial refugia' and post-glacial recolonization. J. Biogeogr 36, 2013–2020.

Steyermark, J.A., 1979. Plant refuge and dispersal centres in Venezuela: their relict and endemic element. In: Larsen, K., Holm-Nielsen, L. (Eds.), Tropical Botany. Academic Press, New York, pp. 185–221.

Stuiver, M., Grootes, P.M., Braziunas, T.F., 1995. The GISP2 ^{18}O climate record of the past 16,500 years and the role of the sun, ocean and volcanoes. Quat. Res. 44, 341–354.

Tate, G.H.H., 1930. Notes on the Mount Roraima region. Geogr. Rev. 20, 53–68.

Tate, G.H.H., 1932. Life zones at Mount Roraima. Ecology 13, 235–257.

Thomas, D.J., 1982. Order without Government: the Society of the Pemon Indians of Venezuela. University of Illinois Press, Illinois.

van der Hammen, T., 1974. The Pleistocene changes in vegetation and climate in tropical South America. J. Biogeogr. 1, 3–26.

Vegas-Vilarrúbia, T., Rull, V., Montoya, E., Safont, E., 2011. Quaternary palaeoecology and nature conservation: a general review and some examples from the Neotropics. Quat. Sci. Rev. 30, 2361–2388.

West, R.G., 1980. Pleistocene forest history in East Anglia. New Phytol. 85, 571–622.

Whitmore, T.C., Prance, G.T., 1987. Biogeography and Quaternary History in Tropical Latin America. Oxford University Press, New York.

Wright, H.E., 1977. Quaternary vegetation history—some comparisons between Europe and America. Annu. Rev. Earth Planet. Sci. 5, 123–158.

Zinck, J.A., 2011. Tropical and subtropical peats: an overview. In: Zink, J.A., Huber, O. (Eds.), Peatlands of the Western Guayana Highlands, Venezuela. Springer, Heidelberg, pp. 5–28.

Peatlands of the Western Guayana Highlands. In: Zink, J.A., Huber, O. (Eds.), Venezuela. Springer, Heidelberg.

Zink, J.A., García, P., van de Plicht, J., 2011. Tepui peatlands: age record and environmental changes. In: Zink, J.A., Huber, O. (Eds.), Peatlands of the Western Guayana Highlands, Venezuela. Springer, Heidelberg, pp. 189–236.

CHAPTER

3

Pantepui as a dynamic biogeographical concept

Valentí Rull[1] and Teresa Vegas-Vilarrúbia[2]

[1]**Institute of Earth Sciences Jaume Almera (ICTJA-CSIC), Barcelona, Spain** [2]**Department of Evolutionary Biology, Ecology and Environmental Sciences, Faculty of Biology, University of Barcelona, Barcelona, Spain**

Introduction

Pantepui has been defined in this book using orographic and biotic criteria. This biogeographical province has been considered to include the high mountain ecosystems of the Guiana region in equatorial South America, found both on sandstone table mountains and on traditional mountains with continuous slopes, at elevations between 1300 and 1500 m up to around 3000 m elevation. Such elevational range includes all upper and high mountain ecosystems of the Guiana region located above the continuous submontane forest belt and has been considered the counterpart of the alpine belt situated immediately above the montane forest belt in other mountain systems (Huber et al., 2018). From a biotic perspective, the typical ecosystems of Pantepui are dominated by broadleaved meadows and sclerophyllous woodlands growing on peat, pioneering saxicolous plant communities, and small patches of low forests growing in depressions and crevasses. The most characteristic plant families of Pantepui are the herbaceous Rapateaceae, the woody Bonnetiaceae, and the pitcher plant family Sarraceniaceae with its endemic pantepuian genus *Heliamphora* (Chapter 1: Definition and characterization of the Pantepui biogeographical province). Thus, the modern concept of Pantepui has been defined on the basis of its present-day vascular flora. As the Pantepui fauna is equally characteristic of this biogeographical province and depends to a large degree on Pantepui plant communities for its life, we use the term "biota" to refer to both flora and fauna. However, some differences may exist in the distribution of some animal groups that are explained in the corresponding chapters (8 to 15). The orographic definition of Pantepui implicitly involves

Biodiversity of Pantepui
DOI: https://doi.org/10.1016/B978-0-12-815591-2.00003-3

a climatic characterization. In the Guiana mountains, climate, especially temperature, is a major environmental driver of vegetation distribution. Indeed, the adiabatic lapse rate (−0.6°C/100 m elevation) has been considered the main factor responsible for vegetation arrangement along elevational gradients (Huber, 1995a). The climate of Pantepui is mesothermic ombrophilous in its lowermost part (1500–2400 m elevation) and submicrothermic ombrophilous between 2400 and ~3000 m. Mesothermic climates are characterized by average annual air temperatures between 12°C and 18°C and high to very high average total rainfall (2500–3500 mm per year). Additional moisture is supplied by the frequent occurrence of dense mists. Submicrothermic climates have an average annual air temperature around 10°C or less and are submitted to heavy rainfalls (>3500 mm) and dense cloud and mist formation almost all year (Huber, 1995b). Therefore, Pantepui is defined by the concurrence in time and space of the particular orographic, climatic, and biotic features that characterize this biogeographical province.

This is the Pantepui we know today, but paleoecology shows us that the present is a transient state as part of long-term ecological processes that occur over a time continuum under the action of external environmental drivers and intra-ecosystem interactions (Rull, 2010, 2014). Therefore the present Pantepui state, in which certain orographic, climatic, and biotic features coincide in time and space, could be a transient situation with no analogs in the past and hardly predictable in the future. This is the scenario to be analyzed in this chapter. In particular, we are interested in whether or not Pantepui as a biogeographical unit has remained constant under the action of Quaternary climate changes and what we can expect for the future of Pantepui under the predicted global warming. These questions are pertinent as, in Chapter 2 (Climatic and ecological history of Pantepui and surrounding areas), we have demonstrated that past climatic changes have affected the Pantepui ecosystems and it is also expected that future climatic changes, especially the predicted global warming of the present century, will affect the Pantepui natural life (Rull and Vegas-Vilarrúbia, 2006; Nogué et al., 2009).

Hypotheses and approaches

As mentioned in the introduction, Pantepui could be considered the spatiotemporal coincidence of three main biogeographical features, namely orography, climate, and biota. In this chapter, the null hypothesis (H_0) is that such a situation has not changed in the past and will remain the same in the future, in spite of the occurrence of climate changes. The alternative hypothesis (H_1) is that climate changes may disrupt the tripartite coincidence, thus questioning the continuity of Pantepui through time. In this case, we could ask how this biogeographical unit has changed through time and whether any of the key features (orography, climate, or biota) is more important in the definition of Pantepui under a long-term perspective. The conceptual analysis will be done by examining the key factors individually and then combining the outputs obtained. For this purpose, we will split the Pantepui concept into three components—the orographic Pantepui, the climatic Pantepui, and the biotic Pantepui—and analyze each of them separately, with emphasis on the potential effect of the Quaternary glacial–interglacial alternation. The outcomes obtained will then be used as past analogs to address the potential consequences of the global warming projected for the present century. We should stress that the focus is on Pantepui as a biogeographical unit, rather than on other ecological and evolutionary aspects, which have already been discussed in Chapter 2.

The Pantepui components

The orographic Pantepui

Orographically, the Guiana mountains have experienced no significant changes during the Quaternary. These mountains were already in place, with their present-day geology, elevation, and topography, since the Plio-Pleistocene. According to Briceño and Schubert (1990), the Auyán-tepui erosion surface, which forms the summit of the tepuis cut into the Precambrian Roraima quartzites, could be Mesozoic or older in age (see Fig. 4.4 from Chapter 4: On the origin of Pantepui biota). Minor differences could be attributed to unknown rates of Pleistocene erosion and possibly in edaphic conditions. Today, huge extensions of the more or less flat surface of the tepuian summits are covered by a peat blanket up to several meters in depth, on which most Pantepui ecosystems develop (Plate 3.1). The maximum age of these peats remains unknown, but the oldest dates obtained thus far are Late Glacial and Holocene (Zink et al., 2011). Therefore we cannot make substantiated inferences before those dates. As peat accumulation is closely linked to climate, this aspect will be further discussed in the next sections. All we can say for now is that the orographic Pantepui has not changed through the Quaternary and is not expected to change in the near future. Therefore the physical scenario for the climatic and the biotic Pantepui has remained constant along the time frame considered in this chapter.

The climatic Pantepui

Pantepui is an interglacial entity, as its climatic context corresponds to the Holocene (11.7 cal kyr BP to the present), the interglacial in which we live following the last glaciation, which occurred between about 115 and 11.7 cal kyr BP, with full glacial conditions during the Last Glacial Maximum (LGM), around 21 cal kyr BP. As we have seen in Chapter 2, in the southern part of Pantepui, average temperatures were about 5°C lower than at present, as deduced from the downward migration of sensitive species from the tepuian slopes to the lowlands (Bush et al., 2004). If we consider the Neotropics in general, the magnitude of the temperature lowering, as deduced from multidisciplinary evidence, was of **5°C–6°C**, which corresponds to a downward migration of bioclimatic conditions of 1000–1100 m (Farrera et al., 1999). This has relevant ecological and evolutionary implications, some of which have already been discussed in Chapter 2. Here we will concentrate on the eventual changes in the shape and extension of the climatic Pantepui during the LGM.

Fig. 3.1 compares the present Pantepui area—the summits above 1500 m elevation—with the expected LGM region with similar climatic conditions, considering a maximum

PLATE 3.1 (A) Vertical cliffs of the Chimantá massif modeled on the quartzites/sandstones of the Roraima Group. The vegetation growing on their top develops on Late Glacial/Holocene peats up to a few meters in depth. (B) Peat coring on the Apakará summit (Chimantá massif). The length of the corer shows the maximum depth of the peats encountered (~3 m). (C and D) Eruoda-tepui (Chimantá massif). Small plant stands growing in favorable microenvironments within rock crevasses (microrefugia). (E) Apaurai-tepui, an example of a totally isolated tepuian summit. (F) Fluvial valley connecting the summit of the Apakará-tepui (Chimantá massif) with the surrounding Gran Sabana uplands. *Source: Photos: V. Rull.*

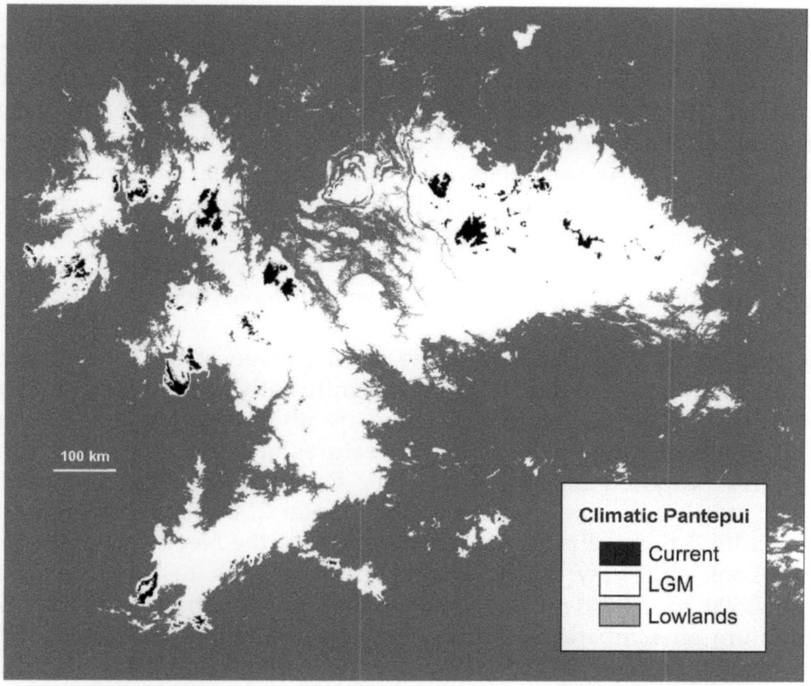

FIGURE 3.1 Present (*black*) and simulated Last Glacial Maximum (LGM) climatic Pantepui (*white*). LGM simulation was performed assuming a downward migration of climatic conditions up to 1100 m below the present lower boundary of Pantepui (1500 m) on a digital elevation model from the Shuttle Radar Topography Mission (USGS/NGA/NASA) of 5 arcsec (90 m) precision. *Source: Modified from Rull, V., Nogué, S., 2007. Potential migration routes and barriers for vascular plants of the neotropical Guyana Highlands during the Quaternary. J. Biogeogr. 34, 1327–1341.*

downward displacement of 1100 m (Rull and Nogué, 2007). In these conditions, the climatic Pantepui would have included all the terrains situated above 400 m. This not only expands significantly the area in which the Pantepui biota would have been able to grow but also replaces the typical fragmentary character of the present Pantepui with a continuous space connecting virtually all the Guiana Highlands. This situation could be extrapolated to the remaining Quaternary glaciations with the pertinent differences in the magnitude of temperature and elevational shifts. In this analysis, due to the lack of suitable meteorological data for the region, only temperature and the adiabatic lapse rate for the Guiana region could be used to infer elevational shifts of the climatic Pantepui, but other analyses using more complete and sophisticated climatic modeling, as for example, maximum entropy-based climate envelope distribution models (CEDMs), yielded similar results (Rödder et al., 2010).

About 50 glacial–interglacial cycles have been recorded during the Quaternary. The pace of these cycles is known to have been controlled by insolation changes generated by the astronomical Milankovitch cycles. The precession cycle (41-kyr period) was dominant until 800 cal kyr BP, and the eccentricity cycle (100-kyr period) set the pace since then (Raymo, 1994). Therefore the climatic Pantepui is believed to have been highly dynamic during this period. To complicate the issue, minor millennial-scale warmings of a few degrees (interstadials) have been recorded during the glaciations, thus increasing the climatic variability (Fig. 3.2). As was mentioned in Chapter 2, glacial climates were dominant during the Quaternary, whereas interglacials can be considered short warmings typically of a 10–20 kyr duration (Willis and Whittaker, 2000; Bush et al., 2001) (Fig. 3.2). Therefore present interglacial conditions could be considered exceptional in the Quaternary context, and therefore the glacial climatic Pantepui would have been the norm during the Quaternary.

Using the present adiabatic lapse rate, the LGM climate of the tepuian summits—that is, the orographic Pantepui—would have been significantly cooler (about 6°C less than the present, in average). In the Guiana region there are no present analogs (i.e., terrains at or above 4000 m elevation) to infer the potential features of these summits under glacial climates. Some have proposed that the orographic Pantepui was arid and devoid of peats (Schubert et al., 1994), but as we have seen in Chapter 2, this is still a matter of speculation. The maximum estimated precipitation decrease for the Neotropics during the LGM is of 25%–35% (Cook and Vizy, 2006), which is not enough to turn the rainy Pantepui climates, with 2500 to >3500 mm of total annual precipitation, into arid or semi-arid climates. Minimum air temperatures measured atop the tepuis are around 1°C–2°C, and frost has not been observed (Huber, 1995b), but an LGM lowering of 6°C on average would have likely promoted ice formation. Eventual glacier formation has been dismissed because of the comparatively low elevation—in the Andean highlands at the same

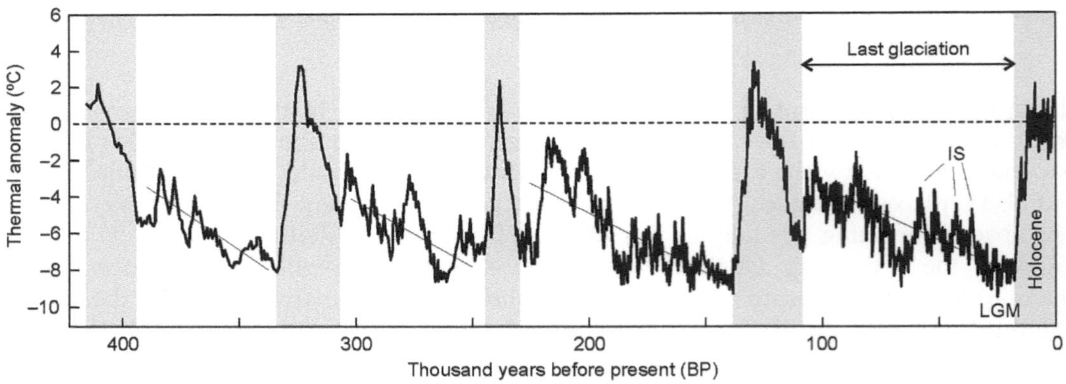

FIGURE 3.2 Temperature anomalies (*solid line*) with respect to the present average (*dotted line*) during the last four glacial cycles as recorded in the Vostok ice core (Antarctica). Interglacials are highlighted in *gray*. Note that the onsets of interglacial warmings are abrupt, whereas glaciations exhibit a gradual temperature decrease spiked by minor interstadial events (*IS*). *LGM*, Last Glacial Maximum. *Source: Redrawn from Petit, J.R., Jouzel, J., Raynaud, D., Barkov, N.I., Barnola, J.-M., Basile, I., et al., 1999. Climate and atmospheric history of the past 420,000 years from the Vostok ice core, Antarctica. Nature, 399, 429–436. Petit et al. (1999).*

latitude, the LGM snowline was at significantly higher elevations (3500–4000 m) than Pantepui (Lachniet and Vazquez-Selem, 2005)—and the absence of geomorphological evidence of glacier activity.

The biotic Pantepui

Paleoecological records from Guiana and adjacent montane regions have shown that not all species responded to temperature shifts by modifying their elevational range. Only the more climatically sensitive species reacted to the temperature shifts by migrating up and down, whereas others with wider tolerance and/or more phenotypic plasticity were able to remain relatively stable (Rull et al., 2013, 2015). Unfortunately, the lack of LGM records for Pantepui species prevents us from knowing the characteristics of the biotic Pantepui during glacial times. The mentioned downward migration of sensitive species from the tepuian slopes to the lowlands during the LGM (Bush et al., 2004) suggests that something similar would have happened at other elevations, for example, between the tepuian summits and their surrounding slopes. This is supported by pollen records from some summits, which have been interpreted in terms of up-and-down migrations of relevant Pantepui taxa coinciding with Late Glacial and Holocene temperature shifts (Rull, 2004a,b; Rull and Montoya, 2017). In this scenario, a significant biotic interchange would be expected in the lowlands and the tepuian slopes during the LGM and the Quaternary glaciations in general. Therefore the biota of the glacial climatic Pantepui (Fig. 3.1) would have been composed of a mixture of the biotic elements we observe today at different elevational levels. The lowlands would have been inhabited by a mixture of lowland-tolerant species and slope-sensitive species, whereas the slopes would have been occupied by slope-tolerant species mixed with summit-sensitive species. In the summits, the situation is less predictable, as local tolerant species would have been present, but there are no higher elevations (>3000 m) from which to migrate downwards. Whether present-like peats existed or not is still speculative, as is the occurrence of potential ecological microrefugia (Rull, 2009) on protected sites (Plate 3.1).

With the available evidence, it is not possible to go beyond such a general picture. We can hypothesize that during the LGM species that characterize the present Pantepui flora would have been irregularly spread in both geographic and topographic terms across the climatic Pantepui (Fig. 3.1), depending on their particular climatic sensitivity. Unfortunately, no detailed autoecological studies exist for the tepuian biota, and any inference should be based on indirect indicators, for example, the elevational range of each species as a proxy for their temperature tolerance. An eventual LGM biogeographer would hardly have described Pantepui as a biogeographical unit, as the biotic Pantepui we observe today was most likely disassembled, eventually fragmented, and partly integrated into other biogeographic units downslope. For long time it was believed that the tabular topography of the tepuis hindered eventual downward displacements of their summit species due to the impossibility of migrating through their huge vertical cliffs, which purportedly acted as biogeographical barriers (Maguire, 1970). However, further floristic and physiographic observations revealed that only a small fraction of the tepuian summits were really isolated, and most of them are connected to the surrounding uplands and lowlands by valleys, ridges, and other topographical features that could have acted as vertical migration corridors (Steyermark and Dunsterville, 1980; Huber, 1988) (Plate 3.1).

The Pantepui oscillator

These observations enable us to hypothesize about the status of Pantepui as a biogeographical unit throughout the Quaternary. For this purpose, the last glaciation and the Holocene will be taken as representatives of the Pleistocene glaciations and interglacials, respectively. Eventual variations in the magnitude of temperature shifts and up-and-down biotic migrations have existed, but they do not significantly change the general picture. As we have seen, only the orographic Pantepui has remained constant, whereas the climatic Pantepui and the biotic Pantepui have been highly variable in terms of area, elevation, and composition. Pantepui, expressed as the coincidence of the orographic, climatic, and biotic components, has been recurrently assembled during interglacials and disassembled during glaciations, a phenomenon that is called here the Pantepui oscillator (Fig. 3.3). Glacial climates have been dominant by far during the Quaternary

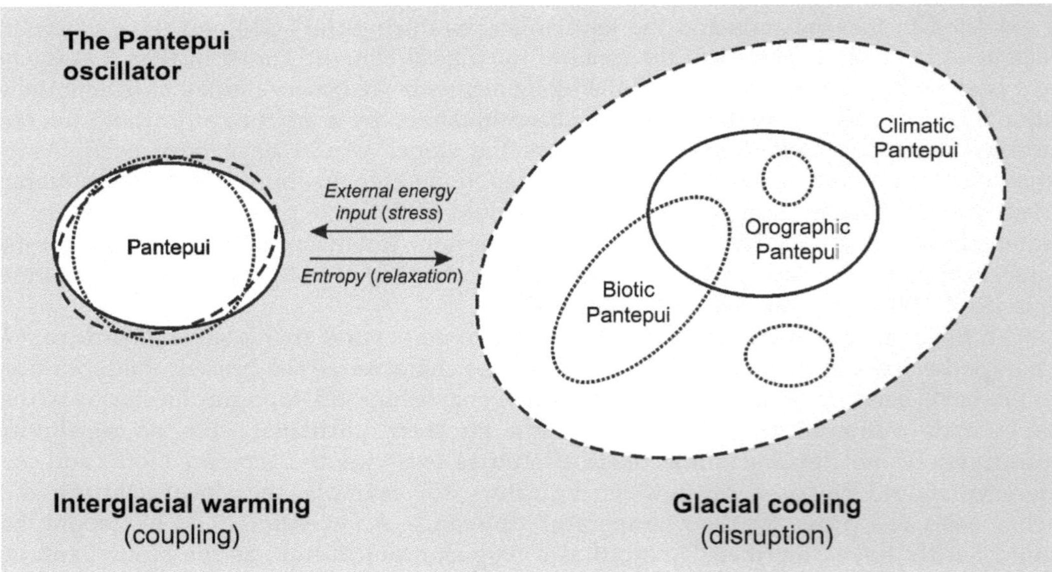

FIGURE 3.3 The Pantepui oscillator. The orographic (*solid lines*), the climatic (*dashed lines*), and the biotic (*dotted lines*) Pantepui assemble during the interglacials and disassemble during glaciations through downslope migration of the last two Pantepui components. The interglacial state is the less stable, as it is maintained by external energy inputs (i.e., temperature maxima), whereas the glacial state is more stable as entropy maximizes.

(Willis and Whittaker, 2000) and have been considered the normal condition for this period (Bush et al., 2001). Therefore Pantepui disruption could be considered the norm and Pantepui assembly the exception.

Fig. 3.2 shows that the glacial–interglacial recurrence is typically asymmetric, with abrupt interglacial warmings and gradual glacial coolings. Warming at the end of glaciations occurs more abruptly than increases in insolation, which has been attributed to the occurrence of positive feedbacks exacerbating the response of the climatic system to solar forcing. The main factor responsible for such feedbacks is changes in ice albedo and atmospheric CO_2 concentration. Ellis and Palmer (2016) explain in detail the mechanism responsible for such climatic forcing, which is beyond the scope of this chapter. What is important here is that interglacials are triggered by the incoming of extra energy inputs to the climatic system causing rapid warmings. When these inputs cease, the system gradually returns to its normal condition, which is characterized by glacial climates. Thus in thermodynamic terms (Hill, 1960), interglacials could be viewed as short unstable states maintained by external energy inputs that, once terminated, enable system relaxation and entropy maximization, which is characteristic of glaciations. Therefore from a thermodynamic point of view, glacial disassembly could be viewed as the more stable state of Pantepui, with maximum entropy (disorder), whereas interglacial assembly (order) would be a transient condition maintained by the incoming of external energy inputs. In other words, interglacials stress the oscillator (Fig. 3.3) leading to the Pantepui coupling by upslope displacement of the climatic and biotic components, whereas glaciations relax the system and maximize entropy by Pantepui disassembly due to downslope migration.

Pantepui assembly and disassembly would have occurred about 50 times during the Quaternary following the glacial–interglacial recurrence. With the available evidence, it is not possible to know the composition of the Pantepui biota from the different interglacial phases, but variations could have existed due to differences in the magnitude and regime of climate changes, the idiosyncratic features of the different species in relation to climate, and to eventual randomness in climate changes and migration processes. Evolution was also active in the Neotropics during the Quaternary (Rull, 2008) and led to the formation of new Pantepui species (e.g., Salerno et al., 2012; Bonaccorso and Guayasamin, 2013). Interglacial warming stress may have eventually led to the extinction of some Pantepui species due to habitat loss atop the tepuis (Rull, 2005). Glacial extinction is far less probable, as downward migration is always possible. Therefore it is likely that during the Quaternary, there have been as many different versions of the biotic Pantepui as interglacials. This biogeographical continuity, in spite of composition differences due to ecological and evolutionary changes, could be an example of the so-called dynamic kinetic stability, which has been considered typical of biological systems in opposition to the physical ones (Wagner and Pross, 2011). Therefore from a thermodynamic perspective, Pantepui could be viewed as a recurrent unstable state maintained by external energy inputs, but from a biological point of view, the Pantepui persistence through time with different compositional states could be considered a manifestation of dynamic stability leading to biogeographical resilience.

The future

The predicted global warming for the end of the present century may represent an additional pressure to the already stressed interglacial climates of the Holocene. In these conditions, the Pantepui oscillator (Fig. 3.3) would be forced beyond the standard for Quaternary interglacials, thus reducing the climatic and the biotic Pantepui inside the orographic Pantepui. It has been estimated that under the current Intergovernmental Panel on Climate Change (IPCC) projections, the climatic Pantepui could experience a surface reduction of almost 90%, which may lead to the extinction of habitat loss of ~80% of its flora, of which nearly half is endemic (Nogué et al., 2009; Safont et al., 2012). Under such projections, several important tepuian summits are set to completely lose the climatic Pantepui and hence the biotic Pantepui (Vegas-Vilarrúbia et al., 2012). Others will retain barely 10% or less of the current Pantepui habitat. Only in three tepuian summits (Roraima-Kukenán, Marahuaka, and the Chimantá massif) will the reduction likely be less severe, though still significant (see details on these estimates in Chapter 17: Pantepui and global warming). This may suggest that if the global warming progresses, the Pantepui biogeographical province might disappear.

A relevant question in our context is whether the current global warming will delay the inception of the next glaciation. According to Tzedakis et al. (2012), the Holocene resembles other interglacials in terms of astronomical forcing but differs in atmospheric CO_2 concentration, because the current anthropogenic increase is delaying the expected temperature decline with respect to former interglacials. Such a delay has been investigated using several types of models, some of which predict that under natural conditions, the next glaciation could start in a couple of millennia, but the anthropogenic CO_2 increase may cause a delay of ~40,000 years (Herrero et al., 2014). Other model outputs are more drastic and prognosticate the interruption of the 100-kyr climatic cycle, which has set the pace of the glacial–interglacial cycles since 800 cal kyr BP to the present (Haqq-Mistra, 2014). According to this, the CO_2 increase would be able to overcome the climatic consequences of astronomical forcing during at least the next million years, thus stopping the glacial–interglacial recurrence. Therefore in natural conditions, the Pantepui would start to disassemble in two millennia, but the current anthropogenic CO_2 increase would stop such disruption for the next 40,000 years or, if the more dramatic models are right, for the next million years. In this case, it would be interesting to know the potential temperature increase predicted by these future time intervals to infer the potential fate of the present Pantepui.

It has been estimated that a temperature increase of 9°C above present values would be needed to completely eliminate the climatic Pantepui and its biota (Rull and Vegas-Vilarrúbia, 2006). Such a temperature increase would lead the Pantepui oscillator to surpass the tipping point, beyond which reversibility is no longer possible, as species extinction is irreversible. Although an average temperature increase of 9°C is beyond the worst expectations for the present century (IPCC, 2014), some believe that even an increase of 2°C, as targeted in the Paris Agreement (UN, 2015), could lead to a cascade effect triggered by nonlinear feedbacks able to destabilize the Earth's system and cease the glacial–interglacial cyclicity (Steffen et al., 2018). These authors suggest that beyond this point, the Earth might enter in a pathway towards a "hothouse Earth" state, characterized by high temperatures, ice-free poles, low pole–equator thermal contrast, high atmospheric CO_2 concentrations, and high sea levels (Kidder and Worsley, 2010). Similar "hothouse" conditions have occurred at least 15 times during geological history, lasting from <1 to several million years and coinciding with biological extinction events (Kidder and Worsley, 2012). If these authors are right, temperature increases sufficient to cause the disappearance of the Pantepui biogeographical province as a whole should not be dismissed. However, empirical data and model simulations to support such hypothesis are still lacking.

Conclusions and further research

The evidence reviewed and discussed in this chapter supports the alternative hypothesis (H_1) that Pantepui as a biogeographical unit has not remained constant through the Quaternary and may significantly change in the future. The available information suggests that Pantepui has recurrently been assembled (interglacials) and disassembled (glacials) by differential up-and-down migration of its climatic and biotic components with respect to the orographic component, which has remained constant. Glacial Pantepui disruption seems to have been the more frequent and stable state, whereas interglacial assembly seems to have been maintained by external energy inputs of shorter duration. Therefore our current understanding of Pantepui—where orographic, climatic, and biotic elements coincide—could be considered a transient and relatively unstable state in the long term, which will be eventually disrupted and stabilized with the incoming of the next glaciation. In the near future, the climatic and the biotic Pantepui are in danger of being significantly reduced inside the orographic Pantepui due to the projected global warming caused by the anthropogenic increase of atmospheric CO_2 concentration. In summary, Pantepui may be considered more than a static present-day biogeographic unit and instead a rather dynamic biogeographic concept—represented by the Pantepui oscillator and including as many potential compositional states as interglacials—in both time and space under the action of Quaternary climate shifts and the ensuing biotic responses.

Future studies should be aimed at testing the hypotheses derived from this chapter and other papers that may emerge in the future. As highlighted in Chapter 2, finding LGM sediments in the tepuian summits and the surrounding uplands and lowlands is critical to obtain direct evidence of the Pantepui glacial climates and biota. Older sediments would ideally provide information on former glacial conditions, including the response of the Pantepui components to the minor millennial-scale interstadial warmings. Sediments corresponding to the interglacial preceding the last glaciation would be crucial to record a Pantepui status likely different from the present Holocene conditions, thus contributing to test the hypothesis of the Pantepui oscillator and the eventual changes in the composition of the Pantepui biota. Regarding the future, estimations of biotic extinction due to habitat loss of the Pantepui biota could be refined with autoecological studies of the involved species. The remoteness and the high Pantepui biodiversity are handicaps to developing this type of study, but a selection of the most representative species could be a useful approach (Safont et al., 2012; Rull and Vegas-Vilarrúbia, 2017).

Acknowledgments

The authors are grateful to Fred Stauffer and José María Fernández-Palacios for their critical revision of the manuscript.

References

Bonaccorso, E., Guayasamin, J.M., 2013. On the origin of Pantepui montane biotas: a perspective based on the phylogeny of *Aulancorhynchus tocanets*. PLoS One 8, 6e67321.

Briceño, H.O., Schubert, C., 1990. Geomorphology of the Gran Sabana, Guayana Shield, southeastern Venezuela. Geomorphology 3, 125–141.

Bush, M., Stute, M., Ledru, M.-P., Behling, H., Colinvaux, P.A., De Oliveira, P.E., et al., 2001. Palaeotemperature estimates for the lowland Americas between 30°S and 30°N at the Last Glacial Maximum. In: Markgraf, V. (Ed.), Interhemispheric Climate Linkages. Academic Press, San Diego, pp. 293–306.

Bush, M.B., De Oliveira, P.E., Colinvaux, P.A., Miller, M.C., Moreno, J.E., 2004. Amazonian paleoecological histories: one hill, three watersheds. Palaeogeogr. Palaeoclimatol. Palaeoecol. 214, 359–393.

Cook, K.H., Vizy, E.K., 2006. South American climate during the Last Glacial Maximum: delayed onset of the South American monsoon. J. Geophys. Res. 11, D02110.

Ellis, R., Palmer, M., 2016. Modulation of ice ages via precession and dust-albedo feedbacks. Geosci. Front. 7, 891–909.

Farrera, I., Harrison, S.P., Prentice, I.C., Ramstein, G., Guiot, J., Bartlein, P.J., et al., 1999. Tropical climates at the Last Glacial Maximum: a new synthesis of terrestrial paleoclimate data. I. Vegetation, lake-levels and geochemistry. Clim. Dyn. 15, 823–856.

Haqq-Mistra, J., 2014. Damping of glacial–interglacial cycles from anthropogenic forcing. J. Adv. Mod. Earth Syst. 6, 950–955.

Herrero, C., García-Olicares, A., Pelegrí, J.L., 2014. Impact of anthropogenic CO_2 on the next glacial cycle. Clim. Change 122, 283–298.

Hill, T.L., 1960. An Introduction to Statistical Thermodynamics. Dover, New York.

Huber, O., 1988. Guayana highlands versus Guayana lowlands, a reappraisal. Taxon 37, 595–614.

Huber, O., 1995a. Vegetation. In: Berry, P.E., Holst, B.K., Yatskievych, K. (Eds.), Flora of the Venezuelan Guayana, vol. 1. Missouri Botanical Garden Press, St. Louis, pp. 97–160. Introduction.

Huber, O., 1995b. Geographical and physical features. In: Berry, P., Holst, B.K., Yaskievych, K. (Eds.), Flora of the Venezuelan Guayana., vol. 1. Missouri Botanical Garden Press, St. Louis, pp. 1–61. Introduction.

Huber, O., Prance, G.T., Kroonenberg, S., Antonelli, A., 2018. The tepuis of the Guiana Highlands. In: Hoorn, C., Perrigo, A., Antonelli, A. (Eds.), Mountains, Climate and Biodiversity. Wiley-Blackwell, Hoboken, pp. 339–353.

IPCC, 2014. Climate Change 2014: Synthesis Report. IPCC, Geneva.

Kidder, D.L., Worsley, T.R., 2010. Phaneorozoic Large Igneous Provinces (LIPs), HEATT (Haline Euxinic Acidic Thermal Transgression) episodes, and mass extinctions. Palaeogeogr. Palaeoclimatol. Palaeoecol. 295, 162–191.

Kidder, D.L., Worsley, T.L., 2012. A human-induced hothouse climate?. GSA Today 22, 4–11.

Lachniet, M.S., Vazquez-Selem, L., 2005. Last Glacial Maximum equilibrium line altitudes in the circum-Caribbean (Mexico, Guatemala, Costa Rica, Colombia, and Venezuela). Quat. Int. 138-139, 129–144.

Maguire, B., 1970. On the flora of the Guayana Highland. Biotropica 2, 85–100.

Nogué, S., Rull, V., Vegas-Vilarrúbia, T., 2009. Modelling biodiversity los by global warming on Pantepui, northern South America: projected upward migration and potential habitat loss. Clim. Change 96, 77–85.

Petit, J.R., Jouzel, J., Raynaud, D., Barkov, N.I., Barnola, J.-M., Basile, I., et al., 1999. Climate and atmospheric history of the past 420,000 years from the Vostok ice core, Antarctica. Nature 399, 429–436.

Raymo, M.E., 1994. The initiation of Northern Hemisphere Glaciations. Annu. Rev. Earth Planet. Sci. 22, 353–383.

Rödder, D., Schlüter, A., Lötters, S., 2010. Is the 'Lost World' lost? High endemism of amphibians and reptiles on South American tepuis in a changing climate. In: Habel, J.C., Assmann, T. (Eds.), Relict Species: Phylogeography and Conservation Biology. Springer, Berlin, pp. 401–416.

Rull, V., 2004a. An evaluation of the lost world and the vertical displacement hypotheses in the Chimantá massif, Venezuelan Guayana. Glob. Ecol. Biogeogr 13, 141–148.

Rull, V., 2004b. Is the 'Lost World' really lost? Palaeoecological insights into the origin of the peculiar flora of the Guayana Highlands. Naturwissenschaften 91, 139–142.
Rull, V., 2005. Biotic diversification in the Guayana Highlands: a proposal. J. Biogeogr. 32, 921–927.
Rull, V., 2008. Speciation timing and neotropical biodiversity: the Tertary–Quaternary debate in the light of molecular phylogenetic evidence. Mol. Ecol. 17, 2722–2729.
Rull, V., 2009. Microrefugia. J. Biogeogr. 36, 481–484.
Rull, V., 2010. Ecology and palaeoecology: two approaches, one objective. Open Ecol. J. 3, 1–5.
Rull, V., 2014. Time continuum and true long-term ecology: from theory to practice. Front. Ecol. Evol. 2, 75. Available from: https://doi.org/10.3389/fevo.2014.00075.
Rull, V., Montoya, E., 2017. Holocene vegetation dynamics on the Apakará summit of the neotropical Guayana Highlands and potential environmental drivers. Rev. Palaeobot. Palynol. 240, 22–32.
Rull, V., Nogué, S., 2007. Potential migration routes and barriers for vascular plants of the neotropical Guyana Highlands during the Quaternary. J. Biogeogr. 34, 1327–1341.
Rull, V., Vegas-Vilarrúbia, T., 2006. Unexpected biodiversity loss under global warming in the neotropical Guayana Highlands: a preliminary appraisal. Glob. Change Biol. 12, 1–9.
Rull, V., Vegas-Vilarrúbia, T., 2017. Potential responses of vascular plants from the pristine "Lost World" of the Neotropical Guayana Highlands to global warming: review and new perspectives. Front. Plant Sci. 8, 81. Available from: https://doi.org/10.3389/fpls.2017.00081.
Rull, V., Montoya, E., Nogué, S., Vegas-Vilarrúbia, T., Safont, E., 2013. Ecological palaeoecology in the neotropical Gran Sabana region: long-term records of vegetation dynamics as a basis for ecological hypothesis testing. Persp. Plant Ecol. Evol. Syst. 15, 338–359.
Rull, V., Vegas-Vilarrúbia, T., Montoya, E., 2015. Neotropical vegetation responses to Younger Dryas climates as analogs for future climate change scenarios and lessons for conservation. Quat. Sci. Rev. 115, 28–38.
Safont, E., Vegas-Vilarrúbia, T., Rull, V., 2012. Use of environmental impact assessment (EIA) tools to set priorities and optimize strategies in biodiversity conservation. Biol. Conserv. 149, 113–121.
Salerno, P.E., Ron, S.R., Señaris, J.C., Rojas-Runjaic, F.J.M., Noonan, B.P., Cannatella, D.C., 2012. Ancient tepui summits harbor young rather than old lineages of endemic frogs. Evolution 66, 3000–3013.
Schubert, C., Fritz, P., Aravena, R., 1994. Late Quaternary paleoenvironmental studies in the Gran Sabana (Venezuelan Guayana Shield). Quat. Int. 21, 81–90.
Steffen, W., Rockström, J., Richardson, K., Lenton, T.M., Folke, C., Liverman, D., et al., 2018. Trajectories of the Earth System in the Anthropocene. Proc. Natl. Acad. Sci. U.S.A. 115, 8252–8259.
Steyermark, J.A., Dunsterville, G.C.K., 1980. The lowland floral element on the summit of Cerro Guaiquinima and other cerros of the Guayana highland of Venezuela. J. Biogeogr. 7, 285–303.
Tzedakis, P.C., Channell, J.E.T., Hodell, D.A., Kleiven, H.F., Skinner, L.C., 2012. Determining the natural length of the current interglacial. Nature Geosci. 5, 138–141.
UN, 2015. Paris Agreement. UN Climate Change, Bonn.
Vegas-Vilarrúbia, T., Nogué, S., Rull, V., 2012. Global warming, habitat shifts and potential refugia for biodiversity conservation in the neotropical Guayana Highlands. Biol. Conserv. 152, 159–168.
Wagner, M., Pross, A., 2011. The nature of stability in replicating systems. Entropy 13, 518–527.
Willis, K.J., Whittaker, R.J., 2000. The refugial debate. Science 287, 1406–1407.
Zink, J.A., García, P., van de Plicht, J., 2011. Tepui peatlands: age record and environmental changes. In: Zink, J. A., Huber, O. (Eds.), Peatlands of the Western Guayana Highlands, Venezuela. Springer, Heidelberg, pp. 189–236.

CHAPTER

4

Origin and evolution of the Pantepui biota

Valentí Rull

Institute of Earth Sciences Jaume Almera (ICTJA-CSIC), Barcelona, Spain

Introduction

The Neotropical region has made great contributions to the sciences of ecology and evolution. Well-known examples are the Galápagos Islands, which in part inspired Darwin's evolutionary theory; the Panama Isthmus, the passage that promoted the Great American Biotic Interchange (GABI); the Amazon basin, one of the largest forests and biodiversity reservoirs on Earth; and the Andes, where Humboldt found inspiration for his theory on altitudinal life zones compared to latitudinal ones. The Guiana Highlands could be considered one of these singular Neotropical regions able to provide significant ecological and evolutionary contributions. However, few synthetic studies exist on the potential ecological and evolutionary mechanisms that generated the exceptional Pantepui biodiversity and endemism patterns, as well as its unique life forms and ecosystems. The origin of the diversity and endemism of Pantepui, which are comparable to those of many oceanic archipelagos (Rull, 2009a), has been a classic biogeographical and evolutionary topic since the early 20th century, when scientific exploration of this archipelago of sky islands began (Huber, 1995a). The Guiana Highlands have been considered a natural laboratory for the study of the origin of the Neotropical biota, but their potential in this sense has not been fully exploited (Rull, 2010).

This chapter reviews the existing evolutionary hypotheses—and the empirical evidence used to support them—on the origin and evolution of the biodiversity and endemism of the Pantepui biota and attempts to provide some clues for an eventual future holistic theory on this topic. The account follows the historical development of the scientific study of Pantepui, largely based on the progress of conceptual and methodological developments over time. In general, the study of Neotropical biodiversity has been characterized by three historical phases, according to the type of evidence used to develop evolutionary

Biodiversity of Pantepui
DOI: https://doi.org/10.1016/B978-0-12-815591-2.00004-5

inferences, namely the biogeographical, paleoecological, and molecular phylogenetic phases (Rull, 2019). The study of Pantepui diversification parallels this trend, and this chapter is organized in the same fashion. This does not mean that these three phases have progressed separately, as they have overlapped over time, but this procedure is considered suitable for understanding the progress of methodological and conceptual advances in the study of the Pantepui biota. It should be clearly stated that this is not an attempt to integrate the information presented in this book, but rather an a priori independent, synthetic attempt based on previous literature.

As in the general study of Neotropical biodiversity, the origin of the Pantepui biota has been addressed from a dual perspective of vicariance versus dispersal, which makes a terminological clarification necessary. Vicariance is a speciation mode well known by most biogeographers, but the term "dispersal" is often used in different senses or as a general word (s.l.) to refer to any type of species displacement or range modification. This chapter follows the terminology of Pielou (1979), who differentiates jump dispersal (also long-distance dispersal)—the movement across inhospitable terrains by active (e.g., flight) or inactive (e.g., wind transport) means—from diffusion or migration, which consists of the progressive colonization of hospitable terrains, thus progressively shifting the distribution area of a species in a given direction. As a geological time framework, the latest update of the Mesozoic and Cenozoic geochronological units from the International Chronostratigraphic Chart is provided (Fig. 4.1). The geographical reference

Era	Period	Epoch	Age (Ma)
			0
Cenozoic	Quaternary	Holocene	0.0117
		Pleistocene	2.58
	Neogene	Pliocene	5.333
		Miocene	23.03
	Paleogene	Oligocene	33.9
		Eocene	56.0
		Paleocene	66.0
Mesozoic	Cretaceous	Upper	100.5
		Lower	145.0
	Jurassic	Upper	163.5
		Middle	174.1
		Lower	201.3
	Triassic	Upper	237
		Middle	247.2
		Lower	251.9

FIGURE 4.1 Mesozoic and Cenozoic geochronological units mentioned in the text (Cohen et al., 2013).

is Fig. 1.2 from Chapter 1: Definition and characterization of the Pantepui biogeographical province.

Evolutionary inferences from biogeography

The pioneers

Mayr and Phelps (1967) summarized the earlier explanations for the origin of the Guiana Highlands' biota in five hypotheses: the Plateau Theory (PT), Cool Climate Theory (CCT), Habitat Shift Theory (HST), Specialized Habitat Theory (SHT), and Distance Dispersal Theory (DDT) (Fig. 4.2). Most of these theories emerged from the study of birds, as happened with the term "Pantepui" itself (Chapter 1). Therefore a significant part of the biogeographical and evolutionary legacy for Pantepui comes from this particular zoological group.

The PT was used by Chapman (1931) for birds and by Tate (1938) for mammals, who suggested that the Pantepui fauna evolved on an extended plateau, of which the present tepuis are erosional remnants. According to Tate (1938), the faunal differences among the tepui summits are due to their physical isolation from one another, and "...the presence of a species on one mountain top and its absence from another suggest only that some

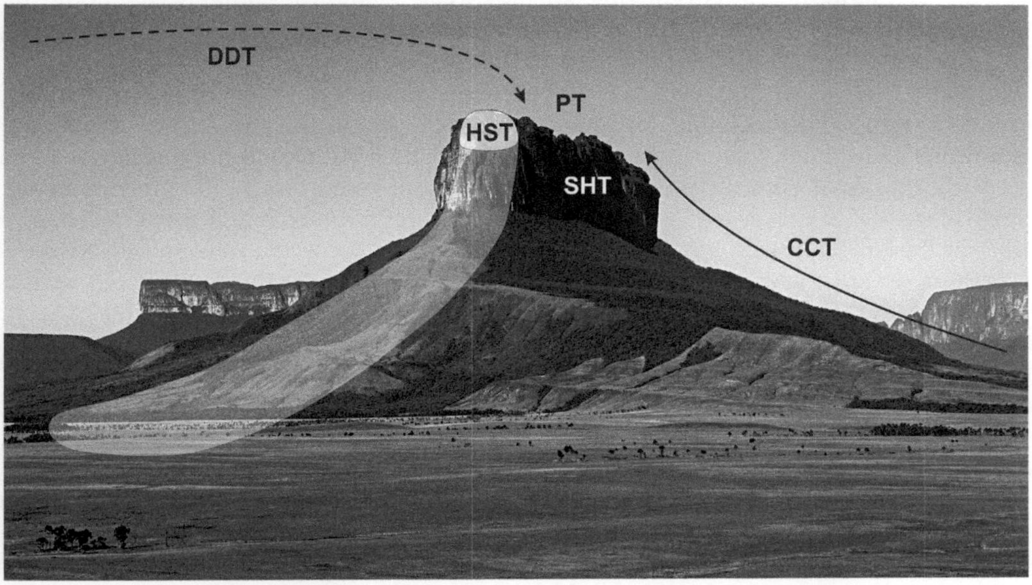

FIGURE 4.2 Graphical representation of the five theories on the origin of the Pantepui biota according to Mayr and Phelps (1967), using the Upuigma-tepui as an example. *CCT*, Cool Climate Theory; *DDT*, Distance Dispersal Theory; *HST*, Habitat Shift Theory; *PT*, Plateau Theory; *SHT*, Specialized Habitat Theory. *Source: Photo by V. Rull.*

inimical condition has arisen on the latter to cause the species to die out". The PT, as originally proposed, was intended to explain not the entire Pantepui fauna, but rather its endemisms. However, according to Mayr and Phelps (1967), the Pantepui endemics are of very unequal age and derived from different sources, which contradicts "...the assumption that the origin of any of the Pantepui forms antedates the dissection of the former plateau into isolated mountain tops". Although the authors do not mention it, it seems clear that the main speciation mode under the PT should have been vicariance.

The CCT contends that the Pantepui fauna colonized the tepuian summits via the surrounding lowlands during the cool Pleistocene phases (i.e., the glaciations). This theory was also suggested by Chapman (1931) and Tate (1938), who considered this the more likely explanation for the origin of the Pantepui fauna. According to these authors, glacial cooling would have facilitated connectivity between the lowlands and summits, thus promoting biotic interchange. However, Mayr and Phelps (1967) criticized this theory, arguing that the cooling needed to produce such an effect would have removed the lowland tropical fauna, which was not the case. These authors also argued that if the CCT is true, the Pantepui fauna should be much more uniform than it actually is. A modification of this theory was proposed by Haffer (1974), who observed that the Pantepui fauna is more similar to the Andean fauna than to the lowland fauna and suggested that glacial cooling could have promoted range expansions in many Andean species, which could have been able to colonize the tepuis. This author believed that Andean species would have been able to disperse through the apparently inhospitable Amazon and Orinoco lowlands by surviving in suboptimal environments, such as low-elevation mountains that could have acted as "stepping stones", until they reached the tepuian summits, which possessed climatic conditions similar to those of their original Andean environments.

Mayr and Phelps (1967) defended the HST, which also proposes that the Pantepui fauna derived from the lowland fauna, but the cause was a shift in the habitat preference of this fauna rather than the occurrence of cooler climates. These authors argue that many species have wide ranges of occurrence and that their eventual restriction to the given climatic or ecological zones is not as rigid as usually thought. If a wide-ranging species becomes extinct in the lowlands and survives on Pantepui, it may become a new species. Although the authors do not mention it, this mechanism is very similar to the concept of parapatric speciation. Mayr and Phelps (1967) called this process "altitudinal segregation" and provided numerous examples of Pantepui species that may have originated in such a way. These authors estimate that the HST may be responsible for the origin of c. 35% of the Pantepui avifauna. According to the same authors, the SHT may be valid for ecological specialists, such as cliff dwellers and others needing very specific habitat requirements that are available only on Pantepui.

The DDT maintains that the Pantepui fauna derived from other regions by island hopping, or jump dispersal, using their flying capacity to cross inhospitable terrains. As logical as this may seem, Chapman (1931) denied such a possibility, arguing that in such a case, the tepuis nearest the Andes should have the richest fauna, which is not true. They also noted that the internal faunistic similarity within Pantepui is not compatible with a theory of chance colonization. Mayr and Phelps (1967) disagreed, arguing that the occurrence of favorable ecological conditions is more decisive than the distance from the source for the establishment of species and that the Pantepui avifauna is not as homogeneous as

Chapman (1931) proposed due to the different dispersal abilities among species. These authors provide abundant examples of this capability for a number of bird species all around the world to demonstrate that 500-km jumps are common for island birds and that "...the inhabitants of Pantepui are simply inhabitants of subtropical islands in a sea of tropical lowlands" (these authors used the terms "tropical" and "subtropical" in climatic terms linked to altitudinal, rather than latitudinal, life zones). Mayr and Phelps (1967) considered that c. 60% of the Pantepui fauna originated by jump dispersal. The potential sources for hopping colonists could have been the Andes (50% of the "hoppers"), the northern Venezuelan coastal ranges (~40%), and the Brazilian highlands south of the Amazon (~10%).

In summary, Mayr and Phelps (1967) emphasized the importance of DDT and HST, accounting for 50% and ~35% of the species, respectively, to explain the composition of the Pantepui avifauna. Of the remaining ~15%, c. 5% would be specialized cliff dwellers (SHT) and ~10% would be "old endemics" that originated by either the DDT or the HST. The PT and the CCT of Chapman (1931) and Tate (1938) were dismissed as potential contributors to the Pantepui fauna. Mayr and Phelps (1967) emphasized that "...although a few of the Pantepui endemics can be called relict, they are not remnants of a formerly more widespread fauna, which is now extinct elsewhere". These authors believed that the colonization of Pantepui was a continuous process that could have elapsed over millions of years, a conclusion that emerged from the fact that "...the degree of differentiation of the Pantepui fauna from the source fauna as well as the differentiation within the fauna indicate strongly an unequal age for the Pantepui elements".

The followers

Hoogmoed (1979) noted that the different Pantepui faunal groups may have had different origins due to differences in geological ages and dispersal abilities. As pointed out by Mayr and Phelps (1967), birds would have benefited from their flying capacity to disperse across inhospitable terrains, but this is not the case for amphibians and reptiles, whose only possibility for range shifting is migration through hospitable lands. Hoogmoed (1979) analyzed the hypotheses of Mayr and Phelps (1967) for the Pantepui herpetofauna and favored the HST. According to this author, the PT could be useful to explain the presence of several relicts, but the CCT, either the original or the modified, and the DDT were incompatible with the limited dispersal ability of amphibians and reptiles.

Almost three decades later, McDiarmid and Donnelly (2005) updated the available information on the Pantepui amphibians and reptiles and revised the theories of Mayr and Phelps (1967). The SHT was considered of little relevance for amphibians and reptiles, as was the DDT because of dispersal limitations, as pointed out by Hoogmoed (1979). The CCT seemed to be contradicted by the high level of endemism among herpetofauna. Indeed, few species occur on more than one tepui, and most of them occur on neighboring tepuis, which seems more consistent with the eventual fragmentation of a continuous plateau population than the transitioning of habitats to the lowlands (in which case the herpetofauna should be more homogeneous among tepuis). The distribution of some amphibians and reptiles seemed to agree with the HST, but in contrast to

Hoogmoed (1979), McDiarmid and Donnelly (2005) considered that the PT may have contributed more significantly to the herpetofaunal diversity of Pantepui. According to these authors, "...vicariance likely has played a more important role in total herpetofaunal diversity than has dispersal", which is explainable by the limited long-distance dispersal capacity of amphibians and reptiles. Local extinction is another factor that should be taken into account to explain the herpetofaunal diversity and endemism patterns (McDiarmid and Donnelly, 2005). However, the extinction patterns are difficult to infer from the distributional data alone.

Mammologists also invoked the theories of Mayr and Phelps (1967) to explain the geographical distribution of this zoological group. A synthesis is provided by Voss et al. (2013), who concluded that the HST and the DDT were the most useful in discussing mammal biogeography.

The outsiders

A few years after the seminal work of Mayr and Phelps (1967), Maguire (1970) summarized the existing botanical data to discuss the origin of the Pantepui flora. This author did not mention the theories summarized by Mayr and Phelps (1967)—a common practice among botanists—and developed his own theoretical framework based exclusively on the geological history accepted at that time and the available information on plant distribution. The starting point of Maguire (1970) was that the Roraima Formation, on which all tepuis are modeled, was deposited during the Cretaceous as a continuous sedimentary cover on the Precambrian Guiana Shield. This would have been the substrate for a more or less continuous vegetation cover that underwent "...floristic development for a great period of time" (Maguire, 1970). The subsequent Miocene uplift would have promoted rapid erosion and fragmentation of the formerly continuous Roraima Formation into isolated table mountains leading to the present tepuis. As a result, the present floras of these tepuis are derived from pre-Miocene floras. In other words, the Pantepui flora was largely established before the Cretaceous and was composed mostly of autochthonous elements and their derivatives. Eventual exchanges with other Guianan floras must have occurred before the Miocene uplift. In summary, the Pantepui flora would have originated during the Cretaceous or earlier (Jurassic), evolved in situ, "...and have been little contaminated by migration or invasion from surrounding territory" (Maguire, 1970). As the tepui summits have progressively reduced their areas since the beginning of the uplift, their floras, composed of species with limited dispersal ability, would have been progressively impoverished. In this way, each tepui is viewed by Maguire (1970) as a "plant microcosmos" full of endemisms due to adaptive radiation. This author ends his paper by suggesting that the antiquity and originality of the Guiana flora "...may compromise a hypothesis of the unique origin of the angiosperms within the region of Malaysia". Although Maguire (1970) did not mention it, his theory largely coincides with the PT.

A decade later, Steyermark and Dunsterville (1980) realized that many species from the summit of Cerro Guaiquinima were from typically lowland genera and deduced that the summit species originated in the lowlands and migrated upwards. These authors provided two explanations for this fact: (1) present-day migration along available dispersal routes (valleys, ridges) by a variety of means (migration, zoochory, anemochory) or (2) upward migration during phases of warmer climates, likely during the Pleistocene. Although not mentioned by the authors, option 1 is similar to the HST, whereas option 2 has similarities with, but also differences from, the CCT. The main resemblance is that both the CCT and option 2 involve vertical migration, and the difference is that the former maintains that upward migration occurred during cool phases, whereas the latter contends that this migration occurred during warmer phases. Later, Steyermark (1986) provided evidence of the same biogeographical pattern, suggesting vertical migration on other tepuian summits (Autana, Jaua, Sarisariñama, Duida). Huber (1988) reinforced this idea by realizing that the species of some emblematic and biodiverse endemic genera of Pantepui, such as *Stegolepis* and *Bonnetia*, were successful in occupying available niches at almost any altitudinal level (coinciding with the HST) and that the number of local endemisms declined as more tepuis were studied, which seemed to contradict Maguire's (1970) previous ideas of an assumedly prolonged isolation through geological time. According to Huber (1988), these biogeographical traits would be the result of repeated vertical migrations of vegetation belts following the Quaternary climatic oscillations. Such climatic and ecological oscillations had already been demonstrated by van der Hammen (1974) in the Andes, and there was no reason to assume that the Guiana Highlands have not been similarly affected (Huber, 1988). In addition, this author confirmed that only a low percentage of tepui summits are truly isolated by vertical cliffs, and the vast majority of them are connected topographically with the surrounding lowlands and uplands by valleys and ridges that provide potential migrational pathways.

Bryophytes seem to show contrasting biogeographical patterns compared to angiosperms. According to Desamorée et al. (2010), the significantly lower bryophyte endemism could be explained by their higher dispersal ability. Based on their biogeographical affinities, these authors dismissed the hypothesis of vertical migration during the Quaternary glacial–interglacial shifts and favored the idea that most endemic bryophytes from Pantepui "...might have originated by dispersal from cold-adapted Andean ancestors" (as proposed by the DDT). This would have occurred by long-distance dispersal across the Orinoco/Amazon lowlands and further allopatric speciation atop the tepuis, a diversification model previously suggested by Noonan and Gaucher (2005) for some amphibian groups on the basis of molecular phylogeographic evidence (see later).

Vicariance versus dispersal

To summarize the discussions on the origin of the Pantepui biota, before the advent of paleoecological studies, the preferred mechanisms for ornithologists seemed to be the DDT and the HST, both involving vertical biotic displacement, whereas herpetologists tended to favor the PT, which is rather static, as dispersal is almost negligible. Botanists did not use the same terms, but they referred to the same or similar processes. For them, the debate was between a long evolution in isolation with no dispersal events, which coincides with the PT, and a vertical migration driven by Quaternary climate shifts, which has similarities with the CCT. In short, the discussion seemed to be reduced to a dual dispute between a static view in which high-elevation vicariance predominates and a more dynamic perspective in which genetic exchange among different elevational levels is possible by jump dispersal, migration, or both. This reproduced the classic biogeographical debate between vicarianism and dispersalism (s.l.) that, in Pantepui, has been expressed as the controversy between the Lost World Hypothesis (LWH) and the Vertical Displacement Hypothesis (VDH) (Rull, 2004). The first was named after the famous novel *The Lost World* by Sir Arthur Conan Doyle, inspired by the Roraima-tepui, to highlight the isolated character of the Pantepui biota as proposed by the PT and Maguire's (1970) static hypothesis. According to the LWH, the biota of each tepuian summit would have evolved in isolation for millions of years due to the erosional breakup of an assumedly continuous initial surface, with the main evolutionary mechanism being vicariance. The VDH, in contrast, accounted for eventual altitudinal migrations following the Pleistocene glacial–interglacial cycles (Steyermark and Dunsterville, 1980; Huber, 1988) and considered other evolutionary mechanisms, including gene flow between highlands and lowlands and among tepuian summits (Rull, 2004). These hypotheses were testable using paleoecological methods.

Paleoecological contributions and geological insights

Quaternary paleoecology

This section will use the available paleoecological and paleoclimatic information that has been explained in detail in Chapter 2: Climatic and ecological history of Pantepui and surrounding areas. In the context of this chapter, the most relevant point is that during the maximum of the last glaciation (Last Glacial Maximum or LGM), some thermosensitive plants migrated ~900 m downward, likely due to a temperature decline of ~5°C below present values. This was consistent with the results obtained in many other Neotropical regions, where the average downward displacement was estimated at ~1100 m (Farrera et al., 1999). For the Guiana Highlands, this means that the Pantepui environments and biota would have reached the surrounding lowlands in the tepuis with a difference of less than 1100 m between their bases and summits, which is the case for nearly half of the tepuis. Therefore it can be estimated that during the LGM, the temperature-sensitive biota of almost half of the tepui summits would have had the opportunity to mix with the surrounding lowlands, whereas the biotas of the other half would have remained isolated (Rull, 2004). Glacial conditions similar to the LGM occurred ~40 times during the Pleistocene in cycles of 40,000–100,000 years (Raymo, 1994); therefore a significant part of the Pantepui biota would have had the opportunity for intense elevational mixing. The potential extant witnesses of such downward migration would be some peculiar peat-forming, tepuian-like communities composed of characteristic tepui taxa that occur in small patches in the Gran Sabana uplands (Huber, 1995b). These results supported the VDH. It should be stressed that although both involve the vertical movements of species due to climate changes, the VDH is substantially different from the CCT. The CCT proposed that the cooler climates would have promoted the upward migration of species from the lowlands to the summits (Mayr and Phelps, 1967). However, paleoecological reconstruction supporting the VDH demonstrated the opposite, that is, the downward migration of sensitive species to the surrounding uplands and lowlands, in agreement with the former ideas of Steyermark (1986) and Huber (1988).

This was the basis for a new hypothesis suggesting that both the LWH and the VDH would be required to explain Pantepui diversification under Quaternary climate changes (Rull, 2005). This suggestion, based on a former diversification model proposed for the Andean highlands (Simpson, 1971), is here called the Quaternary Vicariance–Migration Hypothesis (VMH) (Fig. 4.3). According to this hypothesis, glacial phases would have been characterized by downward spreading of the Pantepui biota across the surrounding uplands and lowlands, thus facilitating floral and faunal mixing and, hence, hybridization and adaptive radiation to the newly colonized environments (Rull, 2005). Glacial conditions have prevailed for ~80% of the Quaternary (Willis and Whittaker, 2000); therefore lowland/upland mixing would have been dominant during this period. The glacial biota of the tepuian summits remains unknown, and it has been speculated that it would have been composed of unknown extinct species, the so-called "glacial ghosts," species with large ranges of temperature tolerances and wide elevational distributions and possibly others surviving within microrefugia (Rull, 2004, 2009b; Rull et al., 1988). Interglacials were shorter warmings (Bush et al., 2001) characterized by upward migration to the tepuian summits, which would have favored vicariance and possibly extinctions due to habitat loss, barring the occurrence of microrefugia. The alternation of glacial and interglacial phases would have favored biotic mixing and gene flow not only between the summits and lowlands but also among all the tepuian summits (Rull, 2005). Indeed, geographic information system (GIS)-based modeling suggested that all tepuian summits would have been connected by lowland migration pathways during the LGM, which is considered representative of the general glacial conditions (Rull and Nogué, 2007; Fig. 3.1 of Chapter 3: Pantepui as a dynamic biogeographical concept). Therefore even species without long-distance dispersal abilities would have experienced up-and-down movements by migration during the Quaternary glacial–interglacial recurrence. This is not incompatible with jump dispersal among tepuian summits and between tepuian summits and other Neotropical mountains for the organisms with a greater dispersal capacity. The combination of all these biotic reorganizations and the associated evolutionary processes would have been a net diversification of the Pantepui biota and the generation of local endemism (Rull, 2005; Rull and Nogué, 2007). The VMH combines vicariance, migration, and jump dispersal processes, showing that the old dichotomy between vicariance and dispersal is not applicable in the Pantepui context.

Pre-Quaternary diversification

The VMH is applicable to the Quaternary, during which the main evolutionary drivers were climate changes, but many Neotropical species emerged during pre-Quaternary times, especially during the Neogene, which was characterized by the occurrence of decisive tectonically-driven regional phenomena, such as the Andean uplift or the formation of the Panama Isthmus, and their important geographical and topographical consequences (Rull, 2008; Hoorn et al., 2010). These drivers did not operate in the

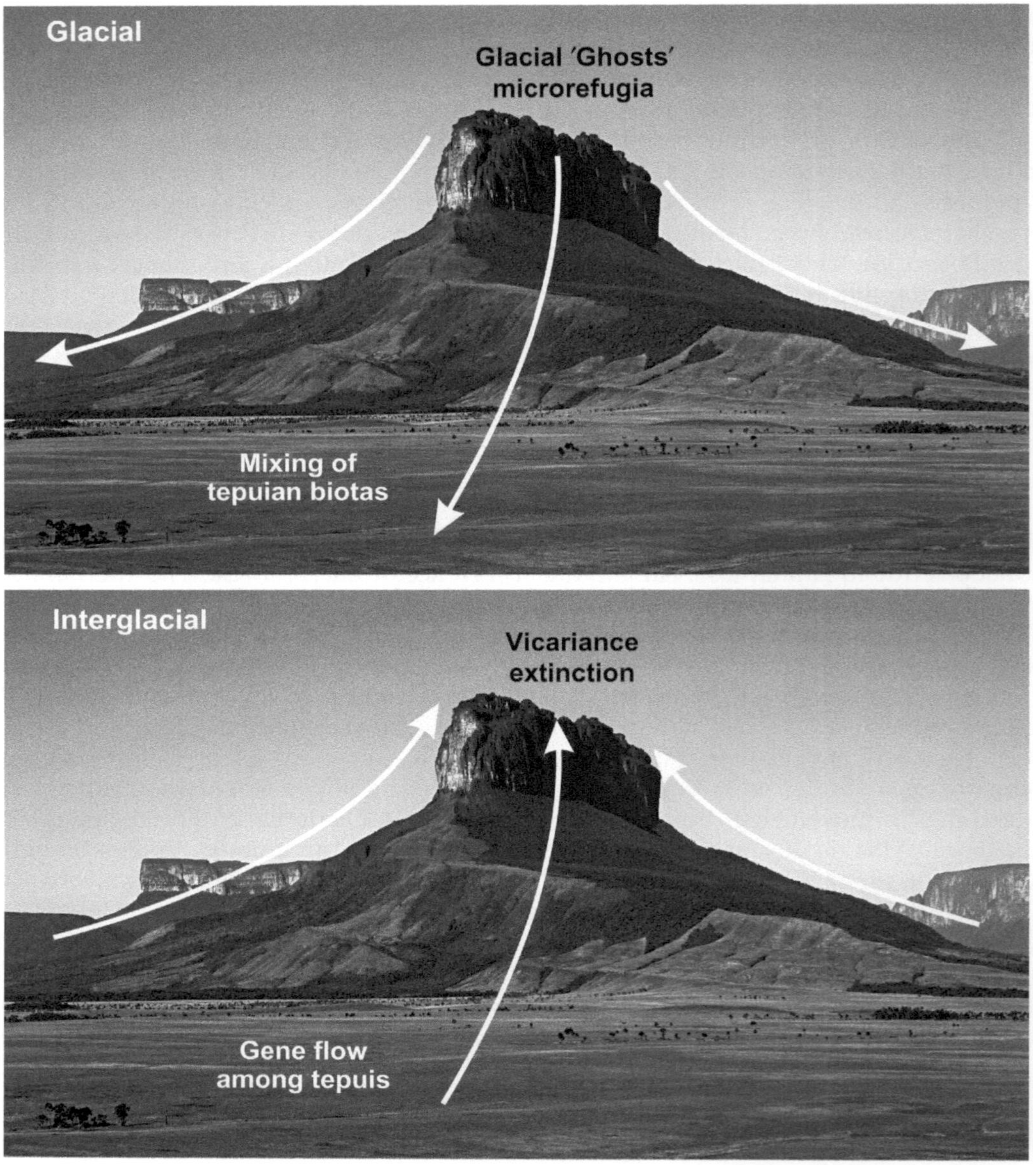

FIGURE 4.3 Graphical representation of the Vicariance–Migration Hypothesis (Rull, 2005), using the Upuigma-tepui, as in Fig. 4.2.

Guiana Shield, which has remained tectonically stable for the last ~600 Ma (late Precambrian to present). During that time, the main geological process has been the continued erosion of the quartzites/sandstones of the 1600- to 1700-year-old Precambrian Roraima Group, formerly the Roraima Formation, resulting in several

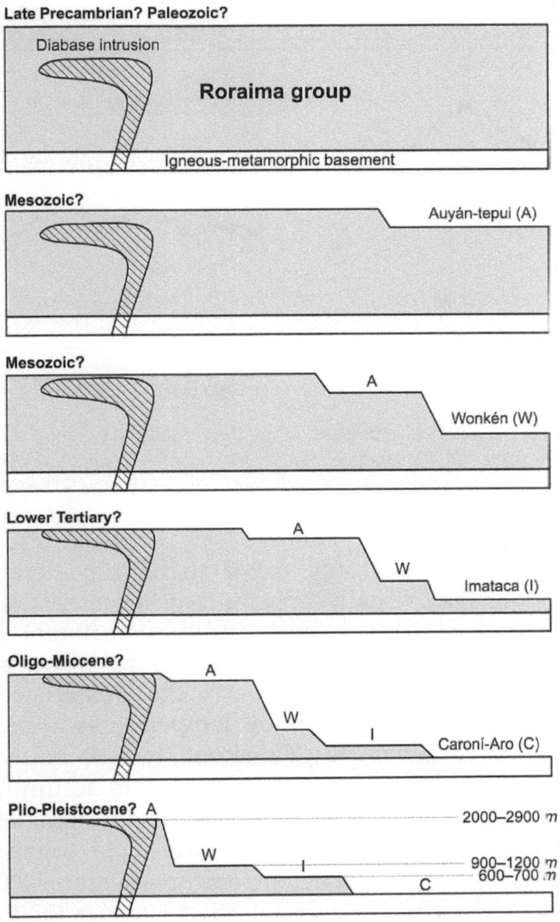

FIGURE 4.4 Formation of the different erosion surfaces (Auyán-tepui, Wonken, and Imataca) on the Precambrian Roraima Group and on the igneous-metamorphic basement (Caroní-Aro erosion surface) since the Precambrian to the Pleistocene. *Source: Redrawn from Briceño, H.O., Schubert, C., 1990. Geomorphology of the Gran Sabana, Guayana Shield, southeastern Venezuela. Geomorphology 3, 125–141.*

erosional surfaces that conferred the typical tabular topography of the Guiana Highlands (Fig. 4.4). Most tepuian summits are the erosional remnants of the oldest of these surfaces, called the Auyán-tepui surface (Briceño and Schubert, 1990). This geological updating, a few decades after the proposal of the PT (Mayr and Phelps, 1967) and Maguire's (1970) static hypothesis, supported the geological scenario considered by these authors. Therefore these hypotheses, condensed here into the LWH, seemed to be favored for pre-Quaternary diversification as far as the geological context is concerned. As in the Quaternary, diversification mechanisms may have been varied, including vicariance, adaptive radiation, and local extinction (Rull, 2005).

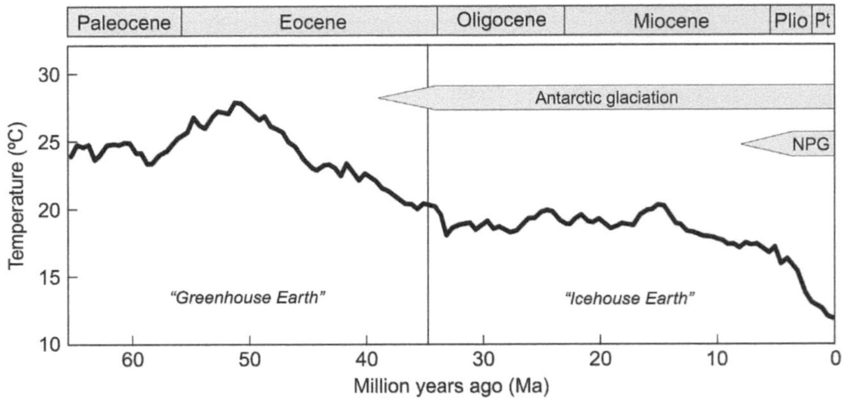

FIGURE 4.5 Global Cenozoic cooling as expressed in the decrease of average surface Earth temperature. *NPG*, North Pole Glaciation; *Plio*, Pliocene; *Pt*, Pleistocene. *Source: Modified from Hansen, J., Sato, M., Russell, G., Kharecha, P., 2013. Climate sensitivity, sea level and atmospheric carbon dioxide. Philos. Trans. R. Soc. A 371, 20120294.*

However, topography is not the only factor to be considered for pre-Quaternary diversification, as climate did not remain constant (an aspect that has not been considered by most scholars), although its trends were radically different from the Quaternary glacial–interglacial oscillations. During the Cenozoic, global average temperatures experienced a maintained decrease until the initiation of the Quaternary glacial cycles in the Plio–Pleistocene boundary (Fig. 4.5). Maximum temperatures approximately 14°C above the present average were recorded in the Paleocene/Eocene boundary, when the poles were ice-free ("greenhouse Earth"). Antarctic ice started to accumulate near the Eocene/Oligocene boundary when average temperatures were approximately 6°C above present, thus initiating an "icehouse Earth" state. The North Pole began to glaciate near the Miocene/Pliocene boundary, when global temperatures were ~4°C above present temperatures (Hansen et al., 2013). Therefore during the Cenozoic, the Pantepui biota evolved under temperatures significantly warmer than those in the Quaternary, even considering the interglacial warmings. Accordingly, in the early Cenozoic, the higher elevations of the ancient plateau, which were only partially eroded (Briceño and Schubert, 1990), were likely occupied by a biota whose temperature requirements were similar to those existing today in the surrounding lowlands. As the erosion progressed and the plateau was being dissected, the temperature decreased, thus creating progressively smaller, more isolated, and cooler highland habitats. This would have favored vicariance and extinction, as proposed by the LWH, but also long-distance colonization by species from climatically similar environments (notably from the Andes), as contended by the DDT. In addition, these proto-Guiana Highlands would have acted as "species pumps" (Rull, 2005) for the lowlands by the downward migration of species from the formerly warmer climates due to the temperature decrease. Effective upward migration, as proposed by the CCT, is not supported due to the decreasing temperature trend. Therefore during pre-Quaternary times, the gene flow among the tepuian summits would have been restricted to the eventual jump dispersal events. The combination of progressive reduction, topographical

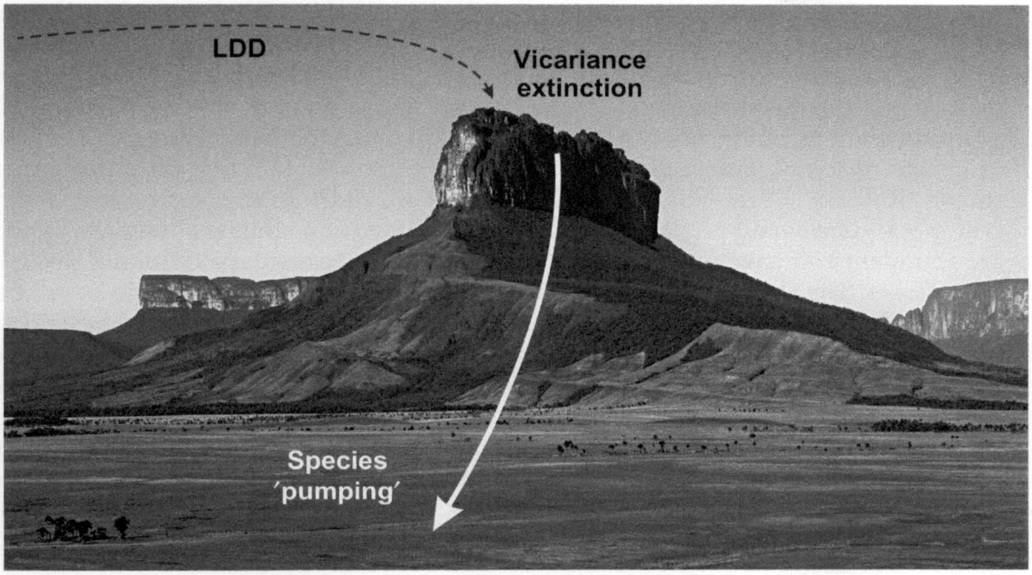

FIGURE 4.6 Graphical representation of the Isolation–Cooling Hypothesis, using the Upuigma-tepui, as in Fig. 4.1. *LDD*, Long-distance dispersal.

isolation, and cooling of the tepuian summits during the Cenozoic, and their eventual evolutionary consequences, is called here the Cenozoic Isolation–Cooling Hypothesis (ICH) (Fig. 4.6).

In theory, this hypothesis is testable, but unfortunately, the lack of Paleogene and Neogene sediments on Pantepui prevents the development of the corresponding paleoecological studies. Indeed, the geology of Pantepui is characterized by a colossal sedimentary gap (hiatus) of at least 1600 million years between the Precambrian Roraima sediments and the overlying Holocene peats from the tepuian summits, whose maximum measured age is less than 13,000 years before the present (Nogué et al., 2009; Zinck et al., 2011). Paleoecologists can hardly overcome this handicap unless an almost miraculous finding of pre-Quaternary sediments occurs in the future. In the meantime, the recently developed methods of molecular phylogenetics and phylogeography can be used, as they have demonstrated their potential to resolve diversification processes and to test evolutionary hypotheses (e.g., Hoorn and Wesselingh, 2010; Richardson and Pennington, 2016; Rull and Carnaval, 2019).

Molecular phylogenetics and phylogeography

Perhaps the most significant advantages of these methods for evolutionary study are related to the possibility of (1) obtaining natural DNA-based phylogenetic relationships among species, (2) estimating the age of speciation events and therefore of the species under study, and (3) mapping the dated phylogenies, thus resolving the spatial component of diversification. The main drawback is the difficulty of measuring extinction (Rull, 2019), but the calibration of molecular mutation rates with fossil evidence is also problematic because of a fragmentary paleontological record. These types of studies are still scarce on Pantepui due to its remoteness and the difficulty of obtaining fieldwork permits for collecting genetic material, although illegal field trips organized for such purposes are not uncommon (Rull and Vegas-Vilarrúbia, 2008; Rull et al., 2009). In the Neotropical context, however, phylogenetic and phylogeographic studies aimed at testing evolutionary hypotheses are abundant and have revolutionized the study of biodiversity patterns and diversification processes. The study of Neotropical diversification has experienced a decisive boost in recent decades due to the exponential increase of molecular phylogenetic and phylogeographic studies. Before this, most hypotheses on Neotropical diversification were based on biogeographical and paleoecological studies or a combination of both. The dual and often excluding debate between the relative importance of either Quaternary or Neogene drivers in the diversification of Neotropical biota became classical and has continued among many researchers to the present day. This debate is still active concerning Pantepui, and previous experiences at a Neotropical scale may be useful for a better understanding and an eventual solution of the problem.

Neotropical diversification

Neotropical diversification is a classical and highly controversial issue that cannot be discussed in depth in this chapter, where only a brief summary is provided. A full review, together with an account of the remaining problems and some potential solutions, can be found in Rull (2019). The idea that Quaternary climatic changes were decisive for the shaping of extant Neotropical biodiversity was most popular during the 1970s and 1980s. This view was championed by defenders of the Refuge Hypothesis (RH), who used paleoclimatic and biogeographical evidence to support the occurrence of refugia that would have allowed extant rainforest species to survive the assumedly arid LGM climates (Haffer, 1969; Prance, 1982; Whitmore and Prance, 1987). This was extrapolated to the entire Quaternary, and glacial–interglacial recurrence was considered the driver of range contraction (glacial) and expansion (interglacial), which would have promoted vicariance (glacial) and gene flow (interglacial). The location of the glacial refugia was inferred from the extant biodiversity and endemism patterns. Using this type of evidence, Pantepui was considered by Steyermark (1979) a refugial center for plants. An implicit assumption of the RH is that most Neotropical species emerged during the Quaternary. The Quaternary hypothesis was contradicted by Raven and Axelrod (1974), also based on extant biogeographical patterns; they argued that pre-Quaternary paleogeographic changes linked to continental drift (e.g., the Andean orogeny or the formation of the Panama Isthmus) were more decisive for the formation of modern biotas. The debate between these two views was explicitly described by Gentry (1982). Further paleoecological studies failed to find support for LGM aridity and forest fragmentation, and the RH was vehemently combated (Colinvaux et al., 1996, 2000). A more complex hypothesis, the Disturbance–Vicariance Hypothesis (DVH), emerged, according to which the biogeographical and paleoecological evidence could be explained by the combination of LGM cooling, moderate precipitation reduction, and atmospheric CO_2 depletion (Bush, 1994; Colinvaux, 1998). The same paleoecological team highlighted the importance of the Miocene Andean uplift and the formation of the Amazon basin for the origin of Neotropical biodiversity and proposed that present-like Amazon rainforests were already in place by the middle Miocene (Colinvaux et al., 2000).

A key aspect of the Quaternary–Neogene debate was the age of Neotropical species whose first estimates emerged with the advent of molecular phylogenetic studies. These first estimates seemed contradictory because a number of species were dated as Quaternary and others as Neogene, which was utilized by different authors to support either one or another hypothesis. An initial meta-analysis performed on >1400 dated species belonging to a wide range of plant and animal groups and covering most Neotropical regions showed that these species had originated in a continual manner since the Neogene until the Quaternary, and no diversification bursts were observed at any particular time. Nearly half of these species emerged during the Neogene, while the other half were of Quaternary origin (Rull, 2008). This suggested that the extant Neotropical biodiversity originated through a diversity of drivers and evolutionary mechanisms and showed that broad-scale generalizations from a few case studies involving small groups of organisms and/or a few geographical settings are not useful to unravel the origin of Neotropical biodiversity as a whole (Rull, 2013, 2015). Therefore it was proposed that the dual simplistic Neogene–Quaternary controversy did not make sense, as the current Neotropical biodiversity was likely the result of a complex interplay of ecological and evolutionary processes across spatial and temporal scales initiated by Neogene tectonic events and continued under the action of Quaternary climatic changes (Rull, 2011). This hypothesis, called here the Continuum Multifactor Hypothesis (CMH), has been supported by further meta-analyses and synthetic reviews on the topic (review in Rull, 2019).

Pantepui

Molecular phylogenetic/phylogeographic studies are still scarce on Pantepui, especially for plants, but these studies are progressing rapidly, and it is expected that they will experience a significant increase in the near future. In addition, the rapid methodological progress in this field of research, especially in reference to dating methods, may suggest that a number of the conclusions related to chronological aspects could eventually change in the future. Therefore an eventual synthesis, similar to that attempted using biogeographical and paleoecological evidence, still seems premature. The following is a brief account of the main molecular phylogenetic/phylogeographic studies available to date, grouped taxonomically, emphasizing the hypotheses favored in each case by the authors of each particular study.

Plants

The first molecular phylogenies from Pantepui corresponded to the vascular plants of the families Bromeliaceae and Rapateaceae. According to these studies, the Rapateaceae likely originated at the end of the Cretaceous in the Guiana lowlands and reached the tepuian summits during the Miocene. *Stegolepis*, an endemic Guianan genus of this family occurring on most tepuian summits with frequent local endemisms, would have invaded Pantepui by 12 Ma and started to diversify at 6 Ma (Givnish et al., 2000). *Stegolepis* diversification atop the tepuis would have resulted from a combination of vicariance for closer tepuis and bird-mediated, long-distance dispersal for more distant tepuis (Givnish et al., 2004). Among Bromeliaceae, *Brocchinia* is also endemic to the Guiana region and well represented on Pantepui, whereas *Lindmannia* is restricted to the tepuian summits. The studied *Brocchinia* species originated between 12 and 17 Ma, and the two *Lindmannia* species analyzed to date diverged at approximately 2.5 Ma, close to the Neogene/Quaternary boundary (Givnish et al., 2007). Although these authors do not discuss the potential diversification drivers and mechanisms for these two genera, all studied species were of Neogene origin, and hence, the ICH would be applicable.

Amphibians

Salerno et al. (2012) studied four of the seven recognized species of the Pantepui frog genus *Tepuihyla* and found that these species originated mostly during the Pliocene and the Quaternary (2–5 Ma). These authors concluded that this dismisses the LWH and discussed other hypotheses such as the DDT, the HST, and the VDH. The first was also disregarded based on the low vagility of *Tepuihyla*, and the most likely explanation was considered to be a combination of the other two hypotheses. Slightly later, the same authors suggested that *Tepuihyla* species may be descendants from a former, now extinct, widespread lowland dweller that climbed the tepuis and speciated on their tops during the Pliocene, possibly due to climate shifts (Salerno et al., 2014). According to these authors, the influence of climate could be indirect, as *Tepuihyla* species live on special microhabitats within the tubular bromeliad *Brocchinia*, which would have been the vehicle of climatically triggered up-and-down migration. Working on the same genus and others that are related (*Oreophrynella*, *Stefania*), Kok et al. (2012) noted that the genetic variability among species from different tepuis was lower than expected under a hypothesis of ancient endemism and favored other mechanisms, such as active (during glaciations) or passive (birds, storms) dispersal coupled with some ensuing extinction, to explain their observations. Later, Kok et al. (2017) used the genus *Stefania* to test whether its Pantepui diversity originated either by Quaternary dispersal, according to the DVH, or by pre-Quaternary vicariance, following the PT. These authors found that the diversification of *Stefania* started in the Oligocene (26 Ma), and most divergence events predated the Quaternary, thus favoring the PT. Therefore the DVH was rejected. According to Kok et al. (2017), *Stefania* originated in the highlands, and its presence on the surrounding Pantepui lowlands and uplands is the result of dispersal events that occurred during the Miocene and the Pleistocene. Differences from other Pantepui genera whose species originated more recently were explained by these authors in terms of repeated vertical migrations induced by climatic instability caused by tectonic events (Andean orogeny, closure of Panama Isthmus) and Quaternary glacial–interglacial oscillations. This is the Neotropical CMH applied to Pantepui.

A recent study suggested that the Pantepui genus *Oreophrynella* diverged from its sister genus *Atelopus* during the Eocene (c. 40 Ma) from a common ancestor that lived on the proto-Andes and reached Pantepui by jump dispersal through the inhospitable Amazon/Orinoco lowlands (Kok et al., 2018). Once in Pantepui, *Oreophrynella* would have diversified according to the PT, which is considered by the authors better suited than other theories to explain the biogeographical patterns of endemic Pantepui amphibians. During the Quaternary, climatic changes would have promoted biotic mixing, leading to massive extinctions and the reduction of genetic diversity that depleted biodiversity. The case of *Oreophrynella* contrasts with the history of the Guianan species of its sister genus *Atelopus*, which, although not on Pantepui, shares common dispersal patterns but at different times. According to Noonan and Gaucher (2005), the ancestor of the extant Guianan *Atelopus* species crossed the Amazon/Orinoco basins during the late Miocene or the early Pleistocene and diversified on the Guiana uplands during the Quaternary, which is consistent with the DVH. Therefore, in contrast to *Oreophrynella* and following the proposals of the DVH, Quaternary glacial–interglacial recurrence seems to have fostered diversification in *Atelopus*.

Birds

An early phylogeographic study of the Neotropical redstars (*Myoborus*) seemed to support a combined dispersal–vicariance hypothesis for the origin of the Pantepui species, which would have diversified in situ after an initial colonization event (Pérez-Emán, 2005). The most likely source would have been either the Andes or the northern coastal range of Venezuela, and the suggested time was the late Pliocene, based on paleoclimatic and paleoecological considerations. Similar conclusions were attained by Bonaccorso and Guayasamin (2013) in studying the toucanets of the genus *Aulacorhynchus* specifically to understand the origin of the Pantepui montane biotas. Phylogeographic evidence supported the Andean origin of the Pantepui endemics, as proposed by the DDT. Although the dated phylogenies of these toucanets indicate that most Neotropical *Aulacorhynchus* species emerged during the Quaternary, the Pantepui species emerged during the Pliocene, and further population differentiation occurred in the Pleistocene. This diversification model—long-distance dispersal followed by in situ allopatric differentiation—has been proposed to be of general utility to explain avian diversity across the entire Neotropical region (Smith et al., 2014). Berv and Prum (2014) studied Neotropical cotingas and found that the two known Pantepui endemic species have different phylogenetic affinities. The red-banded fruiteater *Pipreola whitelyi* is of Andean origin, whereas the rose-collared piha *Lipaugus streptophorus* has its sister group in the surrounding lowlands. The authors concluded that the tepuian avifauna "...has complex evolutionary origins and cannot be explained by a single generalized biogeographic mechanism".

Mammals

The tepuian broad-nosed bat *Platyrrhinus aurarius*, endemic to the Guiana Shield, and its sister species, *Platyrrhinus infuscus*, from western Amazonia, have their more recent common ancestor in the Andes, which suggests Quaternary or older range expansion from this cordillera (Velazco and Patterson, 2008). A similar pattern was found for the short-tailed opossum, *Monodelphis reigi*, whose Andean emigration was first proposed to have occurred during the Miocene, but no dated phylogenies were provided (Lim et al., 2010).

A further phylogeographic analysis of the whole genus placed the origin of *M. reigi* in the Quaternary (1.2–2.4 Ma) and suggested that its more recent ancestor was an Amazonian species that eventually climbed to the tepuis, although the mechanism is not specified (Pavan et al., 2016). Voss et al. (2013) described a new Pantepui endemic species of the didelphid marsupial, *Marmosops pakaraimae*, and found that its sister species lived in the adjacent lowlands. This finding, together with other biogeographic evidence, was used by these authors to suggest that the Pantepui endemic mammals are "...neither ancient relicts of tepui vicariance nor descendants of long-distance-dispersing Andean progenitors" but "...evolved from lowlands species in the late Cenozoic". *Podoxymys roraimae* is a mouse restricted to the Roraima-tepui (Leite et al., 2015). Based on its geographical distribution and its molecular phylogenetic relationships with other genera of the same tribe (Akodontini), the CCT and the HST were considered unlikely by Leite et al. (2015). The DDT was favored, but the more likely source area would have been not the Andes—as originally proposed by the DDT—but the Brazilian Shield, located to the southeast, which is topographically isolated from the Guiana Shield and of much lower elevation than Pantepui. According to Leite et al. (2015), *Podomyxys* diverged from its closest relative at the end of the Pliocene (2.5–3.7 Ma), but when and how this monotypic genus reached Pantepui is still very speculative. Pavan et al. (2016) compared the origin of *P. roraimae* with the evolutionary history of *M. reigi* and concluded that "...the mammalian fauna of Pantepui is descendent from ancestors with heterogeneous distributions and habitat preferences".

General conclusions

As in the study of Neotropical diversification in general, knowledge about the origin of the Pantepui biota has progressed in three successive steps according to the type of evidence used: biogeography, paleoecology, and molecular phylogenetics. Most of the hypotheses on Pantepui diversification emerged during the biogeographical phase, and further methodological developments were used mainly to test these hypotheses. A clear differentiation has been observed between zoologists, who developed the five main theories on Pantepui diversification, and botanists, who rarely mentioned these zoological theories, though they sometimes used some of them to explain the origin of the Pantepui flora. Ornithologists were the pioneers and proposed five main theories for Pantepui diversification, which have survived through time, namely PT, CCT, DDT, HST, and SHT. Botanists, on the other hand, have discussed their evidence using a dual and often mutually exclusive framework between the LWH and the VDH. In summary, these theories and hypotheses reproduce the classical biogeographical debate between vicariance and dispersal to explain extant biodiversity and endemism patterns.

Paleoecology provided further evidence to test these hypotheses for the Quaternary, but the lack of pre-Quaternary sediments on Pantepui prevented researchers from doing the same for the Neogene, where many Neotropical species are known to have emerged. Using this evidence, two combined vicariance–dispersal models emerged, the DVH, formerly proposed for Amazonian diversification, and the VMH, inspired by a previous model for highland Andean diversification. These hypotheses combine elements of the LWH and the VDH, thus suggesting that the old vicarianism–dispersalism debate is fruitless. The LWH almost coincides with the former PT, but the CCT and the VDH are very different, although they are often confused in the literature. Both highlight the potential role of glacial–interglacial cycles but use radically different mechanisms and processes. It should also be noted that many studies use the term "dispersal" in a wide sense, including both long-distance dispersal (also jump dispersal) and migration (also diffusion), which are very different concepts for species range shifts. For pre-Quaternary times, the main diversification mechanism that has been invoked in Pantepui is the PT and its equivalent LWH, both based on vicariance due to long-term isolation without considering dispersal. However, the evidence for gradual and maintained global cooling during the Cenozoic added the possibility of downward migration, and a new hypothesis, the ICH, which is also a combination of vicariance and dispersal (in this case, migration), emerged.

Molecular phylogenetic/phylogeographic evidence provided support for some of the former hypotheses and theories, but there are clear differences according to the different organisms used. These types of studies have been carried out mainly for animal groups, and only a few plant families have been studied from this perspective. There is a clear difference between animal groups with significant long-distance dispersal abilities, for which dispersal theories are favored, and those that cannot fly or be easily transported across inhospitable terrains, for which vicariance seems to be the preferred diversification mechanism. Therefore the ecological and behavioral features of each particular taxonomic group greatly influence its diversification patterns, and applying a single hypothesis to all of them seems unrealistic. As a consequence, a number of authors have proposed that both vicariance and dispersal theories are needed to explain the origin of the Pantepui biota. Molecular phylogenetic/phylogeographic studies have provided evidence for pre-Quaternary diversification that was impossible to obtain from paleoecological records. This has been used to discuss previous biogeographical hypotheses in a wider temporal context encompassing the whole Cenozoic. In this way, some studies have emphasized the importance of either Quaternary or pre-Quaternary diversification drivers and mechanisms or have recognized the need for a more complex perspective that includes a variety of them.

An overview of all the hypotheses and theories reviewed here leads to the conclusion that there is no single explanation for the origin of the Pantepui biota. Different taxonomic groups seem to have acquired their extant biodiversity and endemism patterns by diverse processes and mechanisms, under the action of different drivers, in different Pantepui areas, and at different times. Once more, it is evident that single case studies based on particular taxonomic groups and/or specific geographical areas cannot be generalized (Rull, 2013). As occurs in the Neotropics in general, proper understanding of the origin of the Pantepui biota requires complex thinking and the consideration of multiple drivers, as well as a diversity of ecological and evolutionary processes and mechanisms acting at the same time across spatial and temporal scales, as formerly proposed by the CMH (Rull, 2011).

Further research

To properly address the Pantepui biotic complexity, it is necessary to combine geological, biogeographical, paleoecological, and molecular phylogenetic evidence. Each discipline has its own limitations and requires the support of others to provide a more holistic view. For example, biogeography provides only a static picture, a present snapshot of a long and continuous evolutionary process developed under environmental conditions different from those of the present day. Paleoecology is unable to provide information on the pre-Quaternary Pantepui conditions, which differed greatly from the Quaternary glacial–interglacial oscillations and cannot inform the potential evolutionary impact of the deep tectonic and paleogeographic reorganizations that occurred in northern South America. As a developing discipline, molecular phylogenetics is constantly advancing, and some critical aspects, especially the dating of speciation events, are undergoing continuous improvements that may change previous chronological inferences. In general, the same conceptual and methodological drawbacks identified in the study of Neotropical diversification (Rull, 2013, 2015, 2019) are applicable to Pantepui. A coordinated multidisciplinary effort is needed to overcome these handicaps and advance toward a more comprehensive and, hence, realistic framework for the origin and evolution of the Pantepui biota. Fig. 4.7 provides a conceptual framework for the study of the origin and evolution of Neotropical diversity that may be applicable to Pantepui.

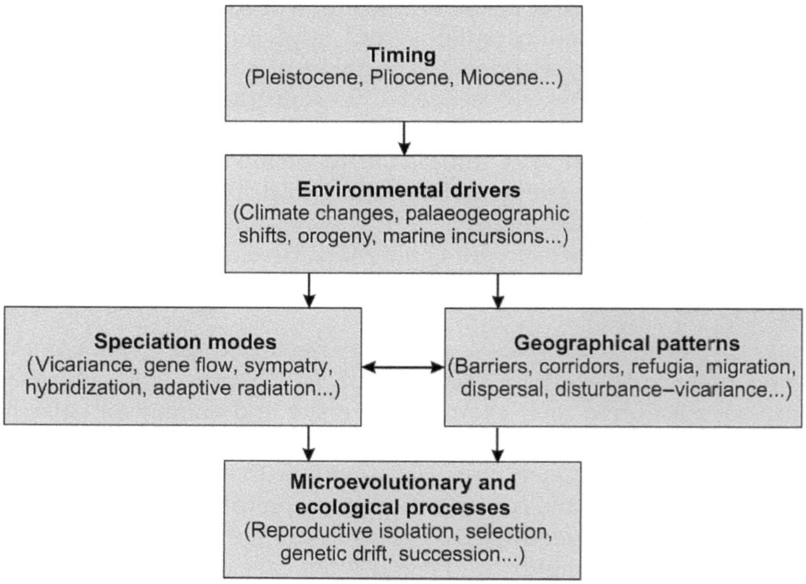

FIGURE 4.7 Schematic view of the conceptual steps proposed by Rull (2015) for making inferences about the origin of Neotropical biodiversity using geological, biogeographical, paleoecological, and molecular phylogenetic evidence. *Source: Modified from Rull, V., 2015. Pleistocene speciation is not refuge speciation. J. Biogeogr. 42, 602–609.*

Acknowledgments

The author is very grateful to Elisa Bonaccorso, Burton Lim, and Celsa Señaris for the critical review of the initial manuscript.

References

Berv, J.S., Prum, R.O., 2014. A comprehensive multilocus phylogeny of the Neotropical cotingas (Cotingidae, Aves) with a comparative evolutionary analysis of breeding system and plumage dimorphism and a revised phyologenetic classification. Mol. Phylogenet. Evol. 81, 120–136.

Bonaccorso, E., Guayasamin, J.M., 2013. On the origin of Pantepui montane biotas: a perspective based on the phylogeny of *Aulancorhynchus tocanets*. PLoS One 8 (6), e67321.

Briceño, H.O., Schubert, C., 1990. Geomorphology of the Gran Sabana, Guayana Shield, southeastern Venezuela. Geomorphology 3, 125–141.

Bush, M., Stute, M., Ledru, M.-P., Behling, H., Colinvaux, P.A., De Oliveira, P.E., et al., 2001. Palaeotemperature estimates for the lowland Americas between 30°S and 30°N at the Last Glacial Maximum. In: Markgraf, V. (Ed.), Interhemispheric Climate Linkages. Academic Press, San Diego, CA, pp. 293–306.

Bush, M.B., 1994. Amazonian speciation: a necessarily complex model. J. Biogeogr. 21, 5–17.

Chapman, J., 1931. The upper zonal bird-life of Mts. Roraima and Duida. Bull. Am. Mus. Nat. Hist. 63, 1–135.

Cohen, K.M., Finney, S.C., Gibbard, P.L., Fan, J.-X., 2013. The ICS International Stratigraphic Chart (v8, 2018). Episodes 36, 199–204.

Colinvaux, P.A., 1998. A new model for Amazon endemics. Glob. Ecol. Biogeogr. Lett. 7, 95–96.

Colinvaux, P.A., De Oliveira, P.E., Bush, M.B., 2000. Amazonian and neotropical plant communities on glacial time-scales: the failure of the aridity and refuge hypothesis. Quat. Sci. Rev. 19, 141–169.

Colinvaux, P.A., De Oliveira, P.E., Moreno, J.E., Miller, M.C., Bush, M.B., 1996. A long pollen record from lowland Amazonia: forest and cooling in glacial times. Science 274, 85–88.

Desamorée, A., Vanderpoorten, A., Laenen, B., Gradstein, S.R., Kok, P.J.R., 2010. Biogeography of the Lost World (Pantepui region, norheastern South America): insights from bryophytes. Phytotaxa 9, 254–265.

Farrera, I., Harrison, S.P., Prentice, I.C., Ramstein, G., Guiot, J., Bartlein, P.J., et al., 1999. Tropical climates at the Last Glacial Maximum: a new synthesis of terrestrial paleoclimate data: I. Vegetation, lake-levels and geochemistry. Clim. Dyn. 15, 823–856.

Gentry, A.H., 1982. Neotropical floristic diversity: phytogeographical connections between Central and South America, Pleistocene climatic fluctuations, or an accident of the Andean orogeny? Ann. Missouri Bot. Gard. 69, 557–593.

Givnish, T.J., Evans, T.M., Zjhra, M.L., Patterson, T.B., Berry, P.E., Sytsma, K.J., 2000. Molecular evolution, adaptive radiation, and geographic diversification in the amphiatlantic family Rapateaceae: edidence from *ndh*F sequences and morphology. Evolution 54, 1815–1937.

Givnish, T.J., Millam, K.C., Evans, T.M., Hall, J.C., Pires, J.C., Berry, P.E., et al., 2004. Ancient vicariance or recent long-distance dispersal? Inferences about phylogeny and SouthAmerican-African disjunctions in Rapateaceae and Bromeliaceae based on *ndh*F sequence data. Int. J. Plant Sci. 165, S35–S54.

Givnish, T.J., Millam, K.C., Berry, P.E., Sytsma, K.J., 2007. Phylogeny, adpative radiation, and historical biogeography of Bromeliaceae inferred from *sdh*F sequence data. Aliso 23, 3–26.

Haffer, J., 1969. Speciation in Amazonian forest birds. Science 165, 131–137.

Haffer, J., 1974. Avian speciation in tropical South America, with a systematic of the toucans (Rhamphastidae) and jacamars (Galbulidae). Publ. Nuttal Ornothol. Club 14, 1–390.

Hansen, J., Sato, M., Russell, G., Kharecha, P., 2013. Climate sensitivity, sea level and atmospheric carbon dioxide. Philos. Trans. R. Soc. A 371, 20120294.

Hoogmoed, M.S., 1979. The herpetofauna of the Guianan region. In: Duellman, W.E. (Ed.), The South American Herpetofauna: Its Origin, Evolution and Dispersal, 7. Monogr. Mus. Nat. Hist, Lawrence, KS, pp. 241–279.

Hoorn, C., Wesselingh, F., 2010. Amazonia: Landscape and Species Evolution. A Look Into the Past. Wiley-Blackwell, Chichester.

Hoorn, C., Wesselingh, F.P., ter Steege, H., Bermudez, M.A., Mora, A., Sevink, J., et al., 2010. Amazonia through time: Andean uplift, climate change, landscape evolution, and biodiversity. Science 330, 927–931.

Huber, O., 1988. Guayana highlands versus Guayan lowlands, a reappraisal. Taxon 37, 595–614.
Huber, O., 1995a. History of botanical exploration. In: Berry, P.E., Holst, B.K., Yatskievych, K. (Eds.), Flora of the Venezuelan Guayana. Introduction, vol. 1. Missouri Botanical Garden Press, St. Louis, MO, pp. 63–95.
Huber, O., 1995b. Vegetation. In: Berry, P.E., Holst, B.K., Yatskievych, K. (Eds.), Flora of the Venezuelan Guayana. Introduction, vol. 1, Missouri Botanical Garden Press, St. Louis, MO, pp. 97–160.
Kok, P.J.R., MacCulloch, R.D., Means, D.B., Roelants, K., Van Bocxlaer, I., Bossuyt, F., 2012. Low genetic diversity in tepui summit vertebrates. Curr. Biol. 22, R589–R590.
Kok, P.J.R., Russo, V.G., Rtaz, S., Means, D.B., MacCulloch, R.D., Lathrop, A., et al., 2017. Evolution in the South American 'Lost World': insights from multilocus phylogeography of stefanias (Anura, Hemiphractidae, *Stefania*). J. Biogeogr. 44, 170–181.
Kok, P.J.R., Ratz, S., MacCulloch, R.D., Lathrop, A., Dezfoulian, R., Aubret, F., et al., 2018. Historical biogeography of the palaeoendemic toad genus Oreophrynella (Amphibia: Bufonidae) sheds a new light on the origin of the Pantepui endemic terrestrial biota. J. Biogeogr. 45, 26–36.
Leite, Y.L.R., Kok, P.J.R., Weksler, M., 2015. Evolutionary affinities of the 'Lost World' mouse suggest a late Pliocene connection between the Guiana and Brzailian shields. J. Biogeogr. 42, 706–715.
Lim, B.K., Engstrom, M.D., Patton, J.C., Bickman, J.W., 2010. Molecular phylogenetics of Reig's short-tailed opossum (*Monodelphis reigi*) and its distributional range extension into Guyana. Mammal Biol 75, 287–293.
Maguire, B., 1970. On the flora of the Guayana Highland. Biotropica 2, 85–100.
Mayr, E., Phelps, W.H., 1967. The origin of the bird fauna of the south Venezuelan highlands. Bull. Am. Mus. Nat. Hist. 136, 269–328.
McDiarmid, R.W., Donnelly, M.A., 2005. The herpetofauna of the Guayana Highlands: amphibians and reptiles of the Lost World. In: Donnelly, M.A., Crother, B.I., Guyer, C., Wake, M.H., White, M.E. (Eds.), Ecology and Evolution in the Tropics: A Herpetological Perspective. University of Chicago Press, Chicago, IL, pp. 461–560.
Nogué, S., Rull, V., Montoya, E., Huber, O., Vegas-Vilarrúbia, T., 2009. Paleoecology of the Guayana highlands (northern South America): holocene pollen record from the Eruoda-tepui, Chimantá massif. Palaeogeogr. Palaeoclimatol. Palaeoecol. 281, 165–173.
Noonan, B.P., Gaucher, P., 2005. Phylogeography and demography of Guyanan harlequin toads (*Atelopus*): diversification within a refuge. Mol. Ecol. 14, 3017–3031.
Pavan, S.E., Jansa, S.A., Voss, R.S., 2016. Spatiotemporal diversification of a low-vagility Neotropical vertebrate clade (short-tailed opossums, Didelphidae: *Monodelphis*). J. Biogeogr. 43, 1299–1309.
Pérez-Emán, J.L., 2005. Molecular phylogenetics and biogeography of the Neotropical redstars (*Myoborus miniatus*). Ornitol. Monogr. 67, 511–528.
Pielou, E.C., 1979. Biogeography. Wiley, New York.
Prance, G.T., 1982. Biological Diversification in the Tropics. Columbia University Press, New York.
Raven, P.H., Axelrod, D.I., 1974. Angiosperm biogeography and past continental movements. Ann. Missouri Bot. Gard. 61, 539–673.
Raymo, M.E., 1994. The initiation of Northern Hemisphere glaciation. Annu. Rev. Earth Planet. Sci. 22, 353–383.
Richardson, J.E., Pennington, R.T., 2016. Origin of Tropical Diversity: From Clades to Communities. Frontiers Media, Lausanne.
Rull, V., 2004. Biogeography of the 'Lost World': a palaeoecological perspective. Earth-Sci. Rev. 67, 125–137.
Rull, V., 2005. Biotic diversification in the Guayana Highlands: a proposal. J. Biogeogr. 32, 921–927.
Rull, V., 2008. Speciation timing and neotropical biodiversity: the Tertary–Quaternary debate in the light of molecular phylogenetic evidence. Mol. Ecol. 17, 2722–2729.
Rull, V., 2009a. Pantepui. In: Gillespie, R., Clague, D. (Eds.), Encyclopedia of Islands. University of California Press, Berkeley, CA, pp. 717–720.
Rull, V., 2009b. Microrefugia. J. Biogeogr. 36, 481–484.
Rull, V., 2010. The Guayana Highlands: a natural laboratory for the biogeographical and evolutionary study of the neotropical flora. In: Sánchez-Villagra, M.R., Aguilera, O.A., Carlini, A.A. (Eds.), Urumaco and Venezuelan Paleontology. The Fossil Record of the Northern Neotropics. Indiana University Press, Bloomington, IN, pp. 84–102.
Rull, V., 2011. Neotropical biodiversity: timing and potential drivers. Trends Ecol. Evol. 26, 508–513.
Rull, V., 2013. Some problems in the study of the origin of neotropical biodiversity using palaeoecological and molecular phylogenetic evidence. Syst. Biodiv. 11, 415–423.

Rull, V., 2015. Pleistocene speciation is not refuge speciation. J. Biogeogr. 42, 602–609.
Rull, V., 2019. Neotropical diversification: historical overview and conceptual insightsIn: Rull, V., Carnaval, A. (Eds.), Neotropical Diversification. Springer, Berlin*PeerJ Preprints* 6: e27294v1 . Available from: https://doi.org/10.7287/peerj.preprints.27294v1.
Rull, V., Carnaval, A., 2019. Neotropical Diversification. Springer, Berlin.
Rull, V., Nogué, S., 2007. Potential migration routes and barriers for vascular plants of the neotropical Guyana Highlands during the Quaternary. J. Biogeogr. 34, 1327–1341.
Rull, V., Vegas-Vilarrúbia, 2008. Biopiracy rules hinder conservation efforts. Nature 453, 26.
Rull, V., Schubert, C., Aravena, R., 1988. Palynological studies in the Venezuelan Guayana Shield: preliminary results. Curr. Res. Pleistocene 5, 54–56.
Rull, V., Vegas, T., Nogué, S., Montoya, E., 2009. Bureaucratic obstruction of conservation science in the Guayana Highlands. Conserv. Biol. 22, 508–509.
Salerno, P.E., Ron, S.R., Señaris, J.C., Rojas-Runjaic, F.J.M., Noonan, B.P., Cannatella, D.C., 2012. Ancient tepui summits harbor young rather than old lineages of endemic frogs. Evolution 66, 3000–3013.
Salerno, P.E., Señaris, J.C., Rojas-Runjaic, F.J.M., Cannatella, D.C., 2014. Recent evolutionary history of Lost World endemics: population genetics, species delimitation, and phylogeography of sky-island treefrogs. Mol. Phylogenet. Evol. 82, 314–323.
Simpson, B., 1971. Pleistocene changes in the flora and fauna of South America. Science 173, 771–780.
Smith, B.T., McCormack, J.E., Cuervo, A.M., Hickerson, M.J., Aleixo, A., Cadena, C.D., et al., 2014. The drivers of tropical speciation. Nature 515, 406–409.
Steyermark, J.A., 1979. Plant refuge and dispersal centres in Venezuela: their relict and endemic element. In: Larsen, K., Holm-Nielsen, L. (Eds.), Tropical Botany. Academic Press, New York, pp. 185–221.
Steyermark, J.A., 1986. Speciation and endemism in the flora of the Venezuelan tepuis. In: Vuilleumier, F., Monasterio, M. (Eds.), High Altitude Tropical Biogeography. Oxford University Press, New York, pp. 317–373.
Steyermark, J.A., Dunsterville, G.C.K., 1980. The lowland floral element of the summit of Cerro Guaiquinima and other cerros of the Guayana Highlands of Venezuela. J. Biogeogr. 7, 285–303.
Tate, G.H.H., 1938. A new 'Lost World'. Nat. Hist. 2, 107–120.
van der Hammen, T., 1974. The Pleistocene changes in vegetation and climate in tropical South America. J. Biogeogr. 1, 3–26.
Velazco, P.M., Patterson, B.D., 2008. Phylogenetics and biogeography of the broad-nosed bats, genus Platyrrhinus (Chiroptera: Phyllostomidae). Mol. Philogenet. Evol. 49, 749–759.
Voss, R.S., Lim, B.K., Díaz-Nieto, J.F., Jansa, S.A., 2013. A new species of *Marmosops* (Marsupialia: Didelphidae) from the Pakaraima Highlands of Guyana, with remarks on the origin of the endemic Pantepui Mammal fauna. Am. Mus. Novitates 3778, 27.
Whitmore, T.C., Prance, G.T., 1987. Biogeography and Quaternary History in Tropical Latin America. Oxford University Press, New York.
Willis, K.J., Whittaker, R.J., 2000. The refugial debate. Science 287, 1406–1407.
Zinck, J.A., García, P., van de Plicht, J., 2011. Tepui peatlands: age record and environmental changes. In: Zink, J.A., Huber, O. (Eds.), Peatlands of the Western Guayana Highlands, Venezuela. Springer, Heidelberg, pp. 189–236.

PART II

The Photosynthetic World

CHAPTER

5

Algae

Jan Kaštovský[1], *Karolina Fučíková*[2], *Jana Veselá*[1], *Charles Brewer Carías*[3] *and Teresa Vegas-Vilarrúbia*[4]

[1]**University of South Bohemia, České Budějovice, Czech Republic** [2]**Assumption College, Worcester, MA, United States** [3]**Venezuelan Society for Natural Sciences, Caracas, Venezuela** [4]**Department of Evolutionary Biology, Ecology and Environmental Sciences, Faculty of Biology, University of Barcelona, Barcelona, Spain**

Introduction

"Algae *sensu latissimo*" is a compendious term denoting a heterogeneous group of several phylogenetic lineages, often comprising eukaryotic "algae" as well as prokaryotic Cyanobacteria (blue-green algae) (Graham et al., 2009). Aside from planktonic and other microorganisms that may be single-celled or colonial, algae also contain large, plant-like forms such as kelp and other seaweeds. These microorganisms and macroorganisms are not all close relatives of each other (Adl et al., 2013) but generally play the same ecological role in nature: primary production. In fact, algae are responsible for nearly half of the world's primary production (Field et al., 1998). However, some algae also live in symbiotic or parasitic relationships with other organisms, and some algal species have even lost the ability to photosynthesize. Clearly, merely formulating a coherent definition of "algae" is a challenge. The biological discipline of algal study is called phycology, although the term algology is sometimes used in the literature as well. This chapter summarizes the results of phycological research on tepuis, mostly based on our own data but also incorporating information from earlier studies.

Algae and extreme environments

The most typical habitats for algae are various aquatic ecosystems (freshwater and marine), but algae can grow terrestrially too (e.g., on soil, wet walls, man-made surfaces, etc.). Essentially, any place reached by sunlight and moisture can be inhabited by algae. Despite the lack of vascular and other highly specialized tissues, algae exhibit remarkable

Biodiversity of Pantepui
DOI: https://doi.org/10.1016/B978-0-12-815591-2.00005-7

physiological flexibility and can withstand extreme environmental conditions. All types of habitats on the summits of tepuis offer extremely low mineral and nutrient levels, low pH, high humic matter content (examples are in Table 5.1), drastically and rapidly changing water availability, and fluctuating light levels. Therefore tepuis can be considered an extreme biotope for photosynthetic microorganisms, which have to adapt to these environmental challenges.

Extreme ecological conditions in most cases support lower species diversity (e.g., Grime and Pierce, 2012). On tepui summits, we recognize several types of habitats with relatively uniform, yet specific, algal floras across all investigated mountains: wet rock walls, rock pools, wetlands (shallow peat bogs), and streams (Fig. 5.1).

Each of these habitats presents its own suite of ecological challenges to algae: for example, light may be scarce under overhangs, while bare rock surfaces may receive extreme irradiation in addition to drying out periodically. Rock pools are shallow (up to 10 cm deep) reservoirs, with minimum or no detritus on the bottom. They are flooded after rain but may dry out within hours. The pools may also warm up rapidly and become oxygen depleted. Algae meet these challenges with a variety of adaptations, which are consistent with the great evolutionary diversity of these organisms. Protection via pigment accumulation, cellular structure, and production of protective lipids and carbohydrates (e.g., Holzinger et al., 2014; Remias et al., 2005, 2012) are just some examples of ways algae handle extreme, and often rapidly changing, environmental conditions.

TABLE 5.1 Water chemistry of several tepuian watersheds and peat habitats as an example of natural conditions on tepuis (see Fig. 1.2 from Chapter 1 for location).

Site	pH	Cond μS/cm	Si mg/L	Ca mg/L	Mg mg/L	Na mg/L	K mg/L	Fe mg/L	PO_4–P mg/L	NO_3-N mg/L	Al mg/L
Akopán-teui peat	3.75	11.5	0.02	0.103	0.064	0.075	≤0.1	N/A	N/A	N/A	N/A
Akopán-tepui brooklet	3.60	7.5	0.13	0.031	0.051	0.022	≤0.1	N/A	N/A	N/A	N/A
Cerro Guaiquinima	3.55	14	0.36	0.760	0.831	0.690	0.047	N/A	N/A	N/A	N/A
Cerro Guaiquinima peat	3.45	25	0.33	0.803	0.350	0.807	0.027	N/A	N/A	N/A	N/A
Auyán-tepui center	3.90	28	0.10	0.107	0.117	0.034	0.023	N/A	N/A	N/A	N/A
Auyán-tepui	3.70	16	0.02	0.09	0.05	0.08	≤0.1	N/A	N/A	N/A	N/A
Churí-tepui creek/river*	3.3–4.58	18–25	0.54–0.77	N/A	N/A	N/A	N/A	0.03–0.22	0.02–1.3	<0.5–0.8	18–92
Churí-tepui swamp*	5.48	25–26	0.41–0.5	N/A	N/A	N/A	N/A	0.06–0.08	0.03	1.10	N/A
Roraima-tepui lake*	N/A	5	0.09	N/A	N/A	N/A	N/A	0.03	0.04	<0.5	N/A
Roraima-tepui creek*	4.76–5.05	19	0.16–1.01	N/A	N/A	N/A	N/A	0.01–0.06	<0.1	<0.5	<20

*Aubrecht et al. (2012), the rest from Rull (unpubl. data).

FIGURE 5.1 The different types of microhabitats inhabited by algae on tepui summits. (A) Wet rock (Akopán-tepui), (B) stream (Apakará-tepui), (C) wetland (Akopán-tepui), and (D) rock pools (Churí-tepui).

Endemism of algae and other microorganisms

Extreme and unique ecological conditions combined with long-term geographic or topographical isolation should lead to high levels of endemism in algae, similar to the patterns observed in macroscopic organisms. It is generally accepted that the diversity and taxonomic composition of local communities of macroorganisms result both from regional-scale processes (allopatric speciation, geographical dispersion) and processes that promote local extinction (predation, competition). However, considerable controversy exists regarding the relative importance of such processes for microorganisms. The "ubiquity hypothesis" proposes that the rate of dispersal of microorganism species is a function of their absolute abundance, resulting in a moderate global richness of species. As the numbers of microbial individuals are astronomically large, limitations on their dispersal would be insignificant, making local extinction and geographic isolation virtually impossible. As a result, allopatric speciation under the ubiquity model would be rare to nonexistent (Fenchel and Finlay, 2004; Finlay, 1999, 2002). In contrast, moderate endemicity models (Chao et al., 2006; Foissner, 2006; Ryšánek et al., 2014) and evidence of allopatric diversification of cyanobacteria and archaea (Darling et al., 2007; Taylor et al., 2006; Whitaker et al., 2003) indicate that microorganisms can exhibit restricted geographic ranges. In the case of algae and cyanobacteria, it has been widely accepted that most species have a widespread or even cosmopolitan distribution that may be explained by the ubiquity hypothesis (Finlay, 2002). However, detailed studies suggest that the geographic distribution of many groups ranges from global to endemic (e.g., diatoms Vanormelingen et al., 2008; cyanobacteria Moreira et al., 2013 and literature therein; and desmids Bestova et al., 2018, Coesel and Krienitz, 2008).

Accurate estimates of species distribution hinge on a taxonomy that reflects genetic and evolutionary diversity and distinctness. This is particularly problematic in morphologically simple organisms such as unicellular algae. Global distributions may be assumed for similar-looking microorganisms that genetically represent multiple species that are, in fact, geographically restricted. Such morphological crypsis has been well documented in coccoid (unicellular and vegetatively nonmotile) green algae, but perhaps more surprisingly even in macroscopic seaweeds (e.g., Fučíková and Lewis, 2012; Schneider et al., 2017). In many cases, DNA sequencing may be a necessary tool for correct species identification, but this tool may not be readily usable for preserved material, let alone in the field. As sequencing technology advances, and especially with the advent of environmental metagenomics, our understanding of microbial endemicity may change significantly.

Phycological studies on Pantepui

Phycological research on the summits of tepuis does not have a very long or rich history, largely due to the difficult access, which makes the transport of equipment like microscopes extremely challenging—our porters and friends from the Yunek village could tell long stories about crossing flooded rivers with a microscope on their back. Thus the best option is often to collect a variety of samples, preserve them in the field, and conduct the actual examination of the material later in a laboratory, at which point returning for more samples (e.g., if some samples contain particularly interesting but insufficient specimens) is impossible. Still, a number of phycological studies do exist, though some are not necessarily focused on floristic surveys (Büdel, 1999; Büdel et al., 1994; Förster, 1963; Lakatos et al., 2001; Pokorný, 1996). These authors have found 17 species of cyanobacteria (*Stigonema ocellatum* and *Gloeocapsa sanguinea* as dominants), six desmid species, plus the green conjugate alga *Zygogonium* sp., from Roraima-tepui, Cerro Parú, Auyán-tepui, and Kukenán-tepui.

Only a subset of tepuis has been explored from the phycological point of view thus far. Our own explorations over the past decade have included several smaller expeditions on Roraima-tepui (2000, 2004, 2006, and 2009; results published in Fučíková and Kaštovský, 2009 and Kaštovský et al., 2011) and bigger expeditions on Churí-tepui in 2012

(Kaštovský et al., 2016); Akopán-tepui in 2015; and Ptari-tepui, Amurí-tepui, and Apakará-tepui in 2017 (results not yet finalized) (for the positions of localities, see Fig. 1.2 from Chapter 1).

Major groups of algae represented on tepuis

Tables 5.2 to 5.5 list the algae species found to date on the tepui summits, organized by major groups. In most types of habitats, in terms of biomass Cyanobacteria are the dominant group (67 species, Fig. 5.2, Table 5.2), followed by conjugates (Zygnemophyceae, 65 species, Figs. 5.3 and 5.4, Table 5.3) and diatoms (approximately 149 species, Fig. 5.3, Table 5.1). Red algae (Rhodophyta) are present in an appreciable biomass in streams (Fig. 5.4), as are

TABLE 5.2 List of diatoms taxa.

Achanthes cf. *oblongella*	R
Actinella sp*	S
*Actinella siolii**	S
Actinella eunitioides Hustedt*	S
Actinella subperonoides Lange-Bertalot*	P, S
Amphora sp.	R
Anomoeoneis sp.	R
Aulacoseira ambigua (Grunow) Simonsen	P, W
Aulacoseira granulata (Ehrenberg) Simonsen	P
Brachysira amoena Lange-Bertalot*	S
Brachysira apiculata (Boyer)Lange-Bertaloti & Moser*	S
Brachysira brebissonii Ross	S, R, W
Brachysira coraliana Metzeltin et Lange-Bertalot*	*ni*
Brachysira lange-bertalotii Metzelin*	R, P, S
Brachysira macroserians Metzeltin et Lange-Bertalot	S, W
Brachysira (aff.) *neoacuta*	S, R, P
Brachysira rostrata (Krasske) Metzeltin et Lange-Bertalot	S, R, P
Brachysira serians (Brébisson ex Kützing) Round et Mann	*W, P, S*
Brachysira simplex Lange-Bertalot*	S
Brachysira steinitziae Metzeltin et Lange-Bertalot*	ni
Brachysira vixapiculata Metzeltin et Lange-Bertalot*	ni
Brachysira wygaschii Lange-Bertalot in Lange-Bertalot et Moser	R
another 6 unidentified species of *Brachysira*	

(Continued)

TABLE 5.2 (Continued)

Caloneis sp.	R
Chamaepinnularia brasilianopsis nov. spec.*	S
cf. *Aulacoseira alpigena*	R
cf. *Chamaepinnularia*	W
cf. *Hantzschia*	S
cf. *Melosira varians*	R
cf. *Pinnularia acoricola*	R
cf. *Stephanocyclus atomus*	R
cf. *Trybrionella*	W
Cyclostephanos dubius (Hustedt) Round	P
Diatoma mesodon (Ehrenberg) Kützing	S
Diploneis sp.	R
Encyonema sp.	R
*Encyonema marginestriatum**	S
Encyonema silesiacum (Bleisch) Mann	R
Encyonema sparsipunctatum Krammer	P, W, R
Encyonopsis buhriana Metzeltin et Lange-Bertalot	R
Encyonopsis cf. *blancheanum*	W
Encyonopsis subminuta Krammer et Reichardt	R
Encyonopsis schneideri Krammer*	S
Eunotia acutinasuta Metzeltin et Lange-Bertalot	S, P, W
Eunotia acutuariola Lange-Bertalot	P, R, W
Eunotia camburnii Metzeltin et Lange-Bertalot*	S
Eunotia cf. *acuticrenulata*	S, R
Eunotia cf. *circumborealis*	W
Eunotia cf. *crassula**	R, W, S
Eunotia cf. dizyga Ehrenberg*	S
Eunotia cf. *fallax*	W
Eunotia cf. *geniculat* *	S, R
Eunotia cf. *noerpeliana**	S, R
Eunotia cf. *tenella*	W, S, R
Eunotia cf. *veneris*	R

(*Continued*)

TABLE 5.2 (Continued)

Eunotia churiensis J. Veselá, Bohunická et Kaštovský	S
Eunotia crassula Metzeltin et Lange-Bertalot	S
Eunotia fennica (Hustedt) Lange-Bertalot	R, P, W
Eunotia gibbosa Grunow	W, R
Eunotia intermedia (Krasske ex Hustedt) Nörpel et Lange-Bertalot	R, P, W
Eunotia multirimoportulata Veselá, Bohunická et Kaštovský	P
Eunotia naegelii Migula	R, P, W
Eunotia napoleonica Metzeltin et Lange-Bertalot*	R, S
Eunotia noerpeliana Metzeltin et Lange-Bertalot*	R, P, W, S
Eunotia paludosa Grunow	R, P, W,
Eunotia rhomboidea Hustedt	R, W
Eunotia triodon Ehreberg	R, S
another approx. 15 unidetified *Eunotia* species	
Eunotioforma sp.	S
Fragilaria capucina Desmazières	R
Fragilarioforma spinulosa (Patrick) Lange-Bertalot	R, S
Frustulia altimontana Metzeltin et Lange-Bertalot	W, R
*Frustulia cf. crassinervia**	S
Frustulia cf. *undosa*	W
Frustulia pararhomboides var. *pararhomboides* Lange-Bertalot	S
Frustulia undosa Metzeltin et Lange-Bertalot	R, P, W, S
Frustulia vulgaris (Thwaites) De Toni	R
Frustulia zizkae Lange-Bertalot	P, W
Gomphonema butantanum Krasske*	S
Gomphonema cf. *bohemicum* ssp. *angustiminus*	R
Gomphonema gracile Ehrenberg	R
Gomphonema micropus Lange-Bertalot	S
Gomphonema parvulum (Kützing) Kützing	S, R
Gomphonema oxycephalum Cleve*	S
Gomphonema riotecense, Lange-Bertalot, Reichart & Metzeltin*	S
Hantzschia amphioxys (Ehrenberg) Grunow	W
Kobayasiella cf. *micropunctata*	R, W

(*Continued*)

TABLE 5.2 (Continued)

Kobayasiella pseudosubtilissima (Manguin) Lange-Bertalot et Reichardt	P, R, W
Kobayasiella sp. 1	W
Kobayasiella sp. 2	R
Kobayasiella subtillisima (Cleve) Lange-Bertalot	S, R
Luticola muticopsis (Van Heurck) Mann	W, R
Luticola sp.	R
Mastogloia sp.	R
Navicula cincta (Ehrenberg) Ralfs	W
Navicula gregaria Donkin	S, R
Navicula rhynchocephala Kützing	S, R, W
Navicula tenelloides Hustedt	R
Nitzschia acicularis (Kützing) W. Smith	R
Nitzschia amphibia Grunow	R
Nitzschia cf. *amphibia*	R
Nitzschia cf. *paleacea*	R
Nitzschia fonticola (Grunow) Grunow	S
Nitzschia hantzschiana Rabenhorst	W, R
Nitzschia palea (Kützing) W. Smith	S
Nitzschia pusilla Grunow	R
Nitzschia sp. 1	R
Nitzschia sp. 2	R
Nupela encyonopsis Metzeltin et Lange-Bertalot	P, R, S, W
Nupela neotropica Lange-Bertalot	S
Nupela zizkae Metzeltin et Lange-Bertalot*	S
Nupela aff. *chlensis*	P
Nupela cf. *astartiella*	R, P
another 4 unidetified *Nupella* species	
Pantocsekiella ocellata (Pantocsek) K.T.Kiss et E.Ács	P
Peronia brasiliensis Hustedt*	R, S
*Peronia cf. fibula, cf. brasiliensis**	S
Peronia sp. 1	R
Peronia sp. 2	R

(*Continued*)

TABLE 5.2 (Continued)

Perinotia diamantina F. Ferrari et C.E. Wetzel	R
Planothidium lanceolatum (Brébisson ex Kützing) Lange-Bertalot	R
Pinnularia cf. *sinistra*	R
Pinnularia cf. *transfversiformis*	W, R
Pinnularia sp. 1	W, R
cf. *Pseudoperonia*	P, S
Stephanocyclus meneghiniana (Kützing) Skabitschevsky	R
Stauroneis phoenicenteron (Nitzsch) Ehrenberg	S, R
Surirella braunii Hustedt*	S
Surirella minuta Brébisson	R
Synedra ulna (Nitzsch) Ehrenberg	R
Thalassiosira sp.	P
Tryblionella sp. 1	P
Tryblionella sp. 2	S, R

Taxa marked by * were found in previous studies of other authors too (Metzeltin and Lange-Bertalot, 1998, 2007).
Typical habitats are marked by *W*, wetlands; *P*, rock pool; *S*, stream; *R*, rock; *ni*, no information available.

TABLE 5.3 List of cyanobacteria taxa.

Albrightia roraimae Kaštovský, Fučíková, Hauer et Bohunická	P, R
Aphanocapsa muscicola (Meneghini) Wille	P, R
Aphanocapsa sp.	P
Aphanothece castagnei (Brebisson) Rabenhorst	R
Aphanothece sp.	W
Anabaena sp. 1	P, W
Anabaena sp. 2	S
Asterocapsa divina group	R, S
Bacularia cf.	S
Calothrix sp.	P, R
Capsosira sp. 1	W
Capsosira sp. 2	R
Cyanodictyon sp.	P, W

(*Continued*)

TABLE 5.3 (Continued)

Cyanothece aeruginosa (Nägeli) Komárek	P, R, W
Cyanothece sp.	W, P
Cyanosarcina sp.	R*
Dichothrix sp. 1	P
Dichothrix sp. 2	P
Enthophysalis arboriformis Kaštovský, Fučíková, Hauer et Bohunická	R
Eucapsis alpina Clements et Schantz	W, P, R
Geitlerinema splendidum (Greville) Anagnostidis	P, R
Gloeocapsa cf. *atrata*	R
Gloeocapsa cf. *biformis*	R
Gloeocapsa kuetzingiana Nägeli	P, R
Gloeocapsa punctata Nägeli	R
Gloeocapsa rupicola Kützing	R
Gloeocapsa sanguinea (C. Agardh) Kützing*	R, P
Gloeothece tepidariorum (Braun) Lagerheim	R
Gloeothece sp.	R
Hapalosiphon cf. *luteolus*	W, R
Heteroleibleinia cf. *pusila*	S
Homeothrix cf. *juliana*	S
Hormoscila sp.	R*
Chroococcus cf. *turgidus*	P, W
Chroococcus turgidus (Kützing) Nägeli	P, W
Chroococcus turicensis (Nägeli) Hansgirg	P
Chroococcus minor (Kützing) Nägeli	P, R
Chroococcus minutus (Kützing) Nägeli	P, S
Komvophoron cf. *schmidlei*	W
Leptolyngbya cf. *boryana*	W, P, S
Leptolyngbya tenuis (Gomont) Anagnostidis et Komárek	W
Leptolyngbya cf. *schmidlei*	P, R, W
Merismopedia elegans Braun ex Kützing	W
Merismopedia glauca (Ehrenberg) Kützing	P
Microcoleus cf. *autumnalis*	R, P, S

(*Continued*)

TABLE 5.3 (Continued)

Nostoc commune Vaucher ex Bornet et Flahault	P, R
Phormidium aerugineo-caeruleum (Gomont) Anagnostidis et Komárek	W
Porphyrosiphon latissimus Kaštovský, Fučíková et Bohunická	R
Pseudophormidium cf. *tenue*	R
Scytonema cf. *multiramosum*	P
Scytonema cf. *myochrous*	P
Scytonema ocellatum (Dillwyn) Lyngbye ex Bornet et Flahault*	R, P
Schizothrix venezuelana Kaštovský, Fučíková et Bohunická*	R
Schizothrix lutea Frémy	P
Stigonema cf. *mesentericum*	P, W
Stigonema cf. *turfaceum*	W
Stigonema crassivaginatum (Geitler) Santanna et al.	R, P, W
Stigonema hormoides (Kützing) Bornet et Flahault	P, R
Stigonema informe Kützing	P, R
Stigionema leprieurii Montagne ex Bornet et Flahault	S
S. ocellatum (Dillwyn) Thuret ex Bornet et Flahault*	R, P, W, S
Synechococcales (5 unidetified taxa with unknown generic level)	R, P, W, S
Xenotholos cf. *kernerii*	P, S

Taxa marked by * were found in previous studies of other authors too (Büdel, 1999; Büdel et al., 1994; Lakatos et al., 2001; Pokorný, 1996).
Typical habitats are marked by *W*, wetlands; *P*, rock pool; *S*, stream; *R*, rock; *R**, rock in caves.

TABLE 5.4 List of desmids taxa.

Actinotaenium cucurbita (Brébisson ex Ralfs) Teiling	P, R
Actinotaenium cf. *cucurbita* var. *cucurbita* f. *minus* (West et West) Teiling	P, R
Actinotaenium cucurbitinum var. *subpolymorphum* (Nordstedt) Teiling	P, R
Actinotaenium cf. *canadense*	W
Actinotaenium cf. *diplosporum*	W, P
Actinotaenium cf. *globosum*	W, P
Actinotaenium cf. *obcuneatum*	P
Actinotaenium cf. *palangula*	R
Actinotaenium cf. *subpalangula*	S, W

(*Continued*)

TABLE 5.4 (Continued)

Bambusina borreri Ralfs	W
Closterium cf. *gracile* Brébisson	W
Cosmarium fontigerum Norstedt	P, S
Cosmarium itatiayae Krieger*	W
Cosmarium laeve Rabenhorst	P
Cosmarium cf. *regneli*	P
Cosmarium obtusatum (Schmidle) Schmidle	W
Cosmarium pentachondrum Børgesen	P, R
Cosmarium subarctoum var. *minutissimum* West	P
Cosmarium subtumidum var. *minutum* (Krieger) Krieger et Gerloff	P
Cosmarium sp.1 "Roraima"	W
Cosmarium sp.3 "Roraima"	W
Cosmarium tinctum var. *intermedium* Nordstedt	W
Croasdalea cf. *marthae* (Grönblad) Bicudo et Mercante	W, P
Cylindrocystis brebissonii (Ralfs) de Bary	W, R, P
Euastrum arciferum var. *goyazense* Förster	P
Euastrum brasiliense Borge*	ni
Euastrum humbertii var. *brasiliense* Förster	W, P, R
Euastrum macrocephalum Förster et Eckert*	P, R
Euastrum sublobatum Brébisson ex Ralfs	W, P, R
Euastrum subbinale Gutwinski	P, W
Euastrum sp. 1 "Roraima"	P
Euastrum sp. 2 "Churi"	P
Ichtyocercus longispinus (Borge) Krieger	W, S
Mesotaenium endlicherianum Nägeli	P, S, W
Mesotaenium chlamydosporum De Bary	W, P
Micrasterias arcuata unidetif. variety	W, P, R
Micrasterias arcuata var. *borgei* Förster	W, P, R
Micrasterias arcuata var. *compacta* Förster	W, P, R
Micrasterias arcuata var. *robusta* f. *goyasensis* Förster et Eckert	W, P, R
Micrasterias simplex Börgesen* (also fac. *triquetra*)	ni
Netrium digitus (Brébisson ex Ralfs) Itzigsohn et Rothe	W, P

(*Continued*)

TABLE 5.4 (Continued)

Onychonema laeve Nordstedt *	ni
Pleurotaenium minutum (Ralfs) Hilse	P
Sphaerozosma luetzelburgianum Krieger	W
Staurastrum binum var. *minus* Borge* (also f. 9-radiata)	ni
Staurastrum cosmarioides Nordstedt	W
Staurastrum elongatum var. *amazonenze* Scott et Croasdale	W
Staurastrum hystrix Ralfs	W
Staurastrum orbiculare (Ehrenberg) Ralfs	P
Staurastrum pseudozonatum var. *minutissimum* Förster	W, P, R
Staurastrum cf. *quadrispinatum*	W
Staurastrum cf. *teliferum*	W
Staurastrum sp.	W, P
Staurodesmus cf. *calyxoides*	P, W
Staurodesmus cf. *wandae*	W
Staurodesmus extensus var. *vulgaris* (B.Eichler et Raciborski) Croasdale	W
Staurodesmus omearae (Archer) Teiling	W, P
Tetmemorus laevis var. *tropicus* Krieger	W
Vincularia roraimae Fučíková et Kaštovský	S
Xanthidium sp.	W

Taxa marked by * were found in previous studies of other authors as well (Förster 1963).
Typical habitats are marked by *W*, wetlands; *P*, rock pool; *S*, stream; *R*, rock; *ni*, no information available.

TABLE 5.5 List of other algal groups.

Cryptophyta	
Cryptomonas sp.	P, W
Dinophyta	
Gymnodinium sp.	W, P
Rufusiella insignis (Hassell) Loeblich	W, R
Synurophyceae	
Synura cf. *sphagnicola*	W
Euglenophyta	
Euglena mutabils Schmitz	R, W, P
Colatium cf. *siderophus*	P, W

(*Continued*)

TABLE 5.5 (Continued)

Rhodophyta	
Audouinella sp.	S
Batrachospermum sp.	S
unidentified member of Batrachospermales family	S
Chlorophyta	
Aphanochaete sp.	P, S
Binuclearia cf. *tectorum*	P
cf. *Quadrigula*	P
Cylindrocapsa geminella Wolle	R, P, W
Chlamydomonas sp.	P, W
Ekerewekia churiensis Kaštovský, Fučíková, Štenclová et Brewer-Carías	S
Microspora cf. *loefgrenii*	S, R, W
Microspora cf. *tumidula*	S, R
Oedogonium sp. steril 5 μm	R
Oedogonium sp. steril 7.5–10 μm	R, P
Oocystis sp.	P
Oocystis cf. *eliptica*	W
Scenedesmus sp.	P
Sphaerococcomyxa sp.	R
Trentepohlia sp.	R
cf. *Treubaria setigera* sensu Ettl 1968	P
another 2 taxa of coccal green algae, propably from Radiococcaceae family	R
Klebsormidiales	
Klebsormidium cf. *flaccidum*	S, P
Zygnematales	
Mougeotia sp. steril. 12–17 μm	W, P, R, S
Mougeotia sp. steril. 25–35 μm	W, P, R, S
Zygnema sp. steril 1	W, P
Zygnema sp. steril 2	W, P
Zygogonium sp. steril*	P, W

Taxon marked by * was found in previous studies of other authors too (Büdel et al., 1994)
Typical habitats are marked by *W*, wetlands; *P*, rock pool; *S*, stream; *R*, rock.

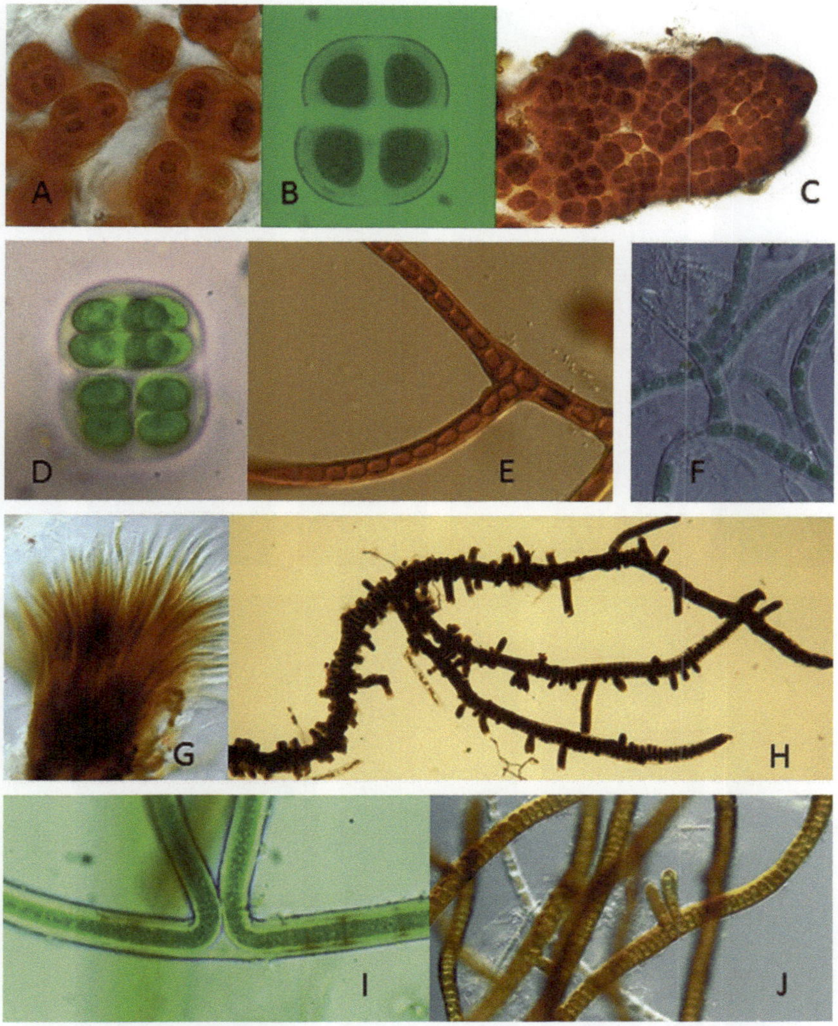

FIGURE 5.2 Examples of cyanobacteria found on tepuis spanning a range of morphologies. (A) *Gloeocapsa sanguinea* (Roraima-tepui), (B) *Chroococcus turicensis* (Churí-tepui), (C) *Enthophysalis arboriformis* (Roraima-tepui), (D) *Eucapsis* cf. *alpina* (Akopán-tepui), (E) *Albrightia roraimae* (Roraima-tepui), (F) *Hapalosiphon* cf. *luteolus* (Churí-tepui), (G) *Homoeothrix* cf. *juliana* (Akopán-tepui), (H) *Stigonema mamillosum* (Roraima-tepui), (I) *Scytonema ocellatum* (Churí-tepui), and (J) *S. ocellatum* (Churí-tepui).

filamentous green algae (Chlorophyta, Fig. 5.4). Coccal green algae, Euglenophyta, Synurophyceae, Cryptophyta, and Dinophyta also occur in various habitats, but they are rare (Table 5.4).

Algal species composition can differ greatly from wetlands to bare rock surfaces—water and light availability obviously plays a big role in community assembly. The

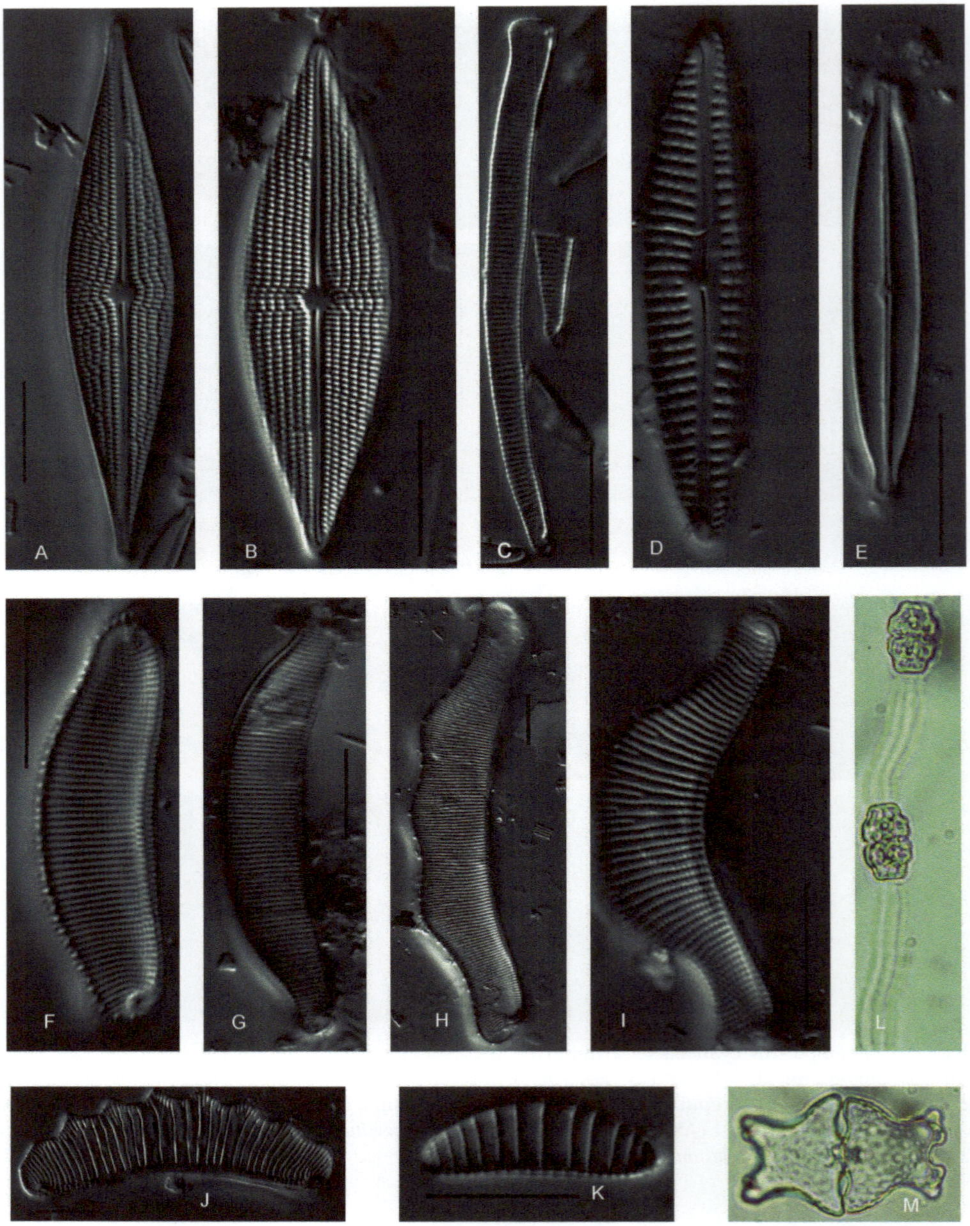

FIGURE 5.3 Examples of diatoms and desmids. (A) *Brachysira* cf. *rostrata*, (B) *Brachysira lange-bertalotii Metzelin*, (C) *Actinella subperonioides* Metzelin et Lange-Bertalot, (D) *Encyonema* sparsipunctatum *Krammer*, (E) *Kobayashiella* cf. *micropunctata*, (F) *Eunotia* noerpeliana *Metzeltin* et Lange-Bertalot, (G) *Eunotia acutuariola* Metzeltin et Lange-Bertalot, (H) *Eunotia triodon* Ehrenberg, (I) *Eunotia napoleonica* Metzeltin et Lange-Bertalot, (J) *Eunotia* cf. *acuticrenulata* K. Perinotia diamantina F. Ferrari & C.E. Wetzel (all pictures are from samples collected on Churí-tepui), (L) yet-unidentified desmids *Euastrum* sp. 1, and (M) yet-unidentified desmids *Euastrum* sp. 2 (Akopán-tepui).

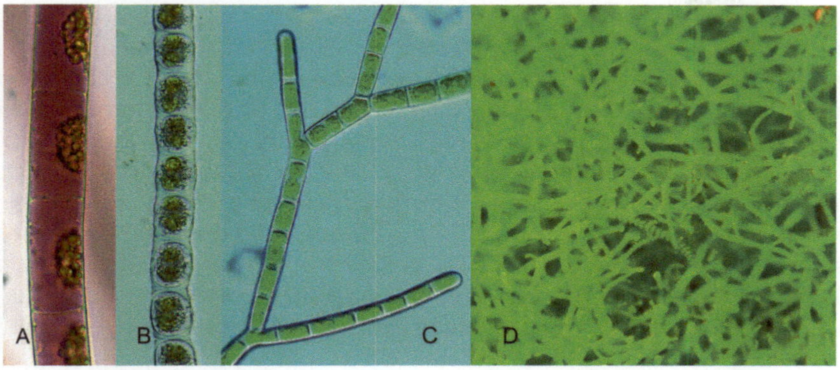

FIGURE 5.4 Important algae from other taxonomic groups. (A) *Zygogonium ericetorum* (Zygnematales, Streptophyta, Churí-tepui), (B) *Cylindrocapsa geminella* (Chlorophyceae, Chlorophyta, Churí-tepui), (C) *Ekerewekia churiensis* (Trebouxiophyceae, Chlorophyta, Churí-tepui), and (D) *Batrachospermum* sp. (Rhodophyta, Apakará-tepui).

unstable water regime of bare rock surfaces especially favors cyanobacteria as a dominant group because of their high resistance to desiccation. This applies not only to vertical rocks and overhangs but also shallow rock pools. However, pools can also support green (chlorophyte and zygnemophyte) algae in significant numbers. The water regime in deeper wetlands and reservoirs (0.5–1 m deep) is more stable that in shallow rock pools, which causes eukaryotic algae (especially desmids and filamentous zygnemophytes such as *Mougeotia* sp.) to dominate. Bogs and other wetlands with standing open water also provide a habitat for an assortment of other common planktonic algae: euglenophytes, cryptophytes, dinophytes, and synurophytes. Streams represent the least homogeneous habitat on Pantepui. They range from small temporary brooklets to rivers. Despite this, the composition of the algal flora is often similar from stream to stream, being heavy in filamentous forms spanning cyanobacteria (e.g., *Stigonema hormoides*, *Homeothrix* cf. *juliana*), rhodophytes (e.g., *Batrachospermum* sp., *Audouinella* sp.), zygnemophytes (e.g., *Mougeotia* spp.), and chlorophytes (e.g., *Microspora* spp.).

Cyanobacteria

Cyanobacteria are responsible for creating Earth's oxygen-rich atmosphere in the Precambrian period, thereby enabling aerobic life forms to evolve (Lyons et al., 2014). Some authors do not include cyanobacteria under the term "algae," mainly because they are prokaryotic. Still, cyanobacteria are typically represented in phycology textbooks (e.g., Graham et al., 2009) and have a critical biological relevance to photosynthetic eukaryotes. It is now widely accepted that the photosynthetic organelles (plastids or chloroplasts) of plants and eukaryotic algae descended from cyanobacteria that were engulfed by ancient eukaryotes (e.g., Falcón et al., 2010; Margulis, 1981; Mereschkowsky, 1905).

As mentioned earlier, cyanobacteria typically dominate in most habitats on tepuis. The original rock coloration on tepuis is pink, but the stone surfaces are covered with black biofilm, which consists mainly of cyanobacteria with their characteristic ultraviolet (UV)−protective pigments accumulated either in extracellular mucilaginous sheaths or inside cells (Lakatos et al., 2001). The dominant organisms here are the coccal cyanobacteria of the genus *Gloeocapsa*, mainly *G. sanguinea* (Fig. 5.2A) or filamentous branching cyanobacteria *Scytonema* sp. (Fig. 5.2I) and *Stigonema* sp. (mainly *S. ocellatum*, Fig. 5.2J). Other important taxa are the simple trichal *Schizothrix venezuelana* and *Porphyrosiphon latissimus* (tepuian endemic, Kaštovský et al., 2016). Overhangs (the shadowed parts of the rocks) are also dominated by *G. sanguinea* but in addition *Gloeocapsa punctata* and *Gloeocapsa kuetzingiana* are also present (Kaštovský et al., 2011). A rarer occurrence is *Enthophysalis arboriformis* (Fig. 5.2C), another endemic tepuian coccal species (Kaštovský et al., 2011). In pools, *Scytonema* sp. dominates, mainly *S. ocellatum* (Fig. 5.2I) or *Scytonema myochrous*, along with *Albrightia roraimae* (Fig. 5.2E), another endemic pantepuian species (Kaštovský et al., 2011), plus the ubiquitous *G. sanguinea* and *S. ocellatum*.

Even though filamentous green algae form a large biomass in wetlands, diverse cyanobacterial species occur here too: *Chroococcus turicensis* (Fig. 5.2B), *Chroococcus turgidus*, *Eucapsis* cf. *alpina* (Fig. 5.2D), *Merismopedia elegans*, and the true-branching *Hapalosiphon* cf. *luteolus* (Fig. 5.2F), presumably another species new to science. Streams harbor filamentous cyanobacteria from, for example, the genera *Stigonema*, *Schizothrix*, *Hapalosiphon*, and *Homeothrix* and, less frequently, coccoid forms such as *Chroococcus* and *Gloeocapsa*.

Diatoms (phylum Bacillariophyta)

Diatoms are a eukaryotic phylum in the SAR supergroup (Fig. 5.5) and are known for their charismatic siliceous cell cover. They have a long history with a detailed taxonomy and are thought to be the species-richest group of algae, with ca. 30,000 species described and an estimated total likely being close 100,000 (Mann and Vanormelingen, 2013).

In their monograph on Neotropical diatoms Metzeltin and Lange-Bertalot (1998, 2007) investigated ca. 300 samples and 1300 species along the South American tropics. In most of the studied locations cosmopolitan or pantropical species were dominant, coinciding with the fact that most sampling sites were near human settlements and therefore affected by humans. Pristine waters were frequently remote and difficult to reach. Metzeltin and Lange-Bertalot (1998, 2007) reported 35 diatom species belonging to 11 genera at Auyán and Roraima. Nonetheless, only 19 and 11 species sampled atop Auyán-tepui and Roraima-tepui, respectively (Table 5.1), can be ascribed to Pantepui (≥1300 m a.s.l.) so far, and among them 10 species were unidentified or described as new. Worth mentioning is the high number of acidophilus *Eunotia* taxa.

South America—the tropics especially—seems to be a speciation hotspot for the diatom group of Eunotiophycidae (which includes genera such as *Actinella*, *Eunotia*, *Peronia*, etc.). Many novel species and multiple new genera within this acidophilic group from the Neotropics have been described recently (e.g., Ferrari et al., 2009, 2014; Wetzel et al., 2012). Diatom flora on tepuis is also dominated by members of Eunotiophycidae. The majority of species belongs to the genus *Eunotia*, which represents one of the most species-rich genera within diatoms in general. A number of unknown and unusual *Eunotia* species were encountered on tepuis. For example, Kaštovský et al. (2016) reported and described two

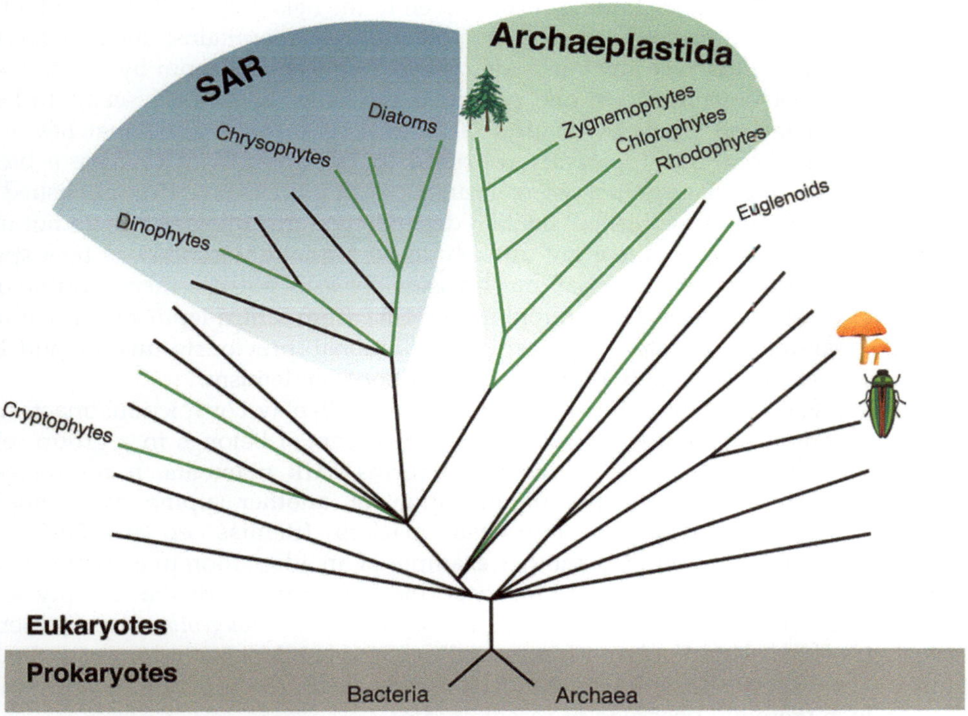

FIGURE 5.5 Simplified phylogenetic tree illustrating eukaryotic diversity and highlighting algal lineages found on tepuis. Green branches mark major photosynthetic lineages of eukaryotes (algae), whether they are documented from tepuis or not. *SAR*, supergroup comprising Stramenopila, Alveolata, and Rhizaria, including several groups of algae (e.g., kelps, diatoms, chrysophytes, synurophytes, and dinoflagellates). *Source: Figure simplified and adapted from Adl et al. (2013).*

new, likely endemic, species of *Eunotia* from Churí-tepui: *Eunotia churiensis* and *Eunotia rimoportulata*. From other genera, members of *Actinella*, *Brachysira*, and *Nupela* were also common and represented by multiple species on tepuis.

Green algae (phyla Charophyta and Chlorophyta)

Desmids (class Zygnemophyceae in Charophyta, Archaeplastida; see Fig. 5.5) tend to be the third most diverse group on tepuis. In the report by Kaštovský et al. (2011), desmids occurred in all sampled habitats with the exception of overhangs, were the most diverse in wetlands and streams (e.g., on Roraima-tepui comprising 27 and 26 species, respectively), and the least diverse in waterfalls (4 species). Only one species occurred in all habitats (except overhangs), *Actinotaenium cucurbita*, which is common worldwide. Other species were mostly restricted to only a few habitat types. Perhaps the most morphologically distinct taxon is *Vincularia roraimae*, a new genus and species described by Fučíková and Kaštovský (2009). This desmid has a unique growth habit with cells attached to long, robust, unbranched stalks that can form bundles visible even to the naked eye—likely an adaptation to stream habitats. Some desmids were found on multiple mountains, such as *Euastrum macrocephalum*, a large *Euastrum* previously described from Roraima-tepui by Förster (1963). This species may not be endemic to one particular peak like *V. roraimae* seems to be but may be specific to several tepuis. A handful of other species found good matches in taxa previously reported from, but apparently restricted to, South America or have a broader tropical distribution (e.g., *Staurastrum pseudozonatum*). Both published and unpublished data by Kaštovský et al. (2011, unpublished) contain desmids that morphologically do not match known species well. However, these are unlikely to be formally described as new species, usually because of insufficient material, but occasionally also because their morphological distinctness is debatable. Regardless, unpublished data from Churí-tepui and preliminary data from Akopán-tepui indicate the presence of additional, previously unseen, and likely new desmid species, quite possibly representing additional endemism.

Other green algae occur on tepuis and in some cases display conspicuous macroscopic growth. The common filamentous alga *Zygogonium ericetorum* belongs to a group related to the desmids (also in Zygnemophyceae) and forms dark filaments. It resembles the recently characterized UV-protective pigmentation of another alpine zygnemophyte, *Mesotaenium berggrenii*, found on European glaciers (Remias et al., 2012). This purpurogallin-derived compound is another example of an adaptation to extreme environments and differs from carotenoids and mycosporine-like amino acids that usually serve a protective purpose in algae and cyanobacteria. *Zygnema* and *Mougeotia*, also filamentous zygnemophytes, occur commonly on tepuis and in some habitats even dominate the algal community. Their species-level identification depends on an examination of reproductive structures, which were not observed in recent studies.

Nonzygnemophyte green algae occur on tepuis as well, but can be taxonomically even more challenging than filamentous zygnemophytes—in coccoid green algae even identification to the genus level may not be possible without genetic data (e.g., Fučíková and Lewis; 2012). Notable filamentous species include *Cylindrocapsa geminella*, a cosmopolitan species, and in stark contrast to a newly described tepui-endemic *Ekerewekia churiensis* (Trebouxiophyceae, Chlorophyta, Archaeplastida), which grows abundantly in Churí-tepui's streams and has a unique combination of morphological traits (Kaštovský et al., 2016). In the language of the indigenous Pemon people, "ekerewek" means "alga" (or any small slimy organism living in the water). In the Bathroom Creek on Churí-tepui summit, a large swath of the stream was covered by this very characteristic alga. Its taxonomic position was supported by molecular methods, and *Ekerewekia* was placed in a phylogenetic cluster, where it is the only branched filamentous form, but is otherwise morphologically consistent with its relatives (Kaštovský et al., 2016). Despite focused efforts, *Ekerewekia* was not found on Roraima-tepui or on Akopán-tepui—the closest neighboring mountains to Churí-tepui.

Other algal groups represented on tepuis

Wetlands (and the traps of larger carnivorous plants such as *Heliamphora* and especially *Brocchinia*) represent the only (semi)stable stagnant water habitat on Pantepui. For this reason, typical planktonic algae such as Euglenophyta inhabit the tepuis' peat bogs and other wetlands. Examples of euglenophytes are the *Euglena mutabilis* or *Colacium* sp., an epizoic alga found worldwide that live in a sessile state on carapaces of microscopic water crustaceans. Euglenophytes (Fig. 5.5) are an ancient group of flagellated algae, characterized by a conspicuous red eyespot and a proteinaceous cell cover (pellicle) made from interlocking bands.

Synurophyceae are a planktonic group related to diatoms, and like diatoms, they typically possess a silica-based cell cover. Unlike diatoms, which have a two-part solid frustule, the shells of synurophytes and chrysophytes consist of siliceous scales, the shape of which is diagnostic for different species. On tepuis, *Synura* cf. *sphagnicola* was encountered (Akopán, unpublished), but without electron microscopy the identification cannot be certain.

Several dinophytes were also found in tepui wetlands but could not be identified more precisely than *Gloeodinium* sp. and *Glenodinium* sp. Dinophyta are a phylum of flagellates that belong to the SAR supergroup (Fig. 5.5) and are of particular interest to evolutionary biologists due to their dynamic endosymbiotic past and many other cellular quirks.

Red algae, or Rhodophyta, are more commonly known as marine organisms, as this group contains many species of macroscopic seaweeds, some of which are of global economic importance. However, freshwater red algae exist too and are most common in streams. On the surveyed tepuis, an alga identified as *Batrachospermum* sp. (Kaštovský et al., 2011) was found in macroscopic tufts. It is possible that this is a new species of *Batrachospermum* or another genus from the same order, which would be consistent with the numerous recent additions to the diversity of Batrachospermales (e.g., Chapuis et al., 2017). A species of *Audouinella*, a less conspicuous but species-rich and common genus, was also observed on Churí-tepui, where another, morphologically enigmatic and unidentified rhodophyte was spotted, but material of the latter was unfortunately not recovered.

Tepuis and algal endemics

As we postulated earlier, tepuis are extreme biotopes because of low nutrients, low pH, fluctuating water availability, and highly variable light conditions. These features and geographical isolation would have been responsible for high levels of endemism in macroorganisms. If we follow the theory of moderate endemism, which in contrast to ubiquity theories allows for approximately one third of microbe species to have restricted distributions (Foissner, 2006, 2008), we should expect a high level of endemism in the microbial realm too. Desmids may be one of the best groups, alongside diatoms, to examine algal endemism on tepuis because their great morphological diversity allows fairly reliable species-level identification. Past discussions of their biogeography (e.g., Coesel et al., 1988; Coesel and Krienitz, 2008) concluded that some species' distributions are restricted by dispersal rather than environmental suitability, with waterfowl being a major dispersing agent—this would also be consistent with high endemism on tepuis, as they are not known for harboring migratory waterfowl. In addition, South American desmids have received quite a bit of taxonomic attention over the past century. A recent study on diatom distribution patterns in the tropical Andean and adjacent Amazon lowlands (Benito et al., 2018) showed that broadscale spatial factors outperformed limnological and geoclimatic variables, suggesting dispersal-assembled communities. Additionally, remoteness decreased species number, and this fact could lead to an expectation of greater endemism on tepuis. Thus it is possible, more so than in other algal groups, to assess which species occur solely on tepuis in contrast to those occurring across the region or throughout the continent. Still, compared to animals and vascular plants, we still know very little about desmid and diatom diversity and distribution in the New World.

Assuming that morphology is a good indicator of species identity, most of the desmids found on tepuis are widely distributed and previously reported from South America (e.g., Araújo et al., 2010; Bicudo and Samanez, 1984; Croasdale et al., 1983; Croasdale and Flint, 1986; Diaz-Olarte and Duque, 2009; Förster, 1963, 1964, 1974; Gonzalez-Gonzalez and Mora-Osejo, 1996; Krieger, 1950; Lenzenweger, 1999; Prescott et al., 1972, 1975, 1977, 1981, 1983; Ramirez et al., 1998). In contrast, some of the tepui-dwelling desmids appear to have more restricted ranges. Two varieties of *Micrasterias arcuata* found on Roraima and Churí (var. *robusta* and var. *goyazensis*) appear to be specific to Brazil. *S. pseudozonatum* was originally described from Guyana, with subspecific taxa also known from Brazil and Sumatra, suggesting a possible pantropical distribution. Somewhat similarly, *Tetmemorus laevis* var. *tropicus* was previously identified in Mississippi, Java, Sumatra, and New Zealand. *Euastrum arciferum* var. *goyazense*, *Cosmarium pentachondrum*, and *Sphaerozosma luetzelburgianum* appear to be restricted to Brazil and were found on Churí-tepui.

So far, two putatively endemic desmids have been described from Roraima: *E. macrocephalum* (Förster, 1963) and the new genus and species *V. roraimae* (Fučíková and Kaštovský, 2009). Further, Förster (1963) suggested multiple new forms (subspecific taxa) of previously described desmid species: *Onychonema laeve* f. *porosa*, *Staurastrum binum* f. *novem-radiata*, *Micrasterias simplex* fac. *triquetra*, and *Cosmarium itatiayae* f. *minor*, though only the *Micrasterias* appears to be formally designated by Förster, and the *Cosmarium* does not appear in the Index Nominum Algarum at all. Similarly, Kaštovský et al. (2011) listed several desmid taxa that did not perfectly match existing species and might therefore represent new forms, varieties, or possibly even species. Preliminary unpublished data from Akopán-tepui suggest additional desmid taxa to be described, including a new *Vincularia*-like alga and a yet-unidentified *Euastrum* (Fig. 5.3L and M).

Two endemic diatoms (*Eu. churiensis* and *Eunotia multirimoportulata*), one green filamentous alga (new genus and species *Ek. churiensis*), and four species of Cyanobacteria (*A. roraimae*, *E. arboriformis*, *P. latissimus*, and *S. venezuelana* were described by Kaštovský et al. (2016). Based on morphological comparisons with previous studies, *S. venezuelana* is not necessarily endemic to tepuis but may occur in other types of aerial habitats in the Neotropics (Golubić, 1967).

Despite the special character of Pantepui, the list of currently known microbial endemics is relatively short. The species list in Tables 5.1–5.4, however, might suggest a different conclusion. Detailed analysis of the list of observed cyanobacteria shows that out of 67 taxa, 30 are identified to the species level (including four newly described species from Pantepui), 32 have an identification on the generic level (sp. or cf.), and 5 other cyanobacterial taxa are impossible to classify even to the level of genera. Fifty-five percent of cyanobacterial taxa are not identified to the species level and therefore cannot be formally recognized as endemic but may turn out to be endemic if examined in more detail. In several cases, a formal circumscription would require more material than is available—the organism may be rare at the sampled localities, or the material is in poor condition and resampling is not possible. In other cases, the material does not contain the life stages with diagnostic traits for the species. For example, for a reliable species determination of *Anabaena* cf. *jonsonii*, the shape and dimensions of akinetes are needed, which were never observed in the tepui material. In other groups, the situation is similar. For example, for *Mougeotia* sp. and *Actinotaenium* cf. *diplosporum*, without examining zygospores (the product of sexual reproduction in zygnemophyte algae), we cannot determine which species the material represents, and all observed *Mougeotia* filaments were sterile. Many algal species from tepuis differ from known taxa only in small morphological details or in size, and the taxonomic importance of such details cannot be verified without additional analyses such as DNA sequencing (e.g., cyanobacteria or green coccal algae) or electron microscopy (e.g., a large amount of diatom species or *Synura*). Such methods are difficult to implement in inaccessible locations such as the tepuis, and transporting live material into laboratories has its own problems. Many taxa reported as "cf." in recent publications may be species or varieties new to science, but given the technical constraints, we can only confidently describe the few taxa that differ substantially from known species—for example, *Vincularia roraimae* by the unique stalk or *E. multirimoportulata* by its specific rimoportulae arrangement (Fučíková and Kaštovský, 2009; Kaštovský et al., 2016).

Another challenge when considering endemism on tepuis is the general paucity of phycological floristic studies in the Neotropics. Thus some species occurring on tepuis may be present in the surrounding areas and regions but have simply not been seen because nobody has looked. In such cases, the conclusion of a tepui-restricted distribution would be erroneous. Unlike Europe, which has an extensive suite of published algal floras (e.g., the multivolume compendium Süßwasserflora von Mitteleuropa), comprehensive taxonomic literature for the Neotropics is lacking. This further complicates microfloristic efforts on tepuis and subsequently makes it difficult to determine the level of microbe endemism on tepuis. Lastly, as discussed earlier, even the most thorough morphological examinations and literature searches may not be enough to accurately describe the biodiversity of algae on tepuis (or elsewhere). Technological advancements and high-throughput sequencing methods will certainly help in future studies, but the ideal would be a combined approach employing a terrain morphological study, live isolated cultures, targeted DNA sequencing, and environmental metagenomics. Clearly, most of the work on algal biodiversity is yet to be done, and tepuis likely still have many new species yet to be discovered.

Acknowledgements

The authors are thankful to Valentí Rull for providing chemistry data.

References

Adl, S.M., Simpson, A.G., Lane, C.E., Lukeš, J., Bass, D., Bowser, S.S., et al., 2013. The revised classification of eukaryotes. J. Eukaryot. Microbiol. 59 (5), 429–493.

Araújo, A., de, Bueno, N.C., Meurer, T., Bicudo, C.E., de, M., 2010. Charophyceae. In: Forzza, R.C. (Ed.), Catálogo de plantas e fungos do Brasil, vol. 1. Andrea Jakobsson Estúdio, Instituto de Pesquisas Jardim Botânico do Rio de Janeiro, Rio de Janeiro, pp. 310–334.

Aubrecht, R., Audy, M., Barrio-Amoros, C., Breure, A., Brewer-Carías, Ch, Derka, T., et al., 2012. Venezuelan Tepuis: Their Caves and Biota. Acta Geologica Slovaca Monograph, Comenius University, Bratislava.

Benito, X., Fritz, S.C., Steinitz-Kannan, M., Tapia, P.M., Kelly, M., Lowell, T.V., 2018. Geo-climatic factors drive diatom community distribution in tropical South American freshwaters. J. Ecol. 106, 1660–1672.

Bestova, H., Munoz, F., Svoboda, P., Skaloud, P., Violle, C., 2018. Ecological and biogeographical drivers of freshwater green algae biodiversity: from local communities to large-scale species pools of desmids. Oecologia 186, 1017–1030.

Bicudo, C.E.M., Samanez, I.M., 1984. Desmidioflorula paulista III. Bibliotheca Phycologica, vol. 68. J. Cramer, Vaduz.

Büdel, B., 1999. Ecology and diversity of rock-inhabiting cyanobacteria in tropical regions. Eur. J. Phycol. 34, 361–370.

Büdel, B., Lüttge, U., Stelzer, R., Huber, O., Medina, E., 1994. Cyanobacteria of rocks and soils of the Orinoco lowlands and the Guyana uplands, Venezuela. Bot. Acta 107, 422–431.

Chao, A., Li, P.C., Agatha, S., Foissner, W., 2006. A statistical approach to estimate soil ciliate diversity and distribution based on data from five continents. Oikos 114, 479–493.

Chapuis, I.S., Necchi Jr., O., Zuccarello, G.C., Xie, S.-L., Aboal, M., Sanchez Castillo, P.M., et al., 2017. A new genus, *Volatus* and four new species of *Batrachospermum sensu stricto* (Batrachospermales, Rhodophyta). Phycologia 56 (4), 454–468.

Coesel, P.F.M., Krienitz, L., 2008. Diversity and geographic distribution of desmids and other coccoid green algae. Biodivers. Conserv. 17, 381–392.

Coesel, P.F.M., Duque, S.R., Arango, G., 1988. Distributional patterns in some neotropical desmid species (Algae, Chlorophyta) in relation to migratory bird routes. Rev. Hydrobiol. Trop. 21 (3), 197–205.

Croasdale, H.T., Flint, E.A., 1986. Flora of New Zealand. Freshwater algae, Chlorophyta, Desmids. V.R. Ward, Wellington.

Croasdale, H.T., Bicudo, C.E., de, M., Prescott, G.W., 1983. A Synopsis of North American Desmids. Part 2: Desmidiaceae: Placodermae. Section 5. University of Nebraska Press, Lincoln and London.

Darling, K.F., Kucera, M., Wade, C.M., 2007. Global molecular phylogeography reveals persistent Arctic circumpolar isolation in a marine planktonic protist. PNAS 104, 5002–5007.

Diaz-Olarte, J., Duque, S.R., 2009. Algal assemblages in freshwater natural microecosystems from the tropical carnivorous plant *Utricularia foliosa* L. Caldasia 31, 319–337.

Falcón, L., Magallón, S., Castillo, A., 2010. Dating the cyanobacterial ancestor of the chloroplast. ISME J. 4, 777–783.

Fenchel, T., Finlay, B.J., 2004. The ubiquity of small species: patterns of local and global diversity. Bioscience 54, 777–784.

Ferrari, F., Wetzel, C.E., Ector, L., Blanco, S., Viana, J.C.C., Silva, E.M., et al., 2009. *Perinotia diamantina* sp. nov., a new diatom species from the Chapada Diamantina, Northeastern Brazil. Diatom Res. 24, 79–100.

Ferrari, F., Wetzel, C.E., Ector, L., Bicudo, D.C., Bicudo, C.E.M., 2014. A new uncommon epilithic Eunotia (Bacillariophyceae, Eunotiaceae) from the Chapada Diamantina region, Northeast Brazil. Phytotaxa 164, 161–174.

Field, C.B., Behrenfeld, M.J., Randerson, J.T., Falkowski, P., 1998. Primary production of the biosphere: integrating terrestrial and oceanic components. Science 281 (5374), 237–240.

Finlay, B.J., 2002. Global dispersal of free-living microbial eukaryote species. Science 296, 1061–1063.

Foissner, W., 2006. Biogeography and dispersal of micro-organisms: a review emphasizing protists. Acta Protozool. 45, 111–136.

Foissner, W., 2008. Protist diversity and distribution: some basic considerations. Biodivers. Conserv. 17 (2), 235–242.

Förster, K., 1963. Desmidiaceen aus Brazilien, 1: Nord Brazilien. Rev. Algologique 7 (1), 38–92.

Förster, K., 1964. Desmidiaceen aus Brasilien, 2: Bahia, Goyaz, Piauhy und Nord-Brasilien. Hydrobiologia 23, 321–505.
Förster, K., 1974. Amazonische Desmidieen 2: Areal Maues-bacaxis. Amazoniana 5, 135–242.
Fučíková, K., Kaštovský, J., 2009. *Vincularia roraimae* (Zygnematophyceae, Desmidiales), gen. et sp. nov. from the top of Mt. Roraima, Venezuela. Nova Hedwigia 88 (1-2), 49–56.
Fučíková, K., Lewis, L.A., 2012. Intersection of *Chorella, Muriella* and *Bracteacoccus*: resurrecting the genus *Chromochloris* Kol & Chodat (Chlorophyceae, Chlorophyta). Fottea 12, 83–93.
Golubić, S., 1967. Die Algenvegetation an Sandsteinfelsen Ost-Venezuelas (Cumana). Int. Rev. Ges. Hydrobiol. 52, 693–699.
Gonzalez-Gonzalez, L.E., Mora-Osejo, L.E., 1996. Desmids from paramo lakes in Colombia. Caldasia 18, 165–202.
Graham, L.E., Graham, J.M., Wilcox, L.W., 2009. Algae, 2nd ed. Pearson Benjamin Cummings, San Francisco.
Grime, J.P., Pierce, S., 2012. The Evolutionary Strategies That Shape Ecosystems. John Wiley & Sons, Chichester.
Holzinger, A., Kaplan, F., Blaas, K., Zechmann, B., Komsic-Buchmann, K., et al., 2014. Transcriptomics of desiccation tolerance in the streptophyte green alga *Klebsormidium* reveal a land plant-like defense reaction. PLoS ONE 9 (10), e110630.
Kaštovský, J., Fučíková, K., Hauer, T., Bohunická, M., 2011. Microvegetation on the top of Mt. Roraima, Venezuela. Fottea 11 (1), 171–186.
Kaštovský, J., Veselá, J., Bohunická, M., Fučíková, K., Štenclová, L., Brewer-Carías, C., 2016. New and unusual species of cyanobacteria, diatoms and green algae, with a description of a new genus *Ekerewekia* gen. nov. (Chlorophyta) from the table mountain Churí-tepui, Chimantá Massif (Venezuela). Phytotaxa 247 (3), 153–180.
Krieger, W., 1950. Desmidiaceen aus der montanen Region Südost Brasiliens. Ber. Deutsch. Bot. Ges. 63, 36–48.
Lakatos, M., Bilger, W., Büdel, B., 2001. Carotenoid composition of terrestrial cyanobacteria: response to natural light conditions in open rock habitats in Venezuela. Eur. J. Phycol. 36, 367–375.
Lenzenweger, R., 1999. Desmidiaceenflora von Österreich. In: Crammer, J. (Ed.), Bibliotheca Phycologica, vol. 104/3. Gebruder Borntraeger Verlagsbuchhandlung, Berlin–Stuttgart.
Lyons, T.W., Reinhard, C.T., Plavansky, N.J., 2014. The rise of oxygen in Earth's early ocean and atmosphere. Nature 506 (7488), 307–315.
Mann, D.G., Vanormelingen, P., 2013. An inordinate fondness? The number, distributions and origins of diatom species. J. Eukaryot. Microbiol. 60, 414–420.
Margulis, L., 1981. Symbiosis in Cell Evolution. WH Freeman, New York.
Mereschkowsky, C., 1905. Über Natur und Ursprung der Chromatophoren im Pflanzenreiche. Biol. Centralbl. 25, 593–604.
Metzeltin, D., Lange-Bertalot, H., 1998. Tropical Diatoms of South America II. Iconographia Diatomologica, vol. 5. Koeltz Scientific Books, Germany.
Metzeltin, D., Lange-Bertalot, H., 2007. Tropical Diatoms of South America II. Iconographia Diatomologica, vol. 18. Gantner Verlag, Rugell.
Moreira, C., Vasconcelos, V., Antunes, A., 2013. Phylogeny and biogeography of cyanobacteria and their produced toxins. Mar. Drugs 11, 4350–4369.
Pokorný, P., 1996. Tepui. Stolové hory venezuelské Guyany aneb život ve druhém patře [Tepui. Table Mountains of the Venezuelan Guyana alias life in the second floor]. Vesmír 75, 557–564 (in Czech).
Prescott, G.W., Croasdale, H.T., Vinyard, W.C., 1972. Desmidiales. Saccodermae, Mesotaeniaceae. North American Flora. Series II: Part 6. New York Botanical Garden, New York.
Prescott, G.W., Croasdale, H.T., Vinyard, W.C., 1975. A Synopsis of North American Desmids. Part 2: Desmidiaceae: Placodermae. Section 1. University of Nebraska Press, Lincoln and London.
Prescott, G.W., Croasdale, H.T., Vinyard, W.C., 1977. A Synopsis of North American Desmids. Part 2: Desmidiaceae: Placodermae. Section 2. University of Nebraska Press, Lincoln and London.
Prescott, G.W., Croasdale, H.T., Vinyard, W.C., 1981. A Synopsis of North American Desmids. Part 2: Desmidiaceae: Placodermae. Section 3. University of Nebraska Press, Lincoln and London.
Prescott, G.W., Croasdale, H.T., Vinyard, W.C., 1983. A Synopsis of North American Desmids. Part 2: Desmidiaceae: Placodermae. Section 4. University of Nebraska Press, Lincoln and London.
Ramirez, J.J.R., Mercante, J., Thais, C., 1998. Temporal and spatial distribution of *Croasdalea marthae* in Jacaré pond, São Paulo–SP, Brazil. Algol. Stud. 89, 23–38.
Remias, D., Lütz-Meindl, U., Lütz, C., 2005. Photosynthesis, pigments and ultrastructure of the alpine snow alga *Chlamydomonas nivalis*. Eur. J. Phycol. 40 (3), 259–268.

Remias, D., Schwaiger, S., Aigner, S., Leya, T., Stuppner, H., Lütz, C., 2012. Characterization of an UV- and vis-absorbing, purpurogallin-derived secondary pigment new to algae and highly abundant in *Mesotaenium berggrenii* (Zygnematophyceae, Chlorophyta), an extremophyte living on glaciers. FEMS Microbiol. Ecol. 79, 638–648.

Ryšánek, D., Hrčková, K., Škaloud, P., 2014. Global ubiquity and local endemism of free-living terrestrial protists: phylogeographic assessment of the streptophyte alga *Klebsormidium*. Environ. Microbiol. 17 (3), 689–698.

Schneider, C.W., Quach, P.K., Lane, C.E., 2017. A case for true morphological crypsis: Pacific *Dasya anastomosans* and Atlantic *D. cryptica* sp. nov. (Dasyaceae, Rhodophyta). Phycologia 56 (4), 359–368.

Taylor, J.W., Turner, E., Townsend, J.P., Dettman, J.R., Jacobson, D., 2006. Eukaryotic microbes, species recognition and the geographic limits of species: examples from the kingdom fungi. Philos. Trans. R. Soc. Lond. B. Biol. Sci. 361, 1947–1963.

Vanormelingen, P., Verleyen, E., Vyverman, W., 2008. The diversity and distribution of diatoms: from cosmopolitanism to narrow endemism. Biodivers. Conserv. 17, 393–405.

Wetzel, C.E., Lange-Bertalot, H., Morales, E., Bicudo, D.C., Hoffmann, L., Ector, L., 2012. *Bicudoa amazonica* gen. nov. et sp. nov. (Bacillariophyta)—a new freshwater diatom from the Amazon basin with a complete raphe loss in the Eunotioid lineage. Phytotaxa 75, 1–18.

Whittaker, R.J., Grogan, D.W., Taylor, J.W., 2003. Geographic barriers isolate endemic populations of hyperthermophilic archaea. Science 24, 976–978.

CHAPTER

6

Vascular plants and bryophytes

Ricarda Riina[1], *Paul E. Berry*[2], *Otto Huber*[3] *and Fabián A. Michelangeli*[4]

[1]Royal Botanical Garden, RJB-CSIC, Madrid, Spain [2]University of Michigan Herbarium, Department of Ecology and Evolutionary Biology, Ann Arbor, MI, United States [3]Botanical Institute of Venezuela Foundation, Caracas, Venezuela [4]The New York Botanical Garden, Bronx, NY, United States

Introduction

Early naturalists and scientists from diverse fields have long been fascinated by the tepuis of the Guiana Shield in northern South America. This fascination remains today due to the mountains' unique biota, largely pristine state, and spectacular landscapes with their sheer vertical slopes and table-top summits. In addition, their remote and isolated location renders accessibility to conduct long-term field explorations and scientific studies very difficult and expensive.

The tepuis are collectively known as Pantepui, a term coined by Mayr and Phelps (1967), which we use here in its strict sense: a phytogeographic province comprising the flora and plant communities occupying areas between 1500 and 3000 m elevation in the Guiana region (Huber, 1987, 1994; Berry et al., 1995; see also Chapter 1: Definition and characterization of the Pantepui biogeographical province). Mountains with summits above 1500 m in elevation are distributed mainly in southern Venezuela, with only a few tepuis located in adjacent areas of northern Brazil, near the border with Venezuela, and in the northwestern part of Guyana (Fig. 1.2 and Table 1.1 in Chapter 1).

Due to its geographic location in one of the earth's cratons, long geological history, and poor substrates, Pantepui is also considered among the three main significant areas of "old climatically buffered infertile landscapes" (OCBILs) in the world, and the

Biodiversity of Pantepui
DOI: https://doi.org/10.1016/B978-0-12-815591-2.00006-9

only OCBIL still largely unaffected by human activities (Hopper, 2009; Hopper et al., 2016). These characteristics, along with its unique biotic diversity and inland island-like configuration, make Pantepui a focus for ecological and evolutionary research (Rull, 2007). Understanding the commonly observed pattern of high species richness and endemism associated with mountain ranges (Schwery et al., 2015; Merckx et al., 2015; Quintero and Jetz, 2018) is still of interest today due to the advent of new methodological approaches in fields like biogeography and phylogenetics (Ronquist and Sanmartín, 2011; O'Meara, 2012). However, besides a strong floristic knowledge, testing hypotheses about the origin and determinants of biodiversity patterns also requires robust, well-sampled, and dated phylogenies, which can be analyzed along with other types of data such as organisms' traits and environmental variables.

Recent studies in several Neotropical lineages show the importance of mountains and their topographic complexity as important drivers of species diversification across space and time (e.g., Madriñán et al., 2013; Givnish et al., 2011, 2014; Hughes and Atchison, 2015; Lagomarsino et al., 2016; Alcantara et al., 2018). Fortunately, molecular phylogenies representing Neotropical lineages are starting to accumulate, but achieving a high coverage of taxonomic sampling for individual plant groups (i.e., families, genera, sections) can be very challenging in the Neotropics, especially for lineages with clades centered in areas of limited access for fieldwork such as Pantepui. For this reason, our understanding of the role of geological history and ecological characteristics in shaping the evolution and diversification of Pantepui-centered plant lineages (Fig. 6.1) and, more importantly, in the overall assemblage of the Neotropical flora, is still in its infancy (see Chapter 4: Origin and evolution of the Pantepui bota).

After more than two decades since the publication of the last volume of the *Flora of the Venezuelan Guayana* (FVG) series (Berry et al., 2005) and the first detailed account of the diversity and biogeography of the Pantepui vascular flora (Berry and Riina, 2005), it is now opportune to provide an update with significant additions to our current knowledge about this flora. However, progress has been slow, and little has been done in terms of new botanical explorations to these remote mountains concentrated in Venezuela. The increasingly problematic political and economic situation of this country during the past two decades has taken its toll on scientific botanical research (Rull et al., 2008; Núñez-Farfán et al., 2017), especially in terms of conducting field expeditions to revisit sites or to explore new areas and gather more floristic data. The main tepuis that have been the subject of botanical exploration in recent years are located on the Brazilian and Guyanan sides (Kelloff et al., 2011; Coelho et al., 2015; Barbosa-Silva et al., 2016). One of the few exceptions in the case of Venezuela is Roraima-tepui, one of the best-known tepuis floristically and one of the few easily accessible for tourists, for which new botanical collections have been recently carried out (Safont et al., 2014).

Additions and changes to the Pantepui flora come from the publication of new species descriptions based on existing herbarium specimens, taxonomic revisions of some plant groups, or new locality/taxon records from the botanical explorations conducted in Guyana and Brazil. There have also been a few plant records and novelties from the exploration of the so-called "Andean tepuis" from Ecuador and Peru (Neill, 2018). Most of these findings primarily affect the geographic (latitudinal and altitudinal) ranges and endemicity patterns of the taxa studied. Of the main changes we have made to our floristic

FIGURE 6.1 Representative taxa from the Pantepui flora. (A) *Pterozonium spectabile* (Pteridaceae), Roraima-tepui. (B) *Chimantaea lanocaulis* (Asteraceae), Murisipán-tepui. (C) *Bonnetia sessilis* (Bonnetiaceae), Sierra de Lema. (D) *Lindmania holstii* (Bromeliaceae) Murisipán-tepui. (E) *Ledothamnus guyanensis* (Ericaceae), Kukenán-tepui. (F) *Tococa obovata* (Melastomataceae), Cerro Duida. (G) *Kunhardtia rhodantha* (Rapateaceae), Cerro Yutajé. (H) *Retiniphyllum scabrum* (Rubiaceae), Roraima-tepui. (I) *Heliamphora tatei* (Sarraceniaceae), Cerro Huachamakari. *Source: Photos: C (Paul E. Berry), A, B, D–I (Fabián I. Michelangeli).*

dataset are those related to classification systems based on new phylogenetic frameworks (e.g., Chase et al., 2016; PPG I, 2016; LPWG The Legume Phylogeny Working Group, 2017). Most of these changes affect patterns of species richness and endemicity at the genus and family levels.

The original and most complete source of information about the flora of Pantepui, including species descriptions and illustrations, is the FVG (Steyermark et al., 1995–2005), which covers both the tepuis and the lowland and middle elevation areas of the Venezuelan portion of the Guiana region (the states of Amazonas, Bolívar, and Delta Amacuro). Basic floristic information from FVG was later integrated into a broader checklist covering the entire Guiana region, including the available information from Venezuela, Guyana, Suriname, and French Guiana (Funk et al., 2007a), with subsequent updates (Feuillet, 2009). The recent explorations of the Pantepui areas in Brazil have also provided new floristic information from Sierra de la Neblina, Serra do Aracá, and other lower-elevation tepuis in northern Brazil (Coelho et al., 2015; Barbosa-Silva et al., 2016; Costa, 2017).

In this chapter we revisit and update our previous attempt to summarize the main patterns of diversity and endemism of the Pantepui flora (Berry and Riina, 2005). We also update our knowledge on the phylogenetics of Pantepui plant lineages and review studies that give insights into the evolutionary and biogeographic history of the flora. We thus hope to provide a new starting point to stimulate future research in botanical exploration, systematics, evolution, biogeography, and conservation of this singular mountain system of the Neotropics.

Main patterns of diversity in vascular plants

Previous accounts of plant diversity patterns in the Pantepui province (Riina, 2003; Berry and Riina, 2005) were primarily based on the FVG family treatments, including those in the last two volumes that were in preparation at the time of those surveys. The publication of these final volumes and subsequent taxonomic revisions of Neotropical groups have resulted in changes in species numbers and geographic ranges. About 55 new taxa of vascular plants (e.g., Scharf et al., 2005; Croat et al., 2010; Hirai et al., 2011; Pruski, 2011; McPherson et al., 2011; Taylor, 2011; Aymard and Grande, 2012; Maciel et al., 2017; Grande, 2018), including three new genera (Struwe et al., 2008; Chase et al., 2009; Romero-González and Carnevali Fernández-Concha, 2010), have been described for the Pantepui area since the publication of the FVG. However, our knowledge about species richness in a given area is also affected by the recognition of new taxonomic combinations and synonymizations, which is the case for several Pantepui taxa (e.g., Zuloaga et al., 2011; Giraldo-Cañas, 2013; Lucas et al., 2016; Delprete and Kirkbride, 2016; Taylor, 2015; Rocha et al., 2018). In this context, we will highlight the most relevant diversity patterns of the Pantepui flora but recognizing that our floristic database is a work in progress and will continue to change as new taxonomic revisions are published.

Overall flora richness and endemism

In this new survey of the Pantepui flora, we report 2579 species of vascular plants known from Pantepui, which are distributed in 673 genera and 157 families (Table 6.1).

TABLE 6.1 Number of vascular plants in Pantepui and levels of endemicity. Taxa follow current classification systems (e.g., APG V as in Chase et al., 2016; PPG I, 2016) and most taxonomic revisions after the publication of the FVG (Steyermark et al., 1995–2005).

	Pantepui	ET (≥1500 m)	ET + NET (≥1200 m)	ET to a single tepui
Species	2579	885 (34%)	1076 (41%)	640 (25%)
Genera	673	22 (7%)	37 (5%)	14 (2%)
Families	157	0	0	0

ET, number of taxa endemic to Pantepui; *NET*, number of taxa nearly endemic to Pantepui.
Note: Nearly endemic taxa refer to those with most of their range in Pantepui but one or a few taxa/populations also occurring outside the studied area (i.e., below 1500 m in the Guiana Shield and/or in a few cases in other disjunct high-elevation areas of the Neotropics).

This high diversity is confined to a relatively small area of about 5000 km^2 (Berry et al., 1995). However, the new figures represent a modest increase in taxon richness—132 species and 47 genera—from our previous report (Berry and Riina, 2005). This increase is mostly due to taxonomic rearrangements based on modern phylogenetic information, description of new taxa, and, to a lesser degree, new botanical explorations. This trend of changes will continue in the future as more taxa from Pantepui and the Guiana Shield start to be included in phylogenetic analyses and subsequent taxonomic revisions.

The percentage of endemic species to Pantepui reported here (34%) is significantly lower (Table 6.1) than that included in our previous report (42%) of endemism for Pantepui. Even if botanical exploration in the Venezuelan area of the Guiana Shield has practically ceased in the last two decades, collecting in other areas of the Guiana Shield (Colombia, Brazil, and the Guianas) and in the Andes has continued. New botanical records from these countries have expanded the geographic range of several Pantepui taxa, which can explain in part this change in endemism levels. Taxonomic updates are also part of the reason for this difference. At the genus and family levels, diversity and endemism levels have remained pretty much the same (Table 6.1) as in our previous work.

Taxonomic diversity across main plant groups

The top five species-rich families in Pantepui, with more than 100 species each, continue to be Orchidaceae (258 species), Melastomataceae (152 species), Rubiaceae (141 species), Asteraceae (139 species), and Bromeliaceae (118 species) (Table 6.2). Only Bromeliaceae and Asteraceae, families that are more diverse at higher elevations, make this list different from the top five families in lowland areas (0–1000 m) of the Amazon basin (including the Guiana Shield lowlands), where Fabaceae, Rubiaceae, Orchidaceae, Melastomataceae, and Araceae are the five most diverse families (Cardoso et al., 2017). The most significant changes in the composition of this new list of the richest Pantepui families (Table 6.2) have to do with taxonomic recircumscriptions in both pteridophytes (PPG I, 2016) and angiosperms (APG V, Chase et al., 2016). One example of an ecologically important family in Pantepui that has changed its circumscription is the former Theaceae, which is now

TABLE 6.2 List of the most species plant families (≥40 species) occurring in Pantepui, ordered by decreasing number of species and including number of endemic species.

Families (19)	Species present in Pantepui	ES (min. alt. ≥1500 m)	ES + NES (min. alt. ≥1200 m)
Orchidaceae	258	52	67
Melastomataceae	152	51	66
Rubiaceae	141	48	66
Asteraceae	139	68	80
Bromeliaceae	118	64	74
Poaceae	89	30	33
Cyperaceae	69	12	15
Polypodiaceae	64	7	11
Xyridaceae	60	30	37
Ericaceae	59	30	34
Hymenophyllaceae	49	4	5
Rapateaceae	48	17	22
Eriocaulaceae	48	30	32
Araliaceae	46	36	41
Gentianaceae	46	28	29
Myrtaceae	46	18	21
Apocynaceae	44	21	26
Clusiaceae	41	13	17
Ochnaceae	40	21	27

ES, number of species endemic to Pantepui; *NES*, number of species nearly endemic to Pantepui.
Note: Nearly endemic species refer to those with most of their ranges in Pantepui but one or a few populations also occurring outside the studied area (i.e., below 1500 m in the Guiana Shield and/or in a few cases in other disjunct high-elevation areas of the Neotropics).

treated as Bonnetiaceae (including the genera *Archytaea* and *Bonnetia*), and Pentaphylacaceae (including *Freziera* and *Ternstroemia*).

The endemism across families (Table 6.2) is not always proportional to species richness, and it is particularly low in the case of the pteridophyte families Polypodiaceae and Hymenophyllaceae, with seven and four endemic species, respectively. Similarly, among angiosperms, Orchidaceae and Cyperaceae are the families with the lowest levels of endemism in Pantepui (20%, with 52 species; 17%, with 12 species, respectively) (Table 6.2). All these groups, especially ferns, are well known for their ability to disperse over long distances due to their minute dispersal units, which may prevent or delay over evolutionary time the process of allopatric speciation. Families with more than 20 species

TABLE 6.3 Families with more than 20 species in Pantepui and over 50% endemism there.

Families	Species present in Pantepui	ES (min. alt. ≥1500 m)	ES + NES (min. alt. ≥1200 m)
Asteraceae	139	68	80
Bromeliaceae	118	64	74
Xyridaceae	60	30	37
Ericaceae	59	30	34
Eriocaulaceae	48	30	32
Araliaceae	46	36	41
Gentianaceae	46	28	29
Apocynaceae	44	21	26
Ochnaceae	40	21	27
Aquifoliaceae	34	25	25
Primulaceae	29	16	19
Bonnetiaceae	27	19	20
Rutaceae	21	12	12

ES, number of species endemic to Pantepui; *NES*, number of species nearly endemic to Pantepui.
Note: Nearly endemic species refer to those with most of their ranges in Pantepui but one or a few populations also occurring outside the studied area (i.e., below 1500 m in the Guiana Shield and/or in a few cases in other disjunct high-elevation areas of the Neotropics).

in Pantepui and with relatively high levels of endemism (≥50%) are listed in Table 6.3. These, along with the families in Table 6.2, give a better picture of the key components of the Pantepui flora, including ecologically important taxa that are dominant elements in different Pantepui plant communities (see Chapter 7: Plant communities).

The most speciose genus in Pantepui is *Navia* (Bromeliaceae), a genus endemic to the Guiana Shield, with about 66 species occurring in Pantepui (Table 6.4) and about 30 additional species in lower areas of the Guiana Shield (Table 6.5; Funk et al., 2007a; Flora do Brasil, 2020). It is followed by *Xyris* (Xyridaceae) and *Schefflera* (Araliaceae), with 47 and 45 species, respectively, and more than half of their species endemic to Pantepui. Although not as rich, genera like *Lindmania* (Bromeliaceae) and *Bonnetia* (Bonnetiaceae) present high levels of endemism, with more than 70% of their species occurring exclusively in Pantepui (Tables 6.4 and 6.5). Similar to the pattern found at the family level, the genera with the lowest levels of endemism are the pteridophytes *Elaphoglossum* (Dryopteridaceae) and *Trichomanes* (Hymenophyllaceae), each with a single species endemic to Pantepui.

Although we focus on Pantepui here, we also tried to update the list of endemic genera for the Guiana Shield, which is important to provide context to our work. Berry et al. (1995) published the first comprehensive list of genera endemic to the Guiana

TABLE 6.4 List of genera of vascular plants with ≥ 25 species in Pantepui ordered by decreasing number of species and including number of endemic species.

Genus	Family	Species in Pantepui	ES (min. alt. ≥1500 m)	ES + NES (min. alt. ≥1200 m)
Navia	Bromeliaceae	66	18	23
Xyris	Xyridaceae	47	23	29
Schefflera	Araliaceae	45	36	41
Stegolepis	Rapateaceae	34	15	18
Ilex	Aquifoliaceae	34	25	25
Clusia	Clusiaceae	34	11	15
Lindmania	Bromeliaceae	32	27	29
Palicourea	Rubiaceae	31	8	12
Cyathea	Cyatheaceae	31	9	13
Myrcia	Myrtaceae	30	14	16
Epidendrum	Orchidaceae	30	9	10
Elaphoglossum	Dryopteridaceae	29	1	1
Selaginella	Selaginellaceae	27	12	13
Miconia	Melastomataceae	27	9	10
Bonnetia	Bonnetiaceae	26	19	20
Paepalanthus	Eriocaulaceae	26	20	22
Byrsonima	Malpighiaceae	25	6	9
Trichomanes	Hymenophyllaceae	25	1	2

ES, number of species endemic to Pantepui; *NES*, number of species nearly endemic to Pantepui.
Note: Nearly endemic species refer to those with most of their ranges in Pantepui but one or a few populations also occurring outside the studied area (i.e., below 1500 m in the Guiana Shield and/or in a few cases in other disjunct high-elevation areas of the Neotropics).

Shield, reporting 138 genera. Although the number of endemic genera reported here remains almost the same (137) (Table 6.5), the two lists are not identical because of taxonomic changes, new species occurrences outside the Guiana Shield, the recent discovery of several new genera (Struwe et al., 2008; Chase et al., 2009; Romero-González and Carnevali Fernández-Concha, 2010; Philbrick et al., 2011), and because we also included a few genera that we consider nearly endemic to the area. More than half of the genera listed are either endemic to Pantepui (17%, Table 6.5A), with all or almost all their species in Pantepui (e.g., *Heliamphora*), or are well represented (46%, Table 6.5B), usually with a significant portion of their diversity confined to Pantepui. In the latter group are genera like *Bonnetia*, *Duidania*, *Ledothamnus*, *Lindmania*, *Pterozonium*, *Stenopadus*, and *Stegolepis*.

TABLE 6.5 Genera (137) endemic or nearly endemic to the Guiana Shield (GS) with total number of species.

A. Genera endemic to Pantepui (23, 17%)		
Achlyphila (Xyridaceae) 1	*Connellia* (Bromeliaceae) 6	*Neblinantha* (Gentianaceae) 2
Achnopogon (Asteraceae) 2	*Coryphothamnus* (Rubiaceae) 1	*Pyrrorhiza* (Haemodoraceae) 1
Adenanthe (Ochnaceae) 1	*Glossarion* (Asteraceae) 2	*Quelchia* (Asteraceae) 5
Adenarake (Ochnaceae) 2	*Heliamphora* (Sarraceniaceae)[a] 23	*Rutaneblina* (Rutaceae) 1
Aracamunia (Orchidaceae) 1	*Huberopappus* (Asteraceae) 1	*Saccifolium* (Gentianaceae) 1
Celiantha (Gentianaceae) 3	*Mallophyton* (Melastomataceae) 1	*Tepuia* (Ericaceae) 7
Chimantaea (Asteraceae) 9	*Marahuacaea* (Rapateaceae) 1	*Tyleropappus* (Asteraceae) 1
Coccochondra (Rubiaceae) 3	*Maguireothamnus* (Rubiaceae) 3	
B. Genera present in Pantepui (63, 46%)		
Amphiphyllum (Rapateaceae) 1	*Guayania* (Asteraceae) 6	*Pagameopsis* (Rubiaceae) 2
Aphanocarpus (Rubiaceae) 1	*Imeria* (Asteraceae) 2	*Phelpsiella* (Rapateaceae) 1
Aratitiyopea (Xyridaceae) 1	*Koyamaea* (Cyperaceae) 1	*Philacra* (Ochnaceae) 4
Archytaea (Bonnetiaceae) 2	*Kunhardtia* (Rapateaceae) 2	*Poecilandra* (Ochnaceae) 2
Blepharandra (Malpighiaceae) 6	*Ledothamnus* (Ericaceae) 7	*Pterozonium* (Pteridaceae) 14
Bonnetia (Bonnetiaceae) 29	*Lindmania* (Bromeliaceae) 47	*Raveniopsis* (Rutaceae) 20
Brewcaria (Bromeliaceae) 6	*Maguireocharis* (Rubiaceae) 1	*Rhynchocladium* (Cyperaceae) 1
Brocchinia (Bromeliaceae) 21	*Maguireanthus* (Melastomataceae) 1	*Rondonanthus* (Eriocaulaceae) 6
Celianella (Phyllanthaceae) 1	*Merumea* (Rubiaceae) 2	*Roraimaea* (Gentianaceae) 2
Cephalocarpus (Cyperaceae) 4	*Mycerinus* (Ericaceae) 3	*Senefelderopsis* (Euphorbiaceae) 2
Chalepophyllum (Rubiaceae) 2	*Myriocladus* (Poaceae) 13	*Sipapoantha* (Gentianaceae) 1
Chorisepalum (Gentianaceae) 5	*Navia* (Bromeliaceae) 96	*Stegolepis* (Rapateaceae) 35
Comoliopsis (Melastomataceae) 3	*Neblinaea* (Asteraceae) 1	*Stenopadus* (Asteraceae) 15
Decagonocarpus (Rutaceae) 2	*Neblinanthera* (Melastomataceae) 1	*Steyerbromelia* (Bromeliaceae) 6
Diacidia (Malpighiaceae) 11	*Neblinathamnus* (Rubiaceae) 2	*Stomatochaeta* (Asteraceae) 6
Digomphia (Bignoniaceae) 3	*Neotatea* (Calophyllaceae) 4	*Tateanthus* (Melastomataceae) 1
Duidaea (Asteraceae) 4	*Nietneria* (Nartheciaceae) 2	*Tepuianthus* (Thymelaeaceae) 6
Duidania (Rubiaceae) 1	*Nohawilliamsia* (Orchidaceae) 1	*Tyleria* (Ochnaceae) 14
Euceraea (Salicaceae) 3	*Notopora* (Ericaceae) 5	*Tylopsacas* (Gesneriaceae) 1
Euphronia (Euphroniaceae) 3	*Ochthephilus* (Melastomataceae) 1	*Weidmannia* (Orchidaceae) 1
Everardia (Cyperaceae) 13	*Orectanthe* (Xyridaceae) 2	*Yanomamua* (Gentianaceae) 1
Gongylolepis (Asteraceae) 15		

(Continued)

TABLE 6.5 (Continued)

C. *Genera absent in Pantepui (51, 37%)*		
Acanthella (Melastomataceae) 2	*Gynocraterium* (Acanthaceae) 1	*Potarophytum* (Rapateaceae) 1
Aguiaria (Malvaceae) 1	*Haematostemon* (Euphorbiaceae) 2	*Pseudephedranthus* (Annonaceae) 2
Autana (Podostemaceae) 1	*Henriquezia* (Rubiaceae) 3	*Rhoogeton* (Gesneriaceae) 2
Aldina (Fabaceae)[b] 22	*Heterostemon* (Fabaceae) 7	*Rhyncholacis* (Podostemaceae) 23
Angostylis (Euphorbiaceae) 2	*Holstianthus* (Rubiaceae) 1	*Schoenocephalium* (Rapateaceae) 4
Apocaulon (Rutaceae) 1	*Jasarum* (Araceae) 1	*Siapaea* (Asteraceae) 1
Asteranthos (Lecythidaceae) 1	*Lembocarpus* (Gesneriaceae) 1	*Sipaneopsis* (Rubiaceae) 7
Astrococcus (Euphorbiaceae) 1	*Maburea* (Olacaceae) 1	*Spirotropis* (Fabaceae) 2
Blastemanthus (Ochnaceae) 2	*Monotrema* (Rapateaceae) 4	*Steyermarkochloa* (Poaceae) 1
Boyania (Melastomataceae) 2	*Neobertiera* (Rubiaceae) 4	*Thysanostemon* (Clusiaceae) 2
Cheiradenia (Orchidaceae) 1	*Pachyloma* (Melastomataceae) 4	*Tryssophyton* (Melastomataceae) 1
Chonocentrum (Phyllanthaceae) 1	*Pakaraimaea* (Cistaceae) 1	*Tuberculocarpus* (Asteraceae) 1
Cyrillopsis (Ixonanthaceae) 2	*Panurea* (Fabaceae) 2	*Uladendron* (Malvaceae) 1
Degranvillea (Orchidaceae) 1	*Pentamerista* (Tetrameristaceae) 1	*Wallacea* (Ochnaceae) 3
Dendrosipanea (Rubiaceae) 2	*Petaladenium* (Fabaceae) 1	*Whittonia* (Peridiscaceae) 1
Duckeanthus (Annonaceae) 1	*Polylychnis* (Acanthaceae) 2	*Windsorina* (Rapateaceae) 1
Guacamaya (Rapateaceae) 1	*Polyotidium* (Orchidaceae) 1	

[a]*Only one species not reaching above 1500 m.*
[b]*Only one species reaching above 1500 m.*
Note: Nearly endemic genera are those with most of their species in the GS but one or a few species/populations also occurring outside the GS area.

Diversity patterns across tepuis

The overall pattern of plant species richness and endemism across tepuis has not changed significantly with our updated floristic Pantepui database, with the exception of the number of species restricted or endemic to a single-tepui summit, which was 37% in our first survey (Berry and Riina, 2005, Figure 3), whereas here we report 25% (Table 6.1). As usually happens, levels of endemism change as new species occurrences are registered with the consequent expansion of taxon geographic ranges.

The distribution of species richness across tepuis for the richest tepuis (those with >200 species) is shown in Fig. 6.2. Not coincidentally, these are also the tepuis with the highest levels of single-tepui endemic species. The remaining tepuis (see list of tepuis in Table 1.1 from Chapter 1) have between 100 and 190 species, but only between 0 and 4 endemic species. As shown in Berry and Riina (2005), the Chimantá massif presents the highest diversity (889 species) among the tepuis, followed closely by Sierra de la Neblina (762 species).

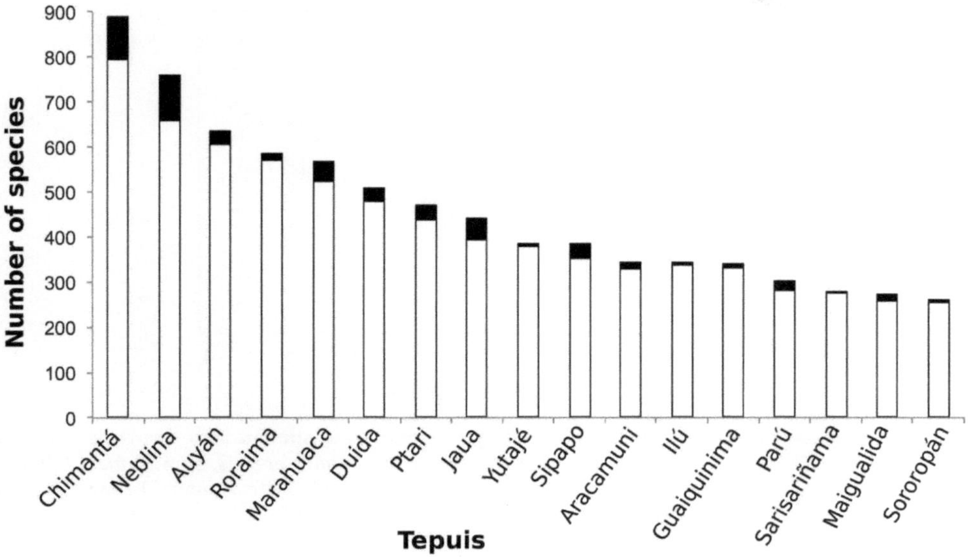

FIGURE 6.2 Tepuis with more than 200 species of vascular plants recorded from their upper slopes and summits (above 1500 m elevation). Black section of bars indicates the number of species endemic to a given tepui (i.e., single-tepui endemic species).

Likewise, the mountain with the highest number of endemic species continues to be Neblina (104 species), followed by Chimantá, with 94 endemic species (Fig. 6.2). Chimantá is a large massif with a summit of about 615 km^2 and a maximum elevation of 2660 m (Huber, 1995). It is similar in size to its neighbor, Auyán-tepui, but it is much more complex in topography, with a summit dissected into several separated sub-tepuis with a common base. Neblina is smaller than Chimantá in summit area (235 km^2, Huber, 1995), but it is higher (3014 m) and much more isolated geographically from other summits. The higher species richness of Chimantá can be explained in part by its larger area, whereas for Neblina, a greater geographic isolation has favored the high level of endemism.

We explored the relationship between species richness and tepui summit area (just the parts of the summits above 1500 m) and found a strong correlation between the two variables (77%) (Fig. 6.3). A similar relationship was reported by Safont et al. (2014) for the summits of the tepuis of the eastern chain (Roraima-tepui to Tramén-tepui). However, a more comprehensive analysis should consider other variables such as habitat diversity, topographic complexity, distance to the closest tepui, and elevational differences that, along with summit area, may better explain the variation in species richness across tepuis. Tepui summits are also quite variable in their type of vegetation coverage. Many of the eastern tepuis have summits with more open rocky areas than the central and western tepuis. Likewise, there are tepuis with more forest cover than herbaceous vegetation, or tepuis with an extraordinary diversity of habitats such as Chimantá and Neblina (Huber, 1992).

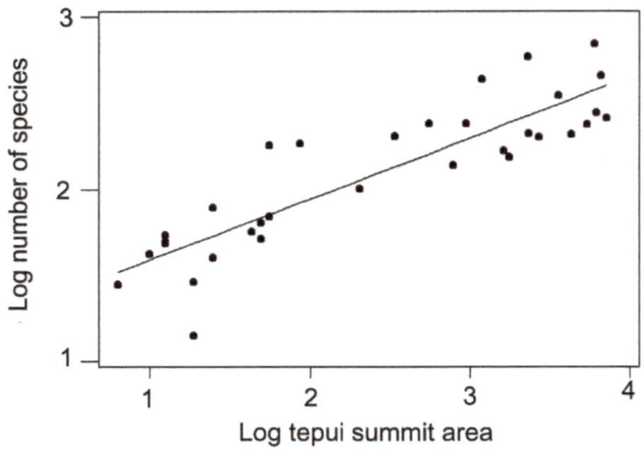

FIGURE 6.3 Relationship between the number of species and tepui summit area (log-transformed) for 33 tepuis of the Guiana Shield ($R^2 = 0.77$, $P < .0001$). Information on summit area is from Huber (1995). Tepuis without data on their summit area (e.g., Sororopán-tepui and Ptari-tepui), with summits below 1500 m (Cerro Autana, Cerro Yapacana), or insufficient botanical knowledge were excluded from this analysis.

Geographic distance between tepuis is an important factor determining their floristic similarities (Fig. 6.4). A cluster analysis of the tepuis based on their floristic composition shows a major division between eastern and western groups of tepuis and two main subgroups within each of them (Fig. 6.4). This general east–west pattern has been pointed out by several authors (e.g., Maguire, 1979; Steyermark, 1986; Funk and Brooks, 1990). Steyermark observed that several genera exhibited an east–west geographic segregation in Pantepui. Similarly, Pruski (1997) showed that the tribe Mutisieae (Asteraceae) has concentrations of actinomorphic-flowered species in the eastern part of the Venezuelan Guiana and bilabiate corollas in the western part of the region. An analogous pattern is seen in *Pentacalia* and other genera of Asteraceae (Pruski, 1997). This strong difference in floristic composition between these two groups of tepuis may be related to the large extension of lowlands separating the two areas, and also the influence of the igneous Sierra Maigualida range that could be working as an effective barrier for the dispersion of some plant groups between the two sides. Other interesting patterns are the tepuis that appear as longer branches in the cluster analysis (Fig. 6.4), such as Maigualida, the Yutajé-Corocoro cluster, Cerro Sipapo, and Aracá massif. This pattern seems to agree with the higher geographic isolation of these tepuis in relation to their closest tepui (Chapter 1, Fig. 1.2). However, the incomplete level of floristic information of the first three (Huber, 1995) could also play a role in the observed pattern. The uneven floristic knowledge and level of botanical exploration across tepuis prevent us from making strong conclusions from these analyses, but it seems likely that the east—-west separation of species distributions is a real distinction for the Pantepui flora.

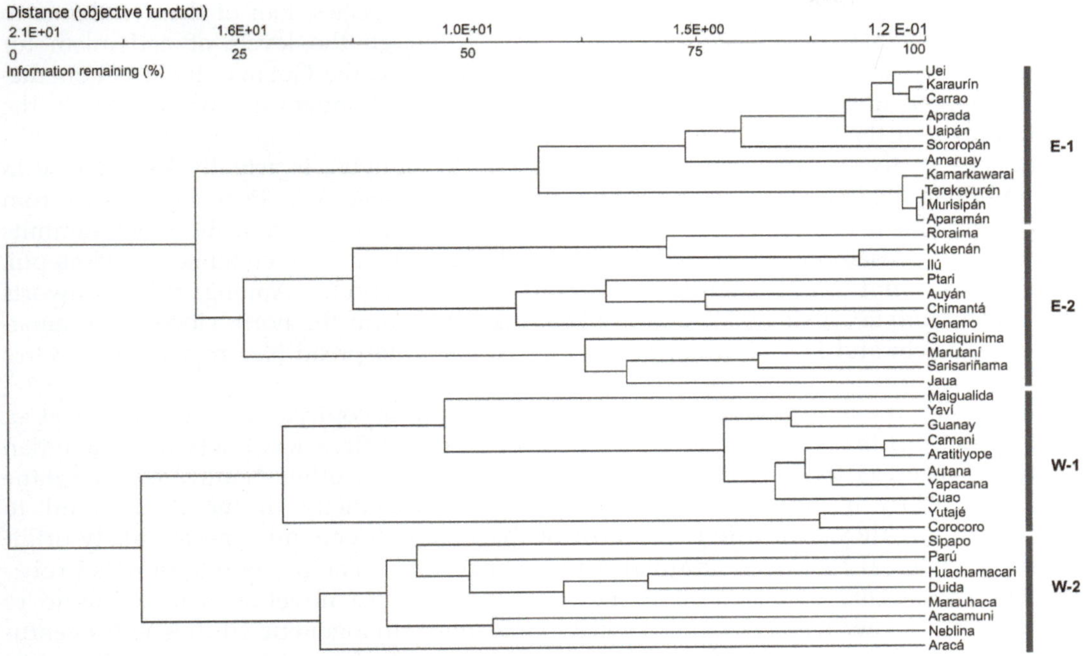

FIGURE 6.4 Cluster analysis based on the floristic composition of 40 tepuis using relative Euclidean distance and Ward's group linkage method. E, eastern tepuis; E1 and E2, subgroups of eastern tepuis; W, western tepuis; W1 and W2, subgroups of western tepuis.

Pantepui bryophytes

In contrast to the vascular plants, there is no complete floristic treatment for the bryophytes of Pantepui or for the broader Guiana Shield region. Instead, there are scattered publications of new taxa (e.g., Gradstein and Costa, 2018); a few inventories of particular tepuis such as Roraima-tepui (Gradstein and Florschütz-de Waard, 1989), Cerro Venamo (Dauphin and Ilkiu-Borges, 2002), and Serra do Aracá (Costa, 2017); and one publication specifically addressing the biogeography of bryophytes from the Pantepui region (Désamoré et al., 2010). From these studies, various conclusions can be made.

Of the three divisions of bryophytes, liverworts (Marchantiophyta) are the most diverse in Pantepui and exhibit the highest degree of endemism, with an estimated 89 genera and 250 species with a 10% endemism at the species level (Désamoré et al., 2010). All of the endemic species are leafy liverworts (Jungermanniidae), and there are seven endemic or nearly endemic genera. Some of these are very differentiated morphologically, such as *Haesselia*, *Odontoseries*, *Trabacellula*, and *Vanaea*, but they are all monotypic with the exception of *Haesselia*, which has two species. The most speciose liverwort genus on the tepuis is the Neotropical *Micropterygium*, which has around 18 species, half of them endemic to the Guiana Highlands (Désamoré et al., 2010). Although the levels of endemism for hepatics are considerably lower than for the vascular flora, the Guiana Shield is nonetheless considered to be the second most important center of endemism for hepatics in the Neotropics after the northern Andes (Gradstein et al., 2001).

The largest division of bryophytes, the mosses (Bryophyta), is actually less diverse in Pantepui than the liverworts. There are an estimated 200 species in 95 genera known from the Guiana Shield region, but only around 70 are known so far from the tepui summits (Buck, 1991; Désamoré et al., 2010). Two of the genera are endemic to Pantepui, *Holomitriopsis* and *Steyermarkiella*, and both are monotypic. Among the hornworts (Anthocerotophyta), a single species has been recorded from the north slope of Roraima-tepui (Gradstein and Florschütz-de Waard, 1989), the widespread Neotropical *Dendroceros crispus*.

Concerning the origins and affinities of the Pantepui bryophyte flora, Désamoré et al. (2010) point out that just 10% of the Pantepui bryophyte flora has lowland Amazonian elements, whereas there is a much stronger affinity with other Neotropical montane floras, including a number of species that occur disjunctly in the Andes and in Pantepui. Thus they conclude that "most of the Pantepui-endemic species likely originated by dispersal from cold-adapted Andean ancestors." The bryophyte flora is probably less thoroughly collected than the vascular flora, so novelties will continue to appear with further exploration. Likewise, molecular phylogenetic studies will eventually provide much more detailed information about the origins of the Pantepui bryophytes.

Current phylogenetic knowledge and biogeographic implications

The proliferation of phylogenetic data has changed the paradigm about how biogeographic research is conducted (Funk and Brooks, 1990; Ronquist and Sanmartín, 2011). In the case of the Pantepui flora, phylogenies should allow us to answer some basic questions, such as whether the flora is composed of early diverging or derived elements and whether plant lineages well represented in the area are the product of a single (or few) incursions into the area, followed or not by adaptive radiation. Here we summarize the rather sparse literature on phylogenetic studies that have included Pantepui taxa. We also highlight their biogeographic implications on the origin and evolution of the Pantepui flora.

Pteridophytes

Pteridaceae—The genus *Pterozonium* has long been recognized as a typical element of Pantepui (Steyermark, 1986). The genus has 14 species, 12 of them endemic to Pantepui, and the other 2 are both present in the Guiana Shield as well as in the Andes, and 1 of them is found in Costa Rica (Lellinger, 1995; Berry and Riina, 2005). Zhang et al. (2015) suggested that the genus is monophyletic, although only three species were included in their phylogeny. More interesting, however, is that it is sister to a clade that includes both Old World (Southeast Asian and Australian) and eastern Brazil elements and that this larger clade is sister to the Andean genera *Eriosorus* and *Jamesonia*.

Angiosperms

Araliaceae—The largest genus of Araliaceae is *Schefflera* with over 900 species (Frodin et al., 2010). The genus has recently been shown to be polyphyletic, but each component clade is largely geographically coherent (Plunkett et al., 2005). Within the Neotropics, all 200 species present there are geographically clustered, with a group of tepui endemics forming a small clade along with one species from lowland Amazonia, another from the sandstone area of the Cordillera del Condor between Ecuador and Peru, and the third species from eastern Brazil (Fiaschi and Plunkett, 2011; Neill et al., 2014). This clade has been informally recognized as the *Crepinella* group and contains about 35 species (Frodin et al., 2010; Fiaschi and Plunkett, 2018). Further analysis of this group using a dated phylogeny will be key to clarifying the tempo and direction of colonization of these three areas in South America.

Asteraceae—Over 300 species of Asteraceae are found in the Guiana Shield (Funk et al., 2007b). Taxa formerly assigned to the Mutisieae tribe are particularly diverse in the region, with many species found at high elevations, although lowland components are also present (Berry and Riina, 2005; Ortiz et al., 2009). Within these species, two distinct groups of genera have been recovered. The first one, the *Gongylolepis* clade of the tribe Stifftieae, includes *Gongylolepis* and *Duidaea* as sisters to the Andean genera *Hyaloseris* and *Dinoseris*, and this group in turn is sister to the Brazilian *Stifftia* (Panero and Funk, 2008). In this same study, the second group recovered was the *Stenopadus* clade, in the tribe Wunderlichieae. This group includes *Stenopadus*, *Stomatochaeta*, and *Chimantaea* as sisters to *Wunderlichia* (from eastern Brazil), and this clade in turn is sister to a clade that contains Andean and Asian taxa (Panero and Funk, 2008; Ortiz et al., 2009). However, this study only included one representative per genus; thus neither the monophyly of the Pantepui genera nor generalizations about biogeography can be addressed with these data.

Bonnetiaceae—The family has three genera, with two of them present in the Guiana Shield. The largest genus is *Bonnetia* with 29 species, 27 of which are endemic to the Guiana Shield and 20 endemic to Pantepui (Table 6.3; Ruhfel et al., 2011, 2016). A large analysis of the clusioid clade of Malpighiales showed that *Bonnetia* is sister to a clade that contains the Guiana Shield-endemic genera *Archytaea* and *Ploiarium*, genera endemic to Southeast Asia (Ruhfel et al., 2011, 2016). Within *Bonnetia*, the sampling is low, but a species endemic to Cuba and another one endemic to eastern Brazil both appear to be derived from Guiana Shield lineages. Although the earliest diverging species is a Pantepui endemic, lowland and highland species are intermixed. One species, *Bonnetia paniculata*, is present in the Guiana Shield and extends into the Andes (Ruhfel et al., 2011). The split between *Bonnetia* and *Archytaea* + *Ploiarium* probably dates to the Paleocene or Eocene (Ruhfel et al., 2016).

Bromeliaceae—The Bromeliaceae is well represented in the Guiana Shield and seems to have originated in the area. The two earliest diverging clades in the family, *Brocchinia* and *Lindmania + Connellia*, are entirely restricted to the Guiana Shield, with most species occurring on the tepui summits and slopes (Givnish et al., 2004, 2011, 2014). The genus *Navia* (including close relatives *Sequencia* and *Brewcaria*) is also centered in Pantepui, but this seems to be a secondary arrival to the region. Additionally, many species from other Bromeliaceae clades are present in Pantepui, but these are often widespread South American species or representatives of widespread genera (Givnish et al., 2004, 2011). The stem age of the Bromeliaceae seems to be c. 100 million years (Ma), and the crown age (and thus of the early diverging Guiana Shield clades) appears to be in the Miocene (Givnish et al., 2011).

Droseraceae—*Drosera* is represented in the Guiana Shield by 16 species, 10 of these endemic (Fleischmann et al., 2018). Although some phylogenetic work has been done in the family, only five of the Guiana species have been included in these analyses (Rivadavia et al., 2003; Rivadavia et al., 2012). Four species endemic to the lowlands of the Guiana Shield belong to one of two clades that contain New World species (Rivadavia et al., 2003). Interestingly, Drosera *meristocaulis*, a species endemic to Neblina and the only Pantepui species sampled, is well nested within a clade otherwise only known from Australia, and long-distance dispersal has been invoked as the most likely explanation for this pattern (Rivadavia et al., 2012).

Ericaceae—This family is represented in the Guiana Shield by 73 species, which tend to be restricted to the summit and upper slopes of the tepuis (Table 6.3; Taylor, 1999; Luteyn, 2007). Four genera are endemic to Pantepui, but they have been represented in molecular phylogenies only by one or two accessions each. *Notopora* is resolved as sister to two species of *Orthaea* (one from the Guiana Shield and one from Trinidad), and this clade is in turn nested within the larger *Vaccinium* complex in a clade that includes taxa from the Andes and the eastern United States (Kron et al., 2002). However, species density sampling is low, especially among *Orthaea* and *Gaylussacea*, to be able to draw any further conclusions. *Ledothamnus*, a genus of seven species, has been resolved as sister to *Bryanthus*, a monotypic genus endemic to Japan (Gillespie and Kron, 2010; Gillespie and Kron, 2012). This Pantepui–Japan disjunction is unique within the Ericacaceae (Gillespie and Kron, 2012) and angiosperms as a whole. Of the seven species of *Tepuia*, two have been included in phylogenetic analyses, and they form a clade either as part of a basal polytomy or along with *Gaultheria procumbens*, a species from eastern North America, as sister to the remaining species of the wintergreen group of Ericaceae (Powell and Kron, 2001; Fritsch et al., 2011). To date, none of the representatives of *Gaultheria* from Pantepui have been included in phylogenetic analyses, and these same analyses show that the genus is polyphyletic (Powell and Kron, 2001; Fritsch et al., 2011).

Eriocaulaceae—This pantropical family is represented in the Guiana Shield by around 100 species (Hensold, 1999; Hakki and Hensold, 2007), 48 of which occur in Pantepui, and 62% of them endemic (Table 6.3). In an analysis of the subfamily Paepalanthoideae, Trovo et al. (2013) showed that *Rondonanthus*, a genus endemic to the Guiana Shield, is sister to the remaining members of the subfamily. This work also showed that a group of *Paepalanthus* species endemic to Pantepui forms a monophyletic group embedded within a clade from the Serra do Espinhaço in eastern Brazil. However, *Rondonanthus* was represented by only 1 of the 6 species in the genus, and only 3 of over 30 species of *Paepalantus* endemic to Pantepui were sampled (Trovo et al., 2013). Within *Comanthera*, the Guiana

Shield-endemic *C. jenmanii* is sister to a species of lowland Amazonia, and this clade in turn is sister to a clade of widespread lowland species and a group of species restricted to Serra do Espinhaço (Echternacht et al., 2014).

Gentianaceae—This family is represented in the Guiana Shield by around 110 species in 23 genera from both lowlands and highlands (Albert et al., 2007). Most of the species and genera are in the Helieae tribe, which has been the subject of several phylogenetic analyses, including molecular and morphological data (Struwe et al., 1998, 2009; Gould and Struwe, 2004; Frasier et al., 2008; Calio et al., 2017). The most comprehensive analysis to date (Calio et al., 2017) shows complex biogeographic patterns. Pantepui species are recovered either as early diverging members of larger clades (*Neblinantha* and *Irlbachia* as successive ancestral clades to the rest of the *Symbolanthus* group; *Symbolanthus argyreus* or *Rogersonanthus* in a basal polytomy that also includes other *Symbolanthus, Tetrapollinia, Calolisianthus*, etc.) or as derived from within widespread Andean or eastern Brazilian clades (*Roraimaea* as sister to Andean species, *Celiantha* within a group of lowland species). This analysis is based on both molecular and morphological data, and some genera critical for our understanding of Pantepui biogeography still lack molecular data (Calio et al., 2017). In a different analysis, the tepuian *Macrocarpaea neblinae* was found to be nested within a northern Andean species of the same genus (Struwe et al., 2009). Lastly, the monotypic genus *Saccifolium*, which used to be placed in its own family, has been shown to be part of the Gentianaceae and is sister to *Gentiana*, an Andean and North American group (Struwe et al., 1998).

Gesneriaceae—This family is represented in the Guiana Shield by 94 species (Berry and Riina, 2005; Feuillet et al., 2007), but only a few are endemic to Pantepui, with many species widely distributed either geographically or along the altitudinal gradient. Recently, Smith and Clark (2013) segregated several small or monotypic genera within the tribe Episceae, many of them endemic to the Guiana Shield. Most of these have been shown to form an early diverging clade within the tribe, sister to a large group that includes Andean, lowland Guiana Shield, and Brazilian species (Mora and Clark, 2016). Unfortunately, endemic Pantepui species of *Centrosolenia* were not included in this study, which would help clarify the biogeographic history of the tribe in the region. Meanwhile, Ferreira et al. (2016) showed that a recently described species of *Lesia* endemic to the summit of Serra do Aracá in Brazil is sister to a species from the lowlands of the Guiana Shield.

Lentibulariaceae—Two genera are present in the Guiana Shield: *Genlisea* and *Utricularia* (Taylor, 1999, 2007). In the case of *Utricularia*, c. 50 species are in the area, with a mix of endemic and widely distributed species, and also a mix of lowland and highland species (Taylor, 1999, 2007). The Guiana Shield-endemic species that have been sampled in phylogenetic studies (*Utricularia humboldtii, U. quelchii*) are located among species that are widespread in South and Central America or present in the Andes, and this clade is in turn sister to a mostly eastern Brazilian clade (Westermeier et al., 2017). *Genlisea* has only seven species in the Guiana Shield, with a mix of endemic and widespread South American species (Taylor, 1999, 2007). The two tepui-endemic species sampled in the most recent phylogeny are not resolved as sisters, but rather as different lineages within a mostly Amazonian/northern South American grade that includes both lowland and highland species (Westermeier et al., 2017).

Melastomataceae—This is a well-represented family in the Guiana Shield with over 500 species spanning over 70 different genera (Berry et al., 2001; Almeda et al., 2007). A number of different phylogenetic studies have included species found in Pantepui. Within the large tribe Miconieae, all basal species of *Tococa* sensu stricto are found in Pantepui, while lowland and ant-bearing species are more recently derived (Michelangeli, 2000; Michelangeli et al., 2004; Goldenberg et al., 2008). Additionally, some species endemic to Pantepui are derived from within Andean taxa (mostly *Miconia* section *Cremanium*) or lowland lineages (other *Miconia* and *Clidemia*) (Goldenberg et al., 2008). Within Merianieae, *Macrocentrum* is centered in the Guiana Shield, but with one species in the Venezuelan Coastal Cordillera and the Andes and a newly described one from the Andes (Michelangeli and Goldenberg, 2018). Neither extra-Guiana species has been sampled in molecular phylogenies, but given that *Macrocentrum* is paraphyletic at the base of the tribe, it seems that the tribe as a whole has a Guiana Shield origin, with later colonization of Amazonia, the Andes, the Antilles, and eastern Brazil (Dellinger et al., 2018). A later return to the Guiana Shield is also inferred: within *Meriania*, a group of two species from Pantepui with two other species present in the Guiana Shield form a clade with a widely distributed low- to mid-elevation species (*Meriania rotundifolia*) and a species from eastern Brazil (*Meriania calophylla*). Within *Graffenrieda*, Pantepui species do not form a clade; rather they are resolved as closely related to Amazonian, Andean, or Venezuelan Coastal Cordillera species (Dellinger et al., 2018). In the case of Marcetieae, the first early diverging pair of species is found in Pantepui (*Comoliopsis*), while the remainder of the clade is mostly found in central and eastern Brazil, in campos rupestres and cerrado areas (Rocha et al., 2016, 2018). Lastly, three enigmatic Guiana Shield genera—*Boyania*, *Phainantha*, and *Tryssophyton*—are resolved at the base of the Sonerileae–Dissochaeteae complex, a mostly Old World clade, where all other early diverging groups are found in Southeast Asia. Interestingly, the three Guiana genera do not form a clade. In the case of *Phainantha*, one species from the Cordillera del Condor in southern Ecuador and northern Peru is clearly related to the Guiana Shield species, but at this point we cannot determine whether it is a basal or a derived member of the genus (Goldenberg et al., 2015; Bacci et al., in press).

Orchidaceae—This is the most species-rich family in several Neotropical biomes, including Pantepui (Table 6.2); however, the number of tepui-endemic species in this family that have been included in phylogenetic studies is disproportionately low. For example, *Epidendrum*, a genus with over 1500 species, has 30 species in Pantepui (Table 6.4), but none of them have been included in phylogenetic analyses (see Pinheiro et al., 2009; Cardoso-Gustavson et al., 2018). However, these same studies have recovered small clades of species that are widespread in both the Andes and the Guiana Shield (i.e., *Epidendrum secundum*, *Epidendrum calanthum*, *Epidendrum ibaguense*, and *Epidendrum incisum*). In the cosmopolitan genus *Habenaria*, with over 600 species, 30 are found in the Guiana Shield, including 4 species occurring on tepui summits and slopes (Carnevali et al., 2003). Batista et al. (2013) showed that *Habenaria roraimensis*, a species mostly from tepui summits, is sister to *Habenaria armata*, a species widespread at low and mid-elevations in South America. Even within closely related genera different patterns arise. For example, in the subtribe Zygopetalinae, *Galeottia burkei*, a species found on tepui slopes and summits, was recovered as sister to *Galeottia ciliata*, which occurs in the lowlands of the Amazon basin (Whitten et al., 2005). In this same analysis, *Zygozepalum tatei* was found as sister to a larger clade that contained species from the southern end of the Atlantic Forest in Brazil (Whitten et al., 2005). In fact, this study showed that *Zygozepalum* was not monophyletic, which prompted the description of the new genus *Weidmannia* (Romero-González and Carnevali Fernández-Concha, 2010). Overall, these would suggest that to understand biogeographic patterns in tepui Orchidaceae, a phylogeographic approach may be more appropriate than a phylogenetic one.

Rapateaceae—With 16 genera and c. 100 species, this family is centered in the Guiana Shield, with one species found in western Africa and a few species extending into Amazonia, Chocó and Panama, and eastern Brazil. Givnish et al. (2000) showed that the origin of the family was likely in the sandy lowlands of the Guiana Shield, with at least two incursions and radiations into the highlands. Extra-Guianan species in Central America, the Brazilian Planalto, and western Africa are all derived from lowland Guiana species as well (Givnish et al., 2004). At this point, species sampling is insufficient to elucidate relationships among the summits of the tepuis (Givnish et al., 2000, 2004).

Rubiaceae—With almost 600 species in the Guiana Shield, Rubiaceae is one of the most diverse families in the region (Anunciaçao et al., 2007). In spite of this diversity, few studies have focused on the endemic elements of the family. The genus *Pagamea*, a group of c. 30 species from northern South America, was studied by Vicentini (2016), and he found that most high-elevation species formed a single clade, derived from within white sand specialists from the lowland Guiana Shield. Andean species found on sandstone in the Andes are also derived from lowland species and are not the closest relatives of Pantepui species (Vicentini, 2016). Although these speciation events are inferred to be relatively recent (less than 2.2 Ma), it should be noted that these estimates are based on secondary calibrations (Vicentini, 2016). The Pantepui-endemic genus *Maguireanthus*, with two species, was found to be sister to *Limnosipanea*, a genus found in flooded campos and lowlands in the Guiana Shield and Llanos (Delprete and Cortés-B, 2004; Cortés-B et al., 2009). The *Maguireanthus* + *Limnosipanea* clade is derived from within a larger group of mostly Guiana Shield-endemic species occurring mostly in the lowlands (Cortés-B and Motley, 2015*).

Sarraceniaceae—The genus *Heliamphora* with 23 species is completely restricted to the Guiana Shield, with almost all species endemic to tepui summits and upper slopes (McPherson et al., 2011; Naczi, 2018a). *Heliamphora* is sister to *Sarracenia*, which is restricted to the eastern United States, and these two genera are in turn sister to the western United States-endemic *Darlingtonia* (Ellison et al., 2012; Naczi, 2018a). With this distribution, it is impossible to establish the origin of the genus, although the alternatives of long-distance dispersal or vicariance have been discussed (Naczi, 2018a). Ellison et al. (2012) hypothesized that the *Heliamphora* + *Sarracenia* clade originated in South America and then dispersed to the eastern United States. However, based on systematics of the obligate pitcher plant mites of the genus *Sarraceniopus* (Histiostomatidae), Naczi (2018b) hypothesized the opposite pattern. The current level of species sampling within *Heliamphora* does not allow us to draw conclusions about area relationships within Pantepui.

In summary, phylogenetic studies that include Pantepui taxa are not only few, but due to poor sampling or to the particular study design, not ideal to answer questions about the origin and relationships of the flora. The majority of the studies only include one or a few Pantepui species as placeholders for lineages putatively endemic to the region, and often, closely related taxa from outside Pantepui are undersampled. Another issue is that depending on the objectives of the study and overall level of sampling, Pantepui taxa are often lumped together during biogeographic analyses with lowland Guiana Shield taxa, Amazonian taxa, or else in a very broad South America area. Moreover, no species-level phylogeny exists yet for any of the larger genera/clades of Pantepui; thus the biogeographic relationships between individual tepuis cannot yet be analyzed in a phylogenetic context. Since most of the groups treated thus far are strictly tropical, the fossil record is too poor to date the origin and diversification of tepui taxa. Some studies have attempted to date the major biogeographic and evolutionary events in their study group, but because these are all secondary calibrations, the dates are often accompanied by large error intervals or several caveats.

Nonetheless, some interesting trends do emerge: the history of Pantepui taxa is complex and idiosyncratic, varying from group to group (and even within groups), resulting in a mosaic of biogeographic patterns. Even though the sampling is small and probably not random, there are examples for both early diverging and later arrivals. Among later arrivals, there are examples of groups that have subsequently diversified in Pantepui and taxa that represent single arrivals. Most of the close biogeographic relationships revealed by these studies are between the surrounding lowlands and Pantepui or between the Andes and Pantepui, although there is also evidence of relationships between eastern Brazil (either the Planalto or the Atlantic Forest) and Pantepui. The relationship between the Andes and Pantepui fall into at least two different categories. Some Pantepui groups are closely related to clades that are widespread throughout the Andes (usually the northern Andes). However, some species present on sandstone areas in southern Ecuador and northern Peru are more closely related to Pantepui-centered clades. Given the similarities in geology and some elements of the flora, these sandstone areas of the Andes have been called "Andean tepuis" (Neill et al., 2014). Based on the few studies that include dates, and based on the overall dates of several families or clades present in Pantepui, most groups seem to have stem ages well before the Holocene (Givnish et al., 2004, 2011; Ellison et al., 2012; Ruhfel et al., 2016). However, with the current low sampling level, it is not possible to tell the crown ages to be able to infer the tempo of the larger radiations. These ages are both in conflict with theories of a Cretaceous origin of the flora in a relatively stable environment (Maguire, 1970) and with theories of diversification mediated by the cyclic climatic changes during the last 2.5 Ma (see reviews in Rull, 2005, 2008). That said, the glaciation cycles of the Cenozoic probably had a significant impact on diversification of groups that had arrived well before that time. In order to discern the importance of climatic adaptation in the origin of the Pantepui biota, we would need considerably denser phylogenetic analyses with both the stem and crown ages of the different clades to be able to differentiate between the age of the origin of the group and the age of the extant species.

Future perspectives on phylogeography

If phylogenetic studies of Pantepui plants are rare and spotty, then the population genetic and phylogeographic literature is practically nonexistent. This represents a major gap because this kind of study would allow us to reconstruct the history of Pantepui species, especially during the Pleistocene. Pantepui represents an ideal system in which to test hypotheses about population genetics, differentiation, and speciation in the Neotropics. Phylogeographic analyses of widespread Pantepui species and comparative analyses of closely related species endemic to one or a few summits would allow us to test the effect of isolation by distance and/or the effect of effective population sizes (using a tepui summit area as a proxy). We could test specific models of speciation in the Neotropics and in the Guiana Shield in particular, such as those outlined by Rull (2005). If during Pleistocene climatic cycles, summit species have come repeatedly into contact during colder periods at lower altitudes and retracted to fragmented populations during warmer periods, we should be able to see signatures of these processes in molecular data, especially while comparing populations of species of closer tepui summits with those of more isolated tepuis. In spite of initiatives to carry out these types of studies, lack of access to genetic material for a representative sampling has made them unattainable so far.

Acknowledgments

We thank Valentí Rull for encouraging us to write this chapter. We are also grateful to Rob Gradstein, Paulo Labiak, Rob Naczi, Tom Philbrick, and Brad Ruhfel for their help clarifying issues about specific plant groups. Two reviewers, Gustavo Romero-González and Bruce K. Holst, made very useful suggestions and comments on an earlier version of this work. Fabián I. Michelangeli provided field photos of a selection of Pantepui species.

References

Albert, V.A., Jansen-Jacobs, M.J., Maas, P.J.M., Pihlar, O., Struwe, L., 2007. Gentianaceae. In: Funk, V., Hollowell, T., Berry, P., Kelloff, C., Alexander, S.N. (Eds.), Checklist of the Plants of the Guiana Shield (Venezuela: Amazonas, Bolivar, Delta Amacuro; Guyana, Surinam, French Guiana). Contributions from the U.S. National Herbarium, vol. 55. Smithsonian Institution, pp. 304–309.

Alcantara, S., Ree, R.H., Mello-Silva, R., 2018. Accelerated diversification and functional trait evolution in Velloziaceae reveal new insights into the origins of the campos rupestres' exceptional floristic richness. Ann. Bot. 122, 165–180.

Almeda, F., Berry, P.E., Freire-Fierro, A., Gröger, A., Holst, B.K., Luckana, N.G., et al., 2007. Melastomataceae. In: Funk, V., Hollowell, T., Berry, P., Kelloff, C., Alexander, S.N. (Eds.), Checklist of the Plants of the Guiana Shield (Venezuela: Amazonas, Bolivar, Delta Amacuro; Guyana, Surinam, French Guiana). Contributions from the U.S. National Herbarium, vol. 55. Smithsonian Institution, pp. 397–417.

Anunciaçao, E.A.D., Costa, C.B., Cortéz, R., Delprete, P.G., Kirkbride, J.H., Persson, C., et al., 2007. Rubiaceae. In: Funk, V., Hollowell, T., Berry, P., Kelloff, C., Alexander, S.N. (Eds.), Checklist of the Plants of the Guiana Shield (Venezuela: Amazonas, Bolivar, Delta Amacuro; Guyana, Surinam, French Guiana). Contributions from the U.S. National Herbarium, vol. 55. Smithsonian Institution, pp. 376–379.

Aymard, G.A., Grande, J.R., 2012. *Duranta neblinensis* (Verbenaceae, Duranteae): a new species from Sierra de la Neblina, Amazonas state, Venezuela. Brittonia 64, 247–251.

Bacci, L.F., Michelangeli, F.A., Goldenberg, R., in press. Revisiting the classification of Melastomataceae: implications for habit and fruit evolution. Bot. J. Linn. Soc.

Barbosa-Silva, R.G., Labiak, P.H., Gil, A.S.B., Goldenberg, R., Michelangeli, F.A., Martinelli, G., et al., 2016. Over the hills and far away: new plant records for the Guayana Shield in Brazil. Brittonia 68, 397–408.

Batista, J.A.N., Borges, K.S., de Faria, M.W.F., Proite, K., Ramalho, A.J., Salazar, G.A., et al., 2013. Molecular phylogenetics of the species-rich genus *Habenaria* (Orchidaceae) in the New World based on nuclear and plastid DNA sequences. Mol. Phylogenet. Evol. 67, 95–109.

Berry, P.E., Riina, R., 2005. Insights into the diversity of the Pantepui flora and the biogeographic complexity of the Guayana Shield. Biologiske Skrifter 55, 145–167.

Berry, P.E., Huber, O., Holst, B.K., 1995. Floristic analysis and phytogeography. In: Berry, P.E., Holst, B.K., Yatskievych, K. (Eds.), Flora of the Venezuelan Guayana, vol. 1. Missouri Botanical Garden/Timber Press, St. Louis/Portland, OR, pp. 161–191.

Berry, P.E., Gröger, A., Holst, B.K., Morley, T., Michelangeli, F.A., Luckana, N.G., et al., 2001. Melastomataceae. In: Berry, P.E., Holst, B.K., Yatskievych, K. (Eds.), Flora of the Venezuelan Guayana, vol. 6. Missouri Botanical Garden Press, St. Louis, pp. 263–528.

Berry, P.E., Yatskievych, K., Holst, B.K. (Eds.), 2005. Flora of the Venezuelan Guayana, vol. 9. Missouri Botanical Garden Press, St. Louis.

Buck, W.R., 1991. Why are there so few mosses on tepui summits? In: Herben, T., McQueen, C.B. (Eds.), Proceedings of the Sixth Meeting of the Central and East European Bryological Working Group (CEBWG). Botanical Institute of the Czechoslovak Academy of Sciences, Průhonice, pp. 46–51.

Calio, M.F., Lepis, K.B., Pirani, J.R., Struwe, L., 2017. Phylogeny of Helieae (Gentianaceae): Resolving taxonomic chaos in a Neotropical clade. Mol. Phylogenet. Evol. 106, 192–208.

Cardoso, D., Särkinen, T., Alexander, S., Amorim, A.M., Bittrich, V., Celis, M., et al., 2017. Amazon plant diversity revealed by a taxonomically verified species list. Proc. Natl Acad. Sci. 114, 10695–10700.

Cardoso-Gustavson, P., Saka, M.N., Pessoa, E.M., Palma-Silva, C., Pinheiro, F., 2018. Unidirectional transitions in nectar gain and loss suggest food deception is a stable evolutionary strategy in *Epidendrum* (Orchidaceae): insights from anatomical and molecular evidence. BMC Plant Biol. 18, 179.

Carnevali, G., Ramírez-Morillo, I.M., Romero-González, G.A., Vargas, C.A., Foldats, E., 2003. Orchidaceae. In: Berry, P.E., Holst, B.K., Yatskievych, K. (Eds.), Flora of the Venezuelan Guayana, vol. 7. Missouri Botanical Garden Press, St. Louis, pp. 200–619.

Chase, M.W., Williams, N.H., de Faria, A.D., Neubig, K.M., Amaral, M., do, C.E., et al., 2009. Floral convergence in Oncidiinae (Cymbidieae; Orchidaceae): an expanded concept of *Gomesa* and a new genus *Nohawilliamsia*. Ann. Bot. 104, 387–402.

Chase, M.W., Christenhusz, M.J.M., Fay, M.F., Byng, J.W., Judd, W.S., Soltis, D.E., et al., 2016. An update of the Angiosperm Phylogeny Group classification for the orders and families of flowering plants: APG IV. Bot. J. Linn. Soc. 181, 1–20.

Coelho, M.A.N., Costa, D.P., Martinelli, G., Moraes, M.A., Forzza, R.C., 2015. Expedições às montanhas da Amazônia. Andrea Jakobsson Estúdio, Rio de Janeiro.

Cortés-B, R., Motley, T.J., 2015. Phylogeny of the Henriquezieae–Posoquerieae–Sipaneeae, a Guayanan-centered clade of Rubiaceae: implications for morphological evolution. Phytotaxa 206, 90–117.

Cortés-B, R., Delprete, P.G., Motley, T.J., 2009. Phylogenetic placement of the tribe Retiniphylleae among the subfamily Ixoroideae (Rubiaceae). Ann. Missouri Bot. Gard. 96, 61–67.

Costa, D.P., 2017. Bryophyte results from a botanical expedition to Serra do Aracá State, Amazonas, Brazil: diversity, distribution, and endemism. Bryologist 120, 45–50.

Croat, T.B., Outman, R., Kostelac, C.V., 2010. New species of Araceae from Venezuela and the Guianas. Willdenowia 40, 137–146.

Dauphin, G., Ilkiu-Borges, A.L., 2002. Hepaticae of Cerro Venamo, Venezuela, collected by J. Steyermark. Trop. Bryol. 22, 115–123.

Dellinger, A.S., Chartier, M., Fernández-Fernández, D., Penneys, D.S., Alvear, M., Almeda, F., et al., 2018. Beyond buzz-pollination—departures from an adaptive plateau lead to new pollination syndromes. New Phytol. 1–14.

Delprete, P.G., Cortés-B, R., 2004. A phylogenetic study of the tribe Sipaneeae (Rubiaceae, Ixoroideae), using *trnL-F* and ITS sequence data. Taxon 53, 347–356.

Delprete, P.G., Kirkbride Jr, J.H., 2016. New combinations and new names in *Palicourea* (Rubiaceae) for species of *Psychotria* subgenus *Heteropsychotria* occurring in the Guianas. J. Bot. Res. Inst. Tex. 10, 409–442.

Désamoré, A., Vanderpooten, A., Laenen, B., Gradstein, S.R., Kok, P.J.R., 2010. Biogeography of the Lost World (Pantepui region, northeastern South America): insights from bryophytes. Phytotaxa 9, 254–264.

Echternacht, L., Sano, P.T., Bonillo, C., Cruaud, C., Couloux, A., Dubuisson, J.-Y., 2014. Phylogeny and taxonomy of *Syngonanthus* and *Comanthera* (Eriocaulaceae): Evidence from expanded sampling. Taxon 63, 47–63.

Ellison, A.M., Butler, E.D., Hicks, E.J., Naczi, R.F.C., Calie, P.J., Bell, C.D., et al., 2012. Phylogeny and biogeography of the carnivorous plant family Sarraceniaceae. PLoS One 7, e39291.

Ferreira, G.E., Chautems, A., Hopkins, M.J.G., Perret, M., 2016. Independent evolution of pouched flowers in the Amazon is supported by the discovery of a new species of *Lesia* (Gesneriaceae) from Serra do Araca tepui in Brazil. Plant Systemat. Evol 302, 1109–1119.

Feuillet, C., 2009. Checklist of the plants of the Guiana Shield 1. An update to the angiosperms. J. Bot. Res. Inst. Tex. 3, 799–814.

Feuillet, C., Skog, L.E., Steyermark, J.A., 2007. Gesneriaceae. In: Funk, V., Hollowell, T., Berry, P., Kelloff, C., Alexander, S.N. (Eds.), Checklist of the Plants of the Guiana Shield (Venezuela: Amazonas, Bolivar, Delta Amacuro; Guyana, Surinam, French Guiana). Contributions from the U.S. National Herbarium, vol. 55. Smithsonian Institution, pp. 309–312.

Fiaschi, P., Plunkett, G.M., 2011. Monophyly and phylogenetic relationships of Neotropical *Schefflera* (Araliaceae) based on plastid and nuclear markers. Syst. Bot. 36, 806–817.

Fiaschi, P., Plunkett, G.M., 2018. Revision of the *Didymopanax* group of Neotropical *Schefflera* (Araliaceae). Ann. Missouri Bot. Gard. 103, 24–105.

Fleischmann, A., Cross, A.T., Gibson, R., Gonella, P.M., Dixon, K.W., 2018. Systematics and evolution of droseraceae. In: Ellison, A.M., Adamec, L. (Eds.), Carnivorous Plants Physiology, Ecology and Evolution. Oxford University Press, Oxford, pp. 45–57.

Flora do Brasil 2020, Under construction. Jardim Botânico do Rio de Janeiro. Available from: <http://floradobrasil.jbrj.gov.br/reflora/floradobrasil/FB34371> (accessed11.11.18).

Frasier, C.L., Albert, V.A., Struwe, L., 2008. Amazonian lowland, white sand areas as ancestral regions for South American biodiversity: biogeographic and phylogenetic patterns in *Potalia* (Angiospermae: Gentianaceae). Organ. Divers. Evol. 8, 44–57.

Fritsch, P.W., Lu, L., Bush, C.M., Cruz, B.C., Kron, K.A., Li, D.Z., 2011. Phylogenetic analysis of the wintergreen group (Ericaceae) based on six genic regions. Syst. Bot. 36, 990–1003.

Frodin, D.G., Lowry, P.P.I., Plunkett, G.M., 2010. *Schefflera* (Araliaceae): taxonomic history, overview and progress. Plant Divers. Evol. 128, 561–595.

Funk, V.A., Brooks, D.R., 1990. Phylogenetic Systematics as the Basis of Comparative Biology. Contributions from the U.S. National Herbarium, vol. 73. Smithsonian Institution, pp. 1–45.

Funk, V., Hollowell, T., Berry, P.E., Kelloff, C., Alexander, S.N., 2007a. Checklist of the plants of the Guiana Shield (Venezuela: Amazonas, Bolivar, Delta Amacuro; Guyana, Surinam, French Guiana). Contributions from the U.S. National Herbarium, vol. 55. Smithsonian Institution, pp. 1–584.

Funk, V.A., Robinson, H., Pruski, J.F., 2007b. Compositae. In: Funk, V., Hollowell, T., Berry, P., Kelloff, C., Alexander, S.N. (Eds.), Checklist of the Plants of the Guiana Shield (Venezuela: Amazonas, Bolivar, Delta Amacuro; Guyana, Surinam, French Guiana). Contributions from the U.S. National Herbarium, vol. 55. Smithsonian Institution, pp. 255–269.

Gillespie, E., Kron, K., 2010. Molecular phylogenetic relationships and a revised classification of the subfamily Ericoideae (Ericaceae). Mol. Phylogenet. Evol. 56, 343–354.

Gillespie, E.L., Kron, K.A., 2012. A new tribe, Bryantheae (Ericoideae: Ericaceae), composed of disjunct genera from South America and Japan. Brittonia 64, 73–75.

Giraldo-Cañas, D., 2013. Las especies del género *Axonopus* (Poaceae: Panicoideae: Paspaleae) en Venezuela. Pittieria 37, 53–114.

Givnish, T.J., Evans, T.M., Zjhra, M.L., Patterson, T.B., Berry, P.E., Sytsma, K.J., 2000. Molecular evolution, adaptive radiation, and geographic diversification in the amphiatlantic family Rapateaceae: evidence from *ndhF* sequences and morphology. Evolution 54, 1915–1937.

Givnish, T.J., Millam, K.C., Evans, T.M., Hall, J.C., Pires, J.C., Berry, P.E., et al., 2004. Ancient vicariance or recent long-distance dispersal? Inferences about phylogeny and South American–African disjunctions in Rapateaceae and Bromeliaceae based on *ndhF* sequence data. Int. J. Plant Sci. 165, S35–S54.

Givnish, T.J., Barfuss, M.H.J., Van Ee, B., Riina, R., Schulte, K., Horres, R., et al., 2011. Phylogeny, adaptive radiation, and historical biogeography in Bromeliaceae: insights from an eight-locus plastid phylogeny, Am. J. Bot., 98. pp. 872–895.

Givnish, T.J., Barfuss, M.H.J., Van Ee, B., Riina, R., Schulte, K., Horres, R., et al., 2014. Adaptive radiation, correlated and contingent evolution, and net species diversification in Bromeliaceae, Mol. Phylogenet. Evol., 71. pp. 55–78.

Goldenberg, R., Penneys, D.S., Almeda, F., Judd, W.S., Michelangeli, F.A., 2008. Phylogeny of *Miconia* (Melastomataceae): patterns of stamen diversification in a megadiverse Neotropical genus. Int. J. Plant Sci. 169, 963–979.

Goldenberg, R., Almeda, F., Sosa, K., Ribeiro, R.C., Michelangeli, F.A., 2015. *Rupestrea*: a new Brazilian genus of Melastomataceae, with anomalous seeds and dry indehiscent fruits. Syst. Bot. 40, 561–571.

Gould, K.R., Struwe, L., 2004. Phylogeny and evolution of *Symbolanthus* and *Wurdackanthus* (Gentianacea-Helteae) in the Guayana Highlands and Andes, based on ribosomal 5S-NTS sequences. Ann. Missouri Bot. Gard. 91, 438–446.

Gradstein, S.R., Costa, D.P., 2018. *Plagiochila lamyana*, a new liverwort species from the Guayana Highland of Brazil. Cryptogam. Bryol. 39, 147–153.

Gradstein, S.R., Florschütz-de Waard, J., 1989. Results of a botanical expedition to Mount Roraima, Guyana. I. Bryophytes. Trop. Bryol. 1, 25–54.

Gradstein, S.R., Churchill, S.P., Salazar, A.N., 2001. Guide to the bryophytes of Tropical America, Memoirs of the New York Botanical Garden, vol. 86. New York Botanical Garden Press, pp. 1–577.

Grande, A.J.R., 2018. Sertulum Ternstroemiacearum II: *Ternstroemia tepuiensis* J.R. Grande, sp. nov. (Ternstroemiaceae), especie nueva del Escudo Guayanés. Anales del Jardín Botánico de Madrid 75, e066.

Hakki, M., Hensold, N., 2007. Eriocaulaceae. In: Funk, V., Hollowell, T., Berry, P., Kelloff, C., Alexander, S.N. (Eds.), Checklist of the Plants of the Guiana Shield (Venezuela: Amazonas, Bolivar, Delta Amacuro; Guyana, Surinam, French Guiana). Contributions from the U.S. National Herbarium, vol. 55. Smithsonian Institution, pp. 104–109.

Hensold, N., 1999. Eriocaulaceae. In: Berry, P.E., Holst, B.K., Yatskievych, K. (Eds.), Flora of the Venezuelan Guayana, vol. 5. Missouri Botanical Garden Press, St. Louis, pp. 1–58.

Hirai, R.Y., Rouhan, G., Labiak, P.H., Ranker, T.A., Prado, J., 2011. *Moranopteris*: a new Neotropical genus of grammitid ferns (Polypodiaceae) segregated from Asian *Micropolypodium*. Taxon 60, 1123–1137.

Hopper, S.D., 2009. OCBIL theory: towards an integrated understanding of the evolution, ecology and conservation of biodiversity on old, climatically buffered, infertile landscapes. Plant Soil 322, 49–86.

Hopper, S.D., Silveira, F.A., Fiedler, P.L., 2016. Biodiversity hotspots and Ocbil theory. Plant Soil 403, 167–216.

Huber, O., 1987. Consideraciones sobre el concepto de Pantepui. Pantepui 2, 2–10.

Huber, O., 1992. La vegetación. In: Huber, O. (Ed.), El Macizo del Chimantá, Escudo de Guayana, Venezuela: Un Ensayo Ecológico Tepuyano. Oscar Todtmann, Caracas, pp. 161–178.

Huber, O., 1994. Recent advances in the phytogeography of the Guayana region, South America. Mémoires de la Société de Biogeographie 4, 53–63.

Huber, O., 1995. Geographical and physical features. In: Berry, P.E., Holst, B.K., Yatskievych, K. (Eds.), Flora of the Venezuelan Guayana, Introduction, vol. 1. Missouri Botanical Garden, St. Louis. Timber Press, Portland, Oregon, pp. 1–51.

Hughes, C.E., Atchison, G.W., 2015. The ubiquity of alpine plant radiations: from the Andes to the Hengduan Mountains. New Phytol. 207, 275–282.

Kelloff, C.L., Alexander, S.N., Funk, V.A., Clarke, H.D., 2011. Smithsonian Plant Collections, Guyana: 1995–2004, H. David Clarke, Smithsonian Contributions to Botany, vol. 97. Smithsonian Institution Scholarly Press, pp. 1–307.

Kron, K.A., Judd, W.S., Stevens, P.F., Crayn, D.M., Anderberg, A.A., Gadek, P.A., et al., 2002. Phylogenetic classification of Ericaceae: molecular and morphological evidence. Bot. Rev. 68, 335–423.

LPWG (The Legume Phylogeny Working Group), 2017. A new subfamily classification of the Leguminosae based on a taxonomically comprehensive phylogeny. Taxon 66, 44–77.

Lagomarsino, L.P., Condamine, F.L., Antonelli, A., Mulch, A., Davis, C.C., 2016. The abiotic and biotic drivers of rapid diversification in Andean bellflowers (Campanulaceae). New Phytol. 210, 1430–1442.

Lellinger, D.B., 1995. Pterozonium. In: Berry, P.E., Holst, B.K., Yatskievych, K. (Eds.), Flora of the Venezuelan Guayana, vol. 2. Missouri Botanical Garden Press, St. Louis, pp. 280–286.

Lucas, E., Wilson, C.E., Lima, D.F., Matsumoto, M.S.K., 2016. A conspectus of *Myrcia* sect. *Aulomyrcia* (Myrtaceae). Ann. Missouri Bot. Gard. 101, 648–698.

Luteyn, J.L., 2007. Ericaceae. In: Funk, V., Hollowell, T., Berry, P., Kelloff, C., Alexander, S.N. (Eds.), Checklist of the Plants of the Guiana Shield (Venezuela: Amazonas, Bolivar, Delta Amacuro; Guyana, Surinam, French Guiana). Contributions from the U.S. National Herbarium, vol. 55. Smithsonian Institution, pp. 285–288.

Maciel, S., Lehnert, M., Hira, R.Y., Prado, J., 2017. Three new species of the *Cyathea "Hymenophyllopsis"* clade (Cyatheaceae) from Venezuela and Brazil. Phytotaxa 329, 159–166.

Madriñán, S., Cortés, A.J., Richardson, J.E. 2013. Páramo is the world's fastest evolving and coolest biodiversity hotspot. Front. Genet. 4: 192.

Maguire, B., 1970. On the Flora of the Guayana highland. Biotropica 2, 85–100.

Maguire, B., 1979. Guayana, region of the roraima sandstone formation. In: Larsen, K., Holm-Nielsen, L.B. (Eds.), Tropical Botany. Academic Press, London, pp. 223–238.

McPherson, S., Wistuba, A., Fleischmann, A., Nerz, J., 2011. Sarraceniaceae of South America. Redfern natural history productions, Poole. Bot. J. Linn. Soc. 170 (1), 562.

Merckx, V.S., Hendriks, K.P., Beentjes, K.K., Mennes, C.B., Becking, L.E., Peijnenburg, K.T., et al., 2015. Evolution of endemism on a young tropical mountain. Nature 524, 347–350.

Michelangeli, F.A., 2000. A cladistic analysis of the genus *Tococa* (Melastomataceae) based on morphological data. Syst. Bot. 25, 211–234.

Michelangeli, F.A., Goldenberg, R., 2018. New and noteworthy Melastomataceae from the Yanachaga-Chemillén National Park and surrounding areas in Oxapampa, Pasco, Peru. Phytotaxa 374, 185–210.

Michelangeli, F.A., Penneys, D.S., Giza, J., Soltis, D., Hils, M.H., Skean Jr., J.D., 2004. A preliminary phylogeny of the tribe Miconieae (Melastomataceae) based on nrITS sequence data and its implications on inflorescence position. Taxon 53, 279–290.

Mora, M.M., Clark, J.L., 2016. Molecular phylogeny of the Neotropical genus *Paradrymonia* (Gesneriaceae), reexamination of generic concepts and the resurrection of *Trichodrymonia* and *Centrosolenia*. Syst. Bot. 41, 82–104.

Naczi, R.F.C., 2018a. Systematics and evolution of Sarracenieae. In: Ellison, A.M., Adamec, L. (Eds.), Carnivorous Plants. Physiology, Ecology and Evolution. Oxford University Press, Oxford, pp. 103–119.

Naczi, R.F.C., 2018b. Systematics and biogeography of the western hemisphere pitcher plants (Sarraceniaceae). Rochester, MN. Botany 2018.

Neill, D.A., 2018. Flora y vegetación de los "tepuyes andinos": endemismo, diversidad y conexiones fitogeográficas con el Escudo Guayanés. Memorias XII Congreso Latinoamericano de Botánica, Quito, Ecuador, p. 18.

Neill, D.A., Paredes, M.R., Montenegro, L.A.T., Vargas, T.J.M., Vriesendorp, C., 2014. Vegetación y flora/Vegetation and flora. In: Pitman, N., Vriesendorp, C., Alvira, D., Markel, J.A., Johnston, M., Inzunza, E.R., Pizango, A.L., Valenzuela, G.S., Álvarez-Loayza, P., Homan, J., Wachter, T., Campo, Á.D., Stotz, D.F., Heilpern, S. (Eds.), Perú: Cordillera Escalera-Loreto. The Field Museum, Chicago, pp. 98–119, 292–311, 408–465.

Núñez-Farfán, J., Simonetti, J.A., Roja, T.M., 2017. Venezuela's botanic heritage in jeopardy. Nature 549, 31.

O'Meara, B.C., 2012. Evolutionary inferences from phylogenies: a review of methods. Annu. Rev. Ecol. Evol. Syst. 43, 267–285.

Ortiz, S., Bonifacino, J.M., Crisci, J.V., Funk, V.A., Hansen, H.V., Hind, D.J.N., et al., 2009. The basal grade of compositae: Mutisieae (sensu Cabrera) and Carduoideae. In: Funk, V., Susanna, A., Stuessy, T., Bayer, R. (Eds.), Systematics, Evolution, and Biogeography of Compositae. IAPT, Vienna, pp. 193–213.

PPG I, 2016. A community-derived classification for extant lycophytes and ferns. J. Syst. Evol. 54, 563–603.

Panero, J.L., Funk, V.A., 2008. The value of sampling anomalous taxa in phylogenetic studies: major clades of the Asteraceae revealed. Mol. Phylogenet. Evol. 47, 757–782.

Philbrick, C.T., Malecki, J., Tippery, N.P., Stevens, H.I., 2011. A new genus of podostemaceae from Venezuela. Novon 21, 475–480.

Pinheiro, F., Koehler, S., Correa, A.M., Salatino, M.L.F., Salatino, A., de Barros, F., 2009. Phylogenetic relationships and infrageneric classification of *Epidendrum* subgenus *Amphiglottium* (Laeliinae, Orchidaceae). Plant Syst. Evol. 283, 165–177.

Plunkett, G.M., Lowry, P.P., Frodin, D.G., Wen, J., 2005. Phylogeny and geography of *Schefflera*: pervasive polyphyly in the largest genus of Araliaceae. Ann. Missouri Bot. Gard. 92, 202–224.

Powell, E.A., Kron, K.A., 2001. An analysis of the phylogenetic relationships in the wintergreen group (*Diplycosia, Gaultheria, Pernettya, Tepuia*; Ericaceae). Syst. Bot. 26, 808–817.

Pruski, J., 1997. Asteraceae. In: Berry, P.E., Holst, B.K., Yatskievych, K. (Eds.), Flora of the Venezuelan Guayana, vol. 3. Missouri Botanical Garden, pp. 177–392.

Pruski, J.F., 2011. Compositae of the Guayana Highland-XIV. Four new species of *Calea* (Neurolaeneae) from tepui summits in Venezuela. Phytoneuron 2011-52, 1–9.

Quintero, I., Jetz, W., 2018. Global elevational diversity and diversification of birds. Nature 555, 246–250.

Riina, R., 2003. Patterns of Plant Species Richness and Endemism on the Tepuis of the Guayana Shield. Master's Dissertation. University of Wisconsin-Madison.

Rivadavia, F., Kondo, K., Kato, M., Hasebe, M., 2003. Phylogeny of the sundews, *Drosera* (Droseraceae), based on chloroplast *rbcL* and nuclear 18S ribosomal DNA sequences. Am. J. Bot. 90, 123–130.

Rivadavia, F., de Miranda, V.F.O., Hoogenstrijd, G., Pinheiro, F., Heubl, G., Fleischmann, A., 2012. Is *Drosera meristocaulis* a pygmy sundew? Evidence of a long-distance dispersal between western Australia and northern South America. Ann. Bot. 110, 11–21.

Rocha, M.J.R., Batista, J.A.N., Guimaraes, P.J.F., Michelangeli, F.A., 2016. Phylogenetic relationships in the *Marcetia* alliance (Melastomeae, Melastomataceae) and implications for generic circumscription. Bot. J. Linn. Soc. 181, 585–609.

Rocha, M.J.R., Guimarães, P.J., Michelangeli, F.A., Nogueira Batista, J.A., 2018. Taxonomy of Marcetieae: a new Neotropical tribe of melastomataceae. Int. J. Plant Sci. 179, 50–74.

Romero-González, G.A., Carnevali Fernández-Concha, G., 2010. *Weidmannia* (Orchidaceae, Zygopetalinae), a new genus from the Guayana Highlands. Harv. Paper Bot. 15, 191–194.

Ronquist, F., Sanmartín, I., 2011. Phylogenetic methods in biogeography. Annu. Rev. Ecol. Evol. Syst. 42, 441–464.

Ruhfel, B.R., Bittrich, V., Bove, C.P., Gustafsson, M.H.G., Philbrick, C.T., Rutishauser, R., et al., 2011. Phylogeny of the Clusioid clade (Malpighiales): evidence from the plastid and mitochondrial genomes. Am. J. Bot. 98, 306–325.

Ruhfel, B.R., Bove, C.P., Philbrick, C.T., Davis, C.C., 2016. Dispersal largely explains the Gondwanan distribution of the ancient tropical clusioid plant clade. Am. J. Bot. 103, 1117–1128.

Rull, V., 2005. Biotic diversification in the Guayana Highlands: a proposal. J. Biogeogr. 32, 921–927.

Rull, V., 2007. The Guayana Highlands: a promised (but threatened) land for ecological and evolutionary science 1. Biotropica 39, 31–34.

Rull, V., 2008. Speciation timing and neotropical biodiversity: the Tertiary–Quaternary debate in the light of molecular phylogenetic evidence. Mol. Ecol. 17, 2722–2729.

Rull, V., Vegas, T., Nogué, S., Montoya, E. 2008. Bureaucratic obstruction of conservation science in the Guayana Highlands. Conserv. Biol. 22: 508-509.

Safont, E., Rull, V., Vegas-Vilarrubia, T., Holst, B.K., Huber, O., Nozawa, S., et al., 2014. Establishing a baseline of plant diversity and endemism on a neotropical mountain summit for future comparative studies assessing upward migration: an approach from biogeography and nature conservation. Syst. Biodivers. 12, 292–314.

Scharf, U., Maas, P.J., Morawetz, W., 2005. Five new species of *Guatteria* (Annonaceae) from the Pakaraima Mountains, Guyana. Blumea-Biodivers. Evol. Biogeogr. Plants 50, 563–573.

Schwery, O., Onstein, R.E., Bouchenak Khelladi, Y., Xing, Y., Carter, R.J., Linder, H.P., 2015. As old as the mountains: the radiations of the Ericaceae. New Phytol. 207, 355–367.

Smith, J.F., Clark, J.L., 2013. Molecular phylogenetic analyses reveal undiscovered monospecific genera in the tribe episcieae (Gesneriaceae). Syst. Bot. 38, 451–463.

Steyermark, J.A., 1986. Speciation and endemism in the flora of the Venezuelan Tepuis. In: Vuilleumier, F., Monasterio, M. (Eds.), High Altitude Tropical Biogeography. Oxford University Press, Oxford, pp. 313–373.

Steyermark, J.A., Berry, P.E., Holst, B.K. (Eds.), 1995–2005. Flora of the Venezuelan Guayana, vol. 9. Missouri Botanical Garden, St. Louis.

Struwe, L., Thiv, M., Kadereit, J.W., Pepper, A.S.R., Motley, T.J., White, P.J., et al., 1998. *Saccifolium* (Saccifoliaceae), an endemic of Sierra de la Neblina on the Brazilian-Venezuelan border, is related to a temperate-alpine lineage of Gentianaceae. Harv. Paper Bot. 3, 199–214.

Struwe, L., Nilsson, S., Lbert, V.A., 2008. *Roraimaea* (Gentianaceae: Helieae)—a new gentian genus from white sand campinas and Cerro de la Neblina of Brazil and Venezuela. Harv. Paper Bot. 13, 35–45.

Struwe, L., Haag, S., Heiberg, E., Grant, J.R., 2009. Andean speciation and vicariance in Neotropical *Macrocarpaea* (Gentianaceae–Helieae). Ann. Missouri Bot. Gard. 96, 450–469.

Taylor, P.G., 1999. Lentibulariaceae. In: Berry, P.E., Holst, B.K., Yatskievych, K. (Eds.), Flora of the Venezuelan Guayana, vol. 5. Missouri Botanical Garden Press, St. Louis, pp. 782–803.

Taylor, P.G., 2007. Lentibulariaceae. In: Funk, V., Hollowell, T., Berry, P., Kelloff, C., Alexander, S.N. (Eds.), Checklist of the Plants of the Guiana Shield (Venezuela: Amazonas, Bolivar, Delta Amacuro; Guyana, Surinam, French Guiana). Contributions from the U.S. National Herbarium, vol. 55. Smithsonian Institution, pp. 376–379.

Taylor, C.M., 2011. The genus *Coccochondra* (neotropical Rubiaceae) expanded. Plant Ecol. Evol. 144, 115–118.

Taylor, C.M., 2015. Rubiacearum Americanarum Magna Hama Pars XXXIV: the new group *Palicourea* sect. *Tricephalium* with eight new species and a new subspecies (Palicoureeae). Novon 24, 55–95.

Trovo, M., Gomes De Andrade, M.J., Sano, P.T., Ribeiro, P.L., Van den Berg, C., 2013. Molecular phylogenetics and biogeography of Neotropical Paepalanthoideae with emphasis on Brazilian *Paepalanthus* (Eriocaulaceae). Bot. J. Linn. Soc. 171, 225–243.

Vicentini, A., 2016. The evolutionary history of *Pagamea* (Rubiaceae), a white-sand specialist lineage in tropical South America. Biotropica 48, 58–69.

Westermeier, A.S., Fleischmann, A., Muller, K., Schaferhoff, B., Rubach, C., Speck, T., et al., 2017. Trap diversity and character evolution in carnivorous bladderworts (*Utricularia*, Lentibulariaceae). Sci. Rep. 7.

Whitten, W.M., Williams, N.H., Dressler, R.L., Gerlach, G., Pupulin, F., 2005. Generic relationships of Zygopetalinae (Orchidaceae: Cymbidieae). Lankesteriana 5, 87–107.

Zhang, L., Rothfels, C.J., Ebihara, A., Schuettpelz, E., Pechon, T.L., Kamau, P., et al., 2015. A global plastid phylogeny of the brake fern genus *Pteris* (Pteridaceae) and related genera in the Pteridoideae. Cladistics 31, 406–523.

Zuloaga, F.O., Morrone, O., Scataglini, M.A., 2011. Monograph of *Trichanthecium* (Poaceae, Paniceae). Syst. Bot. Monogr. 94.

CHAPTER

7

Plant communities

Otto Huber[1] and Valentí Rull[2]

[1]Botanical Institute of Venezuela Foundation, Caracas, Venezuela [2]Institute of Earth Sciences Jaume Almera (ICTJA-CSIC), Barcelona, Spain

Introduction

It is well known that the high-mountain tropical biomes harbor a noticeable richness of ecosystems with remarkable levels of endemic plants and growth forms. Therefore it is not surprising that the Neotropical country of Venezuela, where four larger mountain systems (Andes, Sierra de Perijá, Coastal Cordillera, and Guiana Highlands) occupy a considerable proportion of its surface, has been considered one of the World's megadiverse areas (Mittermeier et al., 1997), especially in terms of plants (Hokche et al., 2008). Whereas the first three mentioned Venezuelan mountain systems are geologically young and exhibit a typically cordilleran physiography with steep peaks and deep valleys, the Guiana Highlands consist of tabular mountains with more or less flat surfaces (locally called tepuis) as the remnants of an ancient, extended sandstone plateau that accumulated on the Guiana Shield during Precambrian times and further dissected, developing a series of fossil erosion surfaces (Briceño and Schubert, 1990). Although the Guiana Highlands are notably lower than the Venezuelan Andes (c. 3000 m vs c. 5000 m elevation) and are 600–1000 km apart, both mountain systems exhibit a roughly similar biotic altitudinal zonation along their flanks. However, such biotic gradients differ in both mountain systems due to their different physiography. In the Andes, the elevational gradient is more or less continuous, but in the Guiana Highlands, the presence of a massive belt of steep walls of rock between the densely forested lower slopes and the flat summits (Plate 7.1A) breaks the pattern and creates a discontinuous environmental gradient for the biota (Huber, 1995a). Therefore it is not surprising to find numerous endemic plant species and vegetation types on the tepuis between approximately 1500 and 3000 m elevation, which form a unique biome and constitute a different biogeographical province called Pantepui (see Chapter 1: Definition and characterization of the Pantepui biogeographical province, for more details).

Biodiversity of Pantepui
DOI: https://doi.org/10.1016/B978-0-12-815591-2.00007-0

The floristic diversity and endemism patterns of Pantepui have already been described in Chapter 6: Vascular plants and bryophytes. Here we classify and briefly describe the vegetation types recognized to date in Pantepui and highlight their special or unique features. The chapter is focused on the Venezuelan part of Pantepui, which contains the majority of the Guianan tepuis and where all vegetation studies have been developed; no similar studies exist for the Guyanan and the Brazilian part of Pantepui (Fig. 1.2 from Chapter 1). Different types of Pantepui plant communities have been reported in several papers. Some vegetation hints were provided by Maguire (1970, 1979) and Vareschi (1980, 1992a,b) as part of their floristic surveys. Huber (1992) described the vegetation types of the Chimantá massif. The *Flora of the Venezuelan Guayana* (Steyermark et al., 1995–2005) contains a chapter on the vegetation types of this region, which includes the most important plant communities of Pantepui (Huber, 1995b). However, a comprehensive synthesis specifically dedicated to Pantepui plant communities is still unavailable. This chapter provides an enhanced and updated view of all Pantepui vegetation types known to date and a succinct diagnostic description of all of them based on qualitative field observations. Quantitative characterizations are still embryonic but are also reported. It is hoped that this will encourage further ecological studies on these plant formations, which are urgently needed not only to understand their origin and development but also to guarantee their conservation.

The table-mountain landscape of the Guiana Shield

Overview

The roughly 70 tepui summits over 1500 m elevation known from the Guiana Shield region are spread over an approximate area of 250,000 km^2, extending mostly into southeastern Venezuela; a smaller, eastern section extends into northwestern Guyana with the tepuis Maringma (approximately 2110 m), Wokomung (1660 m), and Ayanganna (2080 m), whereas the adjacent Brazilian section only includes the headwater area of the Rio Cauaburí and the southern slopes of the Cerro de la Neblina massif (see Table 1.1. and Fig. 1.2 from Chapter 1). These numerous table mountains, with their astonishing diversity in terms of appearance and their impressive shape, are undoubtedly one of the major scenic hotspots on Earth, although they barely reach maximum elevations near 3000 m. The entire surface of the flat tepui summits above 1500 m elevation is estimated at approximately 5000 km^2. Conical summit peaks, so common in many other mountain landscapes of the world, are very rare in the Guiana Highlands and are mostly related to more recent magmatic intrusions (Huber 1995a).

In most cases, the tepuis are typical table-top mountains, well recognizable from far away; their inclined base starts generally between 100 and 500 m elevation in the lowlands of Guiana; this stump is usually surmounted by a massive elevation with vertical, exposed walls of rock, mostly between 600 and 1500 m high. The average elevation of the single table-top mountains or mountain groups oscillates roughly between 1000 and 3000 m, with the Cerro Autana being one of the lowest (1220 m) and the Cerro de la Neblina the highest (2994 m) mountain on the entire Guiana Shield; therefore they are much lower than the Andean Cordillera running from north to south along the western rim of South

America. This also implies that none of the Guianan tepui summits reaches the snow line (approximately at 4500 m in these equatorial latitudes) and are therefore not affected by larger or extensive frost phenomena in their summit region. Only very occasionally one might observe on tepui summits above 2500 m very thin, overnight ice formation in shallow ponds during the colder dry season (November to March or April) (Huber, 1995a).

The substrate found on most tepui summits is predominantly bare rock with numerous depressions temporarily filled with shallow waters. The rock type consists almost exclusively of sedimentary strata of fine- to medium-grain sandstone belonging to the Roraima group (Briceño and Schubert, 1990). Since the thin sandy soil layer deriving from the erosion of this rock type is permanently washed away, the high-tepui plant communities (as well as those surviving on the steep walls) belong evidently to a veritable tropical *peinobiome* with strong environmental and nutrient limitations. Perhaps for that reason, nearly all autochthonous plants of Pantepui have either thick, leathery, and strong leaves or small, stiff leaflets as, for example, in the common shrub or treelet *Bonnetia roraimae* (Bonnetiaceae). Besides the prevailing sedimentary sandstone geology, other geological formations occur in the tepui landscape of the Guiana region, although to a lesser degree: most important among these are Paleo- and Mesozoic intrusive rocks, mainly diabases and granites, found in many places of the Guiana region. Because they are lower in silica and contain several minerals, including calcium, such diabase intrusions favor a better nutrient supply to the surrounding biotopes than the virtually sterile sandstone substrates (Huber, 1995a).

Main plant habitats

Whereas the principal vegetation type growing at the basal slopes of the Guiana tepuis is univariably represented by dense submontane to lower montane tropical forests of different floristic and physiognomic characteristics, the vegetation of the upper tepui sections (c. 1200/1500–3000 m elevation) consists of two fundamentally different habitat types: the mainly vertical walls of rock and the more or less flat summits. The rock face habitats are colonized almost exclusively by saxicolous plants belonging mainly to epilithic lichens, mosses, and low phanerogamic epiphytes. Only in deeper crevasses or cracks found occasionally in the steep walls, can small or low ligneous plants manage to thrive out a meager life there. Because of the very difficult accessibility of these rock faces, general floristic inventories have not yet been produced. Tepui-summit habitats are quite variable depending on several environmental factors, such as elevation, inclination, or physiography: their local topography varies from nearly flat surfaces to little or highly broken rocky terrain, sometimes crossed by deep and narrow rock fissures, but more often consisting of chaotic piles of large boulders and rock fragments.

Depending essentially on the size of the summit surface, many of these plateaus are covered by a great variety of vegetation types. The main components of Pantepui-summit vegetation vary from tiny lichen crusts sticking on the rock surface to extensive *Stegolepis* or bromeliad fields on peat, and up to 15-m tall cloud forests with gnarled trees and a dense, almost impenetrable understory. In several tepuis, however, the summit region is connected to the lower base of the mountain by more or less steep ledges or narrow slope habitats; frequently one finds there taller plant communities, ranging from a few low epipetric cushions to small, chasmophytic shrubbery and, eventually, to gnarled low woods with small crowns (e.g., *Bonnetia roraimae*). Therefore not all tepui summits are totally isolated from their lowland forests or savannas at the base (Huber, 1988). So far, however, a detailed botanical and ecological study of these interesting and unique lowland–upland–highland connections in the Guianan tepui landscape has not been made.

Vegetation types

In this section, a general overview of the principal vegetation types found thus far in the upper tepui zone (submontane and montane, respectively) of Venezuela is presented. These results are mainly based on field observations made in Pantepui by Vareschi (1980, 1992a,b), Maguire (1970, 1979), and Huber (1992, 1995b). A total of 40 vegetation types are described, of which 12 correspond to forest formations, 10 to shrublands, and 18 are herbaceous communities. Each vegetation type is characterized by its elevational range in meters, its physiognomy, and the most characteristic (diagnostic) plant components. The geographical reference is Fig. 1.2 from Chapter 1. The taxonomy follows *The Plant List*, Version 1.1, 2013 (http://www.theplantlist.org/).

Forest formation (types 1–12)

1. Submontane ombrophilous evergreen forests in Sierra de Lema, starting at approximately 1400 m (Plate 7.1A). Dense montane forests cover the slopes from Cerro Venamo in the northeastern Sierra de Lema westwards to the tepui chain Ptari-tepui to Aparamán-tepui, up to 1600 m. Important trees: *Dimorphandra macrostachya* (Leguminosae), *Byrsonima stipulacea* (Malpighiaceae), *Platycarpum rugosum* (Rubiaceae), *Sterigmapetalum guianense* (Rhizophoraceae), and *Moronobea ptaritepuiana* (Clusiaceae).
2. Montane cloud forests of the northern Gran Sabana (Plate 7.1B). Between 1600 and 2000 m, in the true cloud forest zone on tepui slopes until reaching the base of the rock walls of the tepuis (Carrao-tepui, Chimantá massif, Auyán-tepui). Important trees: *Bonnetia tepuiensis*, *B. roraimae* (Bonnetiaceae), *Magnolia ptaritepuiana* (Magnoliaceae), and *Podocarpus* sp. (Podocarpaceae).
3. Montane low forests on high-tepui plateaus (Auyán-tepui, Chimantá massif) (Plate 7.1C). Between 2000 and 2600 m. Low (5–10 m) and dense tepui forests growing on extensive, peat-covered tepui summit plateaus, frequently in depressions or along rivers. Dominant low trees: *Bonnetia tepuiensis*, *B. roraimae*, *B. wurdackii* (Bonnetiaceae), and *Stenopadus chimantensis* (Asteraceae).
4. Montane forests on high-tepui summits (Jaua-Sarisariñama). At 1800–2200 m, on high tepui plateaus: low to medium-high, dense, humid forests with dense understory. Dominant low trees: *Sloanea* sp. (Elaeocarpaceae), *Matayba* sp. (Sapindaceae), *Vochysia* sp. (Vochysiaceae), *Podocarpus* sp. (Podocarpaceae), *Perissocarpa* sp. (Ochnaceae), *Bonnetia jauaensis* (Bonnetiaceae), *Ilex retusa* (Aquifoliaceae), and *Palicourea jauaensis* (Rubiaceae). Understory grass: bambusoid *Neurolepis glomerata* (Poaceae).
5. Montane low forests on high mountains (Sierra de Maigualida). At 1800–2100 m, covered with mosses and epiphytes, on uppermost slopes near high mountain

PLATE 7.1 Forest and shrubland formations. (A) Broadleaved forest on slopes of the Chimantá massif. (B) Tepuian forest dominated by *Bonnetia roraimae* (Chimantá massif). (C) Riparian forest with *B. roraimae* (Chimantá massif). (D) Extensive colonies of *Chimantaea mirabilis* (Chimantá massif). Source: *Photos: O. Huber*.

summit. Dominant low trees: *Cyrilla racemiflora* (Cyrillaceae), *Bonnetia jauaensis* (Bonnetiaceae), *Ilex retusa* (Aquifoliaceae), and *Perissocarpa umbellifera* (Ochnaceae).

6. Upper montane low forests on high tepuis (Yaví–Yutajé massif). At 1500–2200 m, dense, wet, evergreen upper montane cloud forests on slopes near high mountain summit. Dominant low trees: *Schefflera hitchcockii* (Araliaceae), *Clusia* sp. (Clusiaceae), *Weinmannia* sp. (Cunoniaceae), *Cyrilla racemiflora* (Cyrillaceae), and *Euterpe* sp. (Arecaceae).
7. Montane cloud forests on high tepuis (Cuao-Sipapo massif). At 1600–1800 m, montane (cloud) forests on slopes near high mountain summit. Dominant low trees: *Graffenrieda fantastica* (Melastomataceae), *Clusia* sp. (Clusiaceae), *Phyllanthus* sp. (Phyllanthaceae), and *Spathelia ulei* (Rutaceae); with dense understory of *Brocchinia tatei* (Bromeliaceae) and *Neurolepis* sp. (Poaceae).
8. Submontane to montane low forests on high tepuis (Parú–Euaja massif). At 1200–1800 m, in tepui uplands/highlands: forest patches in protected areas. Dominant trees: *Richeria grandis* (Phyllanthaceae), *Clusia* sp. (Clusiaceae), *Simarouba amara* (Simaroubaceae), *Podocarpus tepuiensis* (Podocarpaceae), *Diacidia* spp. (Malpighiaceae), and *Euterpe* sp. (Arecaceae). Occasionally, large clumps of two species of *Saxofridericia* (Rapateaceae) in the understory.
9. Montane low tepui forests on high-tepui summits (Duida–Marahuaka massif). At 1500–2200 m, in tepui highlands: extensive open and low forests on peat substrate. Dominant trees: *Tyleria floribunda*, *T. spathulata* (Ochnaceae), *Neotatea longifolia* (Calophyllaceae), *Gongylolepis* sp. (Asteraceae), and *Schefflera* sp. (Araliaceae).
10. Upper montane forest on high tepuis (Marahuaka). At 2600 m, in high tepui summit region, small tree islands in depressions. Dominant trees: *Podocarpus roraimae* (Podocarpaceae), *Schefflera umbellata* (Araliaceae), *Daphnopsis steyermarkii* (Thymelaeaceae), *Psychotria jauaensis* (Rubiaceae), *Befaria sprucei* (Ericaceae), and *Weinmannia velutina* (Cunoniaceae). In the dense forest floor herbs of *Saxofridericia duidae* and *Stegolepis grandis* (Rapateaceae), *Everardia* sp. (Cyperaceae), and *Brocchinia* sp. (Bromeliaceae).
11. Montane low forests on tepui highlands (Arakamuni). At 1500 m, in tepui uplands/highlands. Low forest patches. Dominant trees and shrubs: *Neblinanthera cumbrensis* (Melastomataceae), *Aegiphila roraimensis* (Lamiaceae), *Clusia* spp. (Clusiaceae), *Diacidia glaucifolia* (Malpighiaceae), *Phyllanthus vacciniifolius* (Phyllanthaceae), *Psychotria duricoria*, *P. jauaensis* (Rubiaceae), and *Tyleria silvana* (Ochnaceae).
12. Upper montane low forests on high tepuis (Neblina). At 1700–2000 m, in tepui highlands. Low forests (“woodlands”). Dominant trees and shrubs: *Bonnetia neblinae*, *B. rubicunda* (Bonnetiaceae), *Neotatea neblinae* (Calophyllaceae), and *Tyleria* spp. (Ochnaceae).

Shrubland formation (types 13–22)

13. Montane scrub on tepui highlands (Auyán-tepui). At 1600–2400 m, colonies of variable size on deep organic soils of tepui summit plateau. Dominant shrubs: *Bonnetia* spp. (Bonnetiaceae), *Notopora auyantepuiensis* (Ericaceae), *Maguireothamnus speciosus*, *Pagameopsis maguirei* (Rubiaceae), and *Tepuianthus auyantepuiensis* (Thymelaeaceae).

14. Montane paramoid shrublands on tepui highlands (Chimantá massif) (Plate 7.1D). At 1900–2600 m, tepui summit plateau, on peat; extensive, 3- to 5-m tall colonies. Dominant shrubs: *Chimantaea mirabilis, C. humilis* (Asteraceae); with *Mallophyton chimantense* (Melastomataceae), *Ledothamnus luteus* (Ericaceae), *Aphanocarpus steyermarkii* (Rubiaceae). Predominant in the herbaceous layer: *Myriocladus steyermarkii* (Poaceae).

15. Montane scrub on tepui highlands (Chimantá massif). At 1900–2600 m, tepui summit plateau. Low dense scrub on peat in broken rocky terrain. Dominant shrubs: *Bonnetia multinervia (fasciculata)* (Bonnetiaceae), *Notopora* sp. (Ericaceae), *Maguireothamnus* sp., *Pagameopsis maguirei* (Rubiaceae), and *Tepuianthus auyantepuiensis* (Thymelaeaceae).

16. Montane scrub on tepui highlands (Jaua-Sarisariñama massif). At 1800–2400 m, southern summit of Cerro Jaua. Low dense scrub on rock and on peat. Dominant shrubs: *Bonnetia jauaensis* (Bonnetiaceae), *Gongylolepis jauaensis* and *G. pedunculata, Stenopadus jauaensis* (Asteraceae), and *Tyleria breweri* (Ochnaceae).

17. Montane scrub on tepui uplands (Guanay). Between 1600 and 2500 m, on summit of Cerro Guanay: very dense, sclerophyllous scrub. Dominant shrubs: *Bonnetia crassa* (Bonnetiaceae), *Graffenrieda* (Melastomataceae), *Ilex* (Aquifoliaceae), *Schefflera* (Araliaceae), and *Stenopadus* and *Gongylolepis* (Asteraceae).

18. Montane scrub on tepui highlands (Yutajé–Coro Coro massif). At approximately 2000–2200 m, on summits of the Yutajé–Coro Coro massif: colonies of stem-rosette shrubs. Dominant shrubs: *Gongylolepis jauaensis* (Asteraceae) and patches of *Bonnetia kathleenae* and *B. celiae* (Bonnetiaceae).

19. Montane scrub on tepui uplands (Duida–Marahuaka massif). At approximately 1500 m, in the south-central section of the Duida plateau, 3–6 m tall, very dense, extensive upland scrub. Dominant shrubs: *Tyleria grandiflora* (Ochnaceae), *Neotatea longifolia* (Calophyllaceae), and *Clusia* spp. (Clusiaceae).

20. Montane scrub on tepui highlands (Duida–Marahuaka massif). At approximately 2200 m, in the southern section of the Duida plateau, up to 3 m tall, extremely dense high-tepui scrub. Dominant shrubs: *Bonnetia crassa, B. tristyla* (Bonnetiaceae), *Neotatea duidae* (Calophyllaceae), *Tyleria linearis* (Ochnaceae), *Blepharandra hypoleuca* (Malpighiaceae), *Duidania montana* (Rubiaceae), *Gongylolepis* spp., *Duidaea* spp., and *Stenopadus* sp. (Asteraceae).

21. Montane scrub on tepui highlands (Huachamacari). At approximately 1800 m, in the western section of Cerro Huachamacari summit, up to 3 m tall, dense high-tepui shrublands. Dominant shrubs: *Bonnetia crassa* (Bonnetiaceae), *Tyleria spathulata* (Ochnaceae), and *Gongylolepis* spp. (Asteraceae).

22. Montane scrub on tepui highlands (Neblina–Avispa–Aracamuni). Between 1600 and 2500 m, on Cerro de la Neblina and adjacent Cerros Avispa and Arakamuni. Dominant shrubs: *Bonnetia maguireorum, B. neblinae* (Bonnetiaceae), and *Rutaneblina pusilla* (Rutaceae).

Herbaceous formation (types 23–40)

In the Guiana Highlands, there are two main types of communities dominated by herbaceous plants. One is the typical grassland dominated by Poaceae species. This is by far the most common herbaceous vegetation type in the lowlands and uplands of the world's tropics, ranging from extensive treeless fields to grasslands, sometimes called savannas, with scattered shrubs, treelets, large trees, or even palms. In the Guiana Highlands, however, this type of vegetation is in the minority, and the more common herbaceous communities are dominated by one or more Rapateaceae (usually broad-leaved species of the genus *Stegolepis*) that confer to these communities a very characteristic physiognomy different from the typical grasslands. These broad-leaved meadows are considered the most characteristic and autochthonous elements in the Pantepui landscape.

Grasslands

23. Upland grasslands (Gran Sabana) (Plate 7.2A). Between 900 and 1400 m, on sandy soils: *Axonopus pruinosus, Trachypogon plumosus, Leptocoryphium lanatum* (Poaceae), *Bulbostylis paradoxa, Rhynchospora globosa*, and *Scleria cyperina* (Cyperaceae).

24. High-tepui grasslands (Auyán-tepui, Chimantá, Sierra de Maigualida, Cerro Marahuaka), at 1600–2650 m, often on water-saturated soils (Plate 7.2B). Dominant grasses and herbs: *Cortaderia roraimensis* (Poaceae), *Rhynchocladium steyermarkii, R. rigidifolia* (Cyperaceae), and occasionally *Orectanthe sceptrum* (Xyridaceae).

Meadows

25. Upland meadows (Gran Sabana). From 950 to 1500 m, extensive broad-leaved meadows on peat. Dominant herbs and low shrubs: *Stegolepis ptaritepuiensis, S. angustata* (Rapateaceae), *Xyris* spp., *Abolboda macrostachya* (Xyridaceae), and *Chalepophyllum guianense* (Rubiaceae).

26. High-tepui meadows (Eastern tepui chain: Ilú to Roraima). Between 2400 and 2750 m, patches of herbaceous and low, subshrubby vegetation on rock are found in depressions and crevasses of the wind-swept summits. Dominant grasses and herbs: patches of rosette meadows with *Orectanthe sceptrum* (Xyridaceae), *Connellia* and *Brocchinia* spp. (Bromeliaceae); small colonies of Eriocaulaceae (*Rondonanthus* spp. and *Syngonanthus* spp.) and Rapateaceae (*Stegolepis guianensis*).

27. High-tepui meadows (Sierra de Lema, Los Testigos and Ptari-tepui group). Around 2400 m, meadows of small rosette plants together with scattered shrubs, mostly on open rock. Dominant herbs and low shrubs: patches of rosette meadows with *Orectanthe sceptrum* (Xyridaceae), *Connellia* and *Brocchinia* spp. (Bromeliaceae); on some plateaus small colonies of *Stegolepis humilis* (Rapateaceae), together with shrubs of *Bonnetia* spp., and a small form of *Chimantaea lanocaulis* (Asteraceae).

28. High-tepui meadows (Auyán-tepui) (Plate 7.2C). Between 1600 and 2400 m, very species-rich, broad-leaved meadows on peat growing on extensive, rocky summit plains. Dominant herbs and low shrubs: *Stegolepis humilis* (Rapateaceae) and *Cladium costatum* (Cyperaceae) and many other species of Cyperaceae, Xyridaceae, and Eriocaulaceae form the herbaceous ground layer; colonies of the large ground rosettes of *Orectanthe sceptrum* (Xyridaceae) are also present, together with shrubs mainly belonging to Rubiaceae, Ericaceae, and Melastomataceae.

29. High-tepui meadows (Chimantá massif). Between 1600 and 2600 m, very species-rich, broad-leaved meadows on peat growing in extensive, rocky summit plains (Plate 7.2D). Dominant and endemic herbs and low shrubs: *Stegolepis ligulata* (Rapateaceae; one of the rare narrow-leaved forms of *Stegolepis*; dominant), *Myriocladus steyermarkii*

PLATE 7.2 Herbaceous formation. (A) Upland grassland in the Gran Sabana (1400 m). Tramén-Ilú massif in the background. (B) High-tepui grassland on Chimantá massif. (C) High-tepui grassland with *Cortaderia roraimensis* on Sierra Maigualida. (D) High-tepui meadow of *Stegolepis ligulata* mixed with low scrub (Chimantá massif). Source: *Photos: O. Huber*.

(Poaceae/Bambusoideae), *Syngonanthus acopanensis* (Eriocaulaceae), *Lindmannia subsimplex* (Bromeliaceae), *Aphanocarpus* sp. (Rubiaceae), *Adenanthe bicarpellata* (Ochnaceae), *Mycerinus chimantensis* (Ericaceae), and many others.

30. Upland meadows (Cerro Guaiquinima). Between 1200 and 1600 m, species-rich, broad-leaved meadows on peat. Dominant herbs and low shrubs: *Stegolepis squarrosa* (Rapateaceae), *Navia ovoidea, Brocchinia acuminata* (Bromeliaceae), *Bonnetia lanceifolia* (Bonnetiaceae), and *Terminalia quintalata* (Combretaceae).

31. High-tepui meadows (Jaua-Sarisariñama–Guanacoco). Between 1800 and 2250 m, species-rich meadows on peat. Dominant herbs and low shrubs: *Brocchinia hechtioides* (Bromeliaceae), *Everardia montana* (Cyperaceae), *Stegolepis* spp. (Rapateaceae), *Orectanthe sceptrum*, (Xyridaceae); *Raveniopsis jauaensis* (Rutaceae), *Bonnetia jauaensis* (Bonnetiaceae), *Thibaudia carrenoi* (Ericaceae), *Chorisepalum ovatum* (Gentianaceae).

32. Meseta de Icutú (2300 m) and Cerro Yudi (N. Maigualida). Not explored botanically.

33. Highland meadows (Sierra de Maigualida). Above 2000 m, upper slopes, on open granitic rock surfaces. Dominant herbs and low shrubs: *Kunhardtia rhodantha* (Rapateaceae), *Brocchinia melanacra* (Bromeliaceae), *Anthurium* sp. (Araceae), several Eriocaulaceae, Cyperaceae, and Xyridaceae species. Low subshrubs are *Tibouchina huberi* and *Leandra gorzulae* (Melastomataceae). Little explored botanically.

34. Pioneer vegetation (Cerro Yaví). At approximately 2400 m, broken rocky tepui summit plateau. Saxicolous vegetation: locally dense mats of low sedges *Cephalocarpus, Everardia,* and *Rhynchospora* (Cyperaceae) on moist sites and large colonies of *Pleurostima celiae* (Velloziaceae) on flat rock surfaces.

35. Highland meadows (Yutajé–Coro Coro massif). Between 1850 and 2400 m, small extensions of floodplains in valley bottoms. Broad-leaved meadows predominantly with *Kundhardtia rhodantha* (Rapateaceae), *Myriocladus* sp. (Poaceae), *Everardia* sp. and *Lagenocarpus* sp. (Cyperaceae), *Brocchinia melanacra* (Bromeliaceae), and *Panicum chnoodes* (Poaceae), as well as other species of Eriocaulaceae, Xyridaceae, Droseraceae, and ferns.

36. Upland and highland meadows (Cuao-Sipapo massif). Between 1300 and 2000 m, mostly on rocks. Large extensions of broad-leaved meadows dominated by *Kundhardtia rhodantha* (Rapateaceae). Little explored botanically.

37. Upland meadows (Parú–Euaja massif). Between 1000 and 1300 m, species-rich, broad-leaved meadows on peat. Dominant herbs and low shrubs: *Saxofridericia grandis* (Rapateaceae), *Xyris* spp. and *Abolboda* sp. (Xyridaceae), *Bonnetia crassa* (Bonnetiaceae), *Celianella montana* (Phyllanthaceae), and an interesting dwarf *Caraipa* sp. (Clusiaceae).

38. High-tepui meadows (Cerro Parú). Between 1300 and 2000 m, open meadows on rocks. Dominant herbs and low shrubs: *Phelpsiella ptericaulis, Stegolepis grandis* (Rapateaceae), and *Xyris* spp. (Xyridaceae). Very little explored botanically.

39. High-tepui meadows (Duida–Marahuaka–Huachamakari massif) (Plate 7.3A). Between 1500 and 2400 m on Cerro Duida, with irregular distribution; dominant herbs and low shrubs: *Brocchinia hechtioides* (Bromeliaceae), *Amphiphyllum rigidum* (Rapateaceae), and *Heliamphora tatei* (Sarraceniaceae). Between 2500 and 2750 m on Cerro Marahuaka, dense meadows of *Marahuacaea schomburgkii* and *Stegolepis terramarensis* (Rapateaceae), *Steyerbromelia discolor*, and *Brewcaria marahuacae* (terrestrial bromeliads) growing on extensive peat soils of the summit plateau.

PLATE 7.3 Herbaceous formation and pioneer vegetation. (A) Broadleaved *Marahuacaea schomburgkii* meadow on Cerro Marahuaka (2800 m). (B) Terrestrial bromeliad colonies in high-tepui swamp (Chimantá massif). (C) *Stegolepis parvipetala* on rock outcrops of the Chimantá massif. (D) Tubiform *Brocchinia* and white-flowering *Lindmannia* on highland Chimantá rocks. Source: *Photos: O. Huber*.

40. Upland and high-tepui meadows (Arakamuni, Neblina). At approximately 1500 m (Arakamuni) and up to 2200 m on Neblina, meadows on peaty soils. Dominant herbs and low shrubs: *Stegolepis* spp. (Rapateaceae), Eriocaulaceae, Xyridaceae spp.; on Neblina *Stegolepis neblinensis* (Rapateaceae), *Navia aloifolia* (Bromeliaceae) and *Heliamphora tatei* subsp. *neblinae* (Sarraceniaceae).

Pioneer formations

Virtually all summits of the Pantepui table mountains offer on their large extensions of open, flat, or level rock surfaces, ideal conditions for the colonization of these areas by a wide selection of lithophytic algae, lichens, and bryophytes. Considering the high amount of annual rainfall (estimated at more than 3000 mm/year) and the low quantity of sunshine hours due to the prevailing damp conditions between the morning and the evening hours, the permanently humid air acts as an effective environmental stimulator for the colonizing agents. Volkmar Vareschi, one of the early botanical tepui explorers, used to insist that "most probably less than one percent of tepui summits are not covered or colonized by cryptogamic crusts or films" (pers. comm. to O. Huber). The first studies on these cryptogam groups were carried out by Ahti (1986, 1987) on the Chimantá massif, and a recent paper by Désamoré et al. (2010) summarizes the degree of diversity and endemism among the Pantepui bryophytes. More information on these groups is provided in Chapter 6. The high level of cryptogam diversification on the nutrient-poor Pantepui sandstones/quartzites is remarkable. Besides these cryptogams, some phanerogams also play a significant role in the initial colonization of bare rock by accumulating organic matter and retaining particulate materials transported by wind and drainage waters. In this way, organic soils progressively accumulate and the rocky substrate is covered by dense vegetation stands. Typical examples of these colonizing plants are several species of Xyridceae, Sarraceniaceae, and Bromeliaceae (Plate 7.3B–D).

Endemic vegetation types in Pantepui

Based on the present state of knowledge concerning the botany of the Guiana Highlands and its regional distribution, some of the larger tepui massifs (e.g., Neblina, Duida, Jaua-Sarisariñama, Chimantá) harbor exceptionally rich plant communities with many endemisms, whereas others of similar size (e.g., Cuao-Sipapo or Guaiquinima) seem to be less diversified floristically; to a certain degree, such unequal distribution patterns can be observed even within some larger high-tepui massifs. On the other hand, the plant life found on smaller tepuis, which often are characterized by mostly bare rock summits, is correspondingly reduced in cover and also less diversified, but often with a relatively higher degree of local endemic taxa. This observation, however, may be due to several factors not yet well understood, such as local differences in microclimatic and/or microedaphic conditions, or in different intrinsic evolutionary driving forces among taxa belonging to different orders or families (see Chapter 4: Origin and evolution of the Pantepui biota).

Probably the most important Pantepui flagship plant families are the Bonnetiaceae in the woody ecosystems and the Rapateaceae in the herbaceous ecosystems. Both families have developed truly autochthonous and noticeable plant communities, such as the widespread *Bonnetia* shrublands alternating with extensive *Stegolepis* herb fields or meadows. The former are growing preferably on deeper, mostly organic, marshy soils with peat, which are found quite frequently in tepui high plains and slightly concave slopes. Herbaceous fields, on the other hand, often cover extensive areas on larger tepui summits with rather dense, up to 1-m-tall meadows, dominated by several species of *Stegolepis* and other rapateaceous genera; several of these may even be endemic to just one mountain.

In addition to these two most representative vegetation types in Pantepui (i.e., the *Bonnetia* scrub and the *Stegolepis* meadows) numerous other geographically or ecologically more restricted members of this peculiar Pantepui flora have been discovered during the past decades. As a result of the increase in helicopter expeditions to the Venezuelan tepuis during the past 30 years, most tepui massifs have now been visited repeatedly, adding new data and images from these hitherto unknown tepui landscapes. Other spectacular but more restricted plant communities discovered recently on remote Venezuelan tepui summits belong to the following plant families.

Asteraceae

Chimantaea colonies

Surely one of the most extraordinary and unparalleled plant communities found in Pantepui at approximately 2400 m on top of the huge Chimantá massif are the extensive *Chimantaea mirabilis* shrublands. These spectacular plants of the Asteraceae family standing together by the millions, forming an almost impenetrable thicket 1–2 m tall and growing on water-saturated peat. *C. mirabilis* (the name was given by Julian A. Steyermark, the botanical discoverer of this unique natural monument) is an endemic species growing exclusively on a few tepuis of the extensive Chimantá massif, which definitely turned out to be one of the top places for the Guiana highland botany and a keystone place for conservation studies (see Chapter 17: Pantepui and global warming). The growth form and habit of *C. mirabilis* are that of a classical paramoid stem rosette with a generally unbranched, single stem partly covered by hundreds of folded, needle-like and stiff leaves; the showy, bright yellow inflorescence sits on or near the apex of the stem and is actively visited by hummingbirds. Although seven more species of *Chimantaea* have been discovered since the last century, *C. mirabilis* and the up-to-8-m-tall, broad-leaved *C. lanocaulis* are among the most spectacular and at the same time graceful and impressive plants of the rich and always surprising Guiana Highlands flora.

Bromeliaceae

Brewcaria colonies

Brewcaria duidensis: terrestrial, acaulescent bromeliads, 0.5–3 m tall in flower, dominating in dense and extensive colonies in swampy open terrains on the summit of Cerro Duida at 1500 m.

Brewcaria marahuacae: terrestrial, acaulescent bromeliad, 0.5 m tall in flower, common to locally dominant on the open rocky tepui summit of Cerro Marahuaka at 2500–2600 m.

Brewcaria hechtioides: terrestrial, acaulescent bromeliad, 0.5–3 m tall in flower, dominating in dense and extensive colonies in montane shrub savannas on the summits of Cerro Autana at about 1200 m and on the summit of nearby Cerro Sipapo at 1500 m.

Brocchinia colonies

B. hechtioides: this ground bromeliad occurs frequently in Guianan uplands and highlands; it is an erect, tubular plant with stiff and glaucous leaves; the paniculate inflorescence reaches up to 2 m or more and has tiny whitish-greenish flowers. It is found preferably in swampy tepui meadows, but also on rock outcrops between 800 and 2500 m.

Sarraceniaceae

Heliamphora colonies

Heliamphora nutans: this unique Neotropical genus of a family mainly distributed in North America now has 11 species in the Venezuelan Guiana Highlands; these Guianan pitcher plant fields are usually found in open areas on tepui slopes and summits at elevations between 1000 and 2950 m. They certainly represent one of the most curious and at the same time most graceful plant communities in the amazingly rich Neotropical plant kingdom.

Phytosociological studies

Phytosociological studies on Pantepui are still in its infancy. Actually, there is only one survey of this nature, developed on the Roraima summit (Safont et al., 2016). This study used the relevé method (Kent, 2012) and defined a series of vegetation plots that were accurately described for species composition and cover using the Wilmanns (1989) scale. In this way, five groups were obtained representing herbaceous (3), shrubland (1), and forest (1) communities. The diagnostic species for herbaceous communities were the sedges *Rhynchospora roraimae* and *Everardia angusta* (Group 1), *E. angusta* (Group 2), and *Cladium costatum* (Group 3). Shrublands were represented by Group 4, dominated by *Bonnetia roraimae* (Bonnettiaceae) and forests corresponding to Group 5, with *Stenopadus connellii* (Asteraceae) and *S. umbellata* (Araliaceae) as the diagnostic species. Safont et al. (2016) also studied the main environmental parameters influencing the different plant community types. Herbaceous communities were associated with flat and frequently flooded terrains, with sandy or peaty soils, whereas shrublands and forests grew on low-pH organic soils on slopes and more irregular topographies.

Final remarks

Numerous other interesting plant communities are surely to be found on many other poorly explored summits of Pantepui. Continuing the scientific exploration of the Guiana Highlands is critical not only to document the impressive diversity and endemicity of their flora and vegetation but also to protect them against the eventual direct and indirect consequences of anthropogenic activities (Rull et al., 2016; Rull and Vegas-Vilarrúbia, 2017; see also Chapters 16: Conservation of Pantepui and 17: Pantepui and global warming). Otherwise, it is possible that a number of the already known plant communities may become extinct in the near future and, eventually, some still-unknown vegetation types will disappear before being discovered. The use of modern quantitative and semi-quantitative methods of vegetation study should be encouraged and the incorporation of remote sensing techniques could be fundamental due to the remoteness and difficulty in accessing tepuian summits.

Acknowledgments

The authors are grateful to Paul E. Berry for his critical review that contributed to improving the manuscript.

References

Ahti, T., 1986. New species and nomenclatural combinations in the lichen genus *Cladonia*. Ann. Bot. Fennici. 23, 205–220.

Ahti, T., 1987. Endemism among *Cladoniaceae* in the table mountains of the Guayana Highland, Venezuela. Progress and problems in lichenology in the eighties. Biblioth. Lichenol. 25, 419–420.

Briceño, H.O., Schubert, C., 1990. Geomorphology of the Gran Sabana, Guayana Shield, southeastern Venezuela. Geomorphology 3, 125–141.

Désamoré, A., Vanderpoorten, A., Laenen, B., Gradstein, S.R., Kok, P.J.R., 2010. Biogeography of the Lost World (Pantepui region, norheastern South America): insights from bryophytes. Phytotaxa 9, 254–265.

Hokche, O., Berry, P.E., Huber, O., 2008. Nuevo catálogo de la flora vascular de Venezuela. Fundación Instituto Botánico de Venezuela, Caracas.

Huber, O., 1988. Guayana highlands versus Guayan lowlands, a reappraisal. Taxon 37, 595–614.

Huber, O. (Ed.), 1992. El macizo del Chimantá, un ensayo ecológico tepuyano. Oscar Todtmann Editores, Caracas.

Huber, O., 1995a. Geographical and physical features. In: Berry, P., Holst, B.K., Yaskievych, K. (Eds.), Flora of the Venezuelan Guayana, Introduction, vol. 1. Missouri Botanical Garden Press, St. Louis, pp. 1–61.

Huber, O., 1995b. Vegetation. In: Berry, P.E., Holst, B.K., Yatskievych, K. (Eds.), Flora of the Venezuelan Guayana, Introduction., vol. 1. Missouri Botanical Garden Press, St. Louis, pp. 97–160.

Kent, M., 2012. Vegetation Description and Data Analysis. A Practical Approach. John Wiley & Sons, Chichester.

Maguire, B., 1970. On the flora of the Guayana Highland. Biotropica 2, 85–100.

Maguire, B., 1979. Guayana, region of the roraima sandstone formation. In: Larsen, K., Holm-Nielsen (Eds.), Tropical Botany, vol. 2. Academic Press, London, pp. 223–238.

Mittermeier, R.A., Mittermeier, C.G., Robles, P., 1997. Megadiversity: Earth's Biologically Wealthiest Nations. CEMEX, Mexico.

Rull, V., Vegas-Vilarrúbia, T., 2017. Potential responses of vascular plants from the pristine "Lost World" of the Neotropical Guayana Highlands to Global Warming: review and new perspectives. Front. Plant Sci. 8, 81.

Rull, V., Vegas-Vilarrúbia, T., Safont, E., 2016. The Lost World's pristinity at risk. Div. Distrib. 22, 995–999.

Safont, E., Vegas-Vilarrúbia, T., Rull, V., Holst, B.K., Huber, O., Nozawa, S., et al., 2016. Plant communities and environmental factors in the Guayana Highlands: monitoring for conservation under future climate change. Syst. Biodivers. 14, 327–344.

Steyermark, J.A., Berry, P.E., Holst, B.K., 1995–2005. Flora of the Venezuelan Guayana, vols. 1–9. Missouri Botanical Garden Press, St. Louis.

Vareschi, V., 1980. Vegetationsökologie der Tropen. Verlag Eugen Ulmer, Stuttgart, p. 293.

Vareschi, V., 1992a. Ecología de la vegetación tropical, con especial atención a investigaciones en Venezuela. Sociedad Venezolana de Ciencias Naturales, Caracas.

Vareschi, V., 1992b. Observaciones sobre la dinámica vegetal en el Macizo del Chimantá. In: Huber, O. (Ed.), El Macizo del Chimantá, Escudo de Guayana, Venezuela: un ensayo ecológico tepuyano. O. Todtmann, Caracas, pp. 179–188.
Wilmanns, O., 1989. Okologische Pflanzensoziologie. Quelle & Meyer, Heidelberg.

PART III

Animal Diversity

CHAPTER

8

Aquatic insects

Tomáš Derka[1], *Carmen Zamora-Muñoz*[2] and *José Manuel Tierno de Figueroa*[2]

[1]Department of Ecology, Faculty of Natural Sciences, Comenius University, Bratislava, Slovakia [2]Department of Zoology, Faculty of Sciences, University of Granada, Granada, Spain

Introduction

Aquatic insects are derived from various terrestrial ancestors that have secondarily invaded aquatic environments and therefore do not represent a distinct taxonomical unit within the class Insecta. Some insect orders contain only species that are aquatic in some life stages (e.g., mayflies, stoneflies, dragonflies, caddisflies, megalopterans), whereas other orders contain both aquatic and terrestrial species (e.g., beetles, bugs, butterflies, neuropterans, orthopterans, and dipterans). Therefore, aquatic insects are a very diverse group with around 76,000 species adapted to all types of freshwater habitats (Balian et al., 2008).

The term Pantepui was introduced by Mayr and Phelps (1955, 1967) to designate the assemblage of sandstone table mountains in Venezuelan Guayana, as well as in adjacent Brazil and Guyana. Although the Pantepui concept has been used in various ways, in this chapter, we have restricted it to high-elevation, mesothermic and microthermic life zones above 1500 m a.s.l., highlands or upper montane, and high-tepui zones, according to Huber (1995a,c). Sites within altitudinal limits are located at the summits, slopes, and foothills of tepuis. Pantepui bears numerous and diverse aquatic habitats occupied by various groups of aquatic insects. The stagnant water habitats found on the summits range from temporary pools in bare rocks (Fig. 8.1A) and phytotelmata to large peat swamps with ponds (Fig. 8.1B–F) and stream pools (Figs. 8.1G and H, 8.2B). Running waters range from cold water springs on summits, slopes, or foothills (Fig. 8.2A) to medium-sized rivers (Figs. 8.1G and H, 8.2E–H). Streams with a bedrock bottom (Figs. 8.1G, 8.2A, D, E, and H), lacking a hyporheic environment, and with alternating cascades and pools (Fig. 8.2C–F) are the most common type of running waters. Sand and gravel accumulate in thin layers in pools (Fig. 8.2B), whereas stretches with stronger currents have a bare bedrock bottom (Fig. 8.2E and H). The low water retention capacity of watersheds and

Biodiversity of Pantepui
DOI: https://doi.org/10.1016/B978-0-12-815591-2.00008-2

FIGURE 8.1 Photographs of Pantepui aquatic habitats (A–H). (A) Temporary pools in bare rocks, Roraima-tepui. (B) Small pond, Roraima-tepui. (C and D) Peat swamp meadow with small ponds, Churí-tepui. (E and F) Ponds, Kukenán-tepui. (G) Pool in stream, Kukenán-tepui. (H) Pool in river, Auyán-tepui.

FIGURE 8.2 Photographs of Pantepui aquatic habitats (A–H). (A) Spring stream with bedrock bottom in a montane forest, Churí-tepui. (B) Pool with sandy bottom and woody debris, Churí-tepui; (C) Black (red) water stream with foam accumulation, Cerro Duida. (D) Bedrock bottom river, Akopán-tepui. (E) River with bedrock bottom, changing to connected pools in a dry period, Auyán-tepui. (F) Small cascade with foam accumulations, Cerro Duida. (G) Pool in river, Akopán-tepui. (H) Bedrock stream in dry period, Auyán-tepui.

heavy rainfalls are responsible for harsh discharge and water temperature fluctuations. During low flows, river pools are converted into ponds, whereas riffles are almost dry (Fig. 8.2E and H) and water temperature increases. The water temperature may range from 12°C to 14°C in springs to over 30°C in ponds and pools (Aubrecht et al., 2012). Aquatic fauna of spring streams are composed mostly of cold stenotherms. On the other hand, fauna of larger streams, swamp ponds, and temporary pools with stagnant water are assumed to be more vagile and adapted to high diurnal thermal fluctuations. The lack of a hyporheic environment limits the refugia for aquatic fauna during low flows and violent floods. The streams are extremely oligotrophic, low in minerals (conductivity usually ranges from 10 to 25 μS/cm), acidic (pH 3.5–5.5), and may contain high concentrations of organic compounds leached from vegetation (Aubrecht et al., 2012). The organic compounds often color the water red (Figs. 8.1H, 8.2B, C, and G) or black and cause foam to coat the surface (Fig. 8.2C, D, and F). Due to the harsh conditions, the aquatic fauna found on the summits of higher tepuis, such as Roraima-tepui, are poor, whereas lower and larger tepuis show higher taxonomical richness (Aubrecht et al., 2012).

The first collections of aquatic insects in Pantepui can be dated to the autumn of 1894, when F.V. McConnell and J.J. Quelch first visited the Roraima-tepui and collected the flora and fauna of the summit. One year later, Waterhouse (1895) described from material collected a new dytiscid, *Rhantus elegans* Waterhouse, 1895, and mentioned damaged fragments of Odonata belonging to Libellulidae and Aeschnidae. In 1927 the American Museum dispatched the Lee Garnett Day Expedition to Roraima-tepui, and the following year, in continuation of the same biological project, the Tyler Duida Expedition was sent to Cerro Duida (Chapman, 1931). As a result of these expeditions, new species from the Roraima-tepui and Cerro Duida have been described (Needham, 1933; Needham and Fischer, 1940; Klots, 1944; Jewett, 1960; Stark, 2011). The most important contributions to the study of aquatic insects of Pantepui were the collections of Paul J. Spangler during the multidisciplinary international expeditions to Sierra de la Neblina in the early 1980s (Brewer-Carías, 1988).

However, considering even the most visited and explored tepuis such as Roraima-tepui, Kukenán-tepui, Auyán-tepui, Cerro Duida, or the tepui summits of the Chimantá massif, there is still a surprising lack of knowledge of aquatic insect groups, especially when compared with intensive botanical exploration (Huber, 1995b). For instance, the dragonfly species *Sympetrum roraimae* De Marmels, 1988, common and easily observable in the stagnant water of the Roraima-tepui, was described as much as a century after the first ascent to Roraima-tepui. There is no hydrobiological study on the structure and functioning of Pantepui aquatic ecosystems, except for the study of aquatic insect communities in two *Heliamphora* pitcher-plant species (Barrera et al., 1989). Information is limited to species descriptions and records that sometimes include characteristics of habitats where the species were collected.

The aim of this study is to review the information about Pantepui aquatic insect fauna, with special attention paid to the groups we have studied: mayflies, stoneflies, and caddisflies. These groups have already been previously reviewed by Derka et al. (2010, 2012) and Derka and Zamora-Muñoz (2012). Aubrecht et al. (2012) added to the previous groups semiaquatic crickets from the genus *Hydrolutos* and beetles from the Elmidae family. In this study, we update previously published reviews with studies published since 2012, and we also provide information about our own new unpublished data. We have also attempted to collect information about other aquatic insect groups recorded in Pantepui, but because we are not experts in these groups, this summary cannot be considered comprehensive. Tepui names, main features and location follow Table 1.1 and Figure 1.2 of Chapter 1: Description and characterization of the Pantepui biogeographical province.

Material collected is deposited in the following institutions: Museum of Agricultural Zoology (MIZA), Faculty of Agronomy, Central University of Venezuela, Maracay, Venezuela; Department of Ecology, Faculty of Natural Sciences, Comenius University, Bratislava, Slovakia; Institute Miguel Lillo, Tucumán, Argentina; Cantonal Museum of Zoology, Lausanne, Switzerland; and Zamora-Muñoz Collection (ZMZUG) of the Department of Zoology, Granada University, Spain.

Mayflies (Ephemeroptera)

The order Ephemeroptera is a small, ancient worldwide-distributed group of hemimetabolous insects. The group comprises more than 3000 described extant species in 42 families and more than 400 genera (Barber-James et al., 2008). South American mayfly fauna is composed of 14 families, more than 100 genera, and more than 450 species (Domínguez et al., 2006). Mayflies have aquatic nymphs and two winged stages: subimago and imago. Mayfly nymphs are common in streams on the Churí-tepui, Akopán-tepui, and Auyán-tepui plateaus; however, they are rare on the Kukenán and Roraima plateaus, where aquatic invertebrates are scarce in general (Aubrecht et al., 2012; Derka et al., 2012). Representatives of three families and 14 genera have been recorded in Pantepui (Table 8.1).

The Baetidae family has a near-cosmopolitan distribution and currently includes almost 100 genera worldwide. In South America this family now encompasses 29 genera (Cruz et al., 2013). Representatives of seven genera were found on the summits, and two additional genera were found on slopes and foothills above 1500 m a.s.l. (Table 8.1). Genera *Baetodes*, *Callibaetis*, *Camelobaetidius*, and *Cloeodes* are widespread in the Neotropics (Domínguez et al., 2006). At least three undescribed species from the genus *Cloeodes* were found at the Churí-tepui and Auyán-tepui, as well as in streams in the Roraima-tepui foothills. The material is currently undergoing description. Material from other genera remains undetermined. The following genera have a more restricted distribution. The genus *Callibaetoides* contains one species of *Callibaetoides caaigua*, described from the Brazilian Amazonia and Mata Atlântica (Cruz et al., 2013). Nymphs were found in a small forested brook in Campamento Militar (1500 m a.s.l.) below the Roraima-tepui. The genus *Zelusia* has a similar distribution containing two morphologically similar species (Salles et al., 2016). Both have been recorded in the Gran Sabana region and in adjacent Brazil (Nieto et al., 2011; Salles et al., 2016). Nymphs of *Zelusia principalis* were recently found in a stream in the Roraima-tepui base camp. The genus *Spiritiops* has two species. The nymphs of *Spiritiops silvudus* have been reported from lower localities in Brazil, Colombia, Surinam, French Guiana, and Venezuela (Domínguez et al., 2006; Nieto et al., 2011). *Spiritiops tepuiensis* has been described from streams at the Auyán-tepui, Churí-tepui (Fig. 8.3B), and Roraima-tepui (Nieto and Derka, 2012). Our unpublished molecular

TABLE 8.1 Review of Ephemeroptera taxa from Pantepui.

Taxa	Locality	Endemic	Source
Baetidae			
Baetodes sp.	Auyán-tepui, Roraima-tepui foothills	?	Derka et al. (2012), Aubrecht et al. (2012)
Callibaetis sp.	Auyán-tepui, Churí-tepui	?	Derka et al. (2012), Aubrecht et al. (2012)
Callibaetoides caaigua Cruz, Sales & Hamada, 2013	8 nymphs, Roraima-tepui foothills—Campamento militar, 7.II.2014	No	New record
Camelobaetidius sp.	Auyán-tepui	?	Derka et al. (2012), Aubrecht et al. (2012)
Cloeodes sp.	Auyán-tepui, Churí-tepui, Roraima-tepui foothills	Yes	Derka et al. (2012), Aubrecht et al. (2012)
Parakari auyanensis Nieto & Derka, 2011	Auyán-tepui	Yes	Nieto and Derka (2011)
Parakari churiensis Nieto & Derka, 2011	Churí-tepui	Yes	Nieto and Derka (2011)
Parakari roraimensis Derka, Nieto & Svitok, 2015	Roraima-tepui foothills	Yes	Derka et al. (2015)
	25 nymphs, Acopán-tepui summit, left side tributary of Río Yunek, 16.XI.2015, 1941 m a.s. l., 05,20548°N, 62,04180°W		New record
Spiritiops tepuiensis Nieto & Derka, 2012	Auyán-tepui, Churí-tepui, Roraima-tepui	No	Nieto and Derka, 2012, Derka et al. (2012), Aubrecht et al. (2012)
	20 nymphs, Kukenán-tepui summit, Kako-Parú River, 28.XI.2015, 2626 m a.s.l., 05,1970833°N, 60,8167167°W; 14 nymphs, unnamed stream in Akopán-tepui summit, 14.XI.2015, 2022 m a.s.l., 5,19413″ N, 62,04478°W, 16 nymphs, Río Yunek, Acopán-tepui summit, 15.XI.2015, 2005 m a.s.l., 05,19662″N, 62,04949°W		New records
Zelusia principalis Lugo-Ortiz & McCafferty, 1998	2 nymphs, Roraima-tepui foothills, a stream in Roraima-tepui base camp, 10.II.2015, 1800 m a.s.l.	No	New record
Leptophlebiidae			
Askola emmerichi Domínguez, Molineri & Mariano, 2009	1 nymph, unnamed right side tributary of Kukenán River via Kukenán-tepui base camp, 25.XI.2015, 1609 m a.s.l., 05,1541°N, 60,82611°W	No	New record

(*Continued*)

TABLE 8.1 (Continued)

Taxa	Locality	Endemic	Source
Hagenulopsis minuta Spieth, 1943	Sierra de la Neblina	No	Peters and Domínguez (2001)
	Auyán-tepui, Churí-tepui, Roraima-tepui foothills		Derka et al. (2012), Aubrecht et al. (2012)
	3 nymphs, unnamed right side tributary of Kukenán River via Kukenán-tepui base camp, 25.XI.2015, 1609 m a.s.l., 05,1541°N, 60,82611°W; 15 males, Akopán-tepui summit, left side tributary of Río Yunek, 16.XI.2015, 1941 m a.s.l., 05,20548°N, 62,04180°W		New records
Massartella devani Derka, 2002	Roraima-tepui	Yes	Derka (2002), Derka et al. (2009, 2012), Aubrecht et al. (2012)
Massartella spp.	Auyán-tepui, Churí-tepui	Yes	Derka et al. (2009, 2012), Aubrecht et al. (2012)
Miroculis bicoloratus Savage, 1987	Sierra de la Neblina	Yes	Savage (1987)
Miroculis spp.	Auyán-tepui, Churí-tepui, Roraima-tepui foothills	?	Derka et al. (2012), Aubrecht et al. (2012)
	3 nymphs, unnamed right side tributary of Kukenán River via Kukenán-tepui base camp, 25.XI.2015, 1609 m a.s.l., 05,1541°N, 60,82611°W	?	New record
Paramaka incognita Domínguez, Grillet, Nieto, Molineri & Guerrero, 2014	4 females, 4 nymphs, unnamed right side tributary of Kukenán River via Kukenán-tepui base camp, 25.XI.2015, 1609 m a.s.l., 05,1541°N, 60,82611°W; 1 female subimago, 20 nymphs, Kukenán River below Kukenán-tepui close to base camp, 25.XI.2015, 1631 m a.s.l., 05,15601°N, 60,82507°W; 11 nymphs, Río Yunek, Acopán-tepui summit, 16.XI.2015, 1941 m a.s.l., 05,20548″N, 62,04180°W	No	New records
Oligoneuriidae			
Fittkauneuria adusta Pescador & Edmunds, 1994	Auyán-tepui, Roraima-tepui foothills	No	Derka et al. (2012), Aubrecht et al. (2012)
Fittkauneuria carina Pescador & Edmunds, 1994	Sierra de la Neblina	No	Pescador and Edmunds (1994)

analyses showed that morphologically identical nymphs from three tepuis represent distinct molecular operational taxonomic units or cryptic species confined to particular massifs. Recently, we discovered new sites at Acopán-tepui and Kukenán-tepui (Table 8.1). Moreover, nymphs that we collected from the streams of La Gran Sabana, where *S. silvidus*

FIGURE 8.3 Photographs of Pantepui aquatic insects (A–H). (A) Mayfly female subimago–undescribed *Massartella*, Churí-tepui. (B) Mayfly imago–*Spiritiops tepuiensis*, Churí-tepui. (C) *Hydrolutos gransabanensis* from Cueva del Tigre, Santa Elena de Uairén. (D) Stonefly nymph–putative Enderleina preclara, Kukenán-tepui foothills. (E) Caddisfly caseless larva–*Atopsyche* sp., Auyán-tepui. (F) Caddisfly larva–*Marilia* sp., Auyán-tepui. (G) Dobsonfly larva of *Corydalus* sp., Akopán-tepui. (H) Larva of snail-case caddisfly *Helicopsyche* sp., Churí-tepui.

is more common, showed that the species is not restricted to Pantepui. The unique baetid genus restricted to Pantepui is the *Parakari* (Nieto and Derka, 2011; Derka et al., 2015). Nymphs tolerate a wider range of temperatures, from cold cave streams with stable temperatures around 14°C to streams with diurnal thermal fluctuations. The genus has three species confined to three massifs. However, *Parakari roraimensis,* described from the foothills of Roraima, has been recently recorded at Akopán-tepui (Table 8.1), where *Parakari churiensis* described from the adjacent Churí-tepui would be expected. This indicates that the evolutionary history of the genus includes vicariance in distinct massifs and recent dispersal events among them.

Leptophlebiidae is the most diverse mayfly family in the Neotropics. In South America it includes approximately 40 genera and more than 150 species (Domínguez et al., 2006). Representatives of five genera have been recorded in Pantepui (Table 8.1). The genera *Askola, Hagenulopsis,* and *Miroculis* have extensive distribution, but the highest diversity is observed in the Guiana and Brazilian shields and Amazonia (Domínguez et al., 2006, 2009). *Askola emmerichi* has been reported from the Gran Sabana (Domínguez et al., 2014), where it can reach higher altitudes, since one nymph of *A. emmerichi* was collected in a small stream in the Kukenán-tepui base camp (Table 8.1). The genus *Hagenulopsis* has five described species (Domínguez et al., 2009)—one of them, *Hagenulopsis minuta,* is known from Sierra de la Neblina (Peters and Domínguez, 2001). It was also published as *Hagenulopsis* sp. from the eastern tepuis (Aubrecht et al., 2012; Derka et al., 2012). *Hagenulopsis minuta* was recently found in a stream in the Kukenán-tepui foothills and on the Akopán-tepui summit (Table 8.1). The genus *Miroculis* currently has 23 described species (Boldrini et al., 2018), but only *Miroculis bicoloratus* is known from Pantepui (Savage, 1987). Unidentified nymphs and undescribed adults of this genus were reported from the eastern tepuis (Aubrecht et al., 2012; Derka et al., 2012). The genus *Paramaka* comprises four species restricted to the Guiana and Brazilian shields. *Paramaka incognita* has been described from the Gran Sabana (Domínguez et al., 2014). Nymphs were found in streams that originated at the Auyán-tepui (e.g., below Salto Angel) and recently have been recorded on the Akopán-tepui plateau and Kukenán-tepui foothills (Table 8.1). The genus *Massartella* (Fig. 8.3A) is the only mayfly genus phylogenetically related to notogeic taxa, which is distributed in the Guiana and Brazilian shields. It has five described species, two from Brazil and three from Venezuela (Pescador and Peters, 1990; Derka et al., 2009; Souto et al., 2013). The only described Pantepui species is *Massartella devani* from Roraima-tepui (Derka, 2002). Nevertheless, undescribed species of the genus have been reported from Churí-tepui and Auyán-tepui (Fig. 8.3A) and recently have been confirmed by our unpublished molecular analyses. One *Massartella* nymph was captured in a cold waterfall below Cerro Cuao (Table 8.1), which confirms that the genus distribution reached as far as the Western Pantepui District (Chapter 1). In Pantepui, *Massartella* nymphs inhabit cold spring streams exclusively with temperatures ranging between 14°C and 18°C.

Among the six recent genera of the family Oligoneuriidae occurring in South America (Domínguez et al., 2006), only the genus *Fittkauneuria,* known from two species distributed in Venezuela and Brazil, has been reported from Pantepui. *Fittkauneuria carina* was sampled in various streams at Sierra de la Neblina at altitudes ranging from 750 to 1820 m a.s.l. (Pescador and Edmunds, 1994). The specimens that we recorded at Cerro Duida at 1109 m a.s.l. indicate that this species has a more extensive distribution (Derka, pers. observ.).

Fittkauneuria adusta is common in small streams in the Gran Sabana (Domínguez et al., 2014; Derka, pers. observ.); in the Auyán-tepui massif and Roraima-tepui foothills, it reaches localities above 1500 m a.s.l. (Table 8.1).

Stoneflies (Plecoptera)

Stoneflies (Plecoptera) are a small order of hemimetabolous insects with about 3500 described extant species in 16 families and 286 genera (Fochetti and Tierno de Figueroa, 2008). Known Neotropical stonefly fauna comprise more than 500 valid species in six families (Fochetti and Tierno de Figueroa, 2008; Stark et al., 2009; Froehlich, 2010). Stonefly nymphs live mainly in cold, well-oxygenated running waters. The ecological requirements of nymphs and the reduced flight capacity of adults are the main factors responsible for their high endemicity (Fochetti and Tierno de Figueroa, 2008).

Data on stoneflies regarding tepuis were published from Cerro Duida, Sierra de la Neblina, and the Auyán-tepui and Churí-tepui and Roraima-tepui foothills (Table 8.2). No stoneflies were recorded in the Roraima plateau streams, probably due to a lack of suitable prey. Four stonefly genera, all belonging to the Perlidae family, have been recorded in Pantepui (Derka et al., 2010; Aubrecht et al., 2012). The genus *Anacroneuria*, with 348 described valid species, is the most diverse stonefly genus in the Neotropics, distributed from Argentina to the South of the United States (Froehlich, 2010; DeWalt et al., 2018). *Anacroneuria shamatari* from Sierra de la Neblina is probably the only species of the genus described from Pantepui (Stark, 1995). Specific recognition of *Anacroneuria* nymphs is very difficult or, in many cases, impossible because many nymphs have not been described. Nymphs have been recorded from the Auyán-tepui and Akopán-tepui plateaus (Table 8.2); new records were published from the foothills of Roraima-tepui and Kukenán-tepui (Derka and Tierno de Figueroa, 2016).

The genus *Enderleina* (Fig. 8.3D) consists of six species known from southern Venezuela and northern Brazil (Derka and Tierno de Figueroa, 2013; Hamada et al., 2016) and one species known from northern Venezuela (Stark, 1989). The holotype of *Enderleina preclara* (the type species of the genus) was collected at Roraima-tepui at 2100 m a.s.l. (Jewett, 1960), probably during the Lee Garnett Day Expedition in 1927. *Enderleina flinti* has been described from Sierra de la Neblina on the Venezuelan-Brazilian border (Stark, 1989). Recently, females and nymphs, which likely belong to *E. preclara*, have been collected on the Akopán-tepui plateau and in the Kukenán-tepui foothills (Derka et al., 2019).

The genus *Kempnyia* includes 38 species known only from Brazil (DeWalt et al., 2018). All published records from Pantepui in Venezuela are based on nymphs (Table 8.2). Because identification of the nymphs at a species level is possible only for a limited number of species (Avelino-Capistrano et al., 2014), Pantepui specimens remain unidentified.

The genus *Macrogynoplax* has 15 described species distributed in Brazil, Guyana, Peru, Surinam, Venezuela, and Colombia (DeWalt et al., 2018). Two Pantepui species have been described: *Macrogynoplax neblina* from Sierra de la Neblina (Stark and Zwick, 1989) and *Macrogynoplax duida* from Cerro Duida (Stark, 2011). Most species' descriptions are based only on adults without the associated immature stages (Ribeiro and Gorayeb, 2015). Therefore, the species identification of nymphs reported from eastern Pantepui was not possible (Table 8.2).

TABLE 8.2 Review of Plecoptera taxa from Pantepui.

Taxa	Locality	Endemic	Source
Perlidae			
Anacroneuria shamatari Stark, 1995	Sierra de la Neblina	Yes	Stark (1995)
Anacroneuria sp.	Auyán-tepui	?	Aubrecht et al. (2012)
	2 females, 15 nymphs, left side tributary of Río Yunek via Akopán-tepui, 12. XI.2015, 1168 m a.s.l., 05,20503″ N, 62,00165°W; 2 nymphs, Río Yunek, Akopán-tepui, 16.XI.2015, 1941 m a.s.l., 05,20548″ N, 62,04180°W; 6 nymph, Roraima-tepui base camp, Roraima-tepui foothills, 7.II.2014	?	New records
Enderleina preclara Jewett, 1960	Roraima-tepui slopes	Yes	Jewett (1960)
Putative *Enderleina preclara*	Akopán-tepui, Kukenán-tepui foothills		Derka et al. (2019)
Enderleina flinti Stark, 1989	Sierra de la Neblina	No	Stark (1989)
Kempnyia sp.	Churí-tepui, Auyán-tepui, Roraima-tepui foothills	?	Derka et al. (2010), Aubrecht et al. (2012)
	3 nymphs, Kukenán River below Kukenán-tepui close to base camp, 25. XI.2015, 1631 m a.s.l., 05,15601°N, 60,82507°W; 1 nymph left side tributary of Río Yunek via Akopán-tepui, 12. XI.2015, 1168 m a.s.l., 05,20503″ N, 62,00165°W; 1 nymph, Río Yunek, Akopán-tepui, 16.XI.2015, 1941 m a.s.l., 05,20548″ N, 62,04180°W	?	New records
Macrogynoplax neblina Stark, 1989	Sierra de la Neblina	Yes	Stark (1989)
Macrogynoplax duida Stark, 2011	Cerro Duida	Yes	Stark (2011)
Macrogynoplax sp.	Churí-tepui, Auyán-tepui	?	Derka et al. (2010), Aubrecht et al. (2012)
	1 nymph, Roraima-tepui base camp, Roraima-tepui foothills, 7.II. 2014	?	New record

Caddisflies (Trichoptera)

The order Trichoptera (caddisflies) is a group of holometabolous insects closely related to butterflies (Lepidoptera). It includes approximately 15,000 extant species within 49 families and approximately 600 genera (Holzenthal et al., 2011, 2015). The immature stages are ubiquitous in freshwaters, but are especially diverse in rivers and streams. In the Neotropics, about 3262 valid species in 25 families and 155 extant genera have been currently recognized (Holzenthal and Calor, 2017). Most species descriptions of South American caddisflies are based only on adults without associated immature stages. Therefore the species identification of larvae (Fig. 8.3E, F, and H) is nearly impossible. Records of Trichoptera in Pantepui are based on material collected at Sierra de la Neblina, Roraima-tepui, Auyán-tepui, Ptari-tepui, Churí-tepui, and Cerro Duida (Aubrecht et al., 2012; Derka and Zamora-Muñoz, 2012; Zamora-Muñoz et al., 2013, 2017). Altogether, 30 species belonging to eight genera and six families have been published from Pantepui, and unidentified specimens of Calamoceratidae, Helicopsychidae (Fig. 8.3H), Hydropsychidae, Hydroptilidae, Leptoceridae, Odontoceridae (Fig. 8.3F), Philopotamidae, Polycentropodidae, and Sericostomatidae belonging to 12 genera have been recorded (Table 8.3).

TABLE 8.3 Review of Trichoptera taxa from Pantepui.

Taxa	Locality	Endemic	Source
Calamoceratidae			
Phylloicus sp.	Churí-tepui	?	Aubrecht et al. (2012), Derka and Zamora-Muñoz (2012)
Helicopsychidae			
Helicopsyche succincta Johanson & Holzenthal, 2004	Sierra de la Neblina	Yes	Johanson and Holzenthal (2004)
Helicopsyche laneblina Johanson & Holzenthal, 2004	Sierra de la Neblina	Yes	Johanson and Holzenthal (2004)
Helicopsyche spp.	Auyán-tepui, Churí-tepui, Roraima-tepui	?	Aubrecht et al. (2012), Derka and Zamora-Muñoz (2012)
Hydrobiosidae			
Atopsyche atahuallpa Schmid, 1989	Ptari-tepui, Churí-tepui, Auyán-tepui	Yes	Schmid (1989), Zamora-Muñoz et al. (2017)
Atopsyche ayacucho Schmid, 1989	Sierra de la Neblina	Yes	Schmid (1989)
Atopsyche ayahuaca Schmid, 1989	Sierra de la Neblina	Yes	Schmid (1989)
Atopsyche calahuaya Schmid, 1989	Ptari-tepui	Yes	Schmid (1989)
Atopsyche cristinae Zamora-Muñoz & Derka, 2017	Churí-tepui	Yes	Zamora-Muñoz et al. (2017)
Atopsyche carmenae Zamora-Muñoz & Derka, 2017	Churí-tepui	Yes	Zamora-Muñoz et al. (2017)

(Continued)

TABLE 8.3 (Continued)

Taxa	Locality	Endemic	Source
Atopsyche chimuru Schmid, 1989	Sierra de la Neblina	Yes	Schmid (1989)
Atopsyche chinchacamac Schmid, 1989	Sierra de la Neblina	Yes	Schmid (1989)
Atopsyche hamata Ross & King, 1952	Roraima-tepui	Yes	Ross and King (1952)
Atopsyche huallaripa Schmid, 1989	Sierra de la Neblina	Yes	Schmid (1989)
Atopsyche iana Mosely, 1949	Roraima-tepui	Yes	Ross and King (1952)
Atopsyche inmae Zamora-Muñoz & Derka, 2017	Auyán-tepui, Churí-tepui, Roraima-tepui	Yes	Zamora-Muñoz et al. (2017)
Atopsyche svitoki Zamora-Muñoz & Derka, 2017	Auyán-tepui	Yes	Zamora-Muñoz et al. (2017)
Hydropsychidae			
Blepharopus sp.	Auyán-tepui	?	Aubrecht et al. (2012), Derka and Zamora-Muñoz (2012)
Leptonema neblinense Flint, McAlpine & Ross, 1987	Sierra de la Neblina	Yes	Flint et al. (1987)
Leptonema amazonense Flint, 1978	Sierra de la Neblina	Yes	Flint et al. (1987)
Leptonema guayanense Flint, McAlpine & Ross, 1987	Auyán-tepui	No	Flint et al. (1987)
Leptonema ramosum Flint, McAlpine & Ross, 1987	Cerro-Duida	No	Flint et al. (1987)
Macrostemum erichsoni Banks, 1920	Churí-tepui	No	Aubrecht et al. (2012), Derka and Zamora-Muñoz (2012)
Hydroptilidae			
Orthotrichia sp.	Auyán-tepui	?	Aubrecht et al. (2012), Derka and Zamora-Muñoz (2012)
Oxyethira sp.	Churí-tepui, Roraima-tepui	?	Aubrecht et al. (2012), Derka and Zamora-Muñoz (2012)
Zumatrichia sp.	Auyán-tepui, Churí-tepui	?	Aubrecht et al. (2012), Derka and Zamora-Muñoz (2012)
Leptoceridae			
Notalina roraima Holzenthal, 1986	Roraima-tepui, Churí-tepui, Auyán-tepui	Yes	Holzenthal (1986), Aubrecht et al. (2012), Derka and Zamora-Muñoz (2012), Zamora-Muñoz et al. (2013)

(*Continued*)

TABLE 8.3 (Continued)

Taxa	Locality	Endemic	Source
Triplectides tepui Holzenthal, 1988	Sierra de la Neblina	Yes	Holzenthal (1988)
Triplectides sp.	Auyán-tepui, Churí-tepui	?	Aubrecht et al. (2012), Derka and Zamora-Muñoz (2012)
Oecetis sp.	Auyán-tepui, Churí-tepui	?	Aubrecht et al. (2012), Derka and Zamora-Muñoz (2012)
Nectopsyche sp.	Churí-tepui	?	Aubrecht et al. (2012), Derka and Zamora-Muñoz (2012)
Odontoceridae			
Marilia sp.	Auyán-tepui	?	Aubrecht et al. (2012), Derka and Zamora-Muñoz (2012)
Polycentropodidae			
Polycentropus neblinensis Hamilton & Holzenthal, 2005	Sierra de la Neblina	Yes	Hamilton and Holzenthal (2005)
Polycentropus sp.	Auyán-tepui, Churí-tepui	?	Aubrecht et al. (2012), Derka and Zamora-Muñoz (2012)
Polyplectropus amazonicus Chamorro & Holzental, 2010	Sierra de la Neblina	Yes	Chamorro and Holzenthal (2010)
Polyplectropus flintorum Chamorro and Holzental, 2010	Sierra de la Neblina	Yes	Chamorro and Holzenthal (2010)
Philopotamidae			
Chimarra ensifera Flint, 1998	Sierra de la Neblina	No	Flint (1998)
Chimarra medioloba Flint, 1971	Sierra de la Neblina	No	Flint (1998)
Chimarra neblina Blahnik, 1997	Sierra de la Neblina	Yes	Blahnik (1997)
Chimarra sp.	Auyán-tepui	?	Aubrecht et al. (2012), Derka and Zamora-Muñoz (2012)
Sericostomatidae	Churí-tepui	?	Aubrecht et al. (2012), Derka and Zamora-Muñoz (2012)

Because comprehensive reviews by Aubrecht et al. (2012) and Derka and Zamora-Muñoz (2012) have been published, only two further studies have been added. Zamora-Muñoz et al. (2013) presented new records of *Notalina roraima* from Roraima-tepui and Churí-tepui and described some morphological variation in the male genitalia. They also described and illustrated the larva of this species and included information on its habitat. Zamora-Muñoz et al. (2017) published descriptions of four new *Atopsyche* species (Fig. 8.3E) from eastern Pantepui: *Atopsyche (Atopsaura) carmenae, Atopsyche (Atopsaura) cristinae, Atopsyche (Atopsaura) inmae, and Atopsyche (Atopsaura) svitoki*. They associated the larvae of two of the species, *Atopsyche cristinae* and *A. inmae*, and performed a phylogenetic tree to assess the molecular validity of the species, establish an evolutionary relation among them, and interpret the historical biogeography of tepuis. Two new *Helicopsyche* species, the first one from Roraima-tepui and the second one from Auyán-tepui and Churí-tepui, are currently undergoing description.

Orthopterans (Orthoptera)

Although orthopterans are usually not considered aquatic insects, some of their members are somehow linked to freshwater habitats. Among them are semiaquatic anostostomatids from the genus *Hydrolutos* (Fig. 8.3C). The genus is known from seven apterous and nocturnal species, five of them restricted to Pantepui (Issa and Jaffe, 1999; Derka and Fedor, 2010, 2012; Derka et al., 2013, 2016). All species have a plastron-like structure on a pleuro-sternal area of the thorax and abdomen, which is generally unique within orthopterans. Their strong legs and tarsal claws enable them to cling to items and to move even against the strong current of streams. Three species are known only from unique tepuis (Issa and Jaffe, 1999): *Hydrolutos aracamuni* Issa & Jaffe, 1999, *Hydrolutos auyan* Issa & Iaffe, 1999, *Hydrolutos chimantea* Issa & Jaffe, 1999, and two species were observed at two adjacent tepuis: *Hydrolutos breweri* Derka & Fedor, 2010 was found at the Churí-tepui and Akopán-tepui, and *Hydrolutos roraimae* Issa & Jaffe, 1999 was found at the Roraima-tepui and Kukenán-tepui and their foothills (Derka and Fedor, 2010; Derka, pers. observ.). *Hydrolutos* used to be considered endemic to Pantepui; however, records and descriptions of new species from the Gran Sabana (Derka et al., 2013) and from the Orinoco lowlands (Derka et al., 2016) showed a more extensive distribution of the genus.

Dragonflies and damselflies (Odonata)

With 5680 extant species, Odonata (dragonflies and damselflies) are a relatively small order of hemimetabolous insects. Their size and color and their diurnal and often conspicuous behavior make them a popular group for entomologists. Nymphs of almost all species are dependent on freshwater habitats. Many species have small distributional ranges and are habitat specialists (Kalkman et al., 2008).

Odonata is probably the first group of aquatic insects described from Pantepui (Needham, 1933; Needham and Fischer, 1940; Klots, 1944). Thanks to numerous studies, many of them published by the Venezuelan odonatologist Jurg Carl De Marmels, dragonflies are known also from tepuis where other groups of aquatic insects have never been studied (De Marmels, 1983, 1985, 1988, 1989, 1992, 1994, 1999, 2007, 2016). Pantepui has a high percentage of endemic Odonata, often confined to a single mountain. The genus *Tepuibasis* includes seven species, all endemic to Pantepui (De Marmels, 2007). However, three species of a new genus, *Austrotepuibasis*, have been described from the low-altitude Amazon region of the Tapajós-Xingu province in Brazil (Machado and Lencioni, 2011). The genus is morphologically very close to *Tepuibasis* and possibly will be synonymized with *Tepuibasis* in the future (De Marmels pers. com.) According to Vivas-Santeliz and De

Marmels (2017), who provided data on the occurrence of dragonflies endemic to Venezuela, the richest endemic fauna are known from the Auyán-tepui, where 13 species belonging to 10 genera have been reported. Cerro Duida houses 10 species from eight genera, Sierra de la Neblina 8 species from eight genera, and Cerro Guaiquinima 7 species from six genera. There are eight other tepuis with known endemic Odonata (Vivas-Santeliz and De Marmels, 2017). Besides endemics, species with larger distributional areas occur in Pantepui as well.

Dobsonflies (Megaloptera)

Megalopteran are a small order of holometabolous insects with 328 described extant species (Cover and Resh, 2008). Larvae have been reported in a wide variety of aquatic habitats. Most Corydalidae are found in fast-flowing riffles under gravel and cobble.

Larvae of dobsonflies from the family Corydalidae (order Megaloptera) are probably the largest aquatic insects inhabiting Pantepui running waters (Fig. 8.3G), where they act as top predators. We collected their larvae from lower tepuis (Auyán-tepui, Churí-tepui, Akopán-tepui) and in the foothills of the Roraima-tepui and Kukenán-tepui. Two genera of dobsonflies (subfamily Corydalinae) are currently recognized in South America: *Chloronia* and *Corydalus*. *Chloronia gaianii* Contreras-Ramos, 2002 has been collected at Cerro Duida but below Pantepui altitudinal limits (Contreras-Ramos, 2002). Three species of *Corydalus* have been described from Pantepui: *Corydalus arpi* Contreras-Ramos, 1998 from Sierra de la Neblina, *Corydalus crossi* Contreras-Ramos, 2002 from Auyán-tepui, and *Corydalus mayri* Contreras-Ramos, 2002 from Cerro Aracamuni (Contreras-Ramos, 1998, 2002).

Beetles (Coleoptera)

Beetles represent the world's largest animal order with about 400,000 species described to date, with about 18,000 species that can be regarded as aquatic. About 30 beetle families have aquatic representatives, and in 25 families, at least 50% of the species can be considered aquatic (Jäch and Balke, 2008).

Elmidae are the best-studied Pantepui aquatic beetle family. This family has approximately 1330 species in 146 genera in two subfamilies: Larainae and Elminae (Kodada and Jäch, 2005). Adults and larvae of all the species are considered aquatic; they appear to be exclusively confined to running water (Jäch and Balke, 2008).

Three Larainae genera are known from Pantepui and all have been described as monotypic (Maier, 2013). *Neblinagena prima* Spangler, 1985 has been described from Sierra de la Neblina, *Neblinagena mira* Čiampor, Čiamporová-Zaťovičová & Kodada, 2017 has been described from the Kukenán-tepui foothills, and other undescribed species were reported from the Auyán-tepui (Spangler, 1985; Čiampor et al., 2017). *Neblinagena doylei* Kodada & Jäch, 1999 seems to prefer lower sites below Pantepui altitudinal limits (Kodada and Jäch, 1999; Čiampor et al., 2017). *Hypsilara royi* Maier & Spangler, 2011, from Sierra de la Neblina, remains a unique Pantepui species of the genus, because two other species have been collected below Pantepui altitudinal limits (Maier and Spangler, 2011; Čiampor et al.,

2013; Laššová et al., 2014). *Roraima carinata* Kodada & Jäch, 1999 from the Roraima-tepui foothills remains the only described representative of the genus, although undescribed new species have been reported from Churí-tepui and Auyán-tepui (Aubrecht et al., 2012).

Elminae are represented by two genera: *Jolyelmis* and *Gyrelmys*. The genus *Jolyelmis* has been established on a single, very distinctive species *Jolyelmis auyana* Spangler & Faitoute, 1991, collected at the Auyán-tepui (Spangler and Faitoute, 1991; Aubrecht et al., 2012). Two additional species, *Jolyelmis derkai* Čiampor & Kodada, 1999 and *Jolyelmis reitmaieri* Čiampor & Kodada, 1999, were discovered in a few small samples from streams in the Roraima-tepui foothills and slopes (Čiampor and Kodada, 1999). *Jolyelmis spangleri* Kodada, Derka & Čiampor, 2012 has been described from the Churí-tepui (Kodada et al., 2012). Two specimens of undetermined *Gyrelmys* Hinton, 1940 were sampled at the Churí-tepui (Aubrecht et al., 2012).

Dytiscids (Dytiscidae), which are the richest family of aquatic beetles, with nearly 4000 described and 5000 expected species (Jäch and Balke, 2008), are represented by two genera in Pantepui. *Rhantus elegans* Waterhouse, 1895 was known only from one type specimen (Waterhouse, 1895) since Spangler (1981) redescribed the species from a new material collected during expeditions organized by Charles Brewer-Carías in 1977 and 1978. He described *Tepuidessus breweri* Spangler, 1981 from Roraima-tepui, where it inhabits temporary rainwater pools (Spangler, 1981). *Tepuidessus grulai* Kodada, Hendrich & Balke, 2018 was found in similar habitats at the Akopán-tepui (Kodada et al., 2018).

Hydrophilidae, the second largest aquatic beetle family with more than 2650 described species (Jäch and Balke, 2008), are represented in Pantepui by *Tropisternus jolyi* Spangler, 1981, described from the Ptari-tepui (Spangler, 1981).

True bugs (Heteroptera)

The Heteroptera, or true bugs, are a group of about 40,000 species of insects in the Hemiptera order. They are hemimetabolous insects, typically developing via a series of five nymphal instars. The aquatic and semiaquatic Heteroptera, consisting of the infraorders Leptopodomorpha, Gerromorpha, and Nepomorpha, comprise a significant component of the world's aquatic insect biota. Within these three infraorders, 20 families, 326 genera, and 4656 species live in freshwater (Polhemus and Polhemus, 2008).

At least two species of two genera are known from Pantepui. Both are members of the family Veliidae. *Microvelia duidana* Drake & Maldonado-Capriles, 1952 has a wide altitudinal range, given that it has been described and reported from Cerro Duida and San Fernando de Atabapo (Drake and Maldonado-Capriles, 1952). *Oiovelia spumicola* Spangler, 1986 has been described from Sierra de la Neblina (Spangler, 1986).

True flies (Diptera)

The order Diptera, or true flies, is a major aquatic insect order representing 43% of known freshwater insect species (Balian et al., 2008). The most species-rich families are

Chironomidae, Culicidae, Simuliidae, and Tipulidae, but other smaller families constitute an important portion of the freshwater Diptera diversity (Balian et al., 2008).

There is scarce information published on Pantepui aquatic dipterans. The summits of tepuis are not free of mosquitoes and blackflies, which are abundant pests in surrounding lowlands and uplands but usually have low population densities on the summits (Derka, pers. observ.). Larvae of mosquitoes (Culicidae) from the genus *Wyeomyia* inhabit phytotelmata of pitcher plants of the genus *Heliamphora* (Sarraceniaceae), together with nonbiting midges from the genus *Metriocnemus* (Chironomidae: Orthocladiinae) and larvae of Ceratopogonidae (Jaffe et al., 1992). Larvae of *Metriocnemus* are known also from phytotelmata of bromeliad species *Brocchinia reducta* (González et al., 1991). Zavortink (1986) described *Wyeomyia* from Roraima-tepui as a new species: *Wyeomyia zinzala* Zavortink, 1986. *Anopheles (Kerteszia) auyantepuiensis* Harbach & Navarro, 1996 is a mosquito species described from bromeliad tanks at the Auyán-tepui summit. An unidentified species of *Wyeomyia* (subgenus *Nunezia*) and another from *Culex* (subgenus *Microculex*) were collected in association with larvae of *A. auyantepuiensis*. Moreover, some unknown *Aedes (Howardina) sexlineatus* Theobald, 1901, *Anopheles (Kerteszia) homunculus* Komp, 1937, and an unidentified species of *Aedes* (subgenus *Ochlerotatus*) were captured at the Auyán-tepui (Harbach and Navarro, 1996). Larvae of blackflies (Simuliidae) were frequently observed in Pantepui streams, with the exception of the Roraima-tepui summit, where they are very rare (Derka, pers. observ.).

Origin and evolution of Pantepui aquatic insects

Due to its ancient age and mountain island topography (Briceño and Schubert, 1990; Aubrecht et al., 2012), the Pantepui area has been assumed to be a mountain counterpart to oceanic islands and a cradle of speciation (Rull, 2005). Pantepui holds very high endemicity for many taxa (McDiarmid and Donnelly, 2005; Huber, 2005; Berry and Riina, 2005), including aquatic insects (Aubrecht et al., 2012). Traditionally, the high endemism has been explained by two major speciation processes: ancient vicariance *versus* recent dispersals. Various hypotheses have been proposed to explain the current distribution of Pantepui plants and animals (see Chapter 4: Origin and evolution of the Pantepui biota). The Lost World hypothesis (Chapman, 1931; Maguire, 1970) predicts that highland species are ancient relicts of formerly widespread plateau taxa, which arose through vicariance following ancient fragmentation of the plateau. Nevertheless, many studies have suggested that vicariance alone cannot explain the current distribution of highland biota and that diversification occurred more recently through various mechanisms such as habitat shifts (Mayr and Phelps, 1967; Huber, 1988; McDiarmid and Donnelly, 2005), vertical displacement (Rull, 2004, 2005; Rull and Nogué, 2006), and island hopping (Chapman, 1931). Palynological findings indicating a downward biotic migration during glacials, and the subsequent interglacial upward shift, in response to colder and warmer climates have inspired Rull (2005) to propose a new diversification model that emphasized the impact of quaternary climatic oscillations on the current distribution of Pantepui biota. According to this model, during glacials, biotic mixing promoting sympatric speciation, hybridization, and polyploidy is expected in the lowlands. At the mountaintops, unknown cold-adapted taxa and páramo-like communities are expected to have occurred, and vicariance prevailed. Subsequently, in the interglacial periods, many taxa have had the opportunity to ascend to the mountains again, allowing genetic interchange among their slopes and summits, while others would have adapted to lowlands. The interglacial highland communities, where vicariance still predominated, experienced some extinction due to habitat loss by upland displacement. Recently, Kok et al. (2017) confirmed that vicariance and dispersal played an important role in the diversification of the frog genus *Stefania* and highlighted a complex pattern implying several multiple processes: (1) Cenozoic vicariance, (2) reorganization of species diversity due to periods of climatic instability, and (3) Pleistocene invasions of widespread upland taxa. However, writing evolutionary scenarios is complicated due to the serious scientific disagreement that exists about the processes that formed the present-day landscapes of the Guiana Shield. The present landscape is considered to be inherited from the Cretaceous (Galan and Lagarde, 1988; Briceño et al., 1991) because quartz dissolution requires a long time. However, this estimation has not been supported by any data (Aubrecht et al., 2011). The theory of silification (Grabert, 1976) or lithification proposed by Aubrecht et al. (2012) implies that the erosional processes that have formed the present landscapes could have been faster and the landscapes could have been formed in the Neogene and older Pleistocene.

Unfortunately, studies of aquatic insects of Pantepui are scarce and mostly limited to species records and descriptions. Moreover, they are available only from a small number of existing tepuis; thus it is difficult to draw conclusions about the evolutionary and ecological processes that have formed its aquatic communities. From a limited dataset of mayflies, stoneflies, caddisflies, crickets of the genus *Hydrolutos*, and elmids that we have available, it seems that Pantepui aquatic insect fauna comprise three groups of taxa. The first one includes taxa confined to nutrient-poor black waters of the Guiana Shield, for example, mayflies from the genera *Parakari* and *Fittkauneuria*; endemic damselfly genus *Tepuibasis*; endemic elmid genera *Jolyelmis*, *Neblinagena*, and *Roraima*; and crickets from the genus *Hydrolutos*. The second group includes taxa with a disjunct distribution in the Guiana and Brazilian shields, for example, the mayfly genera *Massartella*, *Paramaka*, and *Spiritiops*. Baetidae species *Callibaetoides caaigua* and *Spiritiops silvudus* show a disjunct distribution; however, most taxa are represented in each area by different congeneric species. Finally, the third group is represented by widespread and ecologically more plastic taxa, for example, the mayfly genera *Baetodes*, *Camelobaetidius*, and *Cloeodes*; the stonefly genus *Anacroneuria*; and most caddisfly genera. Nevertheless, even in this group, the species recorded in Pantepui are mostly endemic to the Guiana region.

Pantepui species, especially cold stenotherms (e.g., *Massartella devani*, elmids from the genera *Jolyelmis* and *Roraima*), are usually confined to higher altitudes of a single tepui or adjacent tepuis and the cold springs at their foothills, whereas species tolerating wider temperature fluctuations (e.g., mayflies from the genus *Fittkauneuria*, *Spiritiops tepuiensis*, *Hagenulopsis minuta*, caddisflies *Atopsyche inmae*, *Leptonema amazonense*) can have a wider distribution. Cold stenotherms usually have congeneric relatives inhabiting the lower areas, which indicates that habitat shifts have most likely played a role in their diversification. It seems that high vertical cliffs do not represent an effective physical barrier for most taxa, but the unsuitable habitat can act as an ecological barrier that limits faunal interchange among tepuis through adjacent lowlands and uplands. This idea has been supported by Zamora-Muñoz et al. (2017), who demonstrated that the distributions of *Atopsyche* species showed high structuring among tepuis. Some species were recorded exclusively at one tepui, whereas others were found at various distant tepuis. Those species found at various tepuis showed intraspecific genetic divergence among respective subpopulations, which evidenced low gene flow among tepuis. Moreover, species from the summits and the savannas around the tepuis had highly supported genetic divergence, which suggests isolation across habitats differing in altitude (and most likely temperature), even though they are located nearby. Hence, each tepui and habitat commonly has its own endemic or isolated species of caddisfly. Nevertheless, the isolation among habitats is not absolute, because the occurrence of the same species on multiple tepuis was recorded. This is also the case for other taxa, for example, caddisfly *Notalina roraima* and mayflies *Spiritiops tepuiensis* and *Parakari roraimensis* (Nieto and Derka, 2012; Zamora-Muñoz et al., 2013; Derka et al., 2015). Moreover, altitudinal isolation was overcome by *Atopsyche inmae* from the savanna close to the Roraima-tepui, which was genetically similar to the specimens located on Roraima-tepui, and also by the two haplotypes of an unidentified *Atopsyche* species, which were found on the Auyán-tepui and in a savanna stream below the Auyán-tepui (Zamora-Muñoz et al., 2017). Similarly, we recorded the mayfly *Spiritiops tepuiensis* in Gran Sabana streams. These observations indicate that limited dispersal among tepui summits and habitat shifts from summits to adjacent savannas, or vice versa, followed by long-term isolation, have been playing important roles in species diversification.

Recommendations for conservation of the aquatic insects of Pantepui and ideas for future research

Regarding the conservation status of the aquatic insects from Pantepui, their high endemicity makes them particularly vulnerable. Despite the inaccessibility of this area, which has allowed it to remain relatively unchanged, some direct environmental damages derived from illegal mining and burning of the forest can lead to disruption of aquatic ecosystems or even destruction at a local or regional scale. However, it is assumed that the greater threat to Pantepui biota is global climate change and projected habitat loss by upward displacement. As reported by Rull and Vegas-Vilarrúbia (2006), Nogué et al. (2009), and Kok et al. (2016), the inability to migrate in response to climate change is the main threat to tepui-summit biota. Therefore many taxa that have adapted to the higher parts would be in danger of habitat loss and irreversible extinction (Vegas-Vilarrúbia et al., 2012). Effective environmental protection for the prevention of habitat loss seems illusory under continuing global change, and *ex situ* strategies for species conservation suitable for plants (Rull and Vegas-Vilarrúbia, 2006) are difficult to imagine for aquatic insects. Local actions, such as stopping fires and illegal mining, might have a positive local impact on the survival of Pantepui aquatic insects, especially for those inhabiting the tepui foothills and slopes. Changes in seasonal patterns of precipitation and runoff as a result of climate change would alter the hydrologic characteristics of aquatic systems. Droughts combined with increases in water temperatures can alter fundamental ecological processes and affect species composition and ecosystem productivity. However, due to a complete lack of knowledge about these processes in Pantepui aquatic habitats, as well as very limited knowledge of species composition and species requirements, most specific predictions of impacts are difficult. Therefore it is necessary to gain more information about species distributions, species traits, community compositions of Pantepui aquatic habitats, and the ecological and evolutionary processes forming Pantepui aquatic communities and ecosystems.

Acknowledgments

This study was supported by projects APVV-0251-07, APVV-0213-10, and APVV-14-0276. Tomáš Derka is grateful to the colleagues from the Faculty of Natural Sciences, Comenius University in Bratislava; Faculty of Ecology and Environmental Sciences, Technical University in Zvolen; and Slovak Academy of Science for their enjoyable company during field trips and help with sampling. He is very indebted to the members of the Venezuelan-Slovakian speleological expeditions for the support during field work, especially to Charles Brewer-Carías who organized expeditions to Churí-tepui. We are very grateful to local Amerindian people who have allowed expeditions to their lands and helped as guides, porters, and collectors. We are grateful to Carolina Nieto and Carlos Molineri from CONICET-Instituto de Biodiversidad Neotropical, Tucumán, Argentina, who helped with identification of mayflies. Thanks also to Cesc Múrria from the Department of Evolutionary Biology, Ecology and Environmental Sciences, University of Barcelona, Spain, for the phylogeographic discussion on caddisflies.

References

Aubrecht, R., Lánczos, T., Gregor, M., Schlögl, J., Liščák, P., Brewer-Carías, C., et al., 2011. Sandstone caves on Venezuelan tepuis: return to pseudokarst? Geomorphology 132 (3–4), 351–365. Available from: https://doi.org/10.1016/j.geomorph.2011.05.023.

Aubrecht, R., Barrio-Amorós, C.L., Breure, A.S.H., Brewer-Carías, C., Derka, T., Fuentes-Ramos, O.A., et al., 2012. Venezuelan tepuis: their caves and biota. Acta Geologica Slovaca–Monograph, Comenius University, Bratislava 1–167.

Avelino-Capistrano, F., Nessimian, J.L., Santos-Mallet, J.R., Takiya, D.M., 2014. DNA based identification and descriptions of immature of *Kempnyi*a Klapálek (Insecta: Plecoptera) from Macaé River Basin, Rio de Janeiro State, Brazil. Mol. Approach. Freshw. Ecol. 33 (1), 325–337.

Balian, E.V., Segers, H., Lévêque, C., Martens, K., 2008. An introduction to the Freshwater Animal Diversity Assessment (FADA) project. Hydrobiologia 595, 3–8. Available from: https://doi.org/10.1007/s10750-007-9235-6.

Barber-James, H.M., Gattolliat, J.L., Sartori, M., Hubbard, M.D., 2008. Global diversity of mayflies (Ephemeroptera, Insecta) in fresh water. Hydrobiologia 595, 339–350.

Barrera, R., Fish, D., Machado-Alison, C.E., 1989. Ecological patterns of aquatic insects communities in two *Heliamphora* pitcher-plant species of Venezuelan highlands. Ecotropicos 2, 31–44.

Berry, P.E., Riina, R., 2005. Insights into the diversity of the Pantepui flora and the biogeographic complexity of the Guayana Shield. Biol. Skr. 55, 145–167.

Blahnik, R.J., 1997. Systematics of *Chimarrita*, a new subgenus of *Chimarra* (Trichoptera: Philopotamidae). Syst. Entomol. 22, 199–243.

Boldrini, R., Lima, L.R.C., De Paiva Ottoni-Boldrini, B.M., 2018. Description of the nymph of *Miroculis* (*Miroculis*) *eldorado* Gama-Neto & Hamada, 2014 (Leptophlebiidae: Ephemeroptera) from Roraima, northern Brazil. Zootaxa 4459 (1), 197–200. Available from: https://doi.org/10.11646/zootaxa.4459.1.12.

Brewer-Carías, C. (Ed.), 1988. Cerro de la Neblina-Resultados de la Expedición 1983–1987. FUDECI, Caracas.

Briceño, H.O., Schubert, C., 1990. Geomorphology of the Gran Sabana, Guayana Shield, southeastern Venezuela. Geomorphology 3, 125–141.

Briceño, H.O., Schubert, C., Paolini, J., 1991. Table-mountain geology and surficial geochemistry: Chimanta massif, Venezuelan Guayana Shield. J. South Am. Earth Sci. 3, 179–194.

Chamorro, M.L., Holzental, R.W., 2010. Taxonomy and phylogeny of New World *Polyplectropus* Ulmer, 1905 (Trichoptera: Psychomyioidea: Polycentropodidae) with the description of 39 new species. Zootaxa 2582, 1–252.

Chapman, F.M., 1931. The upper zonal bird—life of Mts. Roraima and Duida. Bull. Am. Mus. Nat. Hist. 63, 1–135.

Čiampor, F., Kodada, J., 1999. Description of two new species of the genus *Jolyelmis* from Mount Roraima, Venezuela (Coleoptera: Elmidae). Entomol. Prob. 30 (2), 55–60.

Čiampor Jr., F., Laššová, K., Čiamporová-Zaťovičová, Z., 2013. *Hypsilara breweri* n.sp. from Venezuela: description of new species with notes on the morphology and phylogenetic relationships of the genus (Coleoptera: Elmidae: Larainae). Zootaxa 3635 (5), 591–597. Available from: https://doi.org/10.11646/zootaxa.3635.5.10.

Čiampor Jr., F., Čiamporová-Zaťovičová, Z., Kodada, J., 2017. A new species of *Neblinagena* Spangler from Kukenán tepui and DNA barcoding of Neblinagena and related genera (Coleoptera: Elmidae). Zootaxa 4286 (2), 176–186. Available from: https://doi.org/10.11646/zootaxa.4286.2.2.

Contreras-Ramos, A., 1998. Systematics of the Dobsonfly Genus *Corydalus* (Megaloptera: Corydalidae). Thomas Say Publications in Entomology: Monographs, Entomological Society of America, Lanham, Maryland.

Contreras-Ramos, A., 2002. Six new species of dobsonflies from Venezuela (Megaloptera: Corydalidae: Corydalinae). Aquat. Insects 24 (1), 55–75. Available from: https://doi.org/10.1076/aqin.24.1.55.4909.

Cover, M.R., Resh, V.H., 2008. Global biodiversity of dobsonflies, fishflies and alderflies (Megaloptera; Insecta) and spongillaflies, nevrorthids, and osmylids (Neuroptera; Insecta) in freshwater. Hydrobiologia 595, 409–417.

Cruz, P.V., Salles, F.F., Hamada, N., 2013. A new genus and species of Baetidae (Insecta: Ephemeroptera) from Brazil. Ann. Limnol. 49, 1–12.

De Marmels, J., 1983. The Odonata of the region of Mount Auyantepui and the Sierra de Lema, in Venezuelan Guyana. 3. Additions to the families Gomphidae, Aeshnidae and Corduliidae, with description of *Progomphus racenisi* spec. nov. Odonatologica 12 (1), 5–13.

De Marmels, J., 1985. *Acanthagrion dichrostigma* sp.n. y *Acanthagrion tepuiense* sp.n. de Venezuela (Odonata: Coenagrionidae). Boletín de Entomología Venezolana 4 (2), 9–16.

De Marmels, J., 1988. *Sympetrum roraimae* spec. nov. vom venezolanischen Guayana-Hochland (Odonata: Libellulidae). Opuscula Zoologica Fluminensia 28, 1–6.

De Marmels, J., 1989. Odonata or dragonflies from Cerro de la Neblina and the adjacent lowland between the Río Baria, the Casiquiare and the Río Negro (Venezuela). I. Adults. Bol. Acad. Cienc. Fis. Mat. Nat. 25, 11–78.

De Marmels, J., 1992. Odonata del Cerro Guaiquinima (Edo. Bolívar) y zonas aledañas. Boletín de Entomología Venezolana 7 (1), 37–47.

De Marmels, J., 1994. *Sympetrum chaconi* spec. nov. from Auyan-Tepui, Venezuela, with notes on a Pantepuyan form of *Tramea binotata* (Rambur) (Anisoptera: Libellulidae). Odonatologica 23 (4), 405–412.

De Marmels, J., 1999. A new species of *Dimeragrion* Calvert, 1913 from Pantepui, Venezuela (Odonata: Megapodagrionidae). Boletín de Entomología Venezolana 14 (1), 27–36.

De Marmels, J., 2007. *Tepuibasis* gen. nov. from the Pantepui region of Venezuela, with descriptions of four new species, and with biogeographic, phylogenetic and taxonomic considerations on the Teinobasinae (Zygoptera: Coenagrionidae). Odonatologica 36 (2), 117–146.

De Marmels, J., 2016. Lista de los Odonata de Venezuela/Checklist of the Odonata of Venezuela (13 Abril/April 2016). Available from: http://odonata.miza-ucv.org.ve/files/2016-04-13-Checklist_Odonata_Venezuela.pdf (accessed 28.08.18).

DeWalt, R.E., Maehr, M.D., Neu-Becker, U., Stueber, G., 2018. Plecoptera Species File Online. Version 5.0/5.0. Available from: http://Plecoptera.SpeciesFile.org (accessed 28.08.18).

Derka, T., 2002. *Massartella devani*, a new mayfly species from Venezuela's Highlands (Ephemeroptera: Leptophlebiidae: Atalophlebiinae). Aquat. Insects 24, 309–316.

Derka, T., Fedor, P., 2010. *Hydrolutos breweri* sp. n., a new aquatic Lutosini species (Orthoptera: Anostostomatidae) from Churí-tepui (Chimantá Massif, Venezuela). Zootaxa 2653, 51–59.

Derka, T., Fedor, P., 2012. Description of *Hydrolutos breweri* (Orthoptera: Anostostomatidae) female from Chimantá Massif (Venezuela). Zootaxa 3247, 65–68.

Derka, T., Tierno de Figueroa, J.M., 2013. *Enderleina khazeni* sp. n. (Plecoptera, Perlidae), a new stonefly from Venezuelan Guayana. Zootaxa 3619 (1), 75–78. Available from: https://doi.org/10.11646/zootaxa.3619.1.5.

Derka, T., Tierno de Figueroa, J.M., 2016. Contribution to the knowledge on habitat, distribution and morphological variability of two species of *Anacroneuria* Klapálek, 1909 (Plecoptera, Perlidae) from Venezuela. Bol. Asoc. Esp. Entomol. 40 (3–4), 341–349.

Derka, T., Zamora-Muñoz, C., 2012. New records of Trichoptera from the Pantepui biogeographical province. Bol. Asoc. Esp. Entomol. 36 (1–2), 7–30.

Derka, T., Svitok, M., Schlögl, J., 2009. *Massartella hirsuta* sp. nov. (Ephemeroptera: Leptophlebiidae: Atalophlebiinae) and new data on mayflies of Guayana Highlands. Aquat. Insects 31 (Suppl. 1), 83–94.

Derka, T., Tierno de Figueroa, J.M., Gamboa, M., 2010. First records of Plecoptera from Pantepui biogeographical province, with the first record of genus *Kempnyia* Klapálek, 1914 for Venezuela (Insecta: Plecoptera). Bol. Asoc. Esp. Entomol. 33 (3-4), 493–502.

Derka, T., Nieto, C., Svitok, M., 2012. Mayflies (Ephemeroptera) of the Pantepui biogeographical province. Bol. Asoc. Esp. Entomol. 36 (1–2), 119–135.

Derka, T., Fedor, P., Svitok, M., Trizna, M., 2013. *Hydrolutos gransabanensis* sp.n. (Orthoptera: Anostostomatidae), a new semiaquatic Lutosini species from Gran Sabana (Venezuela). Zootaxa 3682 (3), 432–440. Available from: https://doi.org/10.11646/zootaxa.3682.3.4.

Derka, T., Nieto, C., Svitok, M., 2015. A new species of *Parakari* (Ephemeroptera: Baetidae) from Guiana Highlands. Zootaxa 4028 (2), 296–300.

Derka, T., Svitok, M., Fedor, P., 2016. *Hydrolutos piaroa* sp. n. (Orthoptera: Anostostomatidae), a new semiaquatic Lutosini species from south-western Venezuelan lowland streams. Zootaxa 4066 (4), 485–492. Available from: https://doi.org/10.11646/zootaxa.4066.4.9.

Derka, T., Tierno de Figueroa, J.M., Čiampor, F., Jr., 2019. *Enderleina preclara* Jewett, 1960 (Plecoptera, Perlidae), from the Venezuelan Guayana—A description of a putative female, nymph and egg. Zootaxa 4585 (2), 332–342. Available from: https://doi.org/10.11646/zootaxa.4585.2.6.

Domínguez, E., Molineri, C., Pescador, M.L., Hubbard, M.D., Nieto, C., 2006. Ephemeroptera of South America. In: Adis, J., Arias, J.R., Rueda-Delgado, G., Wantzen, K.M. (Eds.), Aquatic Biodiversity in Latin America (ABLA), vol. 2. Pensoft, Sofia-Moscow.

Domínguez, E., Molineri, C., Mariano, R., 2009. Revision of the South American species of *Hagenulopsis* Ulmer and *Askola* Peters (Ephemeroptera: Leptophlebiidae) with description of six new species. Zootaxa 2142, 29–44.

Domínguez, E., Grillet, M.E., Nieto, C., Molineri, C., Guerrero, E., 2014. Ephemeroptera from the Venezuelan Guayanas's Uplands: families Leptophlebiidae, Euthyplociidae and Oligoneuriidae. Zootaxa 3827 (3), 301–317. Available from: https://doi.org/10.11646/zootaxa.3827.3.1.

Drake, C.J., Maldonado-Capriles, J., 1952. Water-striders from Territorio Amazonas of Venezuela (Hemiptera: Hydrometridae, Veliidae). Great Basin Nat. 12, 47–54. Available from: https://doi.org/10.5962/bhl.part.21968.

Flint Jr., O.S., 1998. Studies of Neotropical caddisflies, LIII: a taxonomic revision of the subgenus *Curgia* of the genus *Chimarra* (Trichoptera: Philopotamidae). Smithsonian Contrib. Zool. 594, 1–131.

Flint Jr., O.S., McAlpine, J.F., Ross, H.H., 1987. A revision of the genus *Leptonema* Guérin (Trichoptera: Hydropsychidae: Macronematinae). Smithsonian Contrib. Zool. 450, 1–193.

Fochetti, R., Tierno de Figueroa, J.M., 2008. Global diversity of stoneflies (Plecoptera; Insecta) in freshwater. Hydrobiologia 595, 365–377.

Froehlich, C.G., 2010. Catalogue of Neotropical Plecoptera. Illiesia 6 (12), 118–205.

Galan, C., Lagarde, J., 1988. Morphologie et evolution des cavernes et formes superficielles dans les quartzites du Roraima. Karstologia 11–12, 49–60.

González, J.M., Jaffe, K., Michelangeli, F., 1991. Competition for prey between the carnivorous Bromeliaceae *Brocchinia reducta* and Sarraceneacea *Heliamphora nutans*. Biotropica 23 (4), 602–604.

Grabert, H., 1976. Die Inselberglandschaft des Roraima in Venezolanisch Guayana. Erde 107, 57–69.

Hamada, N., Silva, J.O., Pedroza, M.K., 2016. A new species of *Enderleina* Jewett (Plecoptera, Perlidae) from Amazonas State, Brazil. Zootaxa 4098 (2), 392–400. Available from: https://doi.org/10.11646/zootaxa.4098.2.11.

Hamilton, S.W., Holzenthal, R.W., 2005. Five new species of Polycentropodidae (Trichoptera) from Ecuador and Venezuela. Zootaxa 810, 1–14.

Harbach, R.E., Navarro, J.C., 1996. A new species of *Anopheles subgenus* Kerteszia (Diptera: Culicidae) from Venezuela. Entomol. Scand. 27, 207–216.

Holzenthal, R.W., 1986. The Neotropical *Notalina*, the southern group of long-horned caddisflies (Trichoptera: Leptoceridae). Syst. Entomol. 11, 61–73.

Holzenthal, R.W., 1988. Systematics of Neotropical *Triplectides* (Trichoptera, Leptoceridae). Ann. Entomol. Soc. Am. 81 (2), 187–208.

Holzenthal, R.W., Calor, A.R., 2017. Catalog of the Neotropical Trichoptera (Caddisflies). ZooKeys 654, 1–566. Available from: https://doi.org/10.3897/zookeys.654.9516.

Holzenthal, R.W., Morse, J.C., Kjer, K.M., 2011. Order Trichoptera Kirby, 1813. In: Zhang, Z.-Q. (Ed.), Animal biodiversity: an outline of higher-level classification and survey of taxonomic richness. Zootaxa 3148, 209–211.

Holzenthal, R.W., Thomson, R.E., Ríos-Touma, B., 2015. Order Trichoptera. In: Thorp, J.H., Rogers, D.C. (Eds.), Ecology and General Biology, vol I. Thorp and Covich's Freshwater Invertebrates, fourth ed. Academic Press, Cambrige, pp. 965–1002. Available from: https://doi.org/10.1016/B978-0-12-385026-3.00038-3.

Huber, O., 1988. Guayana highlands versus Guayana lowlands, a reappraisal. Taxon 37, 595–614.

Huber, O., 1995a. Geographical and physical features. In: Berry, P.E., Holst, B.K., Yatskievych, K. (Eds.), Flora of the Venezuelan Guayana, I. Introduction. Missouri Botanical Garden Press, St. Louis, pp. 1–61.

Huber, O., 1995b. History of botanical exploration. In: Berry, P.E., Holst, B.K., Yatskievych, K. (Eds.), Flora of the Venezuelan Guayana, I. Introduction. Missouri Botanical Garden Press, St. Louis, pp. 63–95.

Huber, O., 1995c. Vegetation. In: Berry, P.E., Holst, B.K., Yatskievych, K. (Eds.), Flora of the Venezuelan Guayana, I. Introduction. Missouri Botanical Garden Press, St. Louis, pp. 97–160.

Huber, O., 2005. Diversity and vegetation types in the Guayana Region, an overview. Biol. Skr. 55, 169–188.

Issa, S., Jaffe, K., 1999. *Hydrolutos*: un género nuevo y cuatro especies nuevas de Lutosini Neotropicales (Orthoptera, Anostostomatidae). Nouvelle Revue d'Entomologie (N. S.) 16 (2), 111–121.

Jäch, M.A., Balke, M., 2008. Global diversity of water beetles (Coleoptera) in freshwater. Hydrobiologia 595, 419–442.

Jaffe, K., Michelangeli, F., González, J.M., Miras, B., Ruiz, M.C., 1992. Carnivory in pitcher plants of the genus Heliamphora (Sarraceniaceae). New Phytol. 122, 733–744.

Jewett, S.G., 1960. Notes and descriptions concerning Brazilian stoneflies (Plecoptera). Arq. Mus. Nac. 50, 167–183.

Johanson, K.A., Holzenthal, R.W., 2004. Thirteen new species and new distribution records of *Helicopsyche* (*Feropsyche*) Johanson from Venezuela (Trichoptera: Helicopsychidae). Zootaxa 711, 1–40.

Kalkman, V.J., Clausnitzer, V., Dijkstra, K.D.B., Orr, A.G., Paulson, D.R., van Tol, J., 2008. Global diversity of dragonflies (Odonata) in freshwater. Hydrobiologia 595, 351–363.

Klots, E.B., 1944. Notes on the Gomphinae (Odonata) with description of new species. Am. Mus. Novit. 1259, 1–11.

Kodada, J., Jäch, M.A., 1999. *Roraima carinata* gen. et sp. nov. and *Neblinagena doylei* sp. nov., two Larainae from Mount Roraima, Venezuela (Coleoptera: Elmidae). Entomological Problems 30 (1), 13–29.

Kodada, J., Jäch, M.A., 2005. 18.2. Elmidae Curtis, 1830. In: Beutel, R.G., Leschen, R.A.B. (Eds.), Handbook of Zoology, vol. IV (Part 38), Coleoptera, Beetles, vol. 1, Morphology and Systematics (Archostemata, Adephaga, Myxophaga, Polyphaga partim). Walter de Gruyter, Berlin, pp. 471–496.

Kodada, J., Derka, T., Čiampor Jr., F., 2012. Description of *Jolyelmis spangleri* a new species from Churí-tepui (Chimantá Massif, Venezuela), with a description of the larva of *J. spangleri* and *J. reitmaieri* (Insecta: Coleoptera: Elmidae). Zootaxa 3223, 1–23.

Kodada, J., Hendrich, L., Balke, M., 2018. *Tepuidessus grulai* sp. nov. from Acopán Tepui in Venezuela (Coleoptera: Dytiscidae: Hydroporinae: Bidessini). Zootaxa 4434 (3), 561–572. Available from: https://doi.org/10.11646/zootaxa.4434.3.10.

Kok, P.J.R., Russo, V.G., Ratz, S., Aubret, F., 2016. On the distribution and conservation of two "Lost World" tepui summit endemic frogs, *Stefania ginesi* Rivero, 1968 and *S. satelles* Señaris, Ayarzagüena, and Gorzula, 1997. Amphib. Reptile Conserv. 10 (1), 5–12.

Kok, P.J.R., Russo, V.G., Ratz, S., Means, D.B., MacCulloch, R.D., Lathrop, A., et al., 2017. Evolution in the South American 'Lost World': insights from multilocus phylogeography of stefanias (Anura, Hemiphractidae, Stefania). J. Biogeogr. 44, 170–181. Available from: https://doi.org/10.1111/jbi.12860.

Laššová, K., Čiampor Jr., F., Čiamporová-Zaťovičová, Z., 2014. Two new Larainae species from Guayana region, Venezuela (Coleoptera: Elmidae). Zootaxa 3753 (2), 187–195. Available from: https://doi.org/10.11646/zootaxa.3753.2.8.

Machado, A.B.M., Lencioni, F.F.A., 2011. *Austrotepuibasis* gen. nov. with descriptions of three new species from Brazil (Zygoptera: Coenagrionidae). Odonatologica 40 (1), 27–37.

Maguire, B., 1970. On the flora of the Guayana Highland. Biotropica 2, 85–100.

Maier, C.A., 2013. A Revision of the Larainae (Coleoptera: Elmidae) of Venezuela, with description of nine new species. ZooKeys 329, 33–91. Available from: https://doi.org/10.3897/zookeys.329.4961.

Maier, C.A., Spangler, P.J., 2011. *Hypsilaria royi* gen. n. and sp. n. (Coleoptera, Elmidae, Larainae) from Southern Venezuela, with a revised key to Larainae of the Western Hemisphere. ZooKeys 116, 25–36.

Mayr, E., Phelps, W.H., 1955. Origin of the bird fauna of Pantepui. In: Portman, A., Sutter, E. (Eds.), Acta XI Congressus Internationalis Ornitologici. Basel. pp. 399–400.

Mayr, E., Phelps, W.H., 1967. The origin of the bird fauna of the South Venezuelan Guayana. Bull. Am. Mus. Nat. Hist. 136, 273–327.

McDiarmid, R.W., Donnelly, M.A., 2005. The herpetofauna of the Guayana Highlands: amphibians and reptiles of the Lost World. In: Donnelly, M.A., Crother, B.I., Guyer, C., Wake, M.H., White, M.E. (Eds.), Ecology and Evolution in the Tropics: A Herpetological Perspective. University of Chicago Press, Chicago, pp. 461–560.

Needham, J.G., 1933. Dragonflies from Mt. Duida and the Venezuelan boder. Am. Mus. Novit. 664, 1–6.

Needham, J.G., Fischer, E., 1940. Two Neotropical Agrionine damselflies (Odonata) from Mts. Duida and Roraima. Am. Mus. Novit. 1081, 1–3.

Nieto, C., Derka, T., 2011. *Parakari* a new genus of the family Baetidae (Insecta: Ephemeroptera) from Guyana Highlands. Zootaxa 3032, 47–59.

Nieto, C., Derka, T., 2012. A new species of the genus *Spiritiops* (Ephemeroptera, Baetidae) from the Pantepui biogeographical province. Zootaxa 3256, 58–63.

Nieto, C., Grillet, M.E., Domínguez, E., Molineri, C., Guerrero, E., 2011. The family Baetidae (Insecta: Ephemeroptera) from Venezuelan Guayana's Uplands. Zootaxa 2808, 1–17.

Nogué, S., Rull, V., Vegas-Vilarrúbia, T., 2009. Modeling biodiversity loss by global warming on Pantepui, northern South America: projected upward migration and potential habitat loss. Clim. Change 94, 77–85. Available from: https://doi.org/10.1007/s10584-009-9554-x.

Pescador, M.L., Edmunds Jr., G.F., 1994. New genus of Oligoneuriidae (Ephemeroptera) from South America. Ann. Entomol. Soc. Am. 87, 263–269.

Pescador, M.L., Peters, W.L., 1990. Biosystematics of the genus *Massartella* Lestage (Ephemeroptera: Leptophlebiidae: Atalophlebiinae) from South America. Aquat. Insects 12, 145–160.

Peters, W.L., Domínguez, E., 2001. The identity of *Hagenulopsis minuta* Spieth (Leptophlebiidae: Atalophlebiinae). In: Domínguez, E. (Ed.), Trends in Research in Ephemeroptera and Plecoptera. Kluwer Academic/Plenum, New York, pp. 353–358.

Polhemus, J.T., Polhemus, D.A., 2008. Global diversity of true bugs (Heteroptera; Insecta) in freshwater. Hydrobiologia 595, 379–391.

Ribeiro, F.J.M., Gorayeb, I.S., 2015. Description of immatures associated with adults of two species of *Macrogynoplax* Enderlein and one species of *Enderleina* Jewett (Plecoptera: Perlidae) of the Brazilian Amazon. Zootaxa 3948 (1), 109–124. Available from: https://doi.org/10.11646/zootaxa.3948.1.7.

Ross, H.H., King, E.W., 1952. Biogeographic and taxonomic studies in *Atopsyche* (Trichoptera, Rhyacophilidae). Ann. Entomol. Soc. Am. 45, 177–204.

Rull, V., 2004. An evaluation of the Lost World and the vertical displacement hypotheses in the Chimanta massif, Venezuelan Guayana. Glob. Ecol. Biogeogr. 13, 141–148.

Rull, V., 2005. Biotic diversification in the Guayana Highlands: a proposal. J. Biogeogr. 32, 921–927.

Rull, V., Nogué, S., 2006. Potential migration routes and barriers for vascular plants of the Neotropical Guyana Highlands during the Quaternary. J. Biogeogr. 34, 1–16. Available from: https://doi.org/10.1111/j.1365-2699.2006.01602.x.

Rull, V., Vegas-Vilarrúbia, T., 2006. Unexpected biodiversity loss under global warming in the neotropical Guayana Highlands. Glob. Change Biol. 12, 1–9.

Salles, F.F., Angeli, K.B., Jacobus, L.M., 2016. Review of *Waltzoyphius* McCafferty & Lugo-Ortiz, 1995 and *Zelusia* Lugo-Ortiz & McCafferty, 1998 (Ephemeroptera: Baetidae). Zoosymposia 11, 174–204. Available from: https://doi.org/10.11646/zoosymposia.11.1.19.

Savage, H.M., 1987. Two new species of Miroculis from Cerro de la Neblina, Venezuela with new distribution records for *Miroculis fittkaui* and *Microphlebia surinamensis* (Ephemeroptera: Leptophlebiidae). Aquat. Insects 9 (2), 97–108.

Schmid, F., 1989. Les Hydrobiosides (Trichoptera, Annulipalpia). Bulletin de l'Institute Royal des Sciences Naturelles de Belgique 59 (Suppl.), 1–154.

Souto, P.M., Da-Silva, E.R., Nessimian, J.L., 2013. Description of the nymph of *Massartella alegrettae* Ulmer, 1943 (Ephemeroptera: Leptophlebiidae: Atalophlebiinae) with a key to *Massartella* Lestage, 1930 nymphs. Zootaxa 3681 (5), 583–588. Available from: https://doi.org/10.11646/zootaxa.3681.5.7.

Spangler, P.J., 1981. New and interesting water beetles from Mt. Roraima and Ptari-tepui, Venezuela (Coleoptera: Dytiscidae and Hydrophilidae). Aquat. Insects 3 (1), 1–11.

Spangler, P.J., 1985. A new genus and species of riffle-beetle, *Neblinagena prima*, from the Venezuelan Tepui Cerro de la Neblina (Coleoptera, Elmidae, Larinae). Proc. Entomol. Soc. Wash. 87 (3), 538–544.

Spangler, P.J., 1986. Two new species of water-striders of the genus *Oiovelia* from the Tepui Cerro de la Neblina, Venezuela (Hemiptera, Veliidae). Proc. Entomol. Soc. Wash. 88, 438–450.

Spangler, P.J., Faitoute, R.A., 1991. A new genus and species of Neotropical water beetle, *Jolyelmis auyana* from a Venezuelan Tepui (Coleoptera, Elmidae). Proc. Entomol. Soc. Wash. 104 (2), 322–327.

Stark, B.P., 1989. The genus *Enderleina* (Plecoptera: Perlidae). Aquat. Insects 11 (3), 153–160. Available from: https://doi.org/10.1080/01650428909361364.

Stark, B.P., 1995. New species and records of *Anacroneuria* (Klapálek) from Venezuela (Insecta, Plecoptera, Perlidae). Spixiana 18, 211–249.

Stark, B.P., 2011. *Macrogynoplax duida*, a new species of stonefly (Plecoptera: Perlidae) from Cerro Duida, Venezuela. Illiesia 7 (7), 86–88.

Stark, B.P., Zwick, P., 1989. New species of *Macrogynoplax* from Venezuela and Surinam (Plecoptera: Perlidae). Aquat. Insects 11 (4), 244–255.

Stark, B.P., Froehlich, C., Zúñiga, M.C., 2009. South American Stoneflies (Plecoptera. In: Adis, J., Arias, J.R., Rueda-Delgado, G., Wantzen, K.M. (Eds.), Aquatic Biodiversity in Latin America (ABLA), vol. 5. Pensoft, Sofia, Moscow.

Vegas-Vilarrúbia, T., Nogué, S., Rull, V., 2012. Global warming, habitat shifts and potential refugia for biodiversity conservation in the neotropical Guayana Highlands. Biol. Conserv. 152, 159–168.

Vivas-Santeliz, J., De Marmels, J., 2017. Current knowledge of Odonata in Venezuela: diversity and distribution of endemic taxa. Odonatologica 46 (1/2), 35–54. Available from: https://doi.org/10.5281/zenodo.572355.

Waterhouse, C.O., 1895. Insects collected by Messrs J.J. Quelch and F. McConnell on the Summit of Mount Roraima. Ann. Mag. Nat. Hist. 16, 494–497.

Zamora-Muñoz, C., Derka, T., Cressa, C., 2013. *Notalina roraima* Holzenthal 1986 (Trichoptera: Leptoceridae), male genitalia variability and larval description. Zootaxa 3702, 450–458. Available from: https://doi.org/10.11646/zootaxa.3702.5.4.

Zamora-Muñoz, C., Derka, T., Múrria, C., 2017. Four new species and new records of *Atopsyche* Banks (Trichoptera: Hydrobiosidae) from Pantepui biogeographical region (Venezuela). Zootaxa 4272 (2), 178–200. Available from: https://doi.org/10.11646/zootaxa.4272.2.2.

Zavortink, T.J., 1986. *Zinzala*, a new subgenus of *Wyeomyia* with two new species from pitcher-plants in Venezuela (Diptera: Culicidae, Sabethini). Wasmann J. Biol. 43, 46–59.

CHAPTER

9

Butterflies

Ángel L. Viloria[1] *and Mauro Costa*[2]

[1]Center of Ecology, Venezuelan Institute for Scientific Research (IVIC), Altos de Pipe, Venezuela [2]Museum of the Institute of Agricultural Zoology (MIZA), Faculty of Agronomy, Central University of Venezuela, Maracay, Venezuela

Introduction

Butterflies, skippers, and the less common moth-like members of the family Hedylidae form a monophyletic group of holometabolous insects: the superfamily Papilionoidea. They are all diurnal, and their imagos are conspicuous by having a coilable proboscis specialized for feeding on liquid substrates (suborder Glossata) and two pairs of broad, membranous wings covered by minute scales (order Lepidoptera). These are fully functional for flight, but their characteristic color patterns account also for interspecific and intraspecific visual communication. With few exceptions, their larvae are phytophagous, a condition that makes their life cycles strongly dependent on associations with particular host plants. Butterflies and skippers are naturally present and widely distributed in most continents and islands. They do not currently occur in Antarctica and the oceanic Easter Island.

Vegetation is an environmental precondition for the occurrence of established populations and communities of diurnal lepidoptera, and because of their predominant host specificity, natural high levels of plant diversity usually correlates well with high butterfly/skipper diversity. Terrestrial tropical latitudes harbor the highest figures of taxa within this group of insects, often regarded as megadiverse. It is generally accepted that vegetation heterogeneity resulting from the combination of environmental factors like physiography, altitude, climate, pedology, or substrate geology and the historical variables affecting them contribute to the increase of local or regional plant and butterfly/skipper richness.

The Papilionoidea are more diverse in tropical America than in any other tropical region. Most families are represented in the American continent, including some of the most spectacular examples of colorful patterns and extreme sizes. Their study has been

Biodiversity of Pantepui
DOI: https://doi.org/10.1016/B978-0-12-815591-2.00009-4

very popular among collectors and professional scientists since the earliest stages of natural history and biology. For instance, the study of the diversity of the lowland Amazonian fauna of butterflies during the 19th century stimulated the development of evolutionary ideas, the natural selection hypothesis, and the concept of mimicry. After the Neotropical mountains began to be explored, butterfly studies in the Andes also contributed to the comprehension of endemism, altitude distributions of life forms, and montane isolation. The last three phenomena are fundamental to the present contribution, which deals with the endemism of the butterfly fauna of the tepuis and its meaning to biogeography.

Tepuis are peculiar table mountains generally separated from each other by long distances. Land expansions in between are often covered by tropical forest and/or low (hilly) savannas. In a macroscopic scale they look like discrete islands in a sea of lowland vegetation. Within this scenario, cloud forests and their fauna only develop in narrow belts around these "islands" occupying especially their side slopes. This feature makes the tepuis essentially different from the tropical Andes, where cloud forest belts, roughly developed between 1000 and 3000 m, might expand horizontally for more than 2500 km without significant interruption. The names, characteristics and location of the tepuis and tepuian massifs follow Table 1.1 and Fig. 1.2 from Chapter 1: Definition and characterization of the Pantepui biogeographical province.

The biogeographic Pantepui

The word Pantepui (*pan* = all in Greek and *tepui* = Pemon-language term applied to the mountains of the Gran Sabana) was introduced by Ernst Mayr and William H. Phelps Jr. in 1954 during the 11th International Congress of Ornithology in Basel, Switzerland, when they presented their "Origin of the bird fauna of Pantepui" (Mayr and Phelps, 1955). Later on, they gave a definition of Pantepui as "the sandstone table mountains located in Venezuela in the Territorio Amazonas and Estado Bolívar and the adjacent bordering regions of Brazil and Guyana" (Mayr and Phelps, 1967, 1971). At first sight, it looks as if it was a purely geographic concept, but the wider definition of Mayr and Phelps Jr. contains both an explicit geographic criterion relative to the surface extension and altitude of the region and an implicit biologic criterion referring to the life conditions in such an area. Thus these authors assigned for the first time a biogeographic meaning to this neologism when considering all the elevations of this region as belonging to a set or unit called Pantepui.

Other researchers such as Julian A. Steyermark (botanist), and the herpetologists Marinus Hoogmoed and Paul Müller offered different interpretations of Pantepui (Müller, 1973; Hoogmoed, 1979; Steyermark, 1979, 1982). Keith S. Brown Jr., who interpreted butterfly race formation in tropical America through Jürgen Haffer's speciation hypothesis of the forest paleoecological refugia (Haffer, 1969), accomplished extensive taxonomic and distribution analyses of certain tribes of Neotropical butterflies, focusing on detecting centers of endemism (Brown, 1977a,b, 1979, 1987). His work became very influential among lepidopterists for at least three decades. Brown recognized several centers of endemism in tropical America, among which he named Pantepui the area corresponding to the southeastern portion of the state of Bolívar in

Venezuela (south of parallel 6°N and east of the Erebato River down to Roraima-tepui), without taking into much consideration the biotic differences imposed by the altitude gradient in this territory (see his discussion related to the high-altitude fauna in Brown, 1987). Brown's areas of endemism were inferred from the distribution of butterflies that belong to taxonomic groups characteristic of the tropical zone (i.e., lowlands). These areas had been delimited by means of extrapolation with localities of low elevation, embracing within their extension several mountain regions where the fauna of the tropical zone gradually rarifies as it ascends in altitude. Many Heliconiini and Ithomiini (Nymphalidae) studied by Brown Jr. are denizens of ecosystems developed at lower levels in the mountains of Pantepui. Instead, they are replaced by other taxa better adapted to the natural conditions of higher-elevation ecosystems, which cover considerable areas, occasionally arranged in disjunct patterns. Hence, Pantepui in the sense employed by Brown Jr. has very little or nothing to do with the original concept introduced by Mayr and Phelps. Nevertheless, Neild (1996, 2008) in his ongoing monograph of the butterflies of Venezuela, notable for the amount of new Ithomiine butterflies described from the montane level of Pantepui, has chosen to use a slightly modified notion of Brown's approach. This is because Neild has so far treated the taxa as essentially representative of the lowland or middle-elevation fauna.

The diversity of conceptual interpretations and the heterodox use of the word Pantepui have introduced considerable confusion; therefore it is essential to clarify what has been meant by this word. Starting from the original definition of Mayr and Phelps, and considering broadly the features that characterize the tepuis, Huber (1987) was able improve the concept of Pantepui in the following way: "there is a mountain complex in the Guiana Region of southern Venezuela, northern Brazil, northwestern Guyana and southern Suriname, which is mostly constituted by sandstones of the Roraima Formation. The upper portions (approximately above 1200 m) of this set of mountains, more or less isolated from each other, display highly differentiated ecosystems from the geological, geomorphological, geochemical and biological points of view. The set of these middle and high mountain ecosystems is designated with the name of 'PANTEPUI'." Subsequent research on the flora of southern Venezuela (Huber, 1995; Berry et al., 1995) highlights even more the importance of a clear definition of Pantepui as a biogeographic unit that essentially brings together the ecosystems of the upper Guiana in its upper montane and high-tepui levels, usually above 1500 m.

A refined concept appears in the Glosario Fitoecológico de las Américas (Huber and Riina, 1997) as follows: "phytogeographic province of the Guiana Region that embraces all the high-montane ecosystems of the Guiana Shield developed in the summits of the sandstone table-mountains ("tepuis") and some granite mountains (Sierra de Maigualida). It is characterized by its specialized montane bushes and grasslands. NEOTROPICS Guyana (Mount Wokomong and Mount Ayanganna), meridional Venezuela, septentrional Brazil; approx. 1500–3014 m. Very high endemism. Representative families: Hymenophyllopsidaceae, Saccifoliaceae, Theaceae, Rapateaceae, Sarraceniaceae, Ericaceae, Asteraceae".

Such a concept of Pantepui complements well the original definition of Mayr and Phelps, emphasizing its altitudinal connotation. However, it corresponds to a notion mainly based on the use of biotic elements as a biogeographic unit (particularly floristic associations like bushes and grasslands). It places the possession of high endemism as an outstanding feature, but it is not implicit in its characterization. Therefore it is a definition essentially chorologic or phythogeographic, which in general would only allow one to explain partly the application of the term Pantepui to zoological groups without losing the original conceptual meaning.

For instance, in the case of the butterflies, the lower limit of 1500 m established for the upper montane plant communities is necessarily approximate. The altitudinal range of the habitat of these flying insects could show a variation of several hundred meters, whereas not all tepuis are identical in their vertical zonation (Fig. 9.1). However, it is just around 1500 m a.s.l. that vegetation switches to upper montane, which is featured by high degrees of endemism in plants. That would probably have a strong effect on the distribution of butterfly taxa, as the larvae of these insects are phytophagous and may be markedly specific to their host plants. In fact, butterfly endemism in the tepuis appears gradually from the middle elevations ($\approx$1000 m a.s.l.), at the montane level, and increases notably with altitude.

It is important to mention that the increase in endemism is not only directly affected by the increasing elevation but also by the size of the ecologically suitable area available above the "line" where endemism begins to appear. It is assumed that area size has something to do with the total number of species present in certain

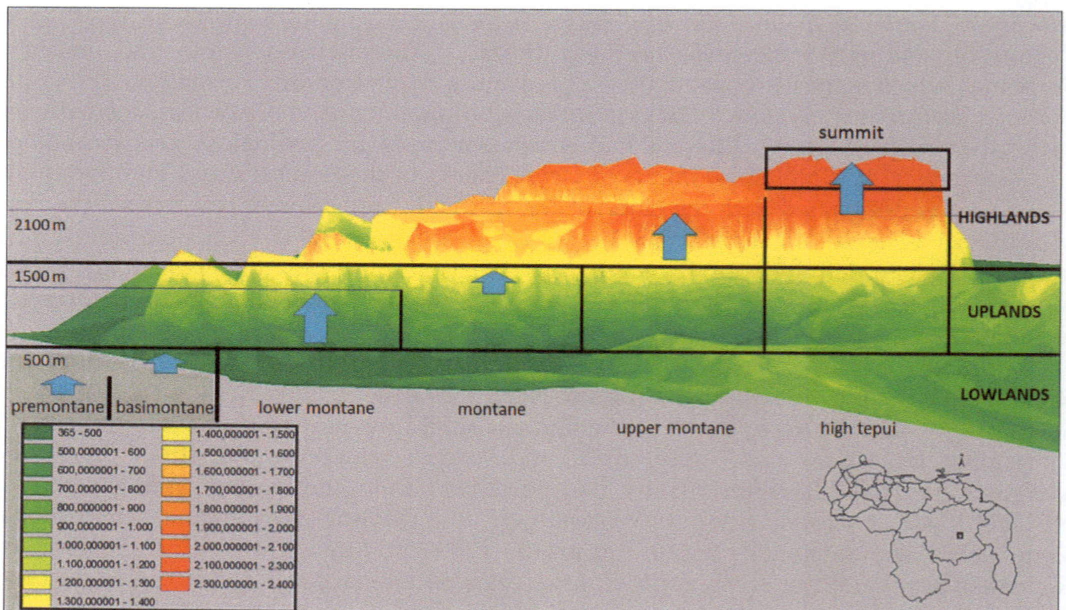

FIGURE 9.1 Physiography and representation of altitudinal levels and life zones on a digital model of Auyán-tepui, Venezuela (based on ALOS World 3D of JAXA). Zonation follows Huber (1995).

biotopes. In general, this phenomenon has been already explored in Andean butterflies (Viloria, 2002). For a better comprehension of the vertical levels of a tepui as treated here, see Fig. 9.1.

Exploration, discovery, and taxonomic studies of butterflies in the tepuis

The evolutionary radiation of the butterflies appears to be correlated with the emergence, proliferation, and diversification of the angiosperms (flowering plants, on which both larvae and adult butterflies generally feed) (Wahlberg et al., 2009; Heikkila et al., 2011). These plant groups are dated to the Mesozoic, over 150 million years ago (Ma) (Bell et al., 2010). The inferred age for the modern butterfly family groups is around or older than 70 Ma (Viloria, 2003; Braby et al., 2006; Wahlberg, 2006; Peña and Wahlberg, 2008; Wahlberg et al., 2009; Simonsen et al., 2011; Espeland et al., 2015; De Jong, 2016). We can assume that butterflies probably existed in the Guiana Shield prior to the uplifting of the Andes. However, this assumption remains a conjecture because there is no known fossil butterfly earlier than Eocene, between 48 and 55 Ma (Tyler et al., 1994; De Jong, 2016).

Scientific exploration of the upper zones of the tepuis began in the middle of the 19th century. Godman et al. (1887) described *Calycopis matho* (Lycaenidae, Theclinae) based on an individual butterfly captured by the British explorer and ornithologist Henry Whitely. This specimen is currently at the Natural History Museum in London (NHMUK), labeled "R. Carimang, British Guiana." It represents an endemic species of the higher elevations of some tepuis, as witnessed by subsequent collecting. Whitely was probably the first naturalist to reach the montane and upper montane levels of a tepui during his pioneer expeditions to Roraima-tepui and Kukenán-tepui, between 1881 and 1883. His collecting trips required great efforts, departing from the Mission of Bartica Grove, along the Essequibo River (Guyana), with the main goal of studying the avifauna of the region. The account of his expeditions (Whitely, 1884) describes in some detail his approaching route to Kukenán-tepui in 1881 along the Mazaruni until reaching its confluence with the Carimang River (or Caramang, according to modern toponymy). Guided by local aborigines, he surmounted the Carimang to the side slopes at the base of Kukenán-tepui, probably reaching no higher than 1500 m. In 1883, Whitely appears to have climbed up above 2000 m while attempting the summit of Roraima-tepui. It is probable that the type specimen of *C. matho* came from the latter expedition, as recent observations by the second author of this chapter (M.C.) indicate that this species flies only between 1500 and 2000 m. Whitely also collected other butterfly species in Roraima, some of which were described later on, such as *Dismorphia crisia roraimae* Hall, 1939, *Heliconius elevatus roraima* Turner, 1966, and *Melinaea lilis kayei* Brown, 1977b.

The success of Whitely quickly encouraged the preparation of further expeditions. E. Im Thurn and H.J. Perkins departed from Great Britain to British Guiana the next year. They left Georgetown in 1884 and reached the summit of Roraima-tepui (Im Thurn, 1885; Perkins, 1885). It is possible, although there is no direct evidence to prove it, that several samples of Lepidoptera currently held in the NHMUK came from that ascent and from others either to Roraima (Anonymous, 1892; or the ornithological expeditions of F.V. McConnell and J.J. Quelch in 1894 and 1898) or to other mountains of the region (Im Thurn, 1882). Some of them represent the types of butterfly subspecies described by the end of the 19th century (i.e., *Dismorphia zathoe proserpina* Grose-Smith and Kirby, 1897) or almost 100 years later (i.e., *Opsiphanes invirae roraimaensis* Bristow, 1991), which to date appear to be endemic to Pantepui. In 1905 Kaye described the ithomian *Napaeogenes silphys potaronus* from the entomological material obtained by C.B. Roberts in 1902 in the upper Potaro River (Guyana), a subspecies belonging to the region but also found flying at lower elevations. The German botanist Ernst Ule, who developed a successful career as an explorer in South America, especially in Brazil (Harms, 1915), had also reached considerable elevations during his exploration of Roraima-tepui (Ule, 1911). Not only did he secure a significant collection of plants (Ule, 1914) but also succeeded in securing zoological specimens. Among the latter, two species of butterflies, *Pedaliodes roraimae* Strand, 1912 and *Antirrhea ulei* Strand, 1912, should be considered endemic to Pantepui (Viloria and Pyrcz, 1995; Orellana, 2004).

Perhaps the most important contributions to the discovery of the butterflies of Pantepui during the first half of the 19th century are due to the efforts of Anglo-American zoologist and explorer George H.H. Tate, who led the Lee Garnett Day Expedition to Roraima-tepui in 1927–1928 (Tate, 1928, 1930, 1932) and the Tyler Expedition to the Cerro Duida between 1928 and 1929 (Tate and Hitchcock, 1930). He also played a leading role in the Phelps Expedition to Auyán-tepui (Tate, 1938). All these enterprises successfully accomplished the objectives of collecting representative zoological specimens in those mountains, as well as reaching their summits. We do not know if Tate managed to collect butterflies on Auyán-tepui, but he certainly did so in Roraima-tepui and Cerro Duida. As a result, several specialists produced the first faunistic lists of butterflies of the region, and a number of taxa were described, particularly hesperids (Bell, 1932) and pierids (Brown, 1932). The riodinids and lycenids collected by Tate in Roraima and Duida were studied by Huntington (1933), who apparently only detected taxa that are widely distributed and are mainly dwellers of the lowlands. Forbes (1942) also published a list of butterflies of the Venezuelan Guayana, basing his results in the study of material captured at lower altitudes. In a later paper, he described *Mechanitis lysimnia bipuncta* from specimens taken by Pablo Anduze in 1941 in the Surukún basin, upper Caroni River (Forbes, 1948).

Reissinger (1970) described several Pantepuian pierids discovered among the entomological material obtained by C. Lindemann in 1964 on the Brazilian slopes of the Sierra de la Neblina. That year coleopterists Jan and Bohumila Bechyné collected insects in the same region (Brown 1977b). Subsequent studies of the butterflies that they found revealed the existence of several unknown and endemic taxa (for instance, the ithomians *Melinaea mnasias neblinae* Brown, 1977b and *Pteronymia alicia* Neild, 2008). A complex multidisciplinary scientific expedition to the Sierra de la Neblina took place between 1983 and 1987 (Brewer-Carías, 1988). During separate campaigns a number of entomologists from different countries visited selected localities, among them were José Clavijo and Jurg De Marmels from the Central University of Venezuela, as well as Robert K. Robbins and colleagues from the Smithsonian Institution (United States). They made representative collections of butterflies in the montane and upper montane zones of this massif. Several endemic taxa were later described based on these expeditions (Viloria, 1995; Neild, 2008).

During the decades of 1950 and 1960 a group of Venezuelan entomologists and collectors (Francisco Fernández-Yépez, Carlos Julio Rosales, Francisco Romero, Harold Skinner, Aníbal Chacón, Carlos Bordón, etc.) started to visit and explore the Sierra de Lema, chiefly because of the opportunity offered by the opening of the road between El Dorado and

Santa Elena de Uairén in the mountain sector known as La Escalera. Butterfly collecting activities intensified in the 1990s (mainly by Mauro Costa, Andrew Neild, Tomasz Pyrcz, Andrés Orellana, and Jürg De Marmels). From that time Mauro Costa has established a continuous and systematic program of exploration in this region, yielding considerable improvement of the knowledge of the butterfly fauna found at the lower-altitude level of Pantepui. Because of its relatively good accessibility, the region of La Escalera has proved to be important for lepidopterological discoveries, including the recent description of species and subspecies of pierids, papilionids, nymphalids, lycaenids, and hesperids peculiar to middle elevations (Pyrcz, 1995; De Marmels et al., 2003; Bollino and Costa, 2004; Pyrcz and Fratello, 2005; Neild, 2008; Neild and Romero Montesino, 2008; Orellana, 2010; Zubek and Pyrcz, 2011; Bálint and Costa, 2012, Costa et al., 2014a, 2016, 2017).

Starting in the 1980s entomologists of the Institute of Agricultural Zoology (MIZA) of the Central University of Venezuela (UCV), such as José Clavijo, Jürg De Marmels, Luis José Joly, John Lattke, and Aníbal Chacón, joined in various expeditions to several tepuis and connected mountain range summits (like Serranía Tapirapecó). These expeditions were organized and funded by Venezuelan institutions like the Foundation for the Development of Physical, Mathematical and Natural Sciences (FUDECI) and Terramar Foundation. Valuable butterfly collections were deposited in the MIZA obtained during these expeditions. Part of these samples are still unique and have been essential as voucher specimens for many of the descriptions of endemic species and subspecies from Roraima-tepui and Kukenán-tepui (Viloria and Pyrcz, 1994, 2000; Neild, 2008), Auyán-tepui (Neild, 1996; Pyrcz and Neild, 1996, 2008; De Marmels and Clavijo, 1998; De Marmels 1999; Viloria and Pyrcz, 2000; Orellana, 2010; Attal and De Marmels, 2012), Serranía Tapirapecó (Viloria, 1998), Serranía Yutajé (Viloria and Pyrcz, 2000; De Marmels et al., 2003; Neild, 2008), Cerro Marahuaka (Viloria and Pyrcz, 2000; Neild, 2008), and Cerro Yaví (Viloria and Pyrcz, 2000).

Between 2007 and 2018, the second author of this chapter (M.C.) conducted 20 lepidopterological expeditions to the mountains of the Eastern Pantepui (Roraima-tepui, Uei-tepui, Ptari-tepui, Sororopán-tepui, Auyán-tepui, Akopán-tepui, Chimantá massif, and Sierra de Lema), discovering an unexpected number of new endemic taxa (Bálint and Costa, 2012; Costa et al., 2014a, 2016, 2017, 2019). Likewise, since 2016 three brief expeditions have been led to the Cerro Cuao (Amazonas state, Venezuela), in the Northwestern Pantepui, some 750 km west of Roraima-tepui. The latter have resulted in the discovery of new biotic elements not detected in the Eastern Pantepui, which are under study and in the process of description (Costa et al., 2019).

On the Guyana side, Steven Fratello and collaborators have been exploring some tepuis such as Roraima-tepui, mounts Wokomong and Ayanganna, and the Acarai Mountains (Fratello 1996a,b, 1999, 2000, 2001a,b, 2003, 2004a,b, 2005). From these localities, only Roraima-tepui and mounts Wokomong and Ayanganna are high enough to be considered part of the biogeographic Pantepui (see later). Samples of butterflies from Mount Ayanganna and the Guyanan slopes of Roraima-tepui have been studied for the description of some endemic taxa: pierids (Bollino and Costa, 2007), riodinids (Hall, 2005, 2006; Costa et al., 2017), and nymphalids (Pyrcz and Fratello, 2005; Neild 2008).

In our opinion, known records of butterflies from the Colombian Serranía de Chiribiquete, which in fact belongs geologically to the Guiana Shield (Constantino and Johnson, 1997; Salazar and Constantino, 2007), do not qualify as records for the biogeographic Pantepui, as such range and table mountains (low tepuis) do not reach a high-enough altitude as to properly develop upper montane and high-tepui ecosystems (maximum elevation 840 m). In the same way, some recently described species of satyrine butterflies of the omnibus genus *Euptychia* Hübner 1818, evidently distributed in the North Amazonian subregion, bordering the area of the tepuis (e.g., *Euptychia alacristata* Neild et al., 2014) or found across the Guiana Shield (e.g., *Euptychia audacia* Brévignon, Fratello and Nakahara, *Euptychia aquila* Fratello, Nakahara and Brévignon, in Fratello et al., 2015), could not be considered for inclusion in our Pantepui list of endemics, as they do not fulfill the criteria established in this contribution (see later) as endemic to the Pantepui province, which differs from the view of a number of previous authors who have used the term Pantepui in a more or less different sense. Another purportedly Pantepuian species, *Euptychia roraima* Nakahara et al., 2014, is considered provisionally in our list of endemics, together with other taxa whose true distribution is still insufficiently known.

Some ranges of mountains of Suriname that are part of the Guiana Shield attain maximum altitudes of 1230 m (Julianatop), 1026 m (Tafelberg), and 986 m (Eilerts de Haan); however, there seems to be no information about endemic butterflies in those heights (Gernaat et al., 2012). On the other hand, in the last decades, numerous species and subspecies of butterflies have been described from French Guiana, some of them apparently endemic, albeit denizens of the lower zones. The highest altitude of the country is 851 m (Mount Bellevue de L'Inini).

In spite of the efforts referred to earlier, current knowledge about the butterflies of Pantepui is still limited, particularly when compared to that of the butterflies of the Venezuelan Coastal Range or the Andes. The most recent expeditions have been tasked with collecting on the forested side slopes of the tepuis, where more diversity is expected. Expeditions aided by helicopter transportation usually go only to the top of certain tepuis. The environment of these places is predominantly rocky, with scarce vegetation; consequently the butterfly diversity, if truly endemic, is definitely poor and limited to biotopes reduced in area: low forests, scrubs, or thickets that grow near deep crevasses or at the base of small heights emerging above the plateau. For example, only three species of diurnal butterflies are so far known in the Roraima-tepui summit (2730 m a.s.l., area 34.4 km^2) in spite of this being probably the best-explored tepui and the one that definitely has received more scientific attention: *Protopedaliodes kukenani*, Viloria and Pyrcz 1994; *Protopedaliodes ridouti* Viloria and Pyrcz, 2000 (Satyrinae, Nymphalidae); and *Catasticta duida* Brown, 1932 (Pieridae). On the other hand, biological studies more or less integral of some tepuian systems have completely excluded the Lepidoptera, insects that do not appear to be the more conspicuous on the tepui summits (Villarreal et al., 2002; Aubrecht et al., 2012).

Some time ago the question emerged as to whether the butterfly fauna of Pantepui was truly poor or just poorly known (Pyrcz and Fratello, 2005). Our later studies have made it evident that it was not only poorly studied and scarcely known but also underestimated, both in richness and in endemicity (Costa et al., 2014a, 2016, 2017, 2019). We refer in particular to the fauna that dwells in the cloud forests that surround the tepuis and cover up their side slopes approximately between 1500 and 2200 m. These forests harbor a fauna that is far more diverse than previously estimated and should be the focus of future long-term studies.

Endemic butterflies of Pantepui

Three genera of the Nymphalidae (Satyrinae: *Protopedaliodes* Viloria and Pyrcz, 1994; *Huberonympha* Viloria and Costa, 2016; and Biblidinae: *Archaeogramma* Attal and Viloria, 2014), as well as a considerable number of species and subspecies of other butterflies and skippers (Papilionoidea), inhabit exclusively the natural forests and scrubs of the middle and high prominences immersed in the mainly lowland area known as the Amazonian subregion (*sensu* Morrone 2000, 2001, 2014a,b, 2017), always in more or less isolated units emerging above the tropical zone (tepuis and some connected mountain ranges) (Viloria, 2000, 2005). Before our first essay on the butterflies of Pantepui (Costa et al., 2014b), the distribution of these taxa had not been coupled satisfactorily with a reasonable level of coherence and precision to the mainstream scheduled system of classification proposed for Latin American and the Caribbean (Morrone, 2001), which proceeds according to the following criterion: areas of endemism are the basic units for biogeographical classification and hierarchization (Harold and Mooi, 1994; Hausdorf, 2002). This notion was adopted later as the fundamental premise for the International Code of Area Nomenclature (ICAN) (Ebach et al., 2008).

The butterfly fauna comprising the set of taxa referred earlier (a checklist is provided later) is taxonomically unitary and cohesive, and its genera, species, and subspecies are inhabitants exclusively associated with the plant formations developed in elevated lands, arranged in a spatial discontinuum, like an archipelago of mounts and table mountains, and spread across the lowlands of the provinces of Guiana, Humid Guiana, Roraima, and Imerí (*sensu* Morrone, 2000, 2006, 2017). As an area of endemism on its own, this collection of land is analogous and probably homologous to the phytogeographic entity previously defined by other authors as Pantepui (sensu Huber, 1987, 1994, 1995; Huber and Riina, 1997; see antecedents in Mayr and Phelps, 1955, 1967, 1971).

We characterize here the biogeographic Pantepui only as an area of endemism for part of the insects of the order Lepidoptera (butterflies and skippers) by the presence of the following list of endemic taxa. It is so far preliminary, although further formal studies on other groups of living organisms have already found congruent evidence confirming our proposal (Morrone, 2017). The geographic information accompanying this taxonomic list has been extracted and compiled from previous publications and has been trascribed literally. Some unpublished records obtained during an ongoing research project led by the second author (M.C.) have also been included.

Note: (*) means possibly not endemic to Pantepui. More sampling across the altitude gradient is needed.

Order **LEPIDOPTERA** Linnaeus 1758
Suborder **Glossata** Fabricius 1775
Superfamily **Papilionoidea** Latreille 1802

Family **Papilionidae** Latreille 1802
Subfamily **Papilioninae** Latreille 1802
Tribe **Troidini** Talbot 1939

**Parides phosphorus laurae* Bollino and Costa: Bollino and Costa, 2004: 9–10, [4] Figs. Type locality: Venezuela, Bolívar, Río Surukum [*sic*], carretera Sta. Elena Icabarú. Additional records: S. Fco. de Yuruaní-Roraima, 900–1200 m (Bolívar).

Family **Hesperiidae** Latreille 1809

Subfamily **Hesperiinae** Latreille 1809

**Cantha roraimae* (Bell): Bell, 1932: 8–9, Fig. 1. Type locality: Arabupu, Brazil (10 miles southeast of Mount Roraima, 4200 ft. altitude).

Eutocus arabupuana Bell: Bell, 1932: 7, Fig. 4. Type locality: Arabupu, Brazil (10 miles southeast of Mount Roraima, 4200 ft. altitude).

Eutocus paulo Bell: Bell, 1932: 7–8, Fig. 6. Type locality: Paulo, Brazil (10 miles southwest of Mount Roraima, 4000 ft. altitude).

Megaleas cervelina Orellana and Costa: Orellana and Costa *in* Costa et al., 2019: 79–82, Figs. 1, 2, 11, 12. Type locality: Venezuela, Edo. Bolívar, Auyán-tepui, entre El Danto y El Peñón, 05°44′24″N, 62°32′18″W, 1750 m.

Thespieus duidensis Bell: Bell, 1932: 14–15, Fig. 2. Type locality: Mount Duida (Provisional Camp), Venezuela (on the summit of the mountain, crest of ridge No. 23, 6000 ft. altitude).

Subfamily **Pyrginae** Burmeister 1878

Tribe **Pyrgini** Burmeister 1878

Pachyneuria duidae (Bell): Bell, 1932: 12, Fig. 5. Type locality: Mount Duida (Middle Camp), Venezuela (at the foot of the mountain, 325 ft. altitude).

Subfamily **Pyrrhopyginae** Mabille 1877

Tribe **Pyrrhopygini** Mabille 1877

Jemadia demarmelsi Orellana: Orellana, 2010: 220, 260, Figs. 113, 114, 278, Fig. 195. Type locality: Venezuela: Bolívar, Río Surukúm [*sic*], 870 m.

Pyrrhopyge caribe camachoi Orellana: Orellana, 2010: 205, 254, Figs. 65, 66, 273, Fig. 183. Type locality: Venezuela: Bolívar: entre Sta. Elena de Uairén e Ikabarú, 850 m.

Pyrrhopyge erazoae Orellana: Orellana, 2010: 205–206, 255, Figs. 67, 68, 288, Fig. 227. Type locality: Bolívar: Auyán-tepui, Guayaraca, 1100 m.

Pyrrhopyge tatei (Bell): Bell, 1932: 10, Fig. 3. Type locality: Mount Duida (Provisional Camp), Venezuela (on the summit of the mountain, crest of ridge No. 23, 6000 ft altitude).

Family **Pieridae** Duponchel 1835

Subfamily **Pierinae** Duponchel 1835

Tribe **Pierini** Duponchel 1835

Catasticta duida F.M. Brown (Fig. 9.2A): F.M. Brown, 1932: 2 (Figs. 1–2, 3 pupa), 3, 6. Type locality: Venezuela, Mount Duida, alt. 6500 ft. Additional records: Brown, 1932: 3, 6: Mount Roraima, Brazil, alt. 7400 ft.; Bollino and Costa, 2007: 13–14, Fig. 10: g–h; Fig. 27. Additional records in Venezuela: Yutajé and Cerro Neblina (Amazonas), Kukenán-tepui, Auyán-tepui and Kamarkawarai-tepui (Bolívar).

Catasticta sisamnus ayanganna Bollino and Costa: Bollino and Costa, 2007: 4, Fig. 4, d, e, f, Fig. 13 (distribution). Type locality: Guiana, Region 7, Mount Ayanganna, Koatse R.

Melete leucadia reyi Neild and Costa: Neild and Costa *in* Costa et al., 2017: 280–284, Figs. 38, 39 (male), 40, 41 (female). Type locality: Venezuela, Bolívar, Sierra de Lema, La Escalera, 5°58′N, 61°24′W, 1400 m. Additional records: Neild and Costa *in* Costa et al., 2017: 280, 284: Venezuela, Bolívar: Sierra de Lema, vía El Dorado-Sta. Elena, km 131.7, 1400 m; Auyán-tepui, entre Guayaraca y El Danto, 5°43′20″N, 62°32′30″W, 1400 m.

Pereute lindemannae lindemannae Reissinger: Reissinger, 1970: 413–414, tfl. 1, 2, Fig. 5. Type locality: Brasilia sept., Serra Neblina, 1250 m.

FIGURE 9.2 (A–F) Selected endemic butterflies of Pantepui in their natural habitat. (A) *Catasticta duida* (Uei-tepui), (B) *Antirrhea ulei* (Uei-tepui), (C) *Hyposcada zarepha bomplandi* (Sororopán-tepui), (D) *Greta clavijoi* (Sororopán-tepui), (E) *Eresia carme judithae* (Uei-tepui), (F) *Stalachtis halloweeni* (Ptari-tepui). (A, B, C, photos Paolo Costa).

FIGURE 9.3 (A–L) Visual sample of selected endemic butterflies (Nymphalidae: Satyrinae) of Pantepui (see list in text for distribution data). (A) *Protopedaliodes kukenani* (♂, ventral, Roraima-tepui); (B) *P. kukenani* (♀, ventral, Roraima-tepui); (C) *Protopedaliodes ridouti* (♂, ventral, Roraima-tepui); (D) *Oxeoschistus romeo* (♂, dorsal, Roraima-tepui); (E) ídem, ventral; (F) *Huberonympha neildi* (♂, ventral, Auyán-tepui); (G) *Pedaliodes roraimae* (♂, ventral, Roraima-tepui); (H) *P. roraimae* (♀, ventral, Uei-tepui); (I) *Eretris agata* (♂, ventral, Sierra de Lema); (J) *Forsterinaria hannieri* (♂, ventral, Ptari-tepui); (K) *Stevenaria yutajensis* (♂, ventral, Cerro Cuao); (L) *Stevenaria nakaharai* (♂, ventral, Ptari-tepui).

FIGURE 9.4 (A–P) Visual sample of selected endemic butterflies of Pantepui (see list in text for taxonomic position and distribution data). (A) *Archaeogramma claritae* (♂, dorsal, Auyán-tepui); (B) idem, ventral; (C) *Perisama tepuiensis* (♂, dorsal, Auyán-tepui); (D) idem, ventral; (E) *Mesotaenia delafuentei* (♂, dorsal, Cerro Cuao); (F) idem, ventral; (G) *Actinote genitrix costae* (♂, dorsal, Sororopán-tepui); (H) *Dismorphia zathoe proserpina* (♂, dorsal, Sierra de Lema); (I) *Lieinix* sp. n. (♂, dorsal, Auyán-tepui. Description in prep.); (J) *Hyposcada dujardini humboldti* (♂, dorsal, Sierra de Lema); (K) *Melanis dulcis* (♀, dorsal, Ptarí-tepui); (L) *Pseudopieris viridula mimaripa* (♂, dorsal, Sierra de Lema); (M) *Callithomia lenea bella* (♂, dorsal, Sororopán-tepui); (N) *Pereute lindemannae pemona* (♂, dorsal, Sierra de Lema); (O) *Ocaria elisa* (♀, dorsal, Ptari-tepui); (P) *Dismorphia crisia roraimae* (♂, dorsal, Roraima-tepui).

Pereute lindemannae pemona De Marmels, Clavijo and Chacín (Fig. 9.4N [male dorsal]): De Marmels et al., 2003: 86–90, Figs. 1–4, 16 (uncus), 18 (distribution). Type locality: Venezuela, Bolívar State, Sierra de Lema, road El Dorado–Santa Elena de Uairén, km 125, 1090 m.

Pereute lindemannae piaroa De Marmels, Clavijo and Chacín: De Marmels et al., 2003: 86–90, Figs. 5–8, 17 (uncus), 18 (distribution), 20–21 (valva). Type locality: Venezuela, Amazonas State, Cerro Yutajé, 1750 m, 5°45′N, 66°08′W.

Subfamily **Dismorphiinae** Schatz 1887

Pseudopieris viridula mimaripa De Marmels, Clavijo and Chacín (Fig. 9.4L [male dorsal]): De Marmels et al., 2003: 87 (Figs. 13–14), 88 (Fig. 19, distribution), 90–91. Type locality: Venezuela, Bolívar State, road El Dorado–Santa Elena de Uairén, km 125, 1090 m. Additional records: road El Dorado–Santa Elena de Uairén, km 131.7, 1400 m (De Marmels et al., 2003: 91).

Lieinix nemesis christa (Reissinger): Reissinger, 1970: 411, tfl. 1, 2, Fig. 7. Type locality: Brasilia sept., Serra Neblina, 1500 m.

Lieinix sp. n. Costa et al., in prep. (Fig. 9.4I [male dorsal]).

Dismorphia crisia roraimae Hall (Fig. 9.4P [male dorsal]): Hall, 1939: 167. Type locality: British Guiana, Roraima.

Dismorphia crisia neblina, Reissinger: Reissinger, 1970: 409–410, tfl. 1, 2, Fig. 1. Type locality: Brasilia sept., Serra Neblina, 1500 m.

Dismorphia zathoe proserpina Grose-Smith and Kirby (Fig. 9.4H [male dorsal]): Grose-Smith and Kirby, 1897: 11, pl. Figs. 1, 2 (male), 3 (female). Type locality: British Guiana, Roraima.

Dismorphia zathoe ssp. n. 1 Costa et al., in prep.

Dismorphia zathoe ssp. n. 2 Costa et al., in prep.

Family **Nymphalidae** Rafinesque 1815

Subfamily **Cyrestinae** Guenée 1865

Tribe **Cyrestini** Guenée 1865

Marpesia pantepuiana Costa, Attal and Viloria (Fig. 9.5C [male dorsal]): Costa, Attal and Viloria *in* Costa et al., 2016: 180–186, Figs. 1, 2 (male), 3, 4 (female), 9–12 (male genitalia). Type locality: Venezuela, Bolívar, talud Ptarí-tepui, Salto Iworé, 1400 m, 5°43′34″N, 61°48′36″ W. Additional records in Venezuela: Costa, Attal and Viloria *in* Costa et al., 2016: 180: Bolívar, Ptarí-tepui, between Punto Phelps and base, 1900 m; Bolívar, Sierra de Lema, La Escalera, 1450 m 5°55′50″N, 61°26′12″W; Amazonas, Cerro Yutajé, 1750 m 5°45′N, 66°8′W.

Subfamily **Biblidinae** Boisduval 1833

Tribe **Callicorini** Orfila 1952

Archaeogramma claritae Costa (Fig. 9.4A [male dorsal], B [male ventral]): Costa *in* Costa et al., 2014a: 40, 43–45, Figs. 1, 5–8, 10. Type locality: El Peñón, 1850 m, Auyán-tepui, Estado Bolívar, Venezuela.

Mesotaenia delafuentei Neild (Fig. 9.4E [male dorsal], F [male ventral]): Neild, 1996: 59, pl. 7, Fig. 356. Type locality: (amended by De Marmels and Clavijo, 1998: 217–219, Fig. 1): Venez.[uela], Amazonas, Cerro Yutajé, 1750 m 5°45′N–66°8′W.

Perisama tepuiensis Attal and De Marmels (Fig. 9.4C [male dorsal], D [male ventral]): Attal and De Marmels, 2012: 111–113, Figs. 1–5. Type locality: Venezuela, Bolívar, El Zanjón, cima, Auyantepuy [*sic*], approx. 2000 m. Additional records: De Marmels, 1999 (as *Perisama* spec.): 155–158, pl. Xa, Figs. 1, 2: Auyán-tepui (Bolívar); Costa et al., 2014a: 45, 48 Figs. 11, 12 (male), 49 Fig. 17 (male genitalia), in press: El Peñón, 1850 m, Auyán-tepui (Bolívar).

FIGURE 9.5 (A–H) Visual sample of selected endemic butterfly species and subspecies of Pantepui (see list in text for taxonomic position and distribution data). (A) *Memphis montesino* (♂, dorsal, Sororopán-tepui); (B) *M. montesino* (♀, dorsal, Ptari-tepui); (C) *Marpesia pantepuiana* (♂, dorsal, Sororopán-tepui); (D) *Heliconius elevatus roraima* (♂, dorsal, Roraima-tepui); (E) *Memphis paulus paulus* (♂, dorsal, Sororopán-tepui); (F) *Heliconius elevatus* ssp. n. (♂, dorsal, Auyán-tepui. Description in prep.); (G) *M. paulus paulus* (♀, dorsal, Sororopán-tepui); (H) *Opsiphanes* sp. n. (♂, dorsal, Uei-tepui. Description in prep).

Subfamily **Charaxinae** Doherty 1886

Tribe **Anaeini** Reuter 1896

Memphis montesino Pyrcz (Fig. 9.5A [male dorsal], B [female dorsal]): Pyrcz, 1995: 11–13, Figs. 1–4, 5 (genitalia). Type locality: Venezuela, Bolívar State, 30 km west of Santa Elena de Uairén, Río Surukum [*sic*], 1000 m; Pyrcz and Neild, 1996: 116, pl. 27, Figs. 1092–1093: "South-eastern Pantepui."

Memphis paulus paulus Orellana and Costa (Fig. 9.5E [male dorsal], G [female dorsal]): Orellana and Costa *in* Costa et al., 2014a: 46–51, Figs. 13, 14 (male), 15, 16 (female), 18 (male genitalia), 19 (wing shape, female), 20 (wing shape, male). Type locality: La Escalera, km 132 sur El Dorado (05°54′50″N; 61°26′10″O), 1450 m, Sierra de Lema, Edo. Bolívar, Venezuela.

Memphis paulus ssp. n. Costa et al., in prep.

Memphis viloriae Pyrcz and Neild: Pyrcz and Neild, 1996: 114, pl. 26, Figs. 1055–1056. Type locality (amended by De Marmels and Clavijo, 1998: 217–219, Fig. 1, distribution): Venez.[uela], Amazonas, Cerro Yutajé, 1750 m, 5°45′N–66°8′W.

Subfamily **Nymphalinae** Rafinesque 1815

Tribe **Melitaeini** Newman 1870

Eresia carme judithae Neild (Fig. 9.2E): Neild, 2008: 85–86, 233. Type locality: Vía El Dorado–Santa Elena, km 122, Sierra de Lema, 1440 m, SE Bolívar, SE Venezuela. Additional records: same road, kms 122–130, 1400–1440 m.

Junonia evarete oscura Neild: Neild, 2008: 59–60, 232, pl. 7, Figs. 191–194. Type locality: Vía Salto Mosquito, approx. 15 km SSW of San Ignacio de Yuruani, SE Bolívar, SE Venezuela, 900 m. Additional records: Forbes, 1929: 312 (as *J. lavinia* ab. *nigralis*): Kaieteur, British Guiana; Neild, 2008: 232: Kavanayen Rd., Canaima N. Park, Bolívar, Venezuela, 1000 m; Bolívar, P.N. Kanaima, Sn. I. Yuruaní a Peray-Tepui, 1300 m; Roraima; La Escalera, 1100 m, Sierra de Lema, Edo. Bolívar, Venezuela; S. Fr. Yuruani, Bolívar; km 7 vía Roraima; km 23 vía Icabarú; Guayaraca, Auyán-tepui, BO, Venezuela, 1100 m; Uonken, BO, Venezuela, 850 m; Kavanayen, Bolívar, Venezuela, ±1300 m; Venezuela, Bolívar, Río Surukum [*sic*], Carretera Sta. Elena-Icabarú, 850 m; Sta. Elena, Venezuela, Guayana.

Junonia genoveva vivida Forbes: Forbes, 1929: 305–320. Type locality: Ireng R. to Roraima, Brazil. Additional records: Neild, 2008: 61–62, pl. 8, Figs. 207–210: Guiana Shield (Eastern Pantepui).

Hypanartia lethe rosamariae Neild: Neild, 2008: 52–53, 232, pl. 6, Figs. 155–158. Type locality: Vía Sta. Elena–Icabarú, km 24.6, SE Bolívar State, 1050 m. Additional records: Neild, 2008: 232: Vía Sta. Elena–Icabarú, km 23, SE Bolívar State; Track to Apoipó, north of km 35 (junction to Betania) on Sta. Elena to Icabarú Road, SE Bolívar State, SE Venezuela, 850–1000 m; 7 km W Waramasén (km 15 vía Icabarú), 1050 m, Estado Bolívar, Venezuela; Alto Orinoco, 300 m, Lat. N2°05′/2°10′ Long. O 63°50′, Venezuela; Alto Orinoco, 500 m, Lat N 2°15′/2°20, Long O 63° 35′; "La Cumbre", T.F.A.

Vanessa myrinna ssp. n. Costa et al., in prep.

Subfamily **Limenitidinae** Behr 1864

Tribe **Limenitidini** Behr 1864

Adelpha irmina arcana Benmesbah and Costa: Benmesbah and Costa *in* Costa et al., 2019: 82–87, Figs. 13, 14 (male), 15, 16 (female), 22 (wing venation), 24 (abdomen), 25, 26 (male genitalia), 27 (distribution). Type locality: Venezuela, Bolívar, Sierra de Lema, La Escalera,

1400 m. Additional records: Venezuela, Bolívar: Sierra de Lema, La Escalera: Salto El Danto, 1200 m; km 122 vía El Dorado-S. Elena, 1400 m; Auyán-tepui: El Danto, 1500 m; Quebrada antes de El Peñón, 1750 m; Sororopán-tepui, 1650–1950 m; Ptarí-tepui: Salto Iworé, 1400 m; entre Salto Iworé y Punto Phelps, 1500 m; entre Salto Iworé y Cumbre, 1800 m; Guiana, Region 7, Mount Ayanganna 05°24.1′N, 59°57.4′W, 3300–4500′.

Subfamily **Satyrinae** Boisduval 1833

Tribe **Morphini** Newman 1834

Antirrhea ulei Strand (Fig. 9.2B): Strand, 1912: 44. Type locality: "Venezuela, Bolívar," "S [ierr]a de Lema. km 125, 1500 m," by subsequent designation (Orellana, 2004: 28). Additional records: Strand, 1912: 44: "Die Lokalität Roraima liegt in Venezuela an der Grenze von Britisch-Guiana, unter 5 Grad nördlicher Breite"; Fratello, 1996a: 4: Guiana, Wokomong Tepui, up to 5000 ft (as *Antirrhea* sp.), 1999: 4: Guiana, Mount Ayanganna, 5000 ft, 2000: 7: Guiana, Mount Ayanganna; Orellana, 2004: Auyán-tepui and Sierra de Lema; Orellana, 2008: "Pantepui highlands of south-eastern Venezuela, west-central Guiana, and perhaps also in adjacent North Brazil."

Tribe **Brassolini** Boisduval 1836

Caligo suzanna ssp. n. Costa et al., in prep.

Opsiphanes invirae roraimaensis Bristow: Bristow, 1991: 258, Fig. 22D, 261, Fig. 23B, 263, Fig. 24 (distribution). Type locality: Guiana, Roraima.

Opsiphanes sp. n. Costa et al., in prep. (Fig. 9.5H [male dorsal]).

Tribe **Satyrini** Boisduval 1836

**Euptychia roraima* Nakahara, Fratello and Harvey: Nakahara et al., 2014: 292–296, Fig. 1 (male), 2, 3 (male genitalia). Type locality: Guiana: N. slope Mount Roraima, 1st Camp, 800 m, 5°17′N, 0°45′W.

Forsterinaria hannieri Zubek and Pyrcz (Fig. 9.3J [male ventral]): Zubek and Pyrcz, 2011: 525–528, Figs. 1, 2, 3 (male genitalia), 4 (female genitalia). Type locality: Venezuela, Estado Bolívar, La Escalera (Santa Elena road) km 130, 1400 m. Additional records: Zubek and Pyrcz, 2011: 525: Venezuela, Estado Bolívar, La Escalera, km 123, 1400–1450 m; La Escalera, km 129, 1400–1450 m; Estado Bolívar, Sororopán, 1700 m; Estado Bolívar, Kavanayén, 1300 m.

Forsterinaria sp. n. Costa et al., in prep.

Huberonympha neildi Viloria, Costa, Fratello and Nakahara (Fig. 9.3F [male ventral]): Viloria et al. *in* Costa et al., 2016: 187, Fig. 17 (wing venation), 188–191 Figs. 20, 22, 23 (male), 24, 25 (female), 26 (male genitalia). Type locality: Venezuela, Bolívar, Auyán-tepui, entre El Danto y El Peñón, 1700 m. Additional records: Viloria et al. *in* Costa et al., 2016: 188, 192: Venezuela, Bolívar, same as type locality, but 1400, 1750, 1850 m; Venezuela, Bolívar, Akopán-tepui, entre Inká y Base, 1900 m; Venezuela, Bolívar, Ptarí-tepui, Base, 2100 m; Guiana, Region 7, Mount Ayanganna 5°24.1′N, 59°57.4′W, 1585 m.

Stevenaria nakaharai Viloria, Costa, Fratello and Neild (Fig. 9.3L [male ventral]: Viloria et al. *in* Costa et al., 2016: 194, Figs. 36, 37 (male), 38, 39 (female), 197–198, Fig. 41 (male genitalia). Type locality: Venezuela, Bolívar, Akopán-tepui, Inká, 1820 m. Additional records: Viloria et al. *in* Costa et al., 2016: 192, 197: Venezuela same as type locality, but 1850 m; Venezuela, Bolívar: Akopán-tepui, Camp. Base, 2000 m; Akopán-tepui, entre Inká y Base, 1900 m; Venezuela, Bolívar: Sierra de Lema, La Escalera km 122, 1440 m; km 123, 1420 m; km 125, 1400 m; km 135,1450 m; Sierra de Lema, Sororopán, Salto Iworé, 1400 m;

Venezuela, Bolívar: Auyán-tepui, El Danto, 1500 m; Auyán-tepui, 1700 m; Auyán-tepui, El Peñón, 1850 m; Auyán-tepui, entre Libertador y El Oso, 2200 m; Venezuela, Bolívar: Ptarí-tepui, entre Punto Phelps y Base, 1900 m; Ptarí-tepui, Punto Phelps, 1800 m; Venezuela, Bolívar, Talud Roraima, 2100 m; Venezuela, Bolívar, San Francisco de Yuruaní, 1200 m; Guiana, Region 7, Mount Ayanganna 5°24.1′N, 59°57.4′W, 1585 m.

Stevenaria yutajensis Viloria and Costa (Fig. 9.3K [male ventral]): Viloria and Costa *in* Costa et al., 2016: 194, Figs. 32, 33 (male), 34, 35 (female), 195–197 Fig. 40 (male genitalia). Type locality: Venezuela, Amazonas, Cerro Yutajé, 5°45′N, 66°8′W, 1750 m.

Eretris agata Pyrcz and Fratello (Fig. 9.3I [male ventral]): Pyrcz and Fratello, 2005: 201–209, Figs. 1, 2, 6 (genitalia), 14. Type locality: Venezuela, Estado Bolívar, Sierra de Lema, road El Dorado–Santa Elena de Uairén km 132, 1350–1400 m.

Eretris cuaoensis Viloria and Attal: Viloria and Attal *in* Costa et al., 2019: 96–100, Figs. 44, 45 (male), 46, 47 (female), 52,53 (male genitalia). Type locality: Venezuela, Amazonas, Serranía del Cuao, Cerro Paraka, Caño Paraka, 05°05′07″N, 67°27′11″W, 1450 m. Additional records: Viloria and Costa *in* Costa et al., 2019: 96: Venezuela, Amazonas, Serranía del Cuao: Valle Caño Negro (Woroi), 5°04′40″N, 67°27′07″W, 1350 m; ídem, 5°05′07″N, 67°26′26″W, 1360 m.

Oxeoschistus romeo Pyrcz and Fratello (Fig. 9.3D [male dorsal], E [male ventral]): Pyrcz and Fratello, 2005: 204–209, Figs. 3, 4, 8 (genitalia). Type locality: Guiana, N. slope Mount Roraima, 2nd Camp, 1300 m 5°16′N, 60°44′W.

Pedaliodes chaconi Viloria: Viloria, 1998: 15–16, 18, Figs. 1, 2 (genitalia). Type locality: Venezuela, T.F. Amazonas, Sierra de Tapirapecó, Cerro Tamakuari, a 1.300 m, 01°13′N, 64°41′30″W.

Pedaliodes demarmelsi Viloria: Viloria, 1995: 526–528, pl. 17a, Fig. 1. Type locality: Venezuela, T.F. Amazonas, Depto. Río Negro, Cerro de La Neblina, foot of Pico Phelps, 2000 mt. Additional records: Viloria, 1995: 526: Venezuela, T.F. Amazonas, Cerro de La Neblina, Camp X, 0°54′N, 60°2′W, 1690 m; Venezuela, T.F. Amazonas, Cerro de La Neblina, Camp II 0°50′N, 65°59′W, 2100 m y 0°49′N, 65°59′W, 2100 m.

Pedaliodes parakana. Viloria and Costa: Viloria and Costa *in* Costa et al., 2019: 90–95, Figs. 32, 33 (male), 34, 35 (female), 40, 41 (male genitalia). Type locality: Venezuela, Amazonas, Serranía del Cuao, Cerro Paraka, Caño Paraka, 05°05′07″N, 67°27′11″W, 1450 m. Additional records: Viloria and Costa *in* Costa et al., 2019: 90: Venezuela, Amazonas, Serranía del Cuao: Valle Caño Negro (Woroi), 5°04′40″N, 67°27′07″W, 1350 m; ídem, 5°05′07″N, 67°26′26″W, 1360 m.

Pedaliodes roraimae Strand (Fig. 9.3G [male ventral], H [female ventral]): Strand, 1912: 48. Type locality: "Die Lokalität Roraima liegt in Venezuela an der Grenze von Britisch-Guiana, unter 5 Grad nördlicher Breite"; Viloria and Pyrcz, 1995: 584–585: Venezuela, Bolívar, Gran Sabana, Roraima, 1800 m; Venezuela, Bolívar, El Dorado–Santa Elena, km 155, 1280 m.

Pedaliodes terramaris Viloria and Pyrcz: Viloria and Pyrcz, 2000: 91–97, Figs. 4, 9 (genitalia). Type locality: Auyán-Tepui, 1500 m, Bolívar, Venezuela, 5°57′N, 62°39′W.

Pedaliodes yutajeana Viloria and Pyrcz: Viloria and Pyrcz, 2000: 93–97, Figs. 5, 6, 10 (genitalia). Type locality: Cerro Marahuaka, 2470 m, Parque Nacional Duida-Marahuaka, Amazonas, Venezuela 3°37′N, 65°22′W. Additional records: Viloria and Pyrcz, 2000: 93:

Cerro Yutajé, 1750 m, Amazonas, Venezuela 5°45′N, 65°08″W; Cerro Yaví, 2200 m, Amazonas, Venezuela, 5°43′N, 65°54′W.

Protopedaliodes kukenani Viloria and Pyrcz (Fig. 9.3A [male ventral], B [female ventral]): Viloria and Pyrcz, 1994: 346–352, Figs. 1–4, 5 (genitalia), 6 (wing venation). Type locality: Cerro Kukenán, Grand [*sic*] Sabana, Bolívar State, Venezuela [2600]–2700 m, 5°19′N, 60°49′W.

Protopedaliodes profauna Viloria and Pyrcz: Viloria and Pyrcz, 2000: 91–97, Figs. 2, 3, 8 (genitalia). Type locality: Auyán-Tepui, 1700 m, Bolívar, Venezuela 5°58′N, 62°32′W. Additional records: Viloria and Pyrcz, 2000: 91: Auyán-Tepui, 1800 m, Bolívar, Venezuela 5°51′N, 62°35′W.

Protopedaliodes ridouti Viloria and Pyrcz (Fig. 9.3C [male ventral]): Viloria and Pyrcz, 2000: 91–97, Figs. 1, 7 (genitalia). Type locality: Mount Roraima, 8000 ft, Venezuela.

Subfamily **Danainae** Boisduval 1833

Tribe **Ithomiini** Godman and Salvin 1879

Greta clavijoi Neild (Fig. 9.2D): Neild, 2008: 180–181, 238, pl. 39, Figs. 1141–1145. Type locality: Venez. Amazonas, Cerro Yutajé, 1750 m, 5°45′N–66°8′W. Additional records: Neild, 2008: 238: Vía El Dorado–Santa Elena, Sierra de Lema, 1400 m, Bolívar, SE Venezuela; La Escalera, kms 119–130, 1100–1400 m, Estado Bolívar, Venezuela; Ptari-Tepui, Bolívar, Venezuela; Campamento V, Cerro Neblina, T.F. Amazonas, Venezuela, 1200 m; Campamento VII [Cerro Neblina], T.F. Amazonas, Venezuela, 1370 m; Venezuela, T.F. Amazonas, Cerro de la Neblina, Camp XI, 1490 m, 0°52′N–65°58W′; Cerro Neblina, Norte Brasil, 1270 m; Brasilia sept. Serra Neblina, 1500 m; Venezuela, T.F. Amazonas, Parque Nacional Duida Marahuaka/Marahuaka–Macizo Central, 3°40′N–65°28′W, 1040 m; Guiana, Region 7, Mount Ayanganna 3300′–4500′, 05°24.1′N–59 57.4′W; idem 4500′–5500′.

Hypothyris ninonia lema Brown: Brown, 1977a: 125 (Figs. 31A, 31B, 31C), 126. Type locality: Venezuela: Bolívar, Carretera El Dorado a Santa Elena, km 125, 5°58′N, 61°28″W, 1100 m [*sic*]. Additional records: Neild, 2008: 119–122, pl. 23: Figs. 653–657: Venezuela, Eastern Bolívar State, Sierra de Lema, 1400 m.

Hypothyris ninonia connexa Hall: Hall, 1939: 29, pl. 8. Type locality: Br. Guiana, Roraima.

Hyposcada dujardini humboldti Neild (Fig. 9.4J [male dorsal]): Neild, 2008: 234. Type locality: km 125, El Dorado–Santa Elena, BO, Venezuela, 1100 m. Additional records: Neild, 2008: 235: Venezuela, Estado Bolívar, Sierra de Lema, La Escalera, km 130, 1400–1450 m; B. Guiana, Roraima.

Hyposcada zarepha bonplandi Neild (Fig. 9.2C): Neild, 2008: 235. Localidad típica: km 125, El Dorado–Santa Elena, BO, Venezuela, 1100 m. Additional records: Neild, 2008: 235: Venezuela, Estado Bolívar, Sierra de Lema, kms 116.8–155 El Dorado–Santa Elena, 1100–1400 m.

Oleria boyeri Neild: Neild, 2008: 235. Type locality: km 125 El Dorado–Santa Elena, BO, Venezuela, 1100 m. Additional records: Neild, 2008: 235: Venezuela, Estado Bolívar, Sierra de Lema, kms 120–133 El Dorado–Santa Elena, 900–1500 m; Guiana, Region 7, Mount Ayanganna, 4500′–5500′.

Pagyris renelichyi Neild: Neild, 2008: 139–142, 235, Fig. 1.16 (male genitalia), pl. 28, Figs. 802–803. Type locality: Caño Tehuri, Base du Marahuaca, Amazonas central, Venezuela.

Eutresis hypereia imeriensis Brown: Brown, 1977a: 113, Fig. 21, 117. Type locality: Brasil: Amazonas, "Cerro Neblina, Norte Brasil" (provavelmente 0°43′N, 66°07′W, num tributáro do alto Rio Cauaburi), 900 m. Additional records: Brown, 1979; Neild, 2008: 95: [Brasil], "upper Rio Tucano."

Melinaea lilis kayei Brown: Brown, 1977b: 167–169, Fig. 24. Type locality: Guiana, Quonga (probably about 6°30′N, 59°W).

Mechanitis lysimnia bipuncta Forbes: Forbes, 1948: 19–20, pl. 1, Fig. 7. Type locality: Surukum [*sic*] basin, upper Caroni River, Venezuelan Guiana.

Callithomia lenea bella Neild (Fig. 9.4M [male dorsal]): Neild, 2008: 146, 235, pl. 30, Figs. 856–859. Type locality: Sierra de Lema, Venezuela, Bolívar, km 112, 1050 m. Additional records: Neild, 2008: 235: Vía El Dorado–Santa Elena, kms 117–132, La Escalera, Sierra de Lema, SE Bolívar, Venezuela, 1100–1400 m; La Escalera, Luepa, 2 km before the end of the forest, Bolívar, Venezuela.

Dircenna adina stevei Neild: Neild, 2008: 152–153, 236, pl. 32, Figs. 902–905. Type locality: km 125, El Dorado–Santa Elena, BO, Venezuela, 1100 m. Additional records: Neild, 2008: 236: Venezuela, Bolívar, kms 123–155, El Dorado–Santa Elena, La Escalera, Sierra de Lema, 1100–1400 m; Brazil, Mount Roraima, Camp Glycor, 6000′.

**Dircenna dero christopheri* Neild: Neild, 2008: 148–151, 235–236, Fig. 1.18 (male genitalia), pl. 30, Figs. 870–873. Type locality: Vía Sta. Elena–Icabarú, km 30.6, SE Bolívar State, SE Venezuela, 900 m. Additional records: Neild, 2008: 235–236: Vía Sta. Elena–Icabarú, km 35–40, 850–1000 m, SE Bolívar State, SE Venezuela; Surukum [*sic*] basin, Upper Caroni, Ven.[ezuela]; Río Surukún, 870 m, Edo. Bolívar; Icabarú Wari-Wantey; Icabarú 950 m; Wariwantey, Bolívar, Venezuela, 900 m; Venezuela, Bolívar, 700 m, Waramasén, Gran Sabana, Bosque Mata Cuchilla; Venezuela, Amazonas, P.N. Parima–Tapirapecó, Parima, 820 m.

**Episcada doto paquito* Neild: Neild, 2008: 160, 236–237, pl. 34, Figs. 970–974. Type locality: Río Surukún, 870 m, Estado Bolívar, Venezuela. Additional records: Neild, 2008: 236–237: Arabupu, Ven.[ezuela]; kms 40–44 vía Icabarú, SE Bolívar State, Venezuela, 900 m; kms 20–35 Santa Elena–Icabarú Road, SE Bolívar, Venezuela, 850–1000 m; vía El Dorado–Sta. Elena, km 254, SE Bolívar State, SE Venezuela, 830 m; vía Roraima, km 7 (1200 m), Edo. Bolívar, Venezuela; Uaiparú, 700 m, Estado Bolívar, Venezuela; 10 km NE Kavanayén, 1350 m, Estado Bolívar, Venezuela.

Pteronymia alissa dorothyae Neild: Neild, 2008: 163–164, 237, pl. 34, Figs. 1005–1008. Type locality: Vía El Dorado–Santa Elena, km 131.7, Sierra de Lema, 1400 m, Bolívar, SE Venezuela. Additional records: Neild, 2008: 237: Vía El Dorado–Santa Elena, kms 123–131.7, La Escalera, Sierra de Lema, 1100–1400 m, Bolívar, SE Venezuela; Venezuela, Arabupu; British Guiana, Roraima.

Pteronymia alissa marjorieae Neild: Neild, 2008: 164–165, 237, pls. 34–35, Figs. 1009–1012. Type locality: Venezuela, T.F. Amazonas, Dpt. Río Negro/Campamento VII, 1870 m [Sierra Neblina]. Additional records: Neild, 2008: 237: Venezuela, T.F. Amaz. Cerro Neblina, Camp VII–1850 m 0°51′N–65°58′W; Cerro Neblina, Norte Brasil, 1500 m; Brasilia sept., Serra Neblina, 1500 m.

Pteronymia alicia Neild: Neild, 2008: 165–167, 237, pl. 35, Figs. 1016–1019. Type locality: km 125 El Dorado–Santa Elena, BO, Venezuela, 1100 m. Additional records: Neild, 2008: 237: kms 125–155, El Dorado–Santa Elena, Sierra de Lema, BO, Venezuela, 1100–1440 m; Venezuela, T.F. Amazonas, Cerro de la Neblina, Camp XI, 1490 m, 0°52′N–65°58′W; Brasilia sept. Serra Neblina, 1250 m.

Pteronymia peteri Neild: Neild, 2008: 237, pl. 35, Figs. 1020–1021. Type locality: Marahuaka–Macizo Central, 3°40′N–65°28′W, 1040 m, T.F. Amazonas, Parque Nacional Duida-Marahuaka, Venezuela.

Subfamily **Heliconiinae** Swainson 1822

Tribe **Acraeini** Boisduval 1833

**Actinote anteas pierrei* Neild and Romero: Neild and Romero *in* Neild, 2008: 27–28, Fig. 1.6, 231, pl. 2, Figs. 41–44. Type locality: km 40 Sta. Elena to Icabarú Road, SE Bolívar, Venezuela. Additional records: Neild, 2008: 231: Vía Sta. Elena–Icabarú kms. 24.6–40, 900–1050 m, SE Bolívar, Venezuela; 10 km east of Sta. Elena de Uairén, Bolívar, Venezuela; Roraima, Br. Guiana.

Actinote genitrix costae Neild (Fig. 9.4G [male dorsal]): Neild, 2008: 39–40 Fig. 1.11, 231, pl. 3, Figs. 73–76. Localidad típica: La Escalera, km 123 (1400 m), Estado Bolívar, Venezuela.

**Actinote pellenea fernandezi* Neild and Romero: Neild and Romero *in* Neild, 2008: 25–27 Fig. 1.5, 231, pl. 1, Figs. 18–21. Type locality: Vía Sta. Elena–Icabarú, km 30.6, SE Bolívar State, SE Venezuela. Additional records: Neild, 2008: Santa Elena–Icabarú, kms 24.6–40, 900–1050 m; Santa Elena–Sampay, 860 m; Sierra de Lema, 1100–1440 m, SE Bolívar State, SE Venezuela; Kavanayén, 1200 m, Bolívar State, Venezuela; El Dorado–Sta. Elena, km 125, 1100 m, Bolívar, Venezuela; Venezuela, Bolívar, Sta. Elena-Peraitepuy, 900 m; Quebrada Jaspe, Estado Bolívar.

**Actinote romeroi* Neild and Costa: Neild and Costa *in* Neild, 2008: 37–39 Fig. 1.10, 231, pl. 3, Figs. 69–72. Type locality: Río Surukún, 870 [m], Estado Bolívar [Venezuela]. Additional records: Neild, 2008: 231: Vía Sta. Elena–Icabarú, km 40, SE Bolívar State, SE Venezuela.

Tribe **Heliconiini** Swainson 1822

Heliconius elevatus roraima Turner (Fig. 9.5D [male dorsal]): Turner, 1966: 104–105 Fig. 2, E, 109–110. Type locality: Roraima, B. Guiana.

Heliconius elevatus ssp. n. Costa et al., in prep. (Fig. 9.5F [male dorsal]).

Eueides procula browni H. Holzinger and R. Holzinger: H. Holzinger and R. Holzinger, 1974: 147–152 Figs. 1–5, map. Type locality: Gran Sabana, Cavanaen [*sic*], Venezuela. Additional records: Holzinger and Holzinger, 1974: 149: El Dorado–Santa Elena highway, km 119, 970 m, Bolívar, Venezuela.

Family **Lycaenidae** Leach 1815

Subfamily **Theclinae** Swainson 1831

Tribe **Eumaeini** Doubleday 1841

Calycopis matho (Godman and Salvin): Godman and Salvin, 1887: 80. Type locality: Guiana Brit., Carimang River. Additional records: K. Johnson, 1993 [as *Argentostriatus matho* (Godman and Salvin): 3–4, photopl. 7, 2B, Figs. 1B, 2A]: Bartica; French Guiana; St. Lauren de Maroni; Surinam, Essequibo River; Christianburg (as *Argentostriatus roraimaevagus* Johnson: 4–5, photopl. 7, 2C, Fig. 1A): Ireng R. to Roraima, Brazil.

Cyanophrys roraimiensis K. Johnson and D.S. Smith: K. Johnson and D.S. Smith, 1993: 7–9, 12 Fig. H, photopl. 1, Fig. 2. Type locality: Brazil: Mount Roraima.

Ocaria elisa Bálint and Costa (Fig. 9.4O [female dorsal]): Bálint and Costa, 2012: 299–310, Figs. 7–10, 11–12 (male genitalia). Type locality: Venezuela, Edo. Bolívar,

La Escalara [*sic*]. Additional records: Bálint and Costa, 2012: 301: Venezuela, Edo. Bolívar, La Escalera 1400 m, km 124 south El Dorado.

Ocaria faurei Bálint, Attal and Costa: Bálint, Attal and Costa *in* Costa et al., 2018: 65–68, Figs. 3, 4 (male), 5, 6 (female), 7 (androconia), 8–12 (male genitalia). Type locality: Venezuela, Bolívar, Auyán-tepui, El Peñón, 1850 m. Additional records: Costa et al., 2018: Venezuela, Bolívar: Auyán-tepui: El Danto, 1500 m; entre El Danto y El Peñón, 1700 m; Akopán-tepui, 1400 m; Ptarí-tepui, Base Camp, 2000 m; Roraima, Base Camp, 2000 m

Gigantofalca duida K. Johnson: K. Johnson, 1991: 17, Figs. 8C, 35. Type locality: Mount Duida, Venezuela.

Strymon auyana Bálint, Benmesbah and Viloria: Bálint, Benmesbah and Viloria *in* Costa et al., 2018: 168–171, Figs. 13, 14 (male), 15, 16 (female), 17 (androconia), 18, 19 (male genitalia), 20 (female genitalia). Type locality: Venezuela, Bolívar, Auyán-tepui, El Danto, 1500 m. Additional records: Costa et al., 2018: Venezuela, Bolívar, Auyán-tepui, entre El Danto y El Peñón, 1700–1750 m.

Family **Riodinidae** Grote 1895

Subfamily **Riodininae** Grote 1895

Tribe **Mesosemiini** Bates 1895

Napaea fratelloi Hall and Harvey: Hall and Harvey *in* Hall, 2005: 123, Figs. 33, 85, 103. Type locality: Guiana: Cuyuni-Mazaruni, Mount Ayanganna, 1120 m, 5°22.22′N 59°57.34′W.

Tribe **Riodinini** Grote 1895

Melanis dulcis Callaghan and Costa (Fig. 9.4K [female dorsal]): Callaghan and Costa *in* Costa et al., 2017: 274–276, Figs. 13, 14 (male), 15, 16 (female), 19, 20 (male genitalia), 21 (female genitalia). Type locality: Venezuela, Bolívar, talud Ptarí-tepui, Salto Iworé, 1400 m. Additional records: Callaghan and Costa *in* Costa et al., 2017: 274: Venezuela, Bolívar: Sororopán-tepui, 5°42′N, 61°47′W, 1650 m; Uei-tepui, 5°00′06″N, 60°37′37″W, 2000 m; Venezuela, T.F. Amazonas, Sierra Tapirapecó, Tamakuari, 1°13′34″N, 64°44′39″W, 1600 m; Guiana, Mount Ayanganna, 5°22′21.5″N, 59°57′33.8″W, 1120 m.

Tribe **Symmachiini** Bates 1859

Pirascca hanneri Gallard, Viloria, Fratello and Costa: Gallard et al. *in* Costa et al., 2017: 270–274, Figs. 1, 2 (male), 3, 4 (female), 9–11 (male genitalia). Type locality: Venezuela, Bolívar, Uei-tepui, 2000 m, vertiente oeste. Additional records: Gallard et al. *in* Costa et al., 2017: 270: Venezuela, Bolívar: Sierra de Lema, La Escalera, 1400 m; Auyán-tepui: El Danto, 1500 m; entre El Danto y El Peñón, 1700 m; El Peñón, 1850 m; Ptarí-tepui, 1900 m; Guiana, Region 7, Mount Ayanganna, 5°24.1′N, 59°57.4′W, 1585 m.

Tribe **Stalachtini** Bates 1861

Stalachtis halloweeni J. Hall (Fig. 9.2F): J. Hall, 2006: 138–142, Figs. 1–2, 3–4 (genitalia). Type locality: Guiana: Cuyuni-Mazaruni, Mount Ayanganna, 1120 m 5°22.22′N, 59°57.34′W.

Stalachtis halloweeni ssp. n. Costa et al., in prep.

Biogeographic significance and conclusions

Using the criterion of a high proportion of endemism, at least three different levels of the taxonomic hierarchy (genus, species, and subspecies), according to the ICAN

(Ebach et al., 2008), Costa et al. (2014b) have proposed to apply the concept of Pantepui not only in the phytogeographic sense conceived of by Huber (1987, 1994, 1995) but also to a disjunct area of endemism whose hierarchic rank stands open to testing through formal analytical methods, at least as a biogeographic province on its own, within the Neotropical region. This proposal has finally been acknowledged by Morrone (2017), who included the Pantepui province in his revised system of the biogeographic regionalization of tropical America.

On nomenclatural grounds, Pantepui is the oldest original available name for this individual entity or area of endemism (Mayr and Phelps 1955), and is herein considered totally or partly equivalent to the following historical denominations in whose proposals their authors separately followed diverse and varied criteria, not necessarily related to endemism but to other aspects of the region, like its biotic element composition, vegetation, geomorphology, physiography, or the combination of two or more of these aspects.

Pantepui province

Synonymy:

Region of Venezuela and Guyana: Good, 1947 [in part]
Pantepui Area: Mayr and Phelps, 1955
Venezuela-Guayana Region: Mattick, 1964
Highlands of Guiana and Brazil area: Sick, 1969 [in part]
Dominio Guayana: Cabrera and Willink, 1973 [in part]
Provincia Guayana: Cabrera and Willink, 1973 [in part]
Pantepui Center: Müller, 1973 [in part]
Distrito Guiano: Ávila-Pires, 1974.
Campos Limpos Province: Udvardy, 1975 [in part]
Dominio Roraima—Guianense—"Gran Sabana": Ab'Sáber, 1977 [in part]
Guiana Centre: Schreiber, 1978 [in part]
Guayana, region of Roraima Sandstone Formation: Maguire, 1979
Duida Subcenter: Cracraft, 1985 [in part]
Gran Sabana Subcenter: Cracraft, 1985 [in part]
Guiana Area: Haffer, 1985 [in part]
Region of the Guayana highlands: Takhtajan, 1986
Provincia de las Tierras Altas de Guayana: Huber and Alarcón, 1988
Setor Setentrional ou Guianense: Fernandes and Bezerra, 1990 [in part]
Provincia de los Tepuis [*sic*]: Rivas-Martínez and Navarro, 1994
Pantepui Province (Provincia Pantepui): Huber 1994, 1997, Berry et al., 1995, Huber and Riina, 1997, Viloria, 2005, Costa et al., 2014b, Morrone, 2017
Área Guiana: Coscarón and Coscarón-Arias, 1995 [in part]
Eco-región del bosque húmedo de alta montaña de la Guiana (Guianan highlands moist forest): Dinerstein et al., 1995 [in part]
Eco-región Tepuyes (tepuis): Dinerstein et al., 1995 [in part]
Eco-región del bosque húmedo de Uatama (Uatama moist forest): Dinerstein et al., 1995 [in part]

Región de las tierras altas de Guayana: Huber and Riina, 1997
Provincia de la Guiana: Morrone, 2000, 2001, 2006 [in part]
Tepuis Zoographic Region (=Pantepui area of endemism): Porzecanski and Cracraft, 2005
Pantepui Region: De Marmels, 2007
Orinoco Guiana Shield ecoregion: Abell et al., 2008
Tepuyan Province (Provincia Tepuyana): Rivas-Martínez et al., 2011

The butterfly fauna of Pantepui shows evident biotic affinities with the traditionally so-called subregion Paramo-Puna of the Andean region and the Amazonian subregion of the Neotropical region (*sensu* Morrone, 2001). The former is equivalent to the more recently defined South American Transition Zone (see Morrone, 2006, 2014a,b, 2015, 2017). Although not yet quantified, these affinity relationships would probably make sense and find plausible explanations in the common geological–biological evolutionary history of the adjacent territories comprising the biogeographic mosaic of the American continent. The first hypotheses on the origin of the biota of Pantepui were based on the detection of its affinities with the Andean (now Transitional South American) biota. These pioneer thoughts began invoking an Andean origin for the birds of Pantepui (Mayr and Phelps, 1955, 1967, 1971). However, almost immediately strong criticisms emerged against this proposal, particularly based on the understanding that, as a geological unit, the Guiana Shield (and its tepuis) is far older than the Andes; therefore the exact opposite would be expected: the Pantepuian biota, or at least an important part of it, should have preceded the Andean one. The latter possibly derived from ancestral elements present in the South American lowlands before the emergence and uplifting of the Andean Cordillera (Croizat-Chaley, 1976; and comments in De Marmels, 1999; Viloria and Pyrcz 2000 and De Marmels et al., 2003). This is a public controversy, reviewed in detail by Pérez-Hernández and Lew (2001) and Marín (2010) (see also Chapter 4: Origin and evolution of the Pantepui biota). Currently, the origin of the fauna and flora of Pantepui continues to be a major subject of paleoecology (see contrasting theories in Pérez-Hernández and Lew, 2001, and the proposals of Rull, 2004, 2005a,b). It is interesting to outline that analyses of endemisms in lower plants (bryophytes) suggested that Pantepui-endemic taxa most probably derived from the dispersal of cold-adapted Andean ancestors (Désamoré et al., 2010). On the other hand investigations of molecular phylogeny in selected groups of birds has yielded the interpretation of an Andean origin for the members of an avian genus (*Aulacorhynchus*), which is well represented in Pantepui (Bonaccorso and Guayasamín, 2013). These opinions are opposed to the idea that a geologically older Pantepui might be the original source of the montane biota of tropical America. Likewise, Salerno et al. (2012), based also on molecular divergences among some amphibian species, have concluded that some lineages of frogs living in tepuian summits are essentially very young (neoendemics). Either way, inquiries about the origin of living beings inhabiting Pantepui will need not only a better taxonomic understanding of microorganisms, fungi, plants, and animals but also the integration of phylogenetic analyses methods coupled with key evidences of the geological history of the region. These topics, however, are beyond the scope of our current lepidopterological studies.

Acknowledgments

We thank Andrés Orellana (Caracas), Stéphane Attal, Patrick Blandin (Paris), Andrew Neild (London), and Zsolt Bálint (Budapest) for their valuable cooperation in building up and correcting the checklist of endemic butterflies of Pantepui. Héctor Suárez, Celsa Señaris, Daniel Lew, Cristóbal Ríos-Málaver (IVIC), Julián Salazar (Manizales), and A. Neild contributed in several ways to locate bibliography and other sources of information, as well as assisting the authors in preparing part of this chapter and its illustrations. We especially acknowledge Grisel Velázquez (UniSIG, IVIC) for elaborating the digital model of Auyán-tepui (Fig. 9.1). C. Señaris, Valentí Rull, and Shinichi Nakahara made valuable comments to help improve the quality of this chapter.

References

Abell, R., Thieme, M.L., Revenga, C., Bryer, M., Kottelat, M., Bogutskaya, N., et al., 2008. Freshwater ecoregions of the world: a new map of biogeographic units for freshwater biodiversity conservation. BioScience 58 (5), 403–414.

Ab'Sáber, A.N., 1977. Espaços ocupados pela expansão dos climas secos na América do Sul, por ocasião dos períodos glaciais quaternários. Paleoclimas 3, 1–19.

Anonymous, 1892. An ascent of Roraima. Proc. R. Geogr. Soc 1892 (4), 242–243.

Attal, S., De Marmels, J., 2012. Le genre *Perisama* Dobleday, 1849, présent à Auyan-Tepui (Venezuela) avec une nouvelle espèce (Lepidoptera, Nymphalidae). Bull. Soc. Ent. France 117 (1), 111–113.

Attal, S., Viloria, Á.L., 2014. *Archaeogramma* Attal & Viloria, n. gen. In: Costa, M., Viloria, Á.L., Attal, S., Orellana, A., Lepidoptera del Pantepui. Parte II. Descripción de nuevos Nymphalidae (Papilionoidea). Bull. Soc. Ent. France 119 (1), 39–43.

Aubrecht, R., Barrio-Amorós, C.L., Breure, A.S.H., Brewer-Carías, C., Derka, T., Fuentes-Ramos, O.A., et al., 2012. *Venezuelan Tepuis: Their Caves and Biota. Acta Geol. Slovaca—Monograph*. Comenius University, Bratislava, 167 pp.

Ávila-Pires, F. de, 1974. Caracterização zoogeográfica de Província Amazônica. II. A familia Callithricidae e a zoogeografía amazônica. An. Acad. Bras. Ciênc. 49, 150–181.

Bálint, Z., Costa, M., 2012. Description of an *Ocaria* species from the Venezuelan Pantepuy [sic] (Lepidoptera: Lycaenidae: Theclinae). Ann. Hist. Nat. Mus. Natl. Hung. 104, 299–310.

Bell, E., 1932. Hesperiidae (Lepidoptera, Rhopalocera) of the Roraima and Duida expeditions, with descriptions of new species. Am. Mus. Nov. 555, 1–16.

Bell, C.D., Soltis, D.E., Soltis, P.S., 2010. The age and diversification of Angiosperms re-revisited. Am. J. Bot. 97 (8), 1296–1303.

Berry, P.E., Huber, O., Holst, B.K., 1995. Floristic analysis and phytogeography. In: Berry, P.E., Holst, B.K., Yatskievych, K. (Eds.), Flora of the Venezuelan Guayana. Introduction, vol. I. Missouri Botanical Garden/Timber Press, St. Louis/Portland, Oregon, pp. 161–191.

Bollino, M., Costa, M., 2004. A new subspecies of *Parides phosphorus* (Bates, 1861) (Lepidoptera: Papilionidae) from South-Eastern Venezuela. Notes Papilionidae 2, 9–10.

Bollino, M., Costa, M., 2007. An illustrated annotated check-list of the species of *Catasticta* (s.l.) Butler (Lepidoptera: Pieridae) of Venezuela. Zootaxa 1469, 1–42.

Bonaccorso, E., Guayasamín, J.M., 2013. On the origin of Pantepui montane biotas: a perspective based on the phylogeny of *Aulacorhynchus* toucanets. PLoS One 8 (6), 1–10.

Braby, M.F., Vila, R., Pierce, N.E., 2006. Molecular phylogeny and systematics of the Pieridae (Lepidoptera: Papilionoidea): higher classification and biogeography. Zool. J. Linn. Soc. 147, 239–275.

Brewer-Carías, C. (Ed.), 1988. Cerro de la Neblina: resultados de la expedición, 1983–1987. Fundación para el Desarrollo de las Ciencias Físicas, Matemáticas y Naturales, Caracas 922 pp.

Bristow, C.R., 1991. A revision of the brassoline genus *Opsiphanes* (Lepidoptera: Rhopalocera). Zool. J. Linn. Soc. 101 (3), 203–293.

Brown, F.M., 1932. Pieridae from the regions of Mt. Duida and Mt. Roraima. Am. Mus. Nov. 572, 1–7.

Brown Jr., K.S., 1977a. Centros de evolução, refúgios quaternários e conservação de patrimônios genéticos na região neotropical: padrões de diferenciação em Ithomiinae (Lepidoptera: Nymphalidae). Acta Amazonica 7 (1), 75–137.

Brown Jr., K.S., 1977b. Geographical patterns of evolution in Neotropical Lepidoptera: differentiation of the species of *Melinaea* and *Mechanitis* (Nymphalidae, Ithomiinae). Syst. Ent. 2 (3), 161–197.
Brown Jr., K.S., 1979. *Ecologia geográfica e evolução nas florestas neotropicais*. Universidade Estadual de Campinas, Campinas, 265 pp. xxi.
Brown Jr., K.S., 1987. Biogeography and evolution of neotropical butterflies. In: Whitmore, T.C., Prance, G.T. (Eds.), Biogeography and Quaternary History of Tropical America. Clarendon Press, Oxford, pp. 66–104.
Cabrera, A.L., Willink, A., 1973. *Biogeografía de América Latina*. Organización de Estados Americanos, Washington, DC, 122 pp.
Constantino, L.M., Johnson, K., 1997. A new species of the *Eumaeus toxana/E. toxea* clade from Tepui's [*sic*] in the Amazon basin of Colombia. Rev. Theclinae Colomb. 1 (8), 1–10 [ii].
Coscarón, S., Coscarón-Arias, C.L., 1995. Distribution of neotropical Simuliidae (Insecta, Diptera) and its areas of endemism. Rev. Acad. Colomb. Cienc. 19 (75), 717–732.
Costa, M., Viloria, Á.L., Attal, S., Orellana, A., Benmesbah, M., 2019. Lepidoptera del Pantepui. Parte VI. Nuevos Hesperiidae Hesperiinae y Nymphalidae Satyrinae, Nymphalinae. Bull. Soc. Ent. France 124 (1), 77–102.
Costa, M., Viloria, Á.L., Attal, S., Orellana, A., 2014a. Lepidoptera del Pantepui. Parte II. Descripción de nuevos Nymphalidae (Papilionoidea). Bull. Soc. Ent. France 119 (1), 39–52.
Costa, M., Viloria, Á.L., Huber, O., Attal, S., Orellana, A., 2014b. Lepidoptera del Pantepui. Parte I. Endemismo y caracterización biogeográfica. Entomotrópica 28 (3), 193–217.
Costa, M., Viloria, Á.L., Attal, S., Neild, A.F.E., Fratello, S., Nakahara, S., 2016. Lepidoptera del Pantepui. Parte III. Nuevos Nymphalidae, Cyrestinae y Satyrinae. Bull. Soc. Ent. France 121 (2), 179–206.
Costa, M., Viloria, Á.L., Attal, S., Neild, A.F.E., Fratello, S., Callaghan, C., et al., 2017. Lepidoptera del Pantepui. Parte IV. Nuevos Riodinidae Riodininae y Pieridae Pierinae. Bull. Soc. Ent. France 122 (3), 269–286.
Costa, M., Viloria, Á.L., Attal, S., Benmesbah, M., Neild, A.F.E., Bálint, Z., 2018. Lepidoptera from the Pantepui. Part V. New Lycaenidae (Theclinae: Eumaeini). Opusc. Zool. Budapest 49 (2), 163–179.
Cracraft, J., 1985. Historical biogeography and patterns of differentiation within the South American avifauna: areas of endemism. Ornith. Monogr. 36, 49–84.
Croizat-Chaley, L., 1976. Biogeografía analítica y sintética ("panbiogeografía") de las Américas, 2 vols. Academia de Ciencias Físicas, Matemáticas y Naturales, Caracas [iv]; 454 pp.; [ii] + [vi]; pp. 455–890; [vi].
De Jong, R., 2016. Reconstructing a 55-million-year-old butterfly (Lepidoptera: Hesperiidae). Eur. J. Ent. 113, 423–428.
De Marmels, J., 1999. First record of the genus *Perisama* Doubleday, 1849 from Pantepui, Venezuela. Atalanta 30 (1/4), 155–158. plate Xa.
De Marmels, J., 2007. *Tepuibasis* gen. nov. from the Pantepui region of Venezuela, with descriptions of four new species and with biogeographic, phylogenetic and taxonomic considerations on the Teinobasinae (Zygoptera: Coenagrionidae). Odonatologica 36, 117–146.
De Marmels, J., Clavijo, J.A., 1998. On the true type localities of *Mesotaenia vaninka delafuentei* Neild and *Memphis viloriae* Pyrcz & Neild (Nymphalidae). J. Lepid. Soc. 52 (2), 217–219.
De Marmels, J., Clavijo, J.A., Chacín, M.E., 2003. Two new subspecies of *Pereute lindemannae* and one of *Pseudopieris viridula* from Pantepui, Venezuela (Pieridae). J. Lepid. Soc. 57 (2), 86–91.
Désamoré, A., Vanderpoorten, A., Laenen, B., Gradstein, S.R., Kok, P.J.R., 2010. Biogeography of the Lost World (Pantepui region, northeastern South America): insights from bryophytes. Phytotaxa 9, 254–265.
Dinerstein, E., Olson, D.M., Graham, D.J., Webster, A.L., Primm, S.A., Bookbinder, M.P., et al., 1995. *Una evaluación del estado de conservación de las eco-regiones terrestres de América Latina y el Caribe*. Fondo Mundial para la Naturaleza (WWF-US). Banco Mundial, Washington, DC, xviii, pp. 135, 1 map.
Ebach, M.C., Morrone, J.J., Parenti, L.R., Viloria, Á.L., 2008. International code of area nomenclature. J. Biogeogr. 35, 1153–1157.
Espeland, M., Hall, J.P.W., DeVries, P.J., Lees, D.C., Cornwall, M., Hsu, Y.F., et al., 2015. Ancient Neotropical origin and recent recolonization: phylogeny, biogeography and diversification of the Riodinidae (Lepidoptera: Papilionoidea). Mol. Phyl. Evol. 93, 296–306.
Fernandes, A.G., Bezerra, P., 1990. *Estudo fitogeográfico do Brasil*. Stylus Comunicaçãos, Fortaleza, p. 205p.
Forbes, W.T.M., 1929. Variation in *Junonia lavinia* (Lepidoptera, Nymphalidae). J.N.Y. Entomol. Soc. 36, 305–320.
Forbes, W.T.M., 1942. Note on the butterflies of Venezuelan Guiana. Bol. Entomol. Venez. 1, 25–36.
Forbes, W.T.M., 1948. A second review of *Melinaea* and *Mechanitis* (Lepidoptera: Ithomiinae). J.N.Y. Entomol. Soc. 56, 1–24. , plates 1–2.
Fratello, S.A., 1996a. Wokomung, a remote Guyana Tepui. Trop. Lepid. News. 1996 (2), 1, 4–5.

Fratello, S.A., 1996b. Additions and corrections. Trop. Lepid. News. 1996 (3), 9.
Fratello, S.A., 1999. The shedding of light. Lepid. News. 1999 (1), 4.
Fratello, S.A., 2000. Guyana montane expedition. Lepid. News. 1999 (4), 4–8.
Fratello, S.A., 2001a. Guyana montane expeditions II. Lepid. News. 2001 (2), 8–9.
Fratello, S.A., 2001b. Falls from paradise. Nat. Hist. 110 (10), 26–28.
Fratello, S.A., 2003. Guyana expeditions (January–April 2001). News. Lepid. Soc. 45 (4), 109–111, 116–121.
Fratello, S.A., 2004a. The female of *Mesosemia phace, M. methion* and other *Mesosemia* mysteries. News. Lepid. Soc. 46 (3), 89, 98–100.
Fratello, S.A., 2004b. New species from Mt. Roraima. News. Lepid. Soc. 46 (4), 122–126.
Fratello, S.A., 2005. An expedition to Guyana's Acarai Mts., including two new and one undescribed butterfly species. News. Lepid. Soc. 47 (1), 29–31, 34–37.
Fratello, S.A., Nakahara, S., Brévignon, C.R., Harvey, D.J., 2015. Two new species of *Euptychia* Hübner, 1818 (Lepidoptera: Nymphalidae: Satyrinae) from the Guiana Shield, with notes on *E. marceli* Brévignon, 2005 and *E. rufocincta* Weymer, 1911. J. Lepid. Soc. 69 (4), 293–306.
Gernaat, H.B.P.E., Beckles, B.G., van Andel, T., 2012. *Butterflies of Suriname: A natural history*. KIT Publishers, Amsterdam, 696 pp. plates 52.
Godman, F., Du, C., Salvin, O., 1887. Biologia Centrali-Americana. Insecta. Lepidoptera–Rhopalocera, vol. 2. Dulau y Co., Bernard Quaritch, London, pp. 65–96, plate 53, no. 61.
Good, R., 1947. *The geography of the flowering plants*. Longmans, Green and Co. Ltd., London, 403 pp. plates 16, 9 col. maps.
Grose-Smith, H., Kirby, W.F., 1897. *Rhopalocera exotica, being illustrations of new, rare, and unfigured species of butterflies. 2:* (Dismorphia *III)*, Gurney & Jackson, London, 11–15 plate 1.
Haffer, J., 1969. Speciation in Amazonian forest birds. Science (n.s.) 165 (3889), 131–137.
Haffer, J., 1985. Avian zoogeography of the Neotropical lowlands. Ornith. Monogr. 36, 113–146.
Hall, A., 1939. Catalogue of the Lepidoptera Rhopalocera (butterflies) of British Guiana. Agric. J. Br. Guiana 10 (1), 25–41. plate 8.
Hall, J.P.W., 2005. *A phylogenetic revision of the Napaeina (Lepidoptera: Riodinidae: Mesosemiini)*. Entomological Society of Washington, Washington, DC [vi]; 235 pp.
Hall, J.P.W., 2006. A remarkable new riodinid species, *Stalachtis halloweeni* (Riodinidae: Stalachtini) from Mount Ayanganna, Guyana. J. Lepid. Soc. 60 (3), 138–142.
Harms, H., 1915. Ernst Ule. *Verhandl.* Bot. Ver. Provinz. Brandenburg 57, 150–184.
Harold, A.S., Mooi, R.D., 1994. Areas of endemism: definition and recognition criteria. Syst. Biol. 43, 261–266.
Hausdorf, B., 2002. Units in biogeography. Syst. Biol. 51 (4), 648–652.
Heikkila, M., Kaila, L., Mutanen, M., Peña, C., Wahlberg, N., 2011. Cretaceous origin and repeated tertiary diversification of the redefined butterflies. Proc. R. Soc. Lond. (B) 279, 1093–1099.
Holzinger, H., Holzinger, R., 1974. *Eueides procula browni*, eine neue subspecies aus Venezuela (Lepidoptera: Nymphalidae). Z. Arbeitsgem. Öster. Ent. 24 (4), 147–152.
Hoogmoed, M.S., 1979. The herpetofauna of the Guianan region. In: Duellman, W.E. (Ed.), The South American Herpetofauna: its Origin, Evolution and Dispersal. Monograph No. 7. University of Kansas, Museum of Natural History, Lawrence, pp. 241–279.
Huber, O., 1987. Consideraciones sobre el concepto de Pantepui. Pantepui 1 (2), 2–10.
Huber, O., 1994. Recent advances in the phytogeography of the Guayana region, South America. Mém. Soc. Biogéogr. (3e. sér.) 4, 53–63.
Huber, O., 1995. Geographical and physical features. In: Berry, P.E., Holst, B.K., Yatskievych, K. (Eds.), Flora of the Venezuelan Guayana. Introduction, vol. I. Missouri Botanical Garden/Timber Press, St. Louis/Portland, Oregon, pp. 1–61.
Huber, O., Alarcón, C., 1988. *Mapa de vegetación de la República de Venezuela. 1: 2,000,000*. Ministerio del Ambiente y de los Recursos Naturales Renovables/The Nature Conservancy, Caracas.
Huber, O., Riina, R. (Eds.), 1997. Glosario fitoecológico de las Américas. América del Sur: países hispanoparlantes. UNESCO/Fundación Instituto Botánico de Venezuela, Caracas vol. I, 500 pp. [ii].
Huntington, E.I., 1933. Erycinidae and Lycaenidae (Lepidoptera, Rhopalocera) from the regions of Mt. Roraima and Mt. Duida. Am. Mus. Nov. 611, 1–5.
Im Thurn, E.F., 1882. A journey to Mount Russell in Guiana. Timehri 1 (2), 216–228.
Im Thurn, E.F., 1885. The ascent of Mount Roraima. Proc. R. Geogr. Soc 7 (8), 497–521. 1 map [p. 560].

Johnson, K., 1991. Neotropical hairstreak butterflies: genera of the "*Calycopis/Calystryma* Grade" of Eumaeini (Lepidoptera, Lycaenidae, Theclinae) and their diagnostics. Rep. Mus. Nat. Hist. Univ. Wisconsin (Stevens Point) 21, 1–128 [iv].

Johnson, K., 1993. Hairstreak butterflies of the genus *Argentostriatus* (Lepidoptera, Lycaenidae, Theclinae). Rep. Mus. Nat. Hist. Univ. Wisconsin (Stevens Point) 36, [i–ii]; 1–19, plate 7.

Kaye, W.J., 1905. Transparency in wings of Lepidoptera. Ent. Rec. J. Var. 17 (4), 83–86 (5): 115–121, plate 3.

Maguire, B., 1979. Guayana, region of the Roraima Sandstone Formation. In: Larsen, K., Holm-Nielsen, L.B. (Eds.), Tropical Botany. Academic Press, London, pp. 223–238.

Marín, G., 2010. Acerca del origen y biogeografía de la avifauna del Pantepuy [sic] o Croizat versus Mayr: una revisión crítica. Biologist 8 (1), 79–108.

Mattick, F., 1964. Übersicht über die Florenreiche und Florengebiete der Erde. In: Engler, A., Melchior, M. (Eds.), Syllabus der Pflanzenfamilien, Bd. 2:12. Gerbruder Borntraeger, Berlin, pp. 626–630.

Mayr, E., Phelps Jr., W.H., 1955. Origin of the bird fauna of Pantepui. In: Portman, A., Sutter, E. (Eds.), Acta XI Congressus Internationalis Ornitologici, 1954. Basel, pp. 399–400.

Mayr, E., Phelps Jr., W.H., 1967. The origin of the bird fauna of the South Venezuelan highlands. Bull. Am. Mus. Nat. Hist. 136 (5), 269–328.

Mayr, E., Phelps Jr., W.H., 1971. Origen de la avifauna de las altiplanicies del sur de Venezuela. Bol. Soc. Venez. Cienc. Nat. 29 (121), 309–401.

Morrone, J.J., 2000. A new regional biogeography of the Amazonian subregion, mainly based on animal taxa. An. Inst. Biol. Ser. Zool. 71 (2), 99–123.

Morrone, J.J., 2001. *Biogeografía de América Latina y el Caribe*. M y T Manuales y Tesis SEA, vol. 3. CYTED/ UNESCO ORCYT/Sociedad Entomológica Aragonesa, Zaragoza, [ii]; 148 pp.; [iv].

Morrone, J.J., 2006. Biogeographic areas and transition zones of Latin America and the Caribbean Islands based on panbiogeographic and cladistic analyses of the entomofauna. Ann. Rev. Ent. 51, 467–494.

Morrone, J.J., 2014a. Perspective. On biotas and their names. Syst. Biodiv. 12 (4), 386–392.

Morrone, J.J., 2014b. Cladistic biogeography of the Neotropical region: identifying the main events in the diversification of the terrestrial biota. Cladistics 30, 202–214.

Morrone, J.J., 2015. Biogeographical regionalisation of the world: a reappraisal. Aust. Syst. Bot. 28, 81–90.

Morrone, J.J., 2017. *Neotropical biogeography: Regionalization and evolution*. CRC Press, Boca Raton, p. 312.

Müller, P., 1973. *The dispersal centres of terrestrial vertebrates in the Neotropical Realm: A study in the evolution of the Neotropical biota and its native landscape*. Junk, The Hague, 244 pp.

Nakahara, S., Fratello, S.A., Harvey, D.J., 2014. A new species of *Euptychia* Hübner, 1818 (Lepidoptera: Nymphalidae: Satyrinae: Satyrini) from Mount Roraima, Guyana. Zootaxa 3881 (3), 291–300.

Neild, A.F.E., 1996. *The Butterflies of Venezuela. Part I: Nymphalidae I (Limenitidinae, Apaturinae, Charaxinae). A Comprehensive Guide to the Identification of Adult Nymphalidae, Papilionidae, and Pieridae*. Meridian Publications, Greenwich, London, 144 pp., plates 32.

Neild, A.F.E., 2008. *The Butterflies of Venezuela. Part 2: Nymphalidae II (Acraeinae, Libytheinae, Nymphalinae, Ithomiinae, Morphinae). A Comprehensive Guide to the Identification of Adult Nymphalidae, Papilionidae, and Pieridae*. Meridian Publications, Greenwich, London, 275 pp., plates 84.

Neild A.F.E., Romero Montesino F., 2008. *Actinote* Hübner [1819]. Species account, In: Neild A.F.E. The Butterflies of Venezuela. Part 2: Nymphalidae II (Acraeinae, Libytheinae, Nymphalinae, Ithomiinae, Morphinae). A Comprehensive Guide to the Identification of Adult Nymphalidae, Papilionidae, and Pieridae, Meridian Publications, Greenwich, London, pp. 25–46.

Neild, A.F.E., Nakahara, S., Fratello, S.A., Harvey, D.J., 2014. A new species of *Euptychia* Hübner, 1818 (Nymphalidae: Satyrinae: Satyrini) from the Amazon basin and the Guaianas. Trop. Lepid. Res. 24 (1), 4–9.

Orellana, A.M., 2004. Descripciones y notas taxonómicas sobre *Antirrhea* Hübner. Entomotrópica 19, 21–29.

Orellana A.M., 2008. Tribe Antirrheini. In: Neild A.F.E. The Butterflies of Venezuela. Part 2: Nymphalidae II (Acraeinae, Libytheinae, Nymphalinae, Ithomiinae, Morphinae). A Comprehensive Guide to the Identification of Adult Nymphalidae, Papilionidae, and Pieridae, Meridian Publications, Greenwich, London, pp. 185–192.

Orellana, A.M., 2010. Pyrrhopyginae de Venezuela (Lepidoptera: Hesperioidea: Hesperiidae). Entomotropica 23 (3), 177–291.

Peña, C., Wahlberg, N., 2008. Prehistorical climate change increased diversification of a group of butterflies. Biol. Lett. 4, 274–278.

Pérez-Hernández, R., Lew, D., 2001. Las clasificaciones e hipótesis biogeográficas para la Guayana venezolana. Interciencia 26 (9), 373–382.

Perkins, H.I., 1885. Notes on a journey to Mount Roraima. Br. Guiana. Proc. R. Geogr. Soc 7 (8), 522–534. 1 map [p. 560].

Porzecanski, A.L., Cracraft, J., 2005. Cladistic analysis of distributions and endemisms (CADE): using raw distributions of birds to unravel the biogeography of the South American aridlands. J. Biogeogr. 32, 261–275.

Pyrcz, T.W., 1995. A new *Memphis* from the Pantepui, Venezuela (Lepidoptera: Nymphalidae: Charaxinae). Trop. Lepid. 6 (1), 11–13.

Pyrcz, T.W., Fratello, S., 2005. Cloud forest butterfly fauna of the Pantepui—poor or poorly known? Description of new species and records of new genera of Pronophilina: *Eretris agata* and *Oxeoschistus romeo* (Nymphalidae: Satyrinae). J. Lepid. Soc. 59 (4), 200–211.

Pyrcz T.W., Neild A.F.E., 1996. Tribe Anaeini. In: Neild A.F.E. The Butterflies of Venezuela. Part I: Nymphalidae I (Limenitidinae, Apaturinae, Charaxinae). A Comprehensive Guide to the Identification of Adult Nymphalidae, Papilionidae, and Pieridae, Meridian Publications, Greenwich, London, pp. 99–116.

Reissinger, E.J., 1970. Neue neotropische Pieriden (Lepidoptera, Dismorphiinae et Pierinae). Acta Ent. Mus. Nat. Pragae 38, 409–414. plates 2.

Rivas-Martínez, S., Navarro, G., 1994. *Mapa biogeográfico de Suramérica*. Servicio de Cartografía, Universidad de León, León.

Rivas-Martínez, S., Navarro, G., Penas, A., Costa, M., 2011. Biogeographic map of South America. A preliminary survey. Int. J. Geobot. Res. 1, 21–41. + map.

Rull, V., 2004. Biogeografía histórica de las tierras altas de Guayana y origen de la biodiversidad neotropical. Orsis 19, 34–48.

Rull, V., 2005a. Biotic diversification in the Guayana highlands: a proposal. J. Biogeogr. 32, 921–927.

Rull, V., 2005b. Vegetation and environmental constancy in the Neotropical Guayana Highlands during the last 6000 years? Rev. Palaeobot. Palynol. 135, 205–222.

Salazar, J.A., Constantino, L.M., 2007. Descripción de nuevas especies de ropaloceros para Colombia (Lepidoptera: Pieridae, Nymphalidae, Satyrinae, Ithomiinae, Riodinidae). Bol. Cient. Mus. Hist. Nat. Univ. Caldas 11, 167–186.

Salerno, P., Ron, S.R., Señaris, C., Rojas-Runjaic, F.J.M., Noonan, B.P., Cannatella, D.C., 2012. Ancient tepui summits harbor young rather than old lineages of endemic frogs. Evolution 66 (10), 3000–3013.

Schreiber, H., 1978. *Dispersal centres of Sphingidae (Lepidoptera) of the neotropical region*, Biogeographica, vol. X. Dr. W. Junk B. V., Publishers, The Hague, Boston, vi; 195 pp.; [iii].

Sick, W.D., 1969. Geographical substance. In: Fittkau, E.J., Illies, J., Klinge, H., Schwabe, H., Sioli, H. (Eds.), Biogeography and Ecology of South America. Monographiae Biologica 19, vol. 2. Dr. W. Junk B. V., Publishers, The Hague, Boston, pp. 449–474.

Simonsen, T.J., Zakharov, E.V., Djernaes, M., Cotton, A.M., Vane-Wright, R.I., Sperling, F.A.H., 2011. Phylogenetics and divergence times of Papilioninae (Lepidoptera) with special reference to the enigmatic genera *Teinopalpus* and *Meandrusa*. Cladistics 27, 113–137.

Steyermark, J.A., 1979. Plant refuge and dispersal centres in Venezuela: their relict and endemic elements. In: Larsen, K., Holm-Nielsen, L.B. (Eds.), Tropical Botany. Academic Press, London, New York, San Francisco, pp. 185–221.

Steyermark, J.A., 1982. Relationships of some Venezuelan forest refuges with lowland tropical floras. In: Whitmore, T.C., Prance, G.T. (Eds.), Biological Diversification in the Tropics. Columbia University Press, New York, pp. 182–220.

Strand, E., 1912. Zwei neue Satyriden von Roraima. Gesammelt von Herrn Botaniker E. Ule. Fauna Exotica 2 (11), 43–44 (12): 47–48.

Takhtajan, A., 1986. *Floristic regions of the World*. University of California Press, Berkeley, p. 522.

Tate, G.H.H., 1928. The "Lost World" of Mount Roraima. Nat. Hist. 28 (3), 318–328.

Tate, G.H.H., 1930. Through Brazil to the summit of Mount Roraima. Nat. Geogr. 58 (5), 585–606.

Tate, G.H.H., 1932. Life zones at Mount Roraima. Ecology 13 (3), 235–257.

Tate, G.H.H., 1938. Auyan-tepui: notes on the Phelps Venezuelan expedition. Geogr. Rev. 28, 452–474.

Tate, G.H.H., Hitchcock, C.B., 1930. The Cerro Duida region of Venezuela. Geogr. Rev. 20 (1), 31–52.

Turner, J.R.G., 1966. A little-recognised species of *Heliconius* butterfly (Nymphalidae). J. Res. Lepid. 5 (2), 97–112.

Tyler, H.A., Brown Jr., K.S., Wilson, K.H., 1994. *Swallowtail butterflies of the Americas. A study in biological dynamics, ecological diversity, biosystematics, and conservation*. Scientific Publishers, Gainesville, p. 376.

Udvardy, M.D.F., 1975. A classification of the biogeographic provinces of the world. IUCN Occas. Paper 18, 1–48.
Ule, E., 1911. Von Amazonenstrom zum Roraima. Süd- und Mittel-Amerika 6, 103–106.
Ule, E., 1914. Die Vegetation des Roraima. Engler's Bot. Jahrb 115 (114), 42–53.
Villarreal, O., Señaris, C., DoNascimiento, C., 2002. Contribución al conocimiento faunístico del Wei-Assipu-Tepui, macizo del Roraima, con énfasis en la anurofauna y opiliofauna. Bol. Soc. Venez. Espel. 36, 46–50.
Viloria, Á.L., 1995. Description of a new species of the genus *Pedaliodes* from the Cerro de La Neblina, Venezuela (Lepidoptera, Satyridae). Atalanta 25 (3–4), 525–529. plate XVIIa.
Viloria, Á.L., 1998. Un nuevo *Pedaliodes* Butler, 1867 de la Serranía de Tapirapecó, Venezuela (Lepidoptera: Nymphalidae, Satyrinae, Pronophilini). SHILAP, Rev. Lepid. 26 (101), 13–18.
Viloria, Á.L., 2000. Estado actual del conocimiento taxonómico de las mariposas (Lepidoptera: Rhopalocera) de Venezuela. In: Martín-Piera, F., Morrone, J.J., Melic, A. (Eds.), Hacia un Proyecto Cyted para el inventario y estimación de la diversidad entomológica en Ibero-América: PrIBES-2000. m3m-Monografías Tercer Milenio, vol. 1. Sociedad Entomológica Aragonesa, Zaragoza, pp. 261–274.
Viloria, Á.L., 2002. Limitaciones que ofrecen distintas interpretaciones taxonómicas y biogeográficas al inventario de lepidópteros hiperdiversos de las montañas neotropicales y a sus posibles aplicaciones. In: Costa, C., Vanin, S.A., Lobo, J., Melic, A. (Eds.), Proyecto de Red Iberoamericana de Biogeografía y Entomología Sistemática PrIBES 2002. m3m-Monografías Tercer Milenio, vol. 2. Sociedad Entomológica Aragonesa, CYTED, Zaragoza, pp. 173–190.
Viloria, Á.L., 2003. Historical biogeography and the origins of the satyrine butterflies of the Tropical Andes (Insecta: Lepidoptera, Rhopalocera. In: Morrone, J.J., Llorente-Bousquets, J. (Eds.), Una perspectiva Latinoamericana de la biogeografía. Las Prensas de Ciencias, Facultad de Ciencias. UNAM, México, DF, pp. 247–261.
Viloria, Á.L., 2005. Las mariposas (Lepidoptera: Papilionoidea) y la regionalización biogeográfica de Venezuela. In: Llorente-Bousquets, J.E., Morrone, J.J. (Eds.), Regionalización Biogeográfica en Iberoamérica y tópicos afines. Primeras Jornadas Biogeográficas de la Red Iberoamericana de Biogeografía y Entomología Sistemática (RIBES XII.I-CYTED). Las Prensas de Ciencias, Facultad de Ciencias. UNAM, México, DF, pp. 441–459.
Viloria, Á.L., Pyrcz, T.W., 1994. A new genus, *Protopedaliodes* and a new species, *Protopedaliodes kukenani*, from the Pantepui, Venezuela (Lepidoptera, Nymphalidae, Satyrinae). Lambillionea 94 (3), 345–352.
Viloria, Á.L., Pyrcz, T.W., 1995. Notes on *Pedaliodes roraimae* Strand, a little known satyrid from south-eastern Venezuela (Lepidoptera: Nymphalidae: Satyrinae). Lambillionea 95 (4), 584–586.
Viloria, Á.L., Pyrcz, T.W., 2000. New pronophiline butterflies from the Venezuelan tepuis (Nymphalidae: Satyrinae). J. Lepid. Soc. 53 (3), 90–98.
Wahlberg, N., 2006. That awkward age for butterflies: insights from the age of the butterfly Subfamily Nymphalinae (Lepidoptera: Nymphalidae). Syst. Biol. 55 (5), 703–714.
Wahlberg, N., Leneveu, J., Kodandaramaiah, U., Peña, C., Nylin, S., Freitas, A.V.L., et al., 2009. Nymphalid butterflies diversity following near demise at the Cretaceous/Tertiaty boundary. Proc. R. Soc. Lond. (B) 276, 4295–4302.
Whitely, H., 1884. Explorations in the neighbourhood of Mounts Roraima and Kukenam, in British Guiana. Proc. R. Geogr. Soc 6 (8), 452–463. plates 2,1 map [p. 488].
Zubek, A., Pyrcz, T.W., 2011. Description of a new species of *Forsterinaria* Gray from the Pantepuy [sic] (Lepidoptera: Nymphalidae: Satyrinae). Genus Int. J. Invert. Taxon. 22 (3), 523–530.

CHAPTER

10

Scorpions

José A. Ochoa[1] *and Fernando J.M. Rojas-Runjaic*[2,3]

[1]Faculty of Sciences, Department of Biology, National University of San Antonio Abad del Cusco, Cusco, Peru [2]Laboratory of Vertebrate Systematics, Zoology Postgraduate Program, School of Sciences, Pontifical Catholic University of Rio Grande do Sul (PUCRS), Porto Alegre, Brazil [3]La Salle Natural History Museum, La Salle Natural Sciences Foundation, Caracas, Venezuela

Introduction

Scorpions are terrestrial predatory arthropods (Order Scorpiones) that inhabit mainly tropical and subtropical zones (Polis, 1990). They are found in different types of terrestrial ecosystems, including deserts, dry forests, savannas, tropical humid forests, mountain regions, paramos and punas, as well as in caves and intertidal zones.

A limited number of scorpion species inhabit high-elevation mountain regions that belong to a few genera and families; in the Alps, the only species of the genus *Euscorpius* Thorell, 1876 (Euscorpiidae) inhabit areas up to 2600 m, while in the Himalayas, scorpions of the genera *Hottentotta* Birula, 1908 (Buthidae); *Himalayotityobuthus* Lourenço, 1997 (Buthidae); *Chaerilus* Simon, 1877 (Chaerilidae); *Scorpiops* Peters, 1861 (Scorpiopidae); and *Tibetiomachus* Lourenço and Qi, 2006 (Hormuridae) can be found in areas up to 4600 m (Beron, 2018; Lourenço, 2018; Lourenço and Qi, 2006; Polis, 1990).

In Southern Africa, the mountain scorpion fauna inhabit areas up to 2570 m in the Brandberg massif (Namibia) and up to 1747 m in the Soutpansberg range (South Africa). These scorpions include members of the families Bothriuridae (genera *Brandbergia* Prendini, 2003 and *Lisposoma* Lawrence, 1928), Buthidae (*Parabuthus* Pocock, 1890; *Pseudolychas* Kraepelin, 1911; and *Uroplectes* Peters, 1861), Hormuridae (*Cheloctonus* Pocock, 1892, *Opisthacanthus* Peters, 1861; and *Hadogenes* Kraepelin, 1894), and Scorpionidae (*Opistophthalmus* Koch, 1837) (Foord et al., 2015; Prendini and Bird, 2008).

The occurrence of extreme ecological conditions in high-mountain regions is possibly the reason why these scorpions spend most of the time in burrows and under rocks, tree bark, logs, interstices, or rock crevices (Lourenço, 2018; Ochoa et al., 2011; Polis, 1990).

Biodiversity of Pantepui
DOI: https://doi.org/10.1016/B978-0-12-815591-2.00010-0

In South America, the most important mountain regions are the Guiana Highlands and the Andes mountain range. The Guiana Highlands are located to the south of the Orinoco River in Venezuela and extend across west-central Guyana and the northern state Roraima in Brazil (Figs. 10.1 and 10.2; see Chapter 1: Definition and characterization of the Pantepui biogeographical province). These highlands, also named Pantepui, are characterized by upland plateaus and tepuis that emerge as high-elevation mountains in the landscape. The biogeographic province of Pantepui includes ecosystems mostly above 1400–1500 m in elevation, whereas the middle elevations (uplands) of the Guiana region that surround the tepuis include forest and savannas located between 800 and 1500 m (Berry and Riina, 2005; Huber, 1987, 1988).

The Andean scorpion fauna are relatively better studied than that of the Guiana Highlands. Bothriuridae, Buthidae, and Chactidae are the most diverse families in the Andan range, but less diverse groups such as Diplocentridae, Hormuridae, Iuridae, and Troglotayosicidae are also present (Acosta and Ochoa, 2002; Botero-Trujillo and Flórez, 2011; Botero-Trujillo et al., 2012; Brito and Borges, 2015; Ceccarelli et al., 2016; Lourenço, 1994a; Lourenço and Duhem, 2010; Ochoa, 2005, Ochoa and Prendini, 2010; Ochoa et al., 2011; Ojanguren-Affilastro, 2005; Ojanguren-Affilastro et al., 2016;

FIGURE 10.1 Habitat of some scorpion species in the Guiana region. (A) Humid forest at La Escalera, Cerro Venamo, Bolívar, Venezuela. (B) Savanna and forest patches near Uruyén, Bolívar, Venezuela. (C) Leaf litter at Guayaraca, Uruyén, middle slopes of Auyán-tepui. Bolívar, Venezuela. (D) Aerial view of the southern part of the Auyán-tepui.

FIGURE 10.2 Habitat of some scorpion species in the Guiana region. (A) Forest near Santa Elena de Uairén, Bolívar, Venezuela. (B) Vegetation around the waterfall Kamá Merú, Bolívar, Venezuela. (C) Canaima lagoon, Canaima National Park, Bolívar, Venezuela.

Rojas-Runjaic and De Sousa, 2007; Ythier and Lourenço, 2017). In the Guiana region, scorpions are represented mainly by two families: Chactidae and Buthidae (González-Sponga, 1984a, 1996, 2001; Rojas-Runjaic and De Sousa, 2007).

Although the first reference to a scorpion species from Pantepui dates from the beginning of the 20th century (Pocock, 1900), it was not until the mid-1970s that a comprehension of the diversity of scorpions in this region was obtained. In 1978, González-Sponga studied the scorpion fauna of the eastern region of the Bolívar state in Venezuela and published the first checklist of the scorpions of Pantepui. Gonzalez-Sponga compiled all the information, which was until then scattered, and described several new taxa, including three endemic genera from the Guiana Shield (González-Sponga, 1978).

The same author had already described some species from the states of Amazonas and Bolívar in Venezuela (González-Sponga, 1972a,b, 1973, 1974a,b, 1975a,b), but it is from 1978 that he made his greatest contributions, describing more than 70% of the known scorpion fauna from the Pantepui province and surrounding areas. This period is considered the beginning of studies on scorpions in the region. At present, the known scorpion fauna for Pantepui comprises only 11 species (distribution above 1400 m); however, when including the surrounding areas, there are 115 species (Figs. 10.3–10.5; Table 10.1; Appendix 10.1).

FIGURE 10.3 Habitus in life. (A) *Tityus venamensis* González-Sponga, 1981, Cerro Venamo, Bolívar, Venezuela. (B) *Tityus filodendron* González-Sponga, 1981, Cerro Autana, Amazonas, Venezuela. (C) *Ananteris venezuelensis* González-Sponga, 1972, Cerro Venamo, Bolívar, Venezuela. (D) *Brotheas libinallyi* González-Sponga, 1978, Santa Elena de Uairén, Bolívar, Venezuela.

First explorations

The first known Pantepui scorpion species were collected by British explorers John J. Quelch and Frederick V. McConnell on their 1898 expedition to the summit of the Roraima-tepui between Brazil, Guyana, and Venezuela. These specimens were deposited at the British Museum of Natural History and later described by Pocock (1900) as *Broteochactas granosus* Pocock, 1900 of the eastern slope of Roraima-tepui (Guyana, 1070 m) and *Broteochactas porosus* (=*Taurepania porosa*) of the summit (2810 m).

In the second half of the 20th century, more than 50 years after the first study, a new study on scorpions of Pantepui and the surrounding areas of the Guiana Shield was carried out. This included the description of *Tityus urbinai* Scorza, 1952, based on scorpions collected in April 1950 by the French entomologist René Lichy (Scorza, 1952), who conducted an expedition to the region La Culebra at the western base of Cerro Duida in the Amazonas state of Venezuela.

FIGURE 10.4 Habitus in life. (A) *Broteochactas efreni* González-Sponga, 1978, Cerro Venamo, Bolívar, Venezuela. (B) *Broteochactas garciai* González-Sponga, 1978, Santa Elena de Uairén, Bolívar, Venezuela. (C) *Broteochactas kjellesvigi* González-Sponga, 1974, Isla Ratón, Amazonas, Venezuela. (D) *Broteochactas macrochelae* González-Sponga, 2004, Uruyén, Bolívar, Venezuela.

A few years later, in December 1952, the La Salle Society of Natural Sciences carried out an expedition to the Orinoco River Delta. In this expedition, several specimens of scorpions were collected, from which José Vicente Scorza two years later described *Broteochactas orinocensis* Scorza, 1954 (Scorza, 1954a).

In that same year *Teuthraustes carmelinae* Scorza, 1954 was described (Scorza, 1954b). This publication also listed several species of chactids from the Guiana Shield and reported the first record of the genus *Chactopsis* Kraepelin, 1912 from Venezuela. It should be noted that the scorpions documented in this publication were collected during the Franco-Venezuelan Expedition to the Orinoco River in 1951.

In 1956, a university expedition to the Auyán-tepui plateau was conducted in which José Vicente Scorza participated. A scorpion he collected in Guayaraca was described a year later as *Broteochactas scorzai* (=*Auyantepuia scorzai*) (Dagert, 1957). The few studies carried out during this period were in no case the result of planned studies on this group of arachnids, and some of them had imprecise geographical citations.

FIGURE 10.5 Habitus in life. (A) *Chactopsoides anduzei* (González-Sponga, 1982), Isla Ratón, Amazonas, Venezuela. (B) *Hadrurochactas machadoi* González-Sponga, 1993, Uruyén, Bolívar, Venezuela. (C) *Megachactops kuemoi* Ochoa, et al., 2013, Cerro Autana, Amazonas, Venezuela. (D) *Taurepania manisapanensis* González-Sponga, 1992, Kamoirán, Bolívar, Venezuela.

Gonzalez-Sponga's contributions

Manuel A. González-Sponga was a Venezuelan arachnologist (1929–2009) who explored the taxonomy, biogeography, and ecology of scorpions and other Venezuelan arachnids between 1970 and 2007. Although his work was not limited to arachnids (he also published some works on myriapods and reptiles), his contribution to scorpions is really remarkable, describing 81% of the species known to Venezuela (Rojas-Runjaic and De Sousa, 2007), which is even higher (85%) if we count only the species of the states of Bolívar and Amazonas. His first manuscripts for the Venezuelan Guayana region were published from 1972 to 1975, when he described seven species of scorpions for states Bolívar and Amazonas: *Brotheas camposi* González-Sponga, 1972; *Brotheas mingueti* González-Sponga, 1973; *Broteochactas kjellesvigi* González-Sponga, 1974; *B. racenisi* González-Sponga, 1975; *B. sanmartini* González-Sponga, 1974; *Ananteris venezuelensis* González-Sponga, 1972 (Fig. 10.3A); and *Teuthraustes adrianae* González-Sponga, 1975 (González-Sponga, 1972a,b, 1973, 1974a,b, 1975a,b), with the latter being the second scorpion species known from Pantepui.

TABLE 10.1 List of scorpion species from Pantepui (> 1400 m) and intermediate tepuian slopes (> 800 m) with the name of the sector or tepui from which they have been recorded. References: 1, González-Sponga, 2006; 2, González-Sponga, 1972a; 3, González-Sponga, 1978; 4, González-Sponga, 1997b; 5, Lourenço, 2008; 6, González-Sponga, 1981; 7, Rojas-Runjaic and De Sousa, 2007; 8, Dagert, 1957; 9, Pocock, 1900; 10, Lourenço et al., 2011; 11, González-Sponga, 1985; 12, González-Sponga, 1991; 13, González-Sponga, 1997a; 14, González-Sponga, 1982; 15, Ochoa et al., 2013, 16, González-Sponga, 1993; 17, González-Sponga, 1992; 18, Lourenço, 2017; 19, Vignoli and Kovařik, 2003; 20, González-Sponga, 1975b; 21, Lourenço, 1994b; 22, Lourenço and Duhem, 2010; 23, unpublished data.

No.	Family	Species	Pantepui	Middle slopes	Altitude (m)	Sector/Tepui	Reference
1	Buthidae	*Ananteris chirimakei* González-Sponga, 2006		X	895	Perai-tepui	1
2	Buthidae	*Ananteris venezuelensis* González-Sponga, 1972		X	800–1140	Cerro Venamo	2, 3, 23
3	Buthidae	*Tityus anduzei* González-Sponga, 1997		X	885	Sierra de Parima	4
4	Buthidae	*Tityus neblina* Lourenço, 2008	X	X	850–2200	Sierra de la Neblina	5
5	Buthidae	*Tityus venamensis* González-Sponga, 1981	X	X	1200–1500	Cerro Venamo	6, 7, 23
6	Chactidae	*Auyantepuia scorzai* (Dagert, 1957)		X	1014	Guayaraca/Auyán-tepui	8, 3
7	Chactidae	*Broteochactas bilbaoi* González-Sponga, 1978		X	810–1100	Cerro Venamo	3
8	Chactidae	*Broteochactas efreni* González-Sponga, 1978		X	960–1150	Cerro Venamo	3, 23
9	Chactidae	*Broteochactas garciai* González-Sponga, 1978		X	700–930	Santa Elena de Uairén	3
10	Chactidae	*Broteochactas granosus* Pocock, 1900		X	1070	Roraima-tepui	9
11	Chactidae	*Broteochactas leoneli* González-Sponga, 1978		X	180–1326	Cerro Venamo	3
12	Chactidae	*Broteochactas niemeyerae* Lourenço et al., 2011		X	1350	Sierra de la Neblina	10
13	Chactidae	*Broteochactas santanai* González-Sponga, 1978		X	800–1200	Yuruani-tepui	3, 23
14	Chactidae	*Broteochactas sarisarinamensis* González-Sponga, 1985	X		1430	Cerro Sarisariñama	11
15	Chactidae	*Brotheas caramaschii* Lourenço et al., 2011		X	1350	Sierra de la Neblina	10
16	Chactidae	*Brotheas libinallyi* González-Sponga, 1978		X	200–1300	Cerro Venamo / Santa Elena de Uairén	3, 7
17	Chactidae	*Brotheas mawarinumensis* González-Sponga, 1991	X	X	140–1690	Sierra de la Neblina	12
18	Chactidae	*Brotheas munozi* González-Sponga, 1997		X	280–980	Cerro Guaiquinima	13
19	Chactidae	*Brotheas sanabriai* González-Sponga, 1997		X	280–1150	Cerro Guaiquinima	13

(*Continued*)

TABLE 10.1 (Continued)

No.	Family	Species	Pantepui	Middle slopes	Altitude (m)	Sector/Tepui	Reference
20	Chactidae	*Chactopsis barajuri* González-Sponga, 1982		X	900–925	Santa Elena de Uairén	14, 15
21	Chactidae	*Chactopsis sujirima* González-Sponga, 1982		X	1200	Serranía Tapirapecó	14, 15
22	Chactidae	*Chactopsoides yanomami* (Lourenço et al., 2011)		X	1350	Serranía Tapirapecó	10
23	Chactidae	*Hadrurochactas machadoi* González-Sponga, 1993		X	455–1014	Auyán-tepui	16
24	Chactidae	*Hadrurochactas odoardoi* González-Sponga, 1985	X		1430	Cerro Sarisariñama	11
25	Chactidae	*Taurepania manisapanensis* González-Sponga, 1992		X	1000	Manisapán-tepui	17
26	Chactidae	*Taurepania mauriciodiasi* (Lourenço, 2017)		X	1050	Serra da Mocidade	18
27	Chactidae	*Taurepania porosus* (Pocock, 1900)	X		2810	Roraima-tepui	9
28	Chactidae	*Taurepania trezzii* Vignoli and Kovarik, 2003	X		1650	Auyán-tepui	19
29	Chactidae	*Taurepania verneti* González-Sponga, 1992	X		2650	Kukenán-tepui	17
30	Chactidae	*Taurepania vestigialis* González-Sponga, 1978		X	1000–1300	Yuruaní-tepui	3, 23
31	Chactidae	*Teuthraustes adrianae* González-Sponga, 1975	X		1550–2200	Sierra de la Neblina	20, 23
32	Chactidae	*Teuthraustes lisei* Lourenço, 1994		X	860	Sierrra de la Neblina	21
33	Chactidae	*Teuthraustes maturaca* González-Sponga, 1991		X	140–860	Sierra de la Neblina	12
34	Chactidae	*Teuthraustes newaribe* Lourenço et al., 2011		X	1350	Sierra de la Neblina	10
35	Chactidae	*Teuthraustes reticulatus* González-Sponga, 1991	X		1800	Sierra de la Neblina	12, 23
36	Chactidae	*Vachoniochactas amazonicus* González-Sponga, 1991	X		1400–1800	Sierra de la Neblina	12, 23
37	Chactidae	*Vachoniochactas ashleeae* Lourenço, 1994		X	400–860	Sierra de la Neblina	21, 23
38	Chactidae	*Vachoniochactas lasallei* González-Sponga, 1978		X	1320	Cerro Venamo	3
39	Chactidae	*Vachoniochactas roraima* Lourenço and Duhem, 2009		X	1380	Roraima-tepui	22

Expeditions in the Guiana region

Subsequent work by González-Sponga includes the publications resulting from specific scientific expeditions in the Guiana region. In 1978 he published "Escorpiofauna de la Región Oriental del Estado Bolívar en Venezuela," which was the result of four expeditions carried out between 1973 and 1977. Two of these expeditions (October to November 1973 and November to December 1974) were led by Maite Bilbao, Edgard García, and collaborators, while the other two expeditions (May 1975 and June 1977) were organized by Abdén R. Lancini of the Museum of Natural Sciences of Caracas. A total of 412 specimens of the families Buthidae and Chactidae were collected (González-Sponga, 1978). These expeditions included collections from several locations on the route from El Dorado to Santa Elena de Uairén, including also the top of the Roraima-tepui. In this work, Gonzalez-Sponga listed 20 species of scorpions from Bolívar state and described several new taxa, including the genera *Auyantepuia* González-Sponga, 1978; *Taurepania* González-Sponga, 1978 (Fig. 10.5D); and *Vachoniochactas* González-Sponga, 1978, and eight new species of Chactidae (González-Sponga, 1978).

A second important publication related to the tepuis corresponds to the results of expeditions carried out between 1983 and 1987 to Sierra de la Neblina, known in Brazil as "Pico da Neblina" (González-Sponga, 1991). These expeditions were sponsored by the Foundation for the Development of Sciences (FUDECI), of the Venezuelan Acadamey of Physical, Mathematical and Natural Sciences under the direction of Charles Brewer-Carías, and included more than 10 fieldtrips and 14 camps located at different altitudes from 140 to 2400 m (Brewer-Carías, 1988). The scorpions were studied by González-Sponga, who listed 11 species, describing 8 new species of Chactidae and 1 of Buthidae (González-Sponga, 1991). Particularly significant are the descriptions of *Teuthraustes reticulatus* González-Sponga, 1991; *Vachoniochactas amazonicus* González-Sponga, 1991; and *Brotheas mawarinumensis* González-Sponga, 1991, all of them endemic species that are distributed in the upper part of the Sierra de la Neblina above 1400 m, as well as *Teuthraustes maturaca* González-Sponga, 1991 that inhabits the slopes and median altitudes (González-Sponga, 1991). From this expedition, collected specimens also resulted in the description of *Chactopsis coriacea* González-Sponga, 1991, which would later be transferred to *Megachactops* Ochoa et al., 2013, an endemic genus of the state of Amazonas in Venezuela (Ochoa et al., 2013).

The Duida-Marahuaka massif and its surrounding areas were explored in 1983 by the Terramar Foundation. Scorpions of the genera *Tityus* C. L. Koch, 1836; *Teuthraustes* Simon, 1878; *Broteochactas* Pocock, 1893; and *Taurepania* and *Chactopsoides* Ochoa et al., 2003 were collected during this expedition and described by González-Sponga in three studies. In 1984, he described *Teuthraustes akananensis* González-Sponga, 1984; *Broteochactas yekuanae* González-Sponga, 1984; and *Brotheas cunucunumensis* González-Sponga, 1984 (González-Sponga, 1984b). In 1994, he described *Tityus culebrensis* González-Sponga, 1994 and confirmed the report of *T. urbinai* Scorza, 1954 (González-Sponga, 1994); finally, in 2004, he described *Chactopsoides marahuacaensis* (González-Sponga, 2004). All these species were collected in the vicinity of La Culebra, a village located at the base of the mountain in the northwest part of Cerro Duida. In addition, a specimen of the genus *Taurepania* was collected at the top of Cerro Marahuaka, which González-Sponga did not describe.

Another important expedition where scorpions were collected was carried out at Cerro Sarisariñama, in the Bolviar state of Venezuela. From this, *Hadrurochactas odoardoi* González-Sponga, 1985 and *Broteochactas sarisarinamensis* González-Sponga, 1985 were described from the lower summit of this tepui, at 1430 m (González-Sponga, 1985), and *Tityus sarisarinamensis* González-Sponga, 2002 was collected at the middle slopes, at 400 m (González-Sponga, 2002).

Other notable species described by González-Sponga from Pantepui were *Brotheas munozi* González-Sponga, 1997 and *Brotheas sanabriai* González-Sponga, 1997, which were collected in the FUDECI expedition at the slopes of Cerro Guaiquinima, from 280 to 1150 m (González-Sponga, 1997a,b). Finally, it is important to highlight the description of *Taurepania verneti* González-Sponga, 1992 collected at the summit of the Kukenán-tepui, at 2650 m (González-Sponga, 1992).

In total, over 37 years, González-Sponga described 87 species of scorpions (23 Buthidae, 63 Chactidae, and one species of the family Hormuridae) from the Venezuelan Guayana region, mainly of the genera *Ananteris* Thorell, 1891 (7 species); *Tityus* (16 species); *Brotheas* C. L. Koch, 1837 (17 species); and *Broteochactas* (28 species).

Recent years

Some additional species have been described from Pantepui and the surrounding areas in recent years, including one species of Buthidae and four species of Chactidae: *Tityus neblina* Lourenço, 2008; *Brotheas caramaschii* Lourenço et al., 2011; *Chactopsoides yanomami* Lourenço et al., 2011; *Broteochactas niemeyerae* Lourenço et al., 2011; and *Teuthraustes newaribe* Lourenço et al., 2011 (Lourenço, 2008; Lourenço et al., 2011), all collected on the slopes of Phelps Peak (known in Brazil as "Pico 31 de Março"), on the border of Brazil and Venezuela near Sierra de la Neblina (Lourenço et al., 2011). In this period, *Vachoniochactas roraima* Lourenço and Duhem, 2009 was also described from the intermediate slopes of the Roraima-tepui, as well as two endemic genera from the Guiana region, *Chactopsoides* and *Megachactops* (Ochoa et al., 2013), and the first reference of a cavernicole scorpion species from Pantepui: *Taurepania trezzii* Vignoli and Kovařik, 2003. This trogloditic scorpion was collected in a cave of the Sima Aonda sector located in the northern part of the Auyán-tepui (1650 m), as part of an expedition carried out in 1992 (Vignoli and Kovařik, 2003).

To date, only 11 described species are distributed in the upper parts of the tepuis at elevations above 1400 m, in the Pantepui biogeographical province. If we consider the slopes and intermediate altitudes of the tepuis ($>$800 m), the list increases to 39 species (Table 10.1).

Pattern of diversity and endemism of scorpions

In many taxa, species diversity declines as altitude increases (Lomolino, 2001). This pattern includes scorpions and all other arachnids, where only a few species are found in high-mountain habitats (Foord et al., 2015; Lourenço and Qi, 2006; Ochoa et al., 2011;

TABLE 10.2 Comparative numbers of families, genera, and species of scorpions from Pantepui and other mountain regions of the World. References: 1, Ochoa et al., 2011; 2, Beron, 2018; 3, Lourenço, 2018; 4, Lourenço and Qi, 2006; 5, Foord et al., 2015; 6, Prendini and Bird, 2008.

	High Andes[1]	Pantepui[a]	Himalaya[2,3,4]	South Africa[5]	Namibia[6]
Families	4	2	5	3	4
Genera	7	7 (11)	7	8	7
Species	39	11 (39)	26	19	20

[a]*The number in parentheses indicates the species of Pantepui and the middle slopes together (>800 m).*

Polis, 1990; Prendini and Bird, 2008). Like in many other mountain regions, the scorpion fauna of Pantepui exhibits a low diversity (Table 10.2). Although the species richness is lower compared to the Andes or the Himalayas, the total area of the discontinuous Pantepui province is considerably smaller (c.5000 km^2) than the Andean highlands and the Old Word mountains. The Andes has the greather scorpion diversity for a high-mountain environment (39 species) recorded above 3000 m, including species from Argentina, Bolivia, Chile, Colombia, Ecuador, and Peru (Ochoa et al., 2011); these include members of the families Bothriuridae (genera *Orobothriurus* Maury, 1976; *Pachakutej* Ochoa, 2004; *Brachistosternus* Pocock, 1893; and *Bothriurus* Peters, 1861), Buthidae (*Tityus*), Iuridae (*Hadruroides* Pocock, 1893), and Chactidae (*Teuthraustes*).

It is important to mention that, recently, research studies on the systematics and biogeography of Andean scorpions have increased significantly (Ceccarelli et al., 2016; Ochoa and Prendini, 2010; Ochoa et al., 2011; Ojanguren-Affilastro et al., 2016), compared to the few studies conducted in Pantepui. Due to the difficulty in terms of accessibility, many tepuis have not been evaluated and, hence, it is very likely that the diversity of Pantepui scorpions is much greater than we know.

The species richness of the scorpion fauna of Pantepui and vicinity (31 species in 11 genera, including the intermediate tepuian slopes >800 m) is comparable to the South African mountain scorpion fauna of the Soutpansberg range (19 species from eight genera; c. 6800 km^2) and the Brandberg massif in Namibia (20 species from seven genera; 650 km^2) (Foord et al., 2015; Prendini and Bird, 2008). The decrease in richness with an increase of altitude, like that of other mountain areas such as the Andes, the Himalayas, and South Africa ranges, is also observed in Pantepui (Table 10.1).

The Guianan forests and savannas surrounding the tepuis are equally remarkable in terms of scorpion diversity, including 115 species, of which 32 species are Buthidae (4 genera), 83 Chactidae (11 genera) and a single representative of Hormuridae (Table 10.3, Appendix 10.1).

The family Chactidae has a remarkable diversity in the Guiana region, with more than 40% of the known species of the family, including most Neotropical genera (except the Andean genus *Chactas* Gervais, 1844), found in this region.

TABLE 10.3 Number of scorpion species from Pantepui (> 1400 m), the intermediate tepuian slopes (> 800 m), and the total number for the Guiana region. References: 1, González-Sponga, 1996; 2, Rojas-Runjaic and De Sousa, 2007; 3, Esposito et al., 2017; 4, Lourenço, 2002; 5, Lourenço, 2008; 6, Lourenço, 2017; 7, Lourenço et al., 2011; 8, Ochoa et al., 2013; 9, Lourenço, 2010; 10, Lourenço and Duhem, 2010; 11, Lourenço, and Duhem, 2009.

Family	Genus	Pantepui (> 1400 m)	Middle tepuian slopes (> 800 m)	Guiana region[a]	Reference
Buthidae	*Ananteris*		2	7	1, 2
	Jaguajir			1	3
	Tityus	2	3	23	2, 4, 5, 6
	Rhopalurus			1	2, 3
Chactidae	*Auyantepuia*		1	1	1, 2
	Broteochactas	1	7	31	1, 2, 7
	Brotheas	1	5	19	1, 2, 4, 7
	Cayooca			1	1, 2
	Chactopsis		2	3	1, 7, 8
	Chactopsoides		1	4	8
	Hadrurochactas	1	1	3	1, 2, 9
	Megachactops			2	8
	Taurepania	3	3	6	1, 2, 6
	Teuthraustes	2	3	7	1, 7, 10
	Vachoniochactas	1	3	5	1, 2, 11
Hormuridae	*Ophistacanthus*			1	2
	Total	11	31	115	

[a]*Including the lowland and upland areas (<800 m) surrounding Pantepui.*

Endemism

Scorpions are generally stenotopic organisms, with a limited ability to disperse great distances (Polis, 1990). As such, in most cases they have a restricted distribution and the degree of endemism is very high in almost all regions of the world where they live. This pattern is due also to the substratum specialization of most scorpions, especially in mountain areas (Prendini, 2001). Therefore it is not surprising that all species of scorpions known from the tepuian summits and their middle slopes are endemic—even 99% of the species of the entire Guiana region have a distribution restricted to this biogeographical region (Appendix 10.1). Only one species, *Rhopalurus laticauda* Thorell, 1876, has a wide distribution that reaches the Venezuelans Caribbean and Atlantic coasts.

However, this pattern of endemism is also due to the region's long geological history, its isolation from other mountain ranges (e.g., the Andes), and the discontinuous distribution of Pantepui summits, where each tepui is practically an island (Steyermark, 1986; Berry and Riina, 2005). The presence of six endemic chactid genera (*Auyantepuia*; *Cayooca* González-Sponga, 1996; *Chactopsoides*; *Megachactops*; *Taurepania*; and *Vachoniochactas*) is a manifestation of the high endemism of this region (Fig. 10.6).

- *Auyantepuia*. It comprises a single described species, *A. scorzai*, known only from Guayaraca, located on the middle slopes of the Auyán-tepui (1014 m) (Fig. 10.6).
- *Cayooca*. It comprises a single described species: *C. venezuelensis*, collected on the base of Sierra de la Neblina (Fig. 10.6).
- *Chactopsoides*. This genus includes four species: *C. anduzei* from the Vichada Department of Colombia and the Amazonas state in Venezuela in the Orinoco River basin, *C. gonzalezspongai* from Municipio Atabapo in Venezuela, *C. marahuacaensis* from Culebra village near the base of Cerro Marahuaka, and *C. yanomami* from Serranía Tapirapecó in Brazil, near the border with Venezuela (Fig. 10.6).
- *Megachactops*. Endemic to the Amazonas state in Venezuela, with two known species: *M. coriaceo* from the base of Sierra de la Neblina and *M. kuemoi* from the base of Cerro Autana (Fig. 10.6).

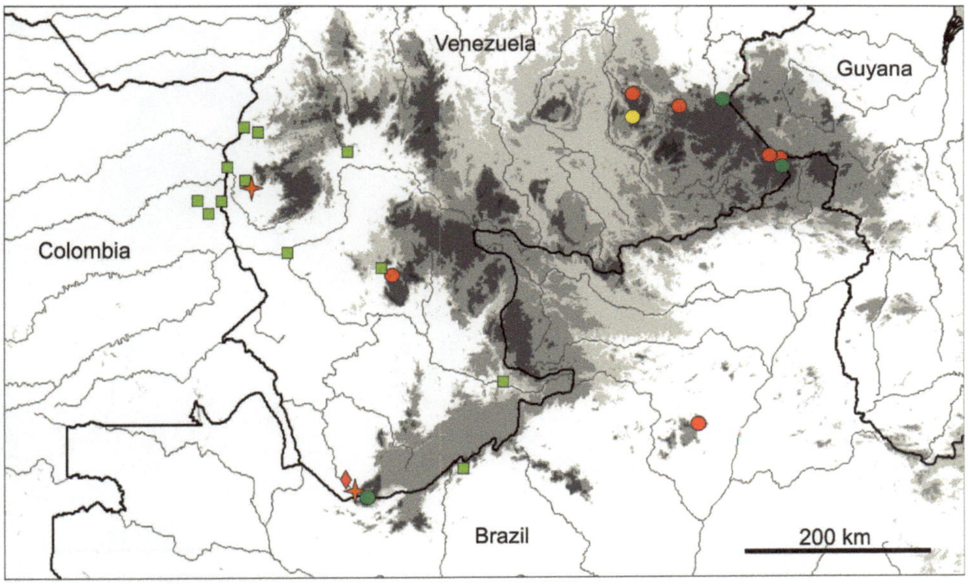

FIGURE 10.6 Distribution of the endemic genera of Pantepui and surrounding areas including records from Brazil, Colombia, and Venezuela. *Auyantepuia* González-Sponga, 1978 (*yellow circle*); *Cayooca* González-Sponga, 1996 (*red diamond*); *Chactopsoides* Ochoa et al., 2013 (*green squares*); *Megachactops* Ochoa et al., 2013 (*orange stars*); *Taurepania* González-Sponga, 1978 (*red circles*); *Vachoniochactas* González-Sponga, 1978 (*green circles*).

- *Taurepania*. This genus was first described by González-Sponga (1978) and includes *T. porosa* Pocock, 1900, known only from the summit of Roraima-tepui (2810 m). *Taurepania* is the only known genus of scorpions that inhabits the top of the tepuis. In addition to *T. porosa*, three other species inhabit these summits: *T. verneti* (Kukenán-tepui, 2650 m), *T. trezzii* (Auyán-tepui, 1650 m), and the unnamed species mentioned by González-Sponga (1984b) for Cerro Marahuaka (Amazonas state), in Venezuela. Three other species can be found the middle slopes of the tepuis: *T. vestigialis* González-Sponga, 1978, described from the Yuruaní River, in the Bolívar state of Venezuela (1000–1300 m); *T. manisapanensis* González-Sponga, 1992, from the base of Cerro Manisapán (1000 m, Bolívar state in Venezuela); and the recently described *T. mauriciodiasi* (Lourenço, 2017), from the Serra da Mocidade National Park, at 1050 m, in the Roraima state of Brazil (Figs. 10.6 and 10.7). It should be mentioned that *Taurepania* was recently considered a subgenus of *Broteochactas* (Lourenço, 2017).
- *Vachoniochactas*. This genus currently includes five species, all related to the Pantepui area (Fig. 10.6). Four of them are located above 1200 m: *V. lasallei* Gonzalez-Sponga, 1978, from Cerro Venamo, in the Bolívar state of Venezuela (1320 m); *V. amazonicus* Gonzalez-Sponga, 1991, from Sierra de la Neblina between, 1390 and 1515 m; *V. ashleeae* Lourenço 1994, from Sierra de la Neblina, although the original description does not indicate the precise altitude (Lourenço, 1994b); and *V. roraima* Lourenço and

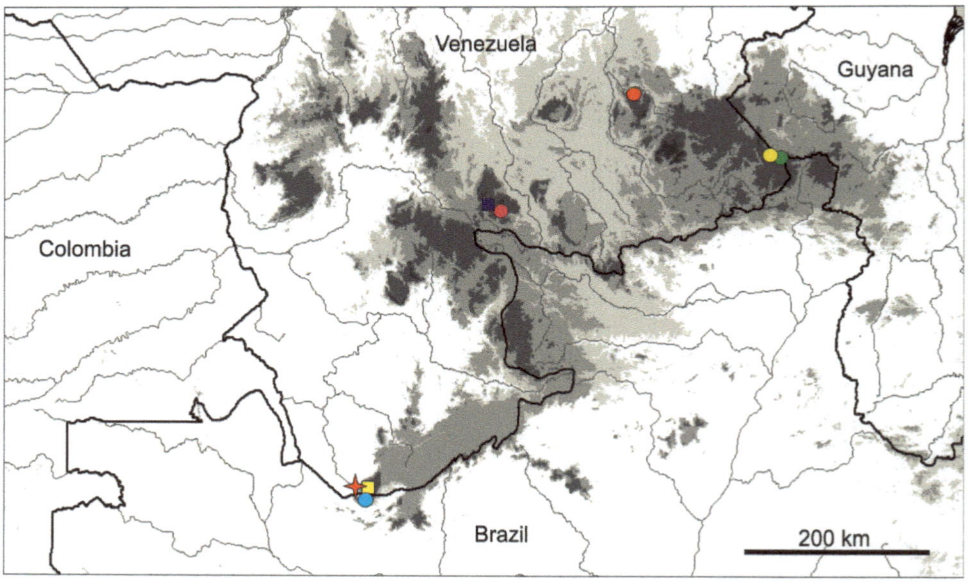

FIGURE 10.7 Distribution of some endemic species of Pantepui including records from Brazil and Venezuela. *Broteochactas sarisarinamensis* González-Sponga, 1985 (*magenta circle*); *Hadrurochactas odoardoi* González-Sponga, 1985 (*blue square*); *Taurepania porosa* Pocock, 1900 (*green circle*); *T. trezzii* Vignoli and Kovarik, 2003 (*red circle*); *T. verneti* González-Sponga, 1992 (*yellow circle*); *Teuthraustes adrianae* González-Sponga, 1975 (*sky-blue circle*); *T. reticulatus* González-Sponga, 1991 (*orange star*); *Vachoniochactas amazonicus* González-Sponga, 1991 (*yellow square*).

Duhem, 2009, from Roraima-tepui (1380 m). *V. humboldti* Flórez et al., 2008 is the only species of the genus distributed in the lowlands (260–300 m), in the Vichada Department of Colombia near the Venezuelan border (Flórez et al., 2008).

Distribution patterns

Most scorpion species of the Guiana region have very limited distribution ranges. From the total number of 39 species that are present above 800 m altitude, 37 are known only from a single sector and/or tepui (Table 10.1). Only *Brotheas libinallyi* has a relatively wider distribution in lowland forested areas from Cerro Venamo to Santa Elena de Uairén, and *Broteochactas garciai* is known from several locations around Santa Elena de Uairén. Presently there are no records of scorpions that occur in two or more tepuis.

Around 76% of the scorpion fauna of Pantepui and the middle slopes (39 species) are only found at an elevation above 800 m, whereas 20% exclusively inhabit regions above 1400 m (Table 10.3).

The elevation pattern of genera distribution is shown in Fig. 10.8. The genus *Auyantepuia* occurs exclusively at an elevation above 800 m. The same pattern occurs in most of the species of the genera *Taurepania* (86%), *Vachoniochactas* (80%), and *Teuthraustes* (57%). Conversely, the genera *Cayooca*, *Megachactops*, *Opisthacanthus*, and *Jaguajir* Esposito et al., 2017 are distributed only in the upland and lowland areas (<800 m) surrounding Pantepui. It is remarkable that seven species (all of the family Chactidae) occur exclusively at the summit of the tepuis (Table 10.3).

Most scorpion species and genera (except *Opisthacanthus autanensis*) are related with Neotropical elements. The genus *Teuthraustes* also includes several species distributed in the

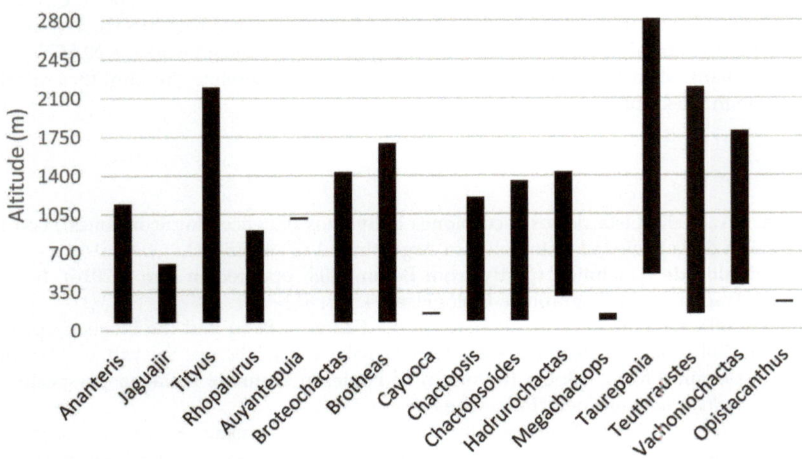

FIGURE 10.8 The elevational pattern of all scorpion genera recorded in the Pantepui province and surrounding areas from the Guiana region based on the references listed in Tables 10.1 and 10.3.

Andes from Ecuador to northern Peru (Lourenço and Duhem, 2010), this distinct distribution pattern from Guiana to the Andes also occurs in several groups of plants (Berry and Riina, 2005) and birds (Bonaccorso and Guayasamin, 2013; Mayr and Phelps, 1967).

Other groups of scorpion species are associated with the lowland rainforest in the Amazonia and Guianan regions, including the genera *Brotheas*, *Broteochactas*, and *Chactopsis*, all of them with a widespread distribution in the Guiana Shield and the Amazonian basin (Lourenço, 2002; Lourenço and Pinto-da-Rocha, 2000; Ochoa et al., 2013; Sissom, 2000). Most of the species of the genus *Hadrurochactas* Pocock, 1893 are also associated with this Guianan–Amazonian rainforest pattern (González-Sponga, 1978, 1985, 1993; Lourenço, 2010).

Contrastingly, the genera *Jaguajir* and *Rhopalurus* Thorell, 1876 and two species of the genus *Hadrurochactas* are associated with open lowland vegetation types such as savannas, caatingas, and the Cerrado (Esposito et al., 2017; Lourenço, 2010).

The Buthidae genera *Ananteris* and *Tityus* have a widespread distribution in the Neotropics and are the most diverse groups of Neotropical scorpions—*Tityus* with more than 200 species and *Ananteris* with more than 80 species (Botero-Trujillo and Flórez, 2011; Fet and Lowe, 2000; Lourenço, 2006). However, few species of these genera inhabit mountain areas (Ochoa et al., 2011). The species of these genera found in the Pantepui province are apparently related to Amazonian elements.

The absence of phylogenetic studies of almost all the groups related to Pantepui does not allow us to know in detail the historical biogeographical relationships of the scorpions, especially of the family Chactidae, which presents the greatest diversity in the region.

Acknowledgments

F.J.M.R.R. was supported by a Ph.D. scholarship from the Brazilian Conselho Nacional de Desenvolvimento Científico e Tecnológico (CNPq, process 142444/2014-6). Fieldwork in the Bolivar and Amazonas states of Venezuela (2008, 2009) was funded in part by a Kalbfleisch Postdoctoral Fellowship from the AMNH to JAO and by U.S. National Science Foundation grant DEB 0228699 to Lorenzo Prendini (AMNH). We are grateful to the González-Sponga family (Caracas, Venezuela) for providing access and facilities to the MAGS private collection in 2009 and 2010. We thank Mika Jormakka for language revision, and Lorenzo Prendini for critical comments on a previous draft of the manuscript.

References

Acosta, L.E., Ochoa, J.A., 2002. Lista de los escorpiones bolivianos (Chelicerata: Scorpiones), con notas sobre su distribución. Revista de la Sociedad Entomológica Argentina 61 (3–4), 15–23.

Beron, P., 2018. High-altitude Arachnida (partly from Beron 2008, updated in Beron 2016). In: Beron, P. (Ed.), Zoogeography of Arachnida. Monographiae Biologicae, 94. pp. 853–891.

Berry, P.E., Riina, R., 2005. Insights into the diversity of the Pantepui Flora and the biogeographic complexity of the Guiana Shield. Biologiske Skrifter—Det Kongelige Danske Videnskbernes Selskab 55, 145–167.

Bonaccorso, E., Guayasamin, J.M., 2013. On the origin of Pantepui montane biotas: a perspective based on the phylogeny of *Aulacorhynchus* toucanets. PLoS One 8 (6), e67321.

Botero-Trujillo, R., Flórez, E., 2011. A revisionary approach of Colombian *Ananteris* (Scorpiones, Buthidae): two new species, a new synonymy, and notes on the value of trichobothria and hemispermatophore for the taxonomy of the group. Zootaxa 2904, 1–44.

Botero-Trujillo, R., Ochoa, J.A., Tovar, O.A., Souza, J.E., 2012. A new species in the scorpion genus *Troglotayosicus* from forest leaf litter in Southwestern Colombia (Scorpiones, Troglotayosicidae). Zootaxa 3506, 63–76.

Brewer-Carías, C., 1988. Geografía del Cerro de la Neblina. In: Brewer-Carias, C. (Ed.), Cerro de la Neblina. Resultados de la Expedición 1983–1987. Fundación para el desarrollo de las Ciencias Físicas, Matemáticas y Naturales, Caracas y Editorial Sucre, Caracas, Venezuela, pp. 1–173.

Brito, A., Borges, A., 2015. A checklist of the scorpions of Ecuador (Arachnida: Scorpiones), with notes on the distribution and medical significance of some species. J. Venom. Anim. Toxins. Incl. Trop. Dis. 21 (23), 1–17.

Ceccarelli, F.S., Ojanguren-Affilastro, A.A., Ramírez, M.J., Ochoa, J.A., Mattoni, C.I., Prendini, L., 2016. Andean uplift drives diversification of the bothriurid scorpion genus *Brachistosternus*. J. Biogeogr. 43, 1942–1954.

Dagert, B.F., 1957. Notas sobre los escorpiones de la región de Auyantepui, Estado Bolívar, Venezuela. Acta Biológica Venezuelica 2, 127–133.

Esposito, L.A., Yamaguti, H.Y., Souza, C.A., Pinto-da-Rocha, R., Prendini, L., 2017. Systematic revision of the Neotropical club-tailed scorpions, *Physoctonus*, *Rhopalurus*, and *Troglorhopalurus*, Revalidation of *Heteroctenus*, and descriptions of two new genera and three new species (Buthidae: Rhopalurusinae). Bull. Am. Mus. Nat. Hist. 415, 1–134.

Fet, V., Lowe, G., 2000. Family Buthidae C.L. Koch, 1837. In: Fet, V., Sissom, W.D., Lowe, G., Braunwalder, M.E. (Eds.), Catalog of the Scorpions of the World (1758–1998). The New York Entomological Society, New York, pp. 54–286.

Flórez, E., Botero-Trujillo, R., Acosta, L.E., 2008. Description of *Vachoniochactas humboldti* sp. nov. from Colombia, with complementary notes on the genus (Scorpiones, Chactidae). Zootaxa 1853, 31–44.

Foord, S.H., Gelebe, V., Prendini, L., 2015. Effects of aspect and altitude on scorpion diversity along a environmental gradient in the Soutpansberg, South Africa. J. Arid Environ. 113, 114–120.

González-Sponga, M.A., 1972a. *Ananteris venezuelensis* (Scorpionida: Buthidae) nueva especie de la Guyana de Venezuela. Memoria de la Sociedad de Ciencias Naturales La Salle. 32 (93), 205–214.

González-Sponga, M.A., 1972b. *Broteas camposi* (Scorpionida: Chactidae) nueva especie para la amazonia colombiana. Memoria de la Sociedad de Ciencias Naturales La Salle. 32 (91), 55–67.

González-Sponga, M.A., 1973. *Broteas mingueti* (Scorpionida: Chactidae). Nueva especie en el Territorio Federal Amazonas, Venezuela Chactidae). Monografías Científicas Augusto Pi Suñer 6, 1–10.

González-Sponga, M.A., 1974a. Una nueva especie de escorpiones (Scorpionida: Chactidae) de la Amazonia venezolana. Acta Biológica Venezuelica 8 (3–4), 299–313.

González-Sponga, M.A., 1974b. *Broteochactas sanmartíni* (Scorpionida: Chactidae) nueva especie del río Caura, en la Guyana de Venezuela. Monografías Científicas Augusto Pi Suñer 7, 1–16.

González-Sponga, M.A., 1975a. *Broteochactas racenisi* (Scorpionida: Chactidae) nueva especie de la Guyana de Venezuela. Caribb. J. Sci. 15 (1–2), 49–55.

González-Sponga, M.A., 1975b. *Teuthraustes adrianae* (Scorpionida: Chactidae) nueva especie en el Tepui La Neblina, Territorio Federal Amazonas. Monografías Científicas Augusto Pi Suñer 8, 1–20.

González-Sponga, M.A., 1978. Escorpiofauna de la Región Oriental del Estado Bolívar, en Venezuela. Roto-Impresos C.A., Caracas.

González-Sponga, M.A., 1981. Seis nuevas especies del género *Tityus* en Venezuela (Scorpionida: Buthidae). Monografías Científicas Augusto Pi Suñer 12, 1–87.

González-Sponga, M.A., 1982. Tres nuevas especies venezolanas del género *Chactopsis* (Scorpionida: Chactidae). Boletín de la Academia de Ciencias Físicas, Matemáticas y Naturales de Venezuela 42 (129–130), 127–146.

González-Sponga, M.A., 1984a. Escorpiones de Venezuela. Cuadernos Lagoven, Cromotip, Caracas.

González-Sponga, M.A., 1984b. Tres nuevas especies de la Amazonía de Venezuela (Scorpionida: Chactidae). Boletín de la Academia de Ciencias Físicas, Matemáticas y Naturales de Venezuela 44 (135–136), 142–165.

González-Sponga, M.A., 1985. Tres nuevas especies de arácnidos de Venezuela (Scorpionida: Chactidae: Buthidae). Memoria de la Sociedad de Ciencias Naturales La Salle 45 (123), 25–45.

González-Sponga, M.A., 1991. Arácnidos de Venezuela. Escorpiones del Tepui "La Neblina", Territorio Federal Amazonas. (Scorpionida: Chactidae: Butidae). Boletín de la Academia de Ciencias Físicas, Matemáticas y Naturales de Venezuela 50 (163–164), 11–62.

González-Sponga, M.A., 1992. Arácnidos de Venezuela. Tres nuevas especies de la familia Chactidae (Escorpiones). Memoria de la Sociedad de Ciencias Naturales La Salle 52 (138), 133–149.

González-Sponga, M.A., 1993. Arácnidos de Venezuela. Cinco nuevas especies del Parque Nacional Canaima, Edo. Bolívar (Scorpionida: Chactidae). Boletín de la Academia de Ciencias Físicas, Matemáticas y Naturales de Venezuela 53 (173–174), 77–100.

González-Sponga, M.A., 1994. Arácnidos de Venezuela. Una nueva especie del género *Tityus* y redescripción de *Tityus urbinai* Scorza, 1952 (Scorpionida: Buthidae). Boletín de la Sociedad Venezolana de Ciencias Naturales 44 (148), 327–342.

González-Sponga, M.A., 1996. Guía para identificar Escorpiones de Venezuela. Cuadernos Lagoven, Editorial Arte, Caracas.

González-Sponga, M.A., 1997a. Arácnidos de Venezuela. Tres nuevas especies del "Tepui" Guaiquinima, Edo. Bolívar (Scorpionida: Chactidae) (Resultado Zoológico de la Expedición de la Fundación para el Desarrollo de las Ciencias de la Academia de Ciencias Físicas, Matemáticas y Naturales de Venezuela). Boletín de la Sociedad Venezolana de Ciencias Naturales 46 (150), 9–33.

González-Sponga, M.A., 1997b. Arácnidos de Venezuela. Tres nuevas especies de escorpiones de la Región Amazónica-Guayanesa (Buthidae: Chactidae). Memoria de la Sociedad de Ciencias Naturales La Salle 57 (148), 55–69.

González-Sponga, M.A., 2001. Arácnidos de Venezuela. Escorpiones y distribución geográfica. Aula y Ambiente 1, 27–60.

González-Sponga, M.A., 2002. Arácnidos de Venezuela. Cuatro nuevas especies del género *Tityus* (Scorpionida: Buthidae). Boletín de la Academia de Ciencias Físicas, Matemáticas y Naturales 62 (2), 49–66.

González-Sponga, M.A., 2004. Arácnidos de Venezuela. Descripción de tres nuevas especies de escorpiones de los géneros *Tityus* (Buthidae), *Chactopsis* y *Broteochactas* (Chactidae). Acta Biológica Venezolana 24 (1), 1–12.

González-Sponga, M.A., 2006. Arácnidos de Venezuela. El género *Ananteris* Thorell 1891, en Venezuela (Scorpionidae: Buthidae). Serie de libros arbitrados del Vicerrectorado de Investigación y Postgrado, UPEL, Caracas.

Huber, O., 1987. Consideraciones sobre el concepto de Pantepui. Pantepui 2, 2–10.

Huber, O., 1988. Guayana highlands versus Guayana lowlands, a reappraisal. Taxon 37, 595–614.

Lomolino, M.V., 2001. Elevation gradients of species-density: historical and prospective views. Glob. Ecol. Biogeogr. 10, 3–13.

Lourenço, W.R., 1994a. Diversity and endemism in tropical versus temperate scorpion communities. Biogeographica 70 (3), 155–160.

Lourenço, W.R., 1994b. Scorpion biogeographic patterns as evidence for a Neblina-São Gabriel endemic center in Brazilian Amazonia. Revista de la Academia Colombiana de Ciencias 19 (72), 181–185.

Lourenço, W.R., 2002. Scorpions of Brazil. Les Éditions de l'If, Paris, France.

Lourenço, W.R., 2006. Nouvelle proposition de découpage sous générique du genre *Tityus* C. L. Koch, 1836 (Scorpiones, Buthidae). Boletín de la Sociedad Entomológica Aragonesa 39, 55–67.

Lourenço, W.R., 2008. Description of *Tityus* (*Atreus*) *neblina* sp.n. (Scorpiones, Buthidae), from the 'Parque Nacional do Pico da Neblina', in Brazil/Venezuela, with comments on some related species. Boletín de la Sociedad Entomológica Aragonesa 43, 75–79.

Lourenço, W.R., 2010. The disrupted pattern of distribution of the genus *Hadrurochactas* Pocock; evidence of past connections between Amazon and the Brazilian Atlantic forest. C. R. Biol. 333, 41–47.

Lourenço, W.R., 2017. Scorpions from Brazilian Amazonia, with a description of two new species from 'Serra da Mocidade' National Park in the State of Roraima (Scorpiones: Buthidae, Chactidae). Arachnida—Rivista Aracnologica Italiana 12, 2–17.

Lourenço, W.R., 2018. Scorpions at high altitudes: a new species of Scorpiops Peters 1861 (Scorpiones: Scorpiopidae) from the Taxkorgan Reserve, Xinjiang, China. C. R. Biol. 341, 362–369.

Lourenço, W.R., Duhem, B., 2009. The genus *Vachoniochactas* González-Sponga (Scorpiones, Chactidae), a model of relictual distribution in past refugia of the Guayana region of South America. C. R. Biol. 332, 1085–1091.

Lourenço, W.R., Duhem, B., 2010. The geographical pattern of distribution of the genus *Teuthraustes* Simon (Scorpiones, Chactidae) in South America and description of a new species. C. R. Biol. 333, 858–863.

Lourenço, W.R., Pinto-da-Rocha, R., 2000. Additions to the knowledge of the Chactidae of Brazilian Amazonia (Arachnida: Scorpiones). Amazoniana 16 (1–2), 259–274.

Lourenço, W.R., Qi, J., 2006. Mountain scorpions: a new genus and species from Tibet (China). C. R. Biol. 329, 289–295.

Lourenço, W.R., Giupponi, A.P.L., Pedroso, D.L., 2011. New species of Chactidae (Scorpiones) from the upper Rio Negro in Brazilian Amazonia. Boletín de la Sociedad Entomológica Aragonesa 49, 65–73.

Mayr, E., Phelps, W.H., 1967. The origin of the bird fauna of the south Venezuelan highlands. Bull. Am. Mus. Nat. Hist. 136, 269–328.

Ochoa, J.A., 2005. Patrones de distribución de escorpiones de la región andina en el sur peruano. Rev. Peru. Biol. 12 (1), 49–68.

Ochoa, J.A., Prendini, L., 2010. The genus *Hadruroides* Pocock, 1893 (Scorpiones: Iuridae), in Peru: New records and descriptions of six new species. Am. Mus. Novit. 3687, 1–56.

Ochoa, J.A., Ojanguren Affilastro, A.A., Mattoni, C.I., Prendini, L., 2011. Systematic revision of the genus *Orobothriurus* Maury, 1976 (Scorpiones: Bothriuridae), with discussion of the altitude record for scorpions. Bull. Am. Mus. Nat. Hist. 359, 1–90.

Ochoa, J.A., Rojas-Runjaic, F.J.M., Pinto-da-Rocha, R., Prendini, L., 2013. Systematic revision of the Neotropical genus *Chactopsis* Kraepelin, 1912 (Chactoidea: Chactidae), with descriptions of two new genera and four new species. Bull. Am. Mus. Nat. Hist. 378, 1–121.

Ojanguren-Affilastro, A.A., 2005. Estudio monográfico de los escorpiones de la República Argentina. Rev. Iber. Aracnol. 11, 75–241.

Ojanguren-Affilastro, A.A., Mattoni, C.I., Ochoa, J.A., Ramírez, M.J., Ceccarelli, F.S., Prendini, L., 2016. Phylogeny, species delimitation and convergence in the South American bothriurid scorpion genus *Brachistosternus* Pocock 1893: integrating morphology, nuclear and mitochondrial DNA. Mol. Phylogenet. Evol. 94, 159–170.

Pocock, R.I., 1900. Myriopoda and Arachnida. II. Report on a collection made by Messrs F.V. McConnell and J.J. Quelch at Mount Roraima in British Guiana. Trans. Linn. Soc. Lond. Zool. 2 (8), 64–71.

Polis, G.A., 1990. Ecology. In: Polis, G.A. (Ed.), The Biology of Scorpions. Stanford University Press, Stanford, CA, pp. 247–293.

Prendini, L., 2001. Substratum specialization and speciation in southern African scorpions: the effect hypothesis revisited. In: Fet, V., Selden, P.A. (Eds.), Scorpions 2001. In Memoriam Gary A. Polis. British Arachnological Society, pp. 113–138.

Prendini, L., Bird, T., 2008. Scorpions of the Brandberg Massif, Namibia: species richness inversely correlated with altitude. Afr. Invertebr. 49, 77–107.

Rojas-Runjaic, F.J.M., De Sousa, L., 2007. Catálogo de los escorpiones de Venezuela (Arachnida: Scorpiones). Boletín de la Sociedad Entomológica Aragonesa 40, 281–307.

Scorza, J.V., 1952. Contribución al estudio de los alacranes venezolanos. *Tityus urbinai* sp. n. (Scorpionida). Novedades Científicas, Contribuciones Ocasionales del Museo Historia Natural La Salle, Serie Zoológica 8, 1–4.

Scorza, J.V., 1954a. Dos especies nuevas de alacranes de Venezuela. Novedades Científicas, Contribuciones Ocasionales del Museo Historia Natural La Salle, Serie Zoológica 12, 3–11.

Scorza, J.V., 1954b. Escorpiones del Alto Orinoco. Boletín de la Sociedad Venezolana de Ciencias Naturales 15 (82), 163–175.

Sissom, W.D., 2000. Family Chactidae Pocock, 1893. In: Fet, V., Sissom, W.D., Lowe, G., Braunwalder, M.E. (Eds.), Catalog of the Scorpions of the World (1758–1998). The New York Entomological Society, New York, pp. 287–322.

Steyermark, J.A., 1986. Speciation and endemism in the Flora of the Venezuelan tepuis. In: Vuilleumier, F., Monasterios, M. (Eds.), High Altitude Tropical Biogeography. Oxford University Press, New York, pp. 317–373.

Vignoli, V., Kovařik, F., 2003. A new troglobitic scorpion of the genus *Taurepania* González-Sponga, 1978 from Venezuela (Scorpiones; Chactidae). Revista Ibérica de Aracnología 7, 127–131.

Ythier, E., Lourenço, W.R., 2017. The geographical patterns of distribution of the genus *Teuthraustes* Simon, 1878 in Ecuador and description of three new species (Scorpiones, Chactidae). ZooKeys 721, 45–63.

Appendix 10.1 Total list of scorpion species from Pantepui and the surrounding areas in the Guiana region, with the name of the state (Venezuela and Brazil) and the countries from which they have been recorded. *Br-Am*, Amazonas (Brazil); *Br-Ro*, Roraima (Brazil); *Co*, Colombia; *Gu*, Guyana; *Ve-Am*, Amazonas (Venezuela); *Ve-Bo*, Bolívar (Venezuela).

			Distribution in the Guayana region					
No.	**Family**	**Species**	**Br-Am**	**Br-Ro**	**Co**	**Gu**	**Ve-Am**	**Ve-Bo**
1	Buthidae	*Ananteris chirimakei* González-Sponga, 2006						X
2	Buthidae	*Ananteris maniapurensis* González-Sponga, 2006						X
3	Buthidae	*Ananteris paoensis* González-Sponga, 2006						X
4	Buthidae	*Ananteris plataensis* González-Sponga, 2006						X
5	Buthidae	*Ananteris riocaurensis* González-Sponga, 2006						X
6	Buthidae	*Ananteris turumbanensis* González-Sponga, 1980						X
7	Buthidae	*Ananteris venezuelensis* González-Sponga, 1972						X
8	Buthidae	*Jaguajir pintoi* (Mello-Leitão, 1932)		X		X		
9	Buthidae	*Rhopalurus laticauda* Thorell, 1876		X		X	X	X
10	Buthidae	*Tityus anduzei* González-Sponga, 1997					X	
11	Buthidae	*Tityus acananensis* González-Sponga, 2009					X	
12	Buthidae	*Tityus breweri* González-Sponga, 1997						X
13	Buthidae	*Tityus caesarbarrioi* González-Sponga, 2001						X
14	Buthidae	*Tiyus clathratus* C. L. Koch, 1844		X				X
15	Buthidae	*Tityus culebrensis* González-Sponga, 1994					X	
16	Buthidae	*Tityus dupouyi* González-Sponga, 1987					X	
17	Buthidae	*Tityus elizabethae* Lourenço and Ramos, 2004		X				
18	Buthidae	*Tityus filodendron* González-Sponga, 1981					X	
19	Buthidae	*Tityus generaltheophiloi* Lourenço, 2017		X				
20	Buthidae	*Tityus manakai* González-Sponga, 2004					X	

21	Buthidae	*Tityus maniapurensis* González-Sponga, 2009				X
22	Buthidae	*Tityus neblina* Lourenço, 2008	X			
23	Buthidae	*Tityus nelsoni* Lourenço, 2005	X			
24	Buthidae	*Tityus riocaurensis* González-Sponga, 1996				X
25	Buthidae	*Tityus rionegrensis* Lourenço, 2006	X			
26	Buthidae	*Tityus romeroi* González-Sponga, 2008				X
27	Buthidae	*Tityus sarisarinamensis* González-Sponga, 2002				X
28	Buthidae	*Tityus shiriana* González-Sponga, 1991			X	
29	Buthidae	*Tityus urbinai* Scorza, 1952			X	
30	Buthidae	*Tityus venamensis* González-Sponga, 1981				X
31	Buthidae	*Tityus ventuarensis* González-Sponga, 2009			X	
32	Buthidae	*Tityus yerenai* González-Sponga, 2009			X	
33	Chactidae	*Auyantepuia scorzai* (Dagert, 1957)				X
34	Chactidae	*Broteochactas cocuyensis* González-Sponga, 2004			X	
35	Chactidae	*Broteochactas parimensis* González-Sponga, 2004			X	
36	Chactidae	*Broteochactas bariensis* González-Sponga, 1991			X	
37	Chactidae	*Broteochactas bilbaoi* González-Sponga, 1978				X
38	Chactidae	*Broteochactas bruzuali* González-Sponga, 1980				X
39	Chactidae	*Broteochactas caroniensis* González-Sponga, 1996				X
40	Chactidae	*Broteochactas colombiensis* González-Sponga, 1976		X	X	
41	Chactidae	*Broteochactas efreni* González-Sponga, 1978				X
42	Chactidae	*Broteochactas eliasilvai* González-Sponga, 1980				X
43	Chactidae	*Broteochactas garciai* González-Sponga, 1978				X
44	Chactidae	*Broteochactas granosus* Pocock, 1900		X		

(*Continued*)

(Continued)

			Distribution in the Guayana region					
No.	**Family**	**Species**	**Br-Am**	**Br-Ro**	**Co**	**Gu**	**Ve-Am**	**Ve-Bo**
45	Chactidae	*Broteochactas guaiquinimensis* González-Sponga, 1997						X
46	Chactidae	*Broteochactas jaspei* Gonzalez-Sponga, 1993						X
47	Chactidae	*Broteochactas josemanueli* González-Sponga, 1992						X
48	Chactidae	*Broteochactas kjellesvigi* González-Sponga, 1974			X		X	
49	Chactidae	*Broteochactas leoneli* González-Sponga, 1978						X
50	Chactidae	*Broteochactas macrochelae* González-Sponga, 2004						X
51	Chactidae	*Broteochactas neblinensis* González-Sponga, 1991					X	
52	Chactidae	*Broteochactas niemeyerae* Lourenço et al., 2011	X					
53	Chactidae	*Broteochactas orinocensis* Scorza, 1954						X
54	Chactidae	*Broteochactas panarei* González-Sponga, 1980						X
55	Chactidae	*Broteochactas paoensis* González-Sponga, 1996						X
56	Chactidae	*Broteochactas racenisi* González-Sponga, 1975						X
57	Chactidae	*Broteochactas riopinensis* González-Sponga, 1992						X
58	Chactidae	*Broteochactas ruizpittoli* González-Sponga, 1993						X
59	Chactidae	*Broteochactas sanmartini* González-Sponga, 1974						X
60	Chactidae	*Broteochactas santanai* González-Sponga, 1978						X
61	Chactidae	*Broteochactas sarisarinamensis* González-Sponga, 1985						X
62	Chactidae	*Broteochactas simarawochensis* González-Sponga, 1980					X	
63	Chactidae	*Broteochactas verai* González-Sponga, 1993						X
64	Chactidae	*Broteochactas yekuanae* González-Sponga, 1984					X	
65	Chactidae	*Brotheas camposi* González-Sponga, 1972			X		X	
66	Chactidae	*Brotheas caramaschii* Lourenço et al., 2011	X					
67	Chactidae	*Brotheas cataniapensis* González-Sponga, 1997					X	
68	Chactidae	*Brotheas cunucunumensis* González-Sponga, 1984					X	

69	Chactidae	*Brotheas dasilvai* González-Sponga, 1978				X
70	Chactidae	*Brotheas humboldti* González-Sponga, 1980				X
71	Chactidae	*Brotheas jourdani* Lourenço, 1997	X			
72	Chactidae	*Brotheas libinallyi* González-Sponga, 1978				X
73	Chactidae	*Brotheas lichyi* González-Sponga, 1980			X	
74	Chactidae	*Brotheas mawarinumensis* González-Sponga, 1991			X	
75	Chactidae	*Brotheas mingueti* González-Sponga, 1973			X	
76	Chactidae	*Brotheas munozi* González-Sponga, 1997				X
77	Chactidae	*Brotheas noguerai* González-Sponga, 1993				X
78	Chactidae	*Brotheas ocamoi* González-Sponga, 2004			X	
79	Chactidae	*Brotheas perezramirezi* González-Sponga, 1996			X	
80	Chactidae	*Brotheas rionegroensis* González-Sponga, 1996			X	
81	Chactidae	*Brotheas sanabriai* González-Sponga, 1997				X
82	Chactidae	*Brotheas wareipai* González-Sponga, 2004				X
83	Chactidae	*Brotheas wilmeri* González-Sponga, 1980			X	
84	Chactidae	*Cayooca venezuelensis* González-Sponga, 1996			X	
85	Chactidae	*Chactopsis barajuri* González-Sponga, 1982				X
86	Chactidae	*Chactopsis siapaensis* González-Sponga, 1991			X	
87	Chactidae	*Chactopsis sujirima* González-Sponga, 1982	X		X	
88	Chactidae	*Chactopsoides anduzei* (Gonzáles-Sponga, 1982)		X	X	
89	Chactidae	*Chactopsoides gonzalezspongai* Ochoa et al., 2013			X	
90	Chactidae	*Chactopsoides marahuacaensis* González-Sponga, 2004			X	
91	Chactidae	*Chactopsoides yanomami* (Lourenço et al., 2011)	X			
92	Chactidae	*Hadrurochactas machadoi* González-Sponga, 1993				X
93	Chactidae	*Hadrurochactas odoardoi* González-Sponga, 1985				X

(*Continued*)

(Continued)

No.	Family	Species	Distribution in the Guayana region					
			Br-Am	Br-Ro	Co	Gu	Ve-Am	Ve-Bo
94	Chactidae	*Hadrurochactas schaumii* (Karsch, 1880)				X		X
95	Chactidae	*Megachactops coriacea* (González-Sponga, 1991)					X	
96	Chactidae	*Megachactops kuemoi* Ochoa et al., 2013					X	
97	Chactidae	*Taurepania manisapanensis* González-Sponga, 1992						X
98	Chactidae	*Taurepania mauriciodiasi* (Lourenço, 2017)		X				
99	Chactidae	*Taurepania porosa* (Pocock, 1900)						X
100	Chactidae	*Taurepania trezzii* Vignoli and Kovarik, 2003						X
101	Chactidae	*Taurepania verneti* González-Sponga, 1992						X
102	Chactidae	*Taurepania vestigialis* González-Sponga, 1978						X
103	Chactidae	*Teuthraustes adrianae* González-Sponga, 1975	X				X	
104	Chactidae	*Teuthraustes akananensis* González-Sponga, 1984					X	
105	Chactidae	*Teuthraustes carmelinae* Scorza, 1954					X	
106	Chactidae	*Teuthraustes lisei* Lourenço, 1994	X					
107	Chactidae	*Teuthraustes maturaca* González-Sponga, 1991	X				X	
108	Chactidae	*Teuthraustes newaribe* Lourenço et al., 2011	X					
109	Chactidae	*Teuthraustes reticulatus* González-Sponga, 1991					X	
110	Chactidae	*Vachoniochactas amazonicus* González-Sponga, 1991	X				X	
111	Chactidae	*Vachoniochactas ashleeae* Lourenço, 1994	X					
112	Chactidae	*Vachoniochactas humboldti* Florez et al., 2008				X	X	
113	Chactidae	*Vachoniochactas lasallei* González-Sponga, 1978						X
114	Chactidae	*Vachoniochactas roraima* Lourenço and Duhem, 2009		X				
115	Hormuridae	*Opisthacanthus autanensis* González-Sponga, 2004					X	

CHAPTER

11

Land snails

Abraham S.H. Breure[1,2]

[1]Naturalis Biodiversity Center, RA Leiden, The Netherlands [2]Royal Belgian Institute of Natural Sciences, Brussels, Belgium

Introduction

This chapter is a slightly updated and extended version of earlier papers (notably Breure, 2009, 2013) from which it borrows a substantial part of the text. Due to the continued inaccessibility of the region, very little new material has been added to recent papers dealing with this region so far (Breure, 2009, 2012, 2013; Breure and Schlögl, 2010; Simone, 2010; Thompson, 2008), which may explain the scanty progress made since then.

Land snails (Mollusca, Gastropoda) have been proposed as good model organisms for the study of complex ecosystems. They are renowned for their limited dispersal abilities, they can be easily collected, and most of them can be identified on the basis of their shells. Land snails are also an important factor in the food web and are vulnerable to predation by other animals, or their shell is used as a calcium source. However, tropical acidic soils are known to support only sparse land snail populations (Schilthuizen, 2011). Although the Neotropical land snail fauna is generally fairly well known, the Pantepui (sensu Huber, 1995) snail fauna is less studied; this is the assemblage of sandstone table mountains in Venezuelan Guayana and in adjacent Brazil and Guyana. According to Huber (1995), the term Pantepui defines the area above 1500 m elevation, but some authors include all the intervening lowlands as well. In this chapter we use Pantepui to refer to the highlands only (sensu Huber, 1988), but we treat the entire land snail species reported from southern Venezuela and adjacent areas in Brazil. Currently, there are a total of 22 identified species (Box 11.1), of which 91% are endemic to the area. In addition, it is noteworthy that the identity of some reported species remains obscure at this time. This particularly refers to a juvenile *Plekocheilus* species (family Amphibulimidae) from near Santa Elena de Uairén (Breure, 2009), a *Happiella* species (family Scolodontidae), and *Euconulus* (?) species (family Euconulidae) from Chimantá massif (Haas, 1955). These species have been reported both from the highlands (*Happiella*, *Euconulus*) and lower elevations (*Plekocheilus*) and are a token of the relatively unexplored malacofauna of this region. Incidentally, all were collected by

Biodiversity of Pantepui
DOI: https://doi.org/10.1016/B978-0-12-815591-2.00011-2

BOX 11.1

Taxonomy of land snails reported from southern Venezuela and adjacent areas in Brazil. Abbreviations refer to tepui or massifs: *CH*, Chimantá massif; *DU*, Duida-Marahuaka massif; *GU*, Cerro Guaiquinima; *MA*, Sierra de Maigualida; *NE*, Sierra de la Neblina; *RO*, Roraima-tepui; *YA*, Cerro Yapacana

Class Gastropoda Cuvier, 1795

Order Stylommatophora A. Schmidt, 1855
- Superfamily Clausilioidea J.E. Gray, 1855
  - Family Clausiliidae J.E. Gray, 1855
    - Genus *Columbinia* Polinski, 1924
      - Subgenus *Columbinia* (*Columbinia*) Polinski, 1924
        - *Columbinia* (*Columbinia*) *exul* Thompson, 2008
- Superfamily Orthalicoidea Martens in Albers and Martens, 1860
  - Family Amphibulimidae P. Fischer, 1873
    - Genus *Plekocheilus* Guilding, 1828
    - Subgenus *Plekocheilus* (*Plekocheilus*) Guilding, 1828
      - *Plekocheilus* (*Plekocheilus*) *linterae* (G.B. Sowerby III, 1890) [RO]
      - *Plekocheilus* (*Plekocheilus*) *alticola* Haas, 1955 [CH]
      - *Plekocheilus* (*Plekocheilus*) *vlceki* Breure and Schlögl, 2010 [CH]
      - *Plekocheilus* (*Plekocheilus*) *philippei* Breure, 2012 [CH]
    - Subgenus *Plekocheilus* (*Eurytus*) Albers, 1850
      - *Plekocheilus* (*Eurytus*) cf. *plectostylus* (L. Pfeiffer, 1848) [MA]
      - *Plekocheilus* (*Eurytus*) *gibber* (Oberwimmer, 1931) [DU]
      - *Plekocheilus* (*Eurytus*) *fusitorsus* (Oberwimmer, 1931) [DU?; GU]
      - *Plekocheilus* (*Eurytus*) *juliani* Haas, 1955 [CH]
      - *Plekocheilus* (*Eurytus*) *mundiperditi* Haas, 1955 [CH]
      - *Plekocheilus* (*Eurytus*) *tatei* Haas, 1955 [DU]
      - *Plekocheilus* (*Eurytus*) *sophiae* Breure, 2009 [RO]
      - *Plekocheilus* (*Eurytus*) *tepuiensis* Breure, 2009 [YA]
      - *Plekocheilus* (*Eurytus*) *huberi* Breure, 2009 [NE]
      - *Plekocheilus* (*Eurytus*) *nebulosus* Breure, 2009 [NE]
      - *Plekocheilus* (*Eurytus*) *breweri* Breure and Schlögl, 2010 [CH]
    - Subgenus *Plekocheilus* (*Eudolichotis*) Pilsbry, 1896
      - *Plekocheilus* (*Eudolichotis*) *sinuatus* (Albers, 1854) [–]
  - Family Bulimulidae Tryon, 1867
    - Genus *Drymaeus* Albers, 1850
    - Subgenus *Drymaeus* (*Drymaeus*) Albers, 1850
      - *Drymaeus* (*Drymaeus*) *steyermarki* (Haas, 1955) [CH]
      - *Drymaeus* (*Drymaeus*) *extraneus* (Haas, 1955) [CH]
      - *Drymaeus* (*Drymaeus*) *yapacanensis* Breure and Eskens, 1981 [YA]
      - *Drymaeus* (*Drymaeus*) *rex* Breure, 2009 [NE]
- Superfamily Helicoidea Rafinesque, 1815

BOX 11.1 (*cont'd*)

Family Solaropsidae H. Nordsieck, 1986
Genus *Olympus* Simone, 2010
Olympus nimbus Simone, 2010 [NE]
Superfamily Scolodontoidea H.B. Baker, 1925
Family Scolodontidae H.B. Baker, 1925
Genus *Happiella* H.B. Baker, 1925
Happiella species [CH]
Superfamily Trochomorphoidea Möllendorff, 1890
Family Euconulidae H.B. Baker, 1928
Genus *Euconulus* Reinhardt, 1883
Euconulus (?) species [CH]

botanists (Otto Huber and Julian A. Steyermark, respectively) or paleontologists (Jan Schlögl), and this illustrates a more generalized pattern: there is generally a paucity of data, with all known snails being collected by nonmalacologists. The expeditions to Sierra de la Neblina (Brewer-Carías, 1988) and Churí-tepui (Brewer-Carías and Audy, 2010), however, are exceptions with also a special interest for malacology and the incorporation of a malacologist (Fred G. Thompson) in the field team of the 1988 expedition.

The aim of this chapter is to provide a comprehensive overview of land snail species and relevant data and to suggest some further avenues for research.

Material and methods

The following abbreviations are used to refer to depositories of material: AMNH, American Museum of Natural History, New York, United States; INPA, Amazon National Research Institute, Manaus, Brazil; MLSC, La Salle Museum of Natural History, Caracas, Venezuela; MZSP, Zoology Museum, University of São Paulo, Sao Paulo, Brazil; NHM, Natural History Museum, London, United Kingdom (label abbreviation NHMUK); RBINS, Royal Belgian Institute of Natural Sciences, Brussels, Belgium; RMNH, Naturalis Biodiversity Center (formerly Rijksmuseum van Natuurlijke Historie, Leiden, the Netherlands); SNMSZ, Slovak National Museum, Bratislava, Slovakia; and UF, Florida State Museum of Natural History, Gainesville, United States.

The principal component analysis (PCA) was done with Past 3.20 (Hammer et al., 2001) using the same methodology as described in Breure (2009). Geographical terminology of the area follows Table 1.1 and Fig. 1.2 from Chapter 1: Definition and characterization of the Pantepui biogeographical province.

Systematics

All known land snails from Pantepui are Stylommatophora, with the majority classified in the superfamily Orthalicoidea. For an overview of the currently known malacofauna,

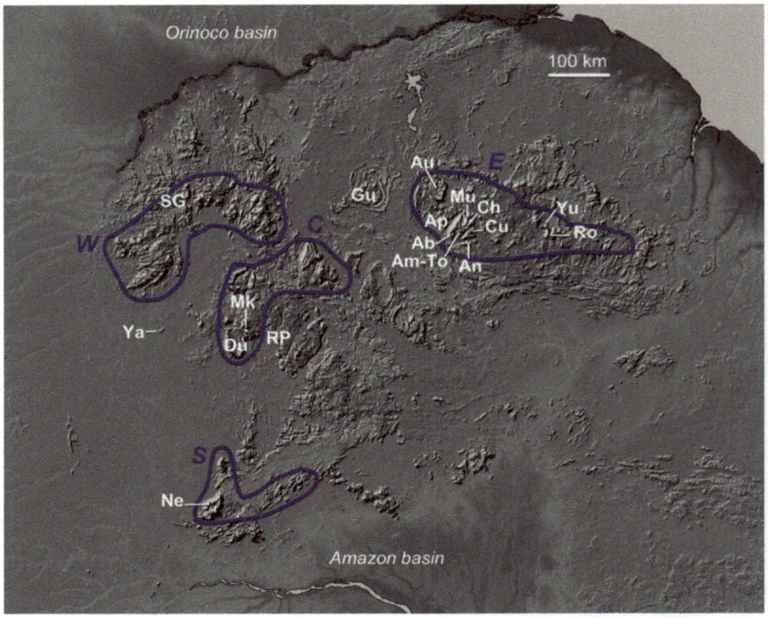

FIGURE 11.1 Map of the Guiana Highlands with snail localities mentioned in the text. Tepuis are sorted by phytogeographic Pantepui districts (Fig. 1.4 of Chapter 1): Eastern District (E): *Ab*, Abakapá-tepui (Chimantá massif); *Am*, Agparamán-tepui (Chimantá massif); *Ap*, Apakará-tepui (Chimantá massif); *Au*, Auyán-tepui; *Ch*, Chimantá massif; *Cu*, Churí-tepui (Chimantá massif); *Mu*, Murey-tepui (Chimantá massif); *Ro*, Roraima-tepui; *To*, Toronó-tepui (Chimantá massif); *Yu*, Yuruaní-tepui. Central District (C; Jaua-Duida subdistrict with outliers): *Du*, Cerro Duida; *Gu*, Cerro Guaiquinama; *Mk*, Cerro Marahuaka; *RP*, Río Padamo; *Ya*, Cerro Yapacana. Southern District (S): *Ne*, Sierra de la Neblina. Western District (W): *SG*, Cerro Guanay. For details of the Chimantá massif see Figs. 1–27 in Huber (1995). Source: *Background image courtesy NASA/JPL.*

see Box 11.1; for a key to families, genera, and species, see Breure (2013, Tables 9–11). The species are summarized next; see Thompson (2008), Breure (2009, 2012, 2013), Simone (2010), and Breure and Schlögl (2010) for more details, especially on the anatomy. The species are presented according to their altitudinal distribution (highlands, respectively, uplands and lowlands), and within each altitudinal stratum are treated according to the different subdivisions of Pantepui province (Berry et al., 1995, p. 173), noted as Eastern, Central (which was called by Berry et al. the Jaua-Duida District), Western, and Southern Pantepui Districts (Fig. 11.1) (see also Chapter 1). This sequence is herein retained to present the species in this chapter.

Species of Pantepui sensu stricto (≥1500 m)

Species from Eastern Pantepui District

Roraima subdistrict

This massif in the Eastern Pantepui chain has only partially been explored in terms of its malacofauna. Until now, only Roraima-tepui and Yuruaní-tepui are known to have endemic snail species.

Plekocheilus (P.) linterae (Sowerby, 1890) (Fig. 11.2B).

Diagnosis. Shell up to 43 mm, 1.8 times as long as wide, elongate-ovate, with a relatively slender last whorl, some irregular, undulating axial reddish-brown stripes may be present, sculptured with axial riblets broken into oblong granules by spiral, incised lines on the last whorl.

Distribution. Known from Roraima-tepui, 2400 m. Type in NHM.

Plekocheilus (Eurytus) sophiae (Breure, 2009) (Fig. 11.2F).

Diagnosis. Shell 44 mm, 1.9 times as long as wide, elongate-ovate, last whorl rather convex, surface shining, sculptured with horizontal treads, partly anastomosing. Animal light brownish, tentacles gray.

Distribution. Described from a single shell in the UF collection. In December 2010, a living specimen was observed at Yuruaní-tepui, 2300 m (Ph. Kok, pers. commun.).

Chimantá subdistrict

On the various tepuis that are part of Chimantá massif, seven identified species of land snails occur.

Plekocheilus (P.) alticola (Haas, 1955) (Fig. 11.2A).

Diagnosis. Shell up to 44.3 mm, 1.75 times as long as wide, elongate-ovate, last whorl relatively slender, dark brown with oblique reddish-brown stripes, sculptured with malleation, a distinct columellar fold visible in the aperture.

Distribution. The sole locality known is Toronó-tepui, where the type material was collected on the slopes bordering Caño Mojado, 2250 m. Type material in FMNH.

Remarks. Breure and Schlögl (2010) expressed doubts on the subspecific status of *Plekocheilus fulminans alticola* and *P. f. linterae* (Sowerby, 1890). After having studied the type of *P. fulminans* (Nyst, 1845) in the Brussels collection (Breure, 2011), these two taxa are treated as separate species (Breure, 2013).

Plekocheilus (P.) vlceki (Breure and Schlögl, 2010) (Fig. 11.2C).

Diagnosis. Shell up to 30.9 mm, 1.75 times as long as wide, with an almost uniform (dark-)yellowish to chestnut-brown color, a finely malleated sculpture crossed by spiral lines on the last whorl and a marked columellar fold. Animal yellowish-beige.

Radula. Rows slightly W-shaped, with the outermost lateromaginals curved distally; radula formula C/1 + LM 60/2 (teeth type C-6, LM-10, LM-11; see Breure, 1978, 1979). The central teeth are monocuspid, triangular; the first lateral teeth are acute, with barely developed ectocones; the next 16 lateromarginal are bicuspid, with rather blunt, spatula-shaped mesocones and more acute, ovate ectocones. The outer 44 lateromarginals are dicuspid and shifted, with rather blunt, spatula-shaped mesocones and blunt deltoid ectocones, which may be bifid in the outermost teeth.

Ecology. Found on the floor of a canyon covered with a dense *Bonnetia* forest and low vegetation. The snails occurred in *Brocchinia tatei* (Bromeliaceae) at the time of collecting.

Distribution. Known only from Churí-tepui, sima noroeste, 2100 m. Type material in MZSP, RMNH (holotype), and UF.

FIGURE 11.2 Shells from the Pantepui area. (A) *Plekocheilus alticola* (Haas, 1955); (B) *Plekocheilus linterae* (Sowerby III, 1890); (C) *Plekocheilus vlceki* (Breure and Schlögl, 2010); (D) *Drymaeus extraneus* (Haas, 1955); (E) *Drymaeus steyermarki* (Haas, 1955); (F) *Plekocheilus sophiae* (Breure, 2009); (G) *Plekocheilus breweri* (Breure and Schlögl, 2010); (H) *Plekocheilus mundiperditi* (Haas, 1955); (I) *Plekocheilus juliani* (Haas, 1955); (J) *Plekocheilus fusitorsus* (Oberwimmer, 1931); (K) *Plekocheilus gibber* (Oberwimmer, 1931); (L) *Plekocheilus tepuiensis* (Breure, 2009); (M) *Plekocheilus nebulosus* (Breure, 2009); (N) *Plekocheilus huberi* (Breure, 2009); (O) *Plekocheilus tatei* (Haas, 1955); (P) *Drymaeus yapacanensis* (Breure and Eskens, 1981); (Q) *Plekocheilus philippei* (Breure, 2012); (R) *Olympus nimbus* (Simone, 2010); (S) *Columbinia exul* (Thompson, 2008); (T) *Drymaeus rex* (Breure, 2009). Scale line equals 2 cm (R), 1.25 cm (S), 1 cm (all other figures).

Remarks. The central teeth of the radula are missing the slight protuberances at the base reported for *P. (Eurytus)* species, and the first lateral teeth are more acute than the second and beyond. This pattern has also been observed in *Plekocheilus (P.) aurissileni* (Born, 1780), the type species of the genus, and a number of other *Plekocheilus* species (Breure, 1978). The mesocones of the first lateral in *P. vlceki* are, however, more developed than those in the group just mentioned and resemble those of *P. (Eurytus)* from Pantepui.

Plekocheilus (P.) philippei (Breure, 2012) (Fig. 11.2Q).

Diagnosis. Shell up to 39.3 mm, 1.84 times as long as wide, with a yellowish- to dark olive-brown color and axial, undulating, reddish-brown streaks. Surface with closely set riblets on upper whorls and fine malleation on the last whorl. Suture slightly descending in front. Animal grayish, with three blackish, longitudinal striped on the head, one median and two on the sides; foot with a dark rim.

Ecology. Found in dwarf forests of *Bonnetia roraimae*, with many *Brocchinia hechtioides* growing on the ground, with snails estivating inside bromeliads.

Distribution. Found only on Angasima-tepui, 2121 m. Type material in RBINS (holotype) and RMNH.

Plekocheilus (Eurytus) juliani (Haas, 1955) (Fig. 11.2I).

Diagnosis. Shell up to 54.2 mm, 2.0 times as long as wide, elongate-ovate, color pattern consisting of obliquely descending dark brown to blackish stripes on a brown background that is darker on the last whorl. Sculpture consisting of short, oblong granules arranged in spiral rows.

Ecology. Julian Steyermark collected the species in a *Bonnetia* forest.

Distribution. Occurring on Chimantá massif: summit of Apakará-tepui, 2100 m, northwest part of the summit of Abapaká-tepui and Auyán-tepui, 1970 m. Specimens tentatively referred to this species have been found on Churí-tepui, 2130 m. Type in FMNH.

Plekocheilus (Eurytus) mundiperditi (Haas, 1955) (Fig. 11.2H).

Diagnosis. Shell up to 55.3 mm, 1.8 times as long as wide, elongate-ovate, last whorl somewhat swollen and saccate, color pattern with blackish to dark brown dots on a (dark-)brown background. Sculptured with puckered spiral bands of oblong granules divided by narrow, smooth areas.

Radula. Additional data to Breure (1978). Rows straight; radula formula C/1 + LM 59/2 (C-6, LM-10, LM-11). The central teeth are monocuspid, with triangular to ovate mesocones and hardly developed protuberances. First 15 lateromarginals are dicuspid, with rather blunt, spatula-shaped mesocones and more acute, ovate ectocones. Outer 44 lateromarginals are dicuspid and shifted, with rather blunt, spatula-shaped mesocones and blunt deltoid ectocones, which may be bifid in the outermost teeth.

Ecology. Found in herbaceous vegetation and in *Bonnetia* forests.

Distribution. Widely distributed on Chimantá massif: Apakará-tepui, 2150 m; Agparamán-tepui, 2120 m; Toronó-tepui, bordering the valley of Caño Mojado, 2250 m; Murey-tepui, 2300 m. Type material in FMNH.

Plekocheilus (Eurytus) breweri (Breure and Schlögl 2010) (Fig. 11.2G).

Diagnosis. Shell up to 41.7 mm, 1.8 times as long as wide, color pattern with oblique stripes of reddish- to blackish-brown, often with a yellowish "shadow" beside the stripes on the last whorl. Sculptured with spiral, puckered bands of oblong granules. Animal dark-gray to blackish.

Radula. Row straight; radula formula C/1 + LM 52/2 (C-6, 20 lateromarginals LM-10, 32 lateromarginals LM-11); description see *P. mundiperditi*.

Ecology. Collected in stands of *Brocchinia hectioides*, up to 1 m high, and in *B. tatei*. See Breure and Schlögl (2010) for more details and discussion of the relation between snails and carnivorous plants in the area.

Distribution. Churí-tepui, both on the summit plateau (2400 m) and in the canyon sima noroeste (2285 m). Type material in MLSC, MZSP, RMNH (holotype), SNMSZ, and UF.

Drymaeus (D.) steyermarki (Haas, 1955) (Fig. 11.2E).

Diagnosis. Shell 27.3 mm, 2.1 times as long as wide, color whitish to brownish with irregular axial, small streaks of chestnut-brown. White lip with a very faint pinkish border lining the inside of the aperture.

Radula. Additional data to Breure and Eskens (1981). Rows slightly curved; radula formula C/1 + LM 53/3 (C-8, LM-12). Central teeth are monocuspid, with relatively small, acute triangular mesocones. The lateromarginals are tricuspid and shifted, with rather blunt ovate mesocones, acute elongate-ovate endocones, and acute deltoid ectocones, which may be bifid in the outermost teeth. Distribution. Single specimen known, collected on plateau below the summit of Apacará-tepui, 1800 m; type in FMNH.

Drymaeus (D.) extraneus (Haas, 1955) (Fig. 11.2D).

Diagnosis. Shell up to 32 mm, 2.2 times as long as wide, unicolored whitish or with spiral color bands of reddish-brown, side slightly convex.

Distribution. Summit of Apakará-tepui, 2100 m, and along trail from Tírica River to lower-summit camp, 1800 m; west side of Abakapá-tepui, 1189 m. One adult and two juvenile specimens known; material in FMNH.

Happiella species.

Haas (1955, p. 384) reported one juvenile specimen identified as belonging to this genus from the Chimantá massif. No subsequent material has been found, and this group needs anatomical material for proper identification.

Euconulus (?) species.

A few juvenile specimens, which are tentatively referred to *Euconulus*, were collected by Jan Schlögl in 2015 on Akopán-tepui. This species is as yet unidentified.

Species from Central Pantepui District

In this region snail species have only been recorded from Cerro Guaiquinima, Cerro Marahuaka, and Cerro Duida.

Plekocheilus (Eurytus) fusitorsus (Oberwimmer, 1931) (Fig. 11.2J).

Diagnosis. Shell up to 63.3 mm, 1.7 times as long as wide, color (yellowish-)brown with darker, undulating stripes which are bordered by a yellowish "shadow" on the last whorl, sculptured with fine granulation, aperture ovate, relatively wide.

Distribution. "Río Padamo" (type locality); Cerro Guaiquinima [no elevation given]. Type in SMF.

Plekocheilus (Eurytus) gibber (Oberwimmer, 1931) (Fig. 11.2K).

Diagnosis. Shell 55.3 mm, 2.0 times as long as wide, color yellowish, turning reddish-brown on the upper whorls, with irregular dots and short, narrow undulating lines of the same color on the last whorl, sculptured with fine granulation, aperture ear-shaped, with a distinct columellar fold.

Distribution. Known only from the holotype, collected on Cerro Marahuaka, 2170 m (SMF).

Plekocheilus (Eurytus) tatei (Haas, 1955) (Fig. 11.2O).

Diagnosis. Shell up to 62 mm, 1.9 times as long as wide, color brownish, lighter on the upper whorls, the last whorl with dark brown, oblique, slightly zig-zag stripes, sculptured with very fine granulation. Aperture ovate, relatively wide, with a weak columellar fold. Animal dark brown.

Radula. Rows slightly V-shaped; radula formula C/1 + LM 42/2 (C-6, 14 lateromarginals LM-10, 28 lateromarginals LM-11); description see *P. mundiperditi*.

Distribution. Cerro Duida (no elevation given). Types in AMNH and FMNH (holotype).

Species from Western Pantepui District

The malacofauna of this extensive region is poorly known, with only one species being reported in the literature.

Plekocheilus (Eurytus) cf. *plectostylus* (L. Pfeiffer, 1848).

Diagnosis. Shell 36 mm, 1.6 times as long as wide, color pale brownish, slightly darker on the upper whorls, sculptured with fine granulation. Aperture ovate, with a strong columellar fold.

Distribution. Known from northeastern Colombia. In the Pantepui area only one worn specimen was reported by Borrero and Breure (2011) from Cerro Guanay, east slope of Parguaza River, ~1700 m (RMNH). More material is needed to verify this tentative identification. Type material in NHM.

Species from Southern Pantepui District

Plekocheilus (Eurytus) huberi (Breure, 2009) (Fig. 11.2N).

Diagnosis. Shell up to 47.9 mm, 2.0 times as long as wide, thin, light chestnut-brown with descending, narrow reddish-brown "lightning stripes"; an almost lusterless surface, very finely granulated on the last whorl.

Radula. Row slightly V-shaped, with the outermost lateromarginals bending distally; radula formula C/1 + LM 44/2 (C-6, 18 lateromarginals LM-10, 26 lateromarginals LM-11); description see *P. mundiperditi*.

Distribution. Venezuela, Amazonas State, Sierra de la Neblina, at the Brazilian–Venezuelan border and near Pico Maguire; at 1800–2000 m elevation. Types in RMNH and UF (holotype).

Plekocheilus (Eurytus) nebulosus (Breure, 2009) (Fig. 11.2M).

Diagnosis. Shell up to 62.5 mm, 2.0 times as long as wide, sculptured on the last whorl with horizontal threads of various length, often anastomosing; color pattern with darker spots on the uniform brownish color; aperture decidedly descending in front, peristome somewhat thickened and reflexed.

Radula. Rows slightly V-shaped; radula formula C/1 + LM.

Distribution. Venezuela, Amazonas State, Sierra de la Neblina, various localities near Pico Phelps and Pico Maguire; at 1800–2100 m elevation. Types in RMNH and UF (holotype). Brazil, Amazonas State, Sierra de la Neblina, Igarape Café (INPA), ibidem, Cumbre do Pico, 3014 m (INPA).

Drymaeus rex (Breure, 2009) (Fig. 11.2T).

Diagnosis. Shell up to 40.7 mm, 2.2 times as long as wide, with a white line below the suture and typically with three interrupted spiral color bands of chestnut- to blackish brown, crossing weaker axial streaks of a less intense color. Aperture with a white, slightly expanded lip, bordered pink inside.

Distribution. Venezuela, Amazonas State, Sierra de la Neblina, various localities near Pico Phelps and Pico Maguire; typically at elevations of 1800–2100 m, but recorded as low as 200 m on the northwestern side of Cerro de la Neblina. Types in RMNH and UF (holotype). Brazil, Amazonas State, Pico de la Neblina, Igarapé Café (INPA).

Species of the uplands and lowlands (below 1500 m)

Only three species from the Eastern and Central Guiana provinces are known to occur in this altitudinal stratum.

Drymaeus (D.) extraneus (Haas, 1955) (Fig. 11.2D).

As mentioned earlier, one specimen was collected on the west side of Abacapá-tepui, 1189 m; material in FMNH.

Plekocheilus (Eurytus) fusitorsus (Oberwimmer, 1931) (Fig. 11.2J).

Remarks. As explained in Breure (2009, p. 35), this species was collected from an imprecise locality in the Rio Padamo area where elevations below 1500 m also occur. The altitudinal stratum thus needs further clarification through additional sampling.

Plekocheilus (Eurytus) tepuiensis (Breure, 2009) (Fig. 11.2L).

Diagnosis. Shell 35 mm, 1.7 times as long as wide, pale yellowish-brown with obliquely descending zig-zag streaks; surface hardly shining, with spiral lines of dot-like granules on the last whorl.

Distribution. Known only from the holotype in RMNH, collected at 800 m elevation on Cerro Yapacana.

Plekocheilus (Eudolichotis) sinuatus (Albers, 1854).

Remarks. This lowland species is mentioned herein for being recorded from the Yapacana-Cerro Tigre subdistrict (sensu Berry et al., 1995, Fig. 4-2), where it was found 10 km south of the Autana River, 100 m, in 1979 by O. Huber (Breure, 2009, p. 39, Fig. 6K).

Drymaeus (D.) yapacanensis (Breure and Eskens, 1981) (Fig. 11.2P).

Diagnosis. Shell up to 31.2 mm, 2.2 times as long as wide, color pink to yellowish, with axial reddish-brown streaks, which may be forked above but are broader below. Aperture with a slightly expanded peristome, pink colored inside.

Distribution. Endemic to Cerro Yapacana; recorded from 1300 m elevation. Types in RMNH.

Species from the Southern Pantepui region

Only the malacofauna of the Neblina massif are known in this region. This massif is partly Venezuelan and partly Brazilian in terms of territory.

Columbinia exul (Thompson, 2008) (Fig. 11.2S).

Diagnosis. Shell slender, up to 24 mm, 4.5 times as long as wide, color brownish, lighter at the blunt apex, which is minutely granulate, teleoconch sculptured with fine, straight thread-riblets, last whorl relatively short. Aperture ovate, about as wide as shell, projecting forward on a short neck, with the palatal plica at a relatively low position.

Distribution. Venezuela, Amazonas State, Sierra de la Neblina, valley north of Pico Phelps, at 1200–1400 m elevation. Types in MLSC, SMF, and UF (holotype).

Ecology. Collected in a dense broadleaf rainforest, at 4–5 m above ground in a dead tree trunk.

Drymaeus rex (Breure, 2009) (Fig. 11.2T).

Remarks. As mentioned earlier, this species has also been collected at altitudes lower than 1500 m on the northwestern side of Cerro de la Neblina.

Olympus nimbus (Simone, 2010) (Fig. 11.2R).

Diagnosis. Shell up to 15 mm wide, height/width ratio 0.9, spire dome-shaped and relatively well developed, sculptured with a regular mosaic of uniformly spaced nodules, each row intercalated with neighboring rows, with about 4–5 spiral rows on the penultimate whorl, last whorl smooth; color beige, with three spiral, narrow bands split into spots, which are fairly uniformly and equidistantly distributed. Aperture semicircular, compressed by penultimate whorl; umbilicus very narrow.

Distribution. Brazil, Edo. Amazonas, São Gabriel da Cachoeira, Cachoeira do Tucano, at 100 m elevation. Types in MZSP.

Ecology

Snails use sources of calcium to form their shells during growth. In the Pantepui area, limestone is not available due to the sandstone substrate; the only calcium sources appear to be the uptake of calcium from plants (Van Bruggen, pers. commun.) and the leaching of the mineral from the sandstone rocks. Although no studies have been carried out on shell formation in these habitats, it can be safely assumed that both the geology of the area and the acidity of the habitat offer a challenge for land snails during their growth. This may partly explain their low abundance. As far as current data allow a conclusion, the occurrence of snails is concentrated on or near the summits of the tepuis. Huber (1988) has made a distinction between highlands (>1500 m), uplands (500–1500 m), and lowlands (<500 m). From the 21 species reported in this area, 14 are known to inhabit only the highlands, 2 are only from the uplands, and 1 species has a broad altitudinal distribution (Table 11.1). As far as is known, the species reported in this study all feed on detritus,

TABLE 11.1 Distribution of snail species among Pantepui phytogeographical districts and altitudinal stratums.

	Pantepui province			
	Eastern District	**Central District**	**Western District**	**Southern District**
Highlands	*Plekocheilus* (*P.*) *linterae* *Plekocheilus* (*P.*) *alticola* *Plekocheilus* (*P.*) *vlceki* *Plekocheilus* (*P.*) *philippei* *Plekocheilus* (*E.*) *sophiae* *Plekocheilus* (*E.*) *juliani* *Plekocheilus* (*E.*) *mundiperditi* *Plekocheilus* (*E.*) *breweri* *Drymaeus* (*D.*) *steyermarki* *Drymaeus* (*D.*) *extraneus* *Happiella* sp. *Euconulus* (?) sp.	*Plekocheilus* (*E.*) *gibber* *Plekocheilus* (*E.*) *fusitorsus* *Plekocheilus* (*E.*) *tatei*	*Plekocheilus* (*E.*) cf. *plectostylus*	*Plekocheilus* (*E.*) *huberi* *Plekocheilus* (*E.*) *nebulosus* *Drymaeus* (*D.*) *rex*
Uplands	*Drymaeus* (*D.*) *extraneus*	*Plekocheilus* (*E.*) *tepuiensis* *Drymaeus* (*D.*) *yapacanensis*		*Drymaeus* (*D.*) *rex* *Columbinia exul*
Lowlands		*Plekocheilus* (*E.*) *sinuatus*		*Drymaeus* (*D.*) *rex Olympus nimbus*
Outside Pantepui		*Plekocheilus* (*E.*) *sinuatus*	*Plekocheilus* (*E.*) *plectostylus*	

Districts and altitudinal ranges from Huber, O., 1995. Geographical and physical features. In: Berry, P.E., Holst, B.K., Yatskievych, K. (Eds.), Flora of the Venezuelan Guayana. Introduction, vol. 1. Missouri Botanical Garden Press, St. Louis, pp. 1–61 and Berry, P.E., Huber, O., Holst, B.K., 1995. Floristic analysis and phytogeography. In: Berry, P.E., Holst, B.K., Yatskievych, K. (Eds.), Flora of the Venezuelan Guayana. Introduction, vol. 1. Missouri Botanical Garden Press, St. Louis, pp. 161–191.

TABLE 11.2 Variables used in principal component analysis (columns 2–12). Column headings

LOC	TOT	TEP	ELE	SUM	SL	NN	FT	VPT	VPE	PPT	PPE
ETC	2	2	2810	70	320	383	1	135	56	112	19
AUY	1	1	2450	668	715	272	4	54	165	76	83
CHI	8	8	2650	615	92	268	4	242	129	156	28
GUQ	1	1	1650	1096	410	260	3	111	40	41	4
GUY	1	1	2080	165	113	367	1	36	22	5	3
YAP	2	1	1300	11	38	421	1	28	13	5	0
DUI	2	2	2800	1219	1100	346	7	220	135	187	28
NEB	4	1	3014	473	1515	538	5	199	118	180	23

Abbreviations: *LOC*, tepui or massif; *TOT*, number of snail species known to occur on each tepui or massif; *TEP*, number of tepuis with snails in each massif; *ELE*, maximum elevation (m); *SUM*, summit area (m^2); *SL*, slope area (m^2); *NN*, average nearest-neighbor distance between each massif (km); *FT*, number of forest/vegetation types on each massif; *VPT*, number of vascular plants known from each massif; *VPE*, number of endemic vascular plants known from each massif; *PPT*, total number of primitive plants known from each massif; *PPE*, number of endemic primitive plants known from each massif. Tepuis or massifs are abbreviated on the rows as follows: *ETC*, Eastern Tepui Chain; *AUY*, Auyán; *CHI*, Chimantá; *GUQ*, Cerro Guaiquinima; *GUY*, Cerro Guanay; *YAP*, Cerro Yapacaná; *DUI*, Duida-Marahuaka; *NEB*, Neblina-Aracamuni.
Data in columns 4–12 adapted from McDiarmid and Donnelly (2005).

lichens, or algae. Although no direct observations are known, it is supposed that at least part of the species are preyed upon by birds (Jan Schlögl, pers. commun.).

Ecologically, with the exception of *O. nimbus*, which is ground-dwelling, all species have been observed in low shrubs or tree-like vegetation (notably *Brocchinia* and *Bonnetia* forests with trees up to 3 m high). In a PCA of different factors related to snail distribution, it appears that endemic plant species play an important role in the primary PCA factor of habitat diversity (Tables 11.2 and 11.3; see also Breure, 2009, for further discussion). Local endemic plant species have mainly been recorded from the slopes and summits of individual tepuis, with the highest numbers recorded at Sierra de la Neblina, Chimantá massif, Cerro Duida, and Cerro Sipapo (Berry et al., 1995; Huber, 1988). Among these plant endemics, *Bonnetia* species are worth mentioning because several endemic snails appear to be associated with them. It is noteworthy that *Plekocheilus* species in the Pantepui area often occur in paramoid scrub (Breure, 2009; Huber, 1988, 1995, p. 37), a habitat that they share with certain Andean species groups in the same genus.

Finally, it is worth mentioning that microgastropods seem to be underrepresented in the faunal overview of this region. The recent find of an unidentified *Euconulid* species suggests that more extensive searches using the right collection methods may yield additional surprises.

Biogeography

Breure (2013) analyzed the Pantepui malacofauna in comparison to the known land snail fauna of Venezuela and found the Pantepui fauna relatively prominent (15% of total

TABLE 11.3 Factor loading scores with the variables used in the PCA and the % variance explained by each factor. Variable abbreviations as in Table 11.2.

	PC1	PC2	PC3
% variance	49.75	20.08	16.50
TOT	0.23136	0.5034	0.19283
TEP	0.19315	0.59616	− 0.047244
ELE	0.3504	− 0.047465	0.21467
SUM	0.25208	− 0.12968	− 0.39839
SL	0.28985	− 0.44995	0.18729
NN	− 0.007687	− 0.24798	0.67239
FT	0.37973	− 0.17183	− 0.11139
VPT	0.37116	0.20325	0.17648
VPE	0.37961	− 0.08171	− 0.20927
PPT	0.40561	0.01325	0.1966
PPE	0.21954	− 0.17455	− 0.37763

number of species in 0.05% of the country area). Bearing in mind that the state of knowledge on the malacofauna in Venezuela is relatively poor, it may be better to compare the Pantepui data to an ecologically similar area. Griffith et al. (2018) listed the Guiana Highlands as one of the ecoregions in the area, which extends into Guyana, Suriname, and French Guiana. The latter country is especially well studied, and Massemin et al. (2009) gave an overview of the 64 species of terrestrial snails confirmed from there. The composition of the snail fauna of Pantepui and these lower parts of the ecoregion differ, however, as *Plekocheilus* species are only known from Pantepui, and the recorded *Drymaeus* species are different. Unfortunately, not enough molecular data have become available in recent years to allow a malacological view on the various hypotheses which have been proposed as explanations of speciation processes in the Guiana Highlands (summarized in Rull, 2010; see also Breure, 2009, p. 48 and Breure, 2012, p. 122; see also Chapter 4: Origin and evolution of the Pantepui biota).

Further avenues for research

The Eastern Pantepui District is comparatively reasonable well studied in malacological terms, although further additions are expected once thorough and prolonged field work is possible. For the other Pantepui lands, the picture is different, with a possible exception for the Neblina area, as these regions have hardly been surveyed in malacological terms, and further research there is strongly encouraged.

To enable phylogenetic analysis of the species in this area, and to make testing of alternative speciation hypotheses possible, it will be necessary to collect additional material with an adequate state of preservation. This will also facilitate phylogeographical analyses if similar material of congeneric species from a broader region in the Neotropics is considered.

References

Berry, P.E., Huber, O., Holst, B.K., 1995. Floristic analysis and phytogeography. In: Berry, P.E., Holst, B.K., Yatskievych, K. (Eds.), Flora of the Venezuelan Guayana. Introduction, vol. 1. Missouri Botanical Garden Press, St. Louis, pp. 161–191.

Borrero, F.J., Breure, A.S.H., 2011. The Amphibulimidae (Mollusca: Gastropoda: Orthalicoidea) from Colombia and adjacent areas. Zootaxa 3054, 1–59.

Breure, A.S.H., 1978. Notes on and descriptions of Bulimulidae (Mollusca, Gastropoda). Zool. Verh. 164, 1–255.

Breure, A.S.H., 1979. Systematics, phylogeny and zoogeography of Bulimulinae (Mollusca). Zool. Verh 168, 1–215.

Breure, A.S.H., 2009. New Orthalicidae (Mollusca, Gastropoda) from Venezuelan Guayana: unravelling secrets from the Lost World. Zootaxa 2065, 25–50.

Breure, A.S.H., 2011. Annotated type catalogue of the Orthalicoidea (Mollusca, Gastropoda) in the Royal Belgian Institute of Sciences, Brussels, with descriptions of two new species. ZooKeys 101, 1–50.

Breure, A.S.H., 2012. Living in isolation: *Plekocheilus* (*P.*) *philippei* spec. nov. from Venezuelan Guayana (Gastropoda, Amphibulimidae). Basteria 76, 101–106.

Breure, A.S.H., 2013. The Pantepui malacofauna: land snails of Churí-tepui and other tepuis in southern Venezuela and adjacent areas. In: Aubrecht, R., Schlögl, J. (Eds.), Venezuelan Tepuis—Their Caves and Biota. Comenius University, Bratislava, pp. 113–124.

Breure, A.S.H., Eskens, A.A.C., 1981. Notes on and descriptions of Bulimulidae (Mollusca, Gastropoda), II. Zool. Verh 186, 1–111.

Breure, A.S.H., Schlögl, J., 2010. Additional notes on Orthalicidae from the Chimantá Massif, Venezuelan Guayana, with descriptions of new species of *Plekocheilus* Guilding, 1828 (Mollusca: Gastropoda). Zootaxa 2416, 51–60.

Brewer-Carías, C., 1988. Cerro de la Neblina. Resultados de la expedición 1983–1987. FUDECI, Caracas.

Brewer-Carías, C., Audy, M., 2010. Entrañas del mundo perdido. Corpo Print, Caracas.

Griffith, G.E., Omernik, J.M., Azevedo, S.H., 2018. Ecoregions of Central and South America. Available from: http://www.ecologicalregions.info/htm/sa_eco.htm (accessed 31.05.18).

Haas, F., 1955. On some small collections of inland shells from South America. Fieldiana Zool. 34, 361–387.

Hammer, O., Harper, D.A.T., Ryan, P.D., 2001. PAST: paleontological statistics software package for education and data analysis. Pal. Electr 41 (1), 9 pp.

Huber, O., 1988. Guayana highlands versus Guayana lowlands, a reappraisal. Taxon 37, 595–614.

Huber, O., 1995. Geographical and physical features. In: Berry, P.E., Holst, B.K., Yatskievych, K. (Eds.), Flora of the Venezuelan Guayana. Introduction, vol. 1. Missouri Botanical Garden Press, St. Louis, pp. 1–61.

Massemin, D., Lamy, D., Pointier, J.P., Gargominy, O., 2009. Coquillages et escargots de Guyane. Seashells and snails from French Guiana. Biotope, Muséum national d'Histoire naturelle, Mèze, Paris.

McDiarmid, R.W., Donnelly, M.A., 2005. The herpetofauna of the Gyayana Highlands: amphibians and reptiles of the Lost World. In: Donnelly, M.A., Crother, B.I., Guyer, C., Marvalee, H., White, M.E. (Eds.), Ecology and evolution in the tropics: a herpetological perspective. University of Chicago Press, Chicago, pp. 461–560.

Rull, V., 2010. The Guayana Highlands: a natural laboratory for the biogeographical and evolutionary study of the neotropical flora. In: Sánchez-Villagra, M., Aguilera, O.A., Carlini, A.A. (Eds.), Urumaco and Venezuelan Palaeontology. The Fossil Record of the Northern Neotropics. Indiana University Press, Bloomington, pp. 84–102.

Schilthuizen, M., 2011. Community ecology of tropical forest snails: 30 years after Solem. Contrib. Zool. 80, 1–15.

Simone, L.R.L., 2010. A new genus and species of camaenid from the Amazon rainforest, Brazil (Pulmonata: Helicoidea). J. Conchol. 40, 149–161.

Thompson, F.G., 2008. The clausiliid landsnail genus *Columbinia* in Venezuela, and the description of a new species (Gastropoda: Pulmonata: Stylommatophora: Clausiliidae: Neniinae). Arch Molluskenkd 137, 127–132.

CHAPTER

12

Amphibians and reptiles

Celsa Señaris[1] *and Fernando J.M. Rojas-Runjaic*[2,3]

[1]Laboratory of Population Genetics, Center of Ecology, Venezuelan Institute for Scientific Research (IVIC), Caracas, Venezuela [2]La Salle Natural History Museum, La Salle Natural Sciences Foundation, Caracas, Venezuela [3]Laboratory of Vertebrate Systematics, Zoology Postgraduate Program, School of Sciences, Pontifical Catholic University of Rio Grande do Sul (PUCRS), Porto Alegre, Brazil

Introduction

The Guiana Shield is a geological unit that comprises about 13% of South America and is located between the Amazon and Orinoco rivers. This region is environmentally, hydrologically, and topographically complex, and the spectacular flat-topped sandstone mountains -locally known as tepuis- that rise abruptly above the surrounding uplands and lowlands are its most characteristic physiographical feature. In addition to the scenic beauty of this area, the diversity of fauna and flora is remarkable, but even more striking is their level of endemism (Hoogmoed, 1979; Berry and Riina, 2005; Hollowell and Reynolds, 2005; Naka, 2011).

In 1979 Marinus Hoogmoed offered the first detailed compendium of the Guiana Shield herpetofauna, including biogeographical analyses, and highlighted the high level of endemism up to 1000 m of altitude (Hoogmoed, 1979). Twenty years later, in a macroanalysis of the distribution patterns of South America amphibians, Duellman (1999) recognized the Guiana Highlands as a natural region—but without establishing precise altitudinal limits—based on its higher number of exclusive taxa and its area/number of species index ratio (ranking first among the 12 regions considered). In the same year, Gorzula and Señaris (1999) summarized the amphibian and reptile fauna of the Venezuelan Guayana based on museum records, published literature, and more than 15 years of fieldwork in 410 localities by the senior author. This contribution is especially noteworthy because most of the *c*. 70 tepuis (Huber, 1995) are concentrated in southern Venezuela and only a few in adjacent Guyana and Brazil (Chapter 1: Definition and characterization of the Pantepui biogeographical province).

Biodiversity of Pantepui
DOI: https://doi.org/10.1016/B978-0-12-815591-2.00012-4

Sometime later, in the framework of the Guiana Shield Priority Setting Workshop held in Paramaribo in 2002 (Conservation International, 2003), the taxonomic lists of amphibians (Señaris and MacCulloch, 2005) and reptiles (Ávila-Pires, 2005) for the Guiana Shield were updated as a basis for the establishment of conservation areas. At the same time McDiarmid and Donnelly (2005) offered a comprehensive review of the Guiana Highlands herpetofauna, including a detailed taxonomic list, identifying patterns of distributions of these taxa, and offering some analysis to explain those patterns. Despite the usefulness of these contributions, they are now outdated due to recent taxonomic changes at the family and genus levels, new species descriptions, splitting of former species, synonyms update, correction of misidentifications, and new data on geographic distribution, especially in the Eastern Guiana Shield. Also, in the last decade, important phylogenetic analyses of Amazonian and Guiana Shield taxa have been developed, as well as several pioneer studies on tepui-endemic clades. A huge cryptic diversity is being revealed in the region (e.g., Gehara et al., 2014; Fouquet et al., 2014, 2016; Kok et al., 2016; Vacher et al., 2017; Rojas et al., 2018), in addition to more complete information about the evolutionary origin and diversification mechanisms of the amphibians and reptiles of the Guiana Shield.

In this chapter, we update the knowledge about the species richness and distribution of amphibians and reptiles of Pantepui, highlighting relevant biogeographical aspects that we consider a priority to define and characterize this region more appropriately. We also identify information gaps that hopefully will be useful to promote and guide future studies on the Guiana Highlands herpetofauna and to define the most effective conservation strategies for this unique region.

Defining Pantepui for amphibians and reptiles

The term *Pantepui* was coined by Mayr and Phelps (1967) to describe the geographical assemblage of tepuis from the Guiana Highlands based on the particular avifauna inhabiting the tepui summits. They did not offer a precise definition, but rather restricted the term to high-elevation, mesothermic, or microthermic life zones (Huber, 1995). Subsequent authors used this term more broadly and extended it to include the adjacent lowlands as well (e.g., Steyermark, 1982; Kok, 2010). Huber (1987, 1995) redefined Pantepui as "a biogeographical province that comprises all high-tepuian ecosystems between 1500 and 3000 m elevation, but better developed between 1800 and 2700 m."

Despite this clear geographical and altitudinal definition of the term by Huber (1995), Pantepui has also been applied to lower elevation areas, depending on and adjusted to the taxonomic groups under study. Additionally, sometimes the term "Guiana Highlands" has been used as synonym of Pantepui. Hoogmoed (1979) called "highland endemics" those amphibians and reptiles known only from localities above 1000 m, and Duellman (1999) probably used that same criterion (McDiarmid and Donnelly, 2005). In updating the reptiles of Venezuela, Rivas et al. (2012) incorporated Pantepui (≥1500 m above sea level) as a different bioregion, following McDiarmid and Donnelly (2005), but they noted that this elevational limit implied some arbitrariness. More recently Kok (2013, 2016) extended the concept and classified Pantepui *sensu lato* into five main landscape types/life zones: *Pantepui lowlands* (*c.* 200–400 m elevation), *Pantepui uplands* (*c.* 400/500–1200 m elevation), *tepui slopes* (the mostly inclined portion of ground separating the tepui summits/cliffs from the uplands or the lowlands), *tepui cliffs* (vertical escarpments separating the tepui summits from the upper slopes and the upper slopes from lower slopes), and *tepui summits* (flat tops, above *c*. 1200 m asl and separated from tepui slopes by vertical walls, or an elevation >1900 m that has a different climate and vegetation from that of the slopes). According to the same author, *Pantepui sensu stricto* corresponded to the upper Pantepui (above *c*. 1200 m elevation, i.e., tepui slopes/tepui summits).

The concept of the Guiana region, Pantepui, or Guiana Highlands as a biogeographic unit changes as new information from taxa distributions becomes available. Likewise, the definition can be different for each biological group depending on its particular attributes, both present and historical. Consequently, it is likely that only one definition is insufficient. Although amphibians and reptiles usually are treated together, they must be analyzed separately because they have not only different life histories and physiological requirements, but also diverse evolutionary histories. In a recent analysis of the Venezuelan Guayana herpetofauna, Señaris and Rojas-Runjaic (in press) noticed differences in the richness, endemism, and taxonomic community assemblages along the elevational gradients in the region. Based on levels of endemism, and partially following Kok (2013), we here consider Pantepui as a biogeographical unit above 1200 m asl. Therefore, our update includes all the amphibians and reptile species with records above this altitudinal limit. Chapter 1 summarizes the general aspects of this region, including the list of all the tepuis and other Guiana Highlands mentioned here.

Amphibians and reptiles of the Guiana Highlands

State of knowledge

The first amphibian and reptile species from the Guiana Highlands were collected by British explorers John J. Quelch and Frederick V. McConnell on Mount Roraima between 1894 and 1898. Shortly after, Boulenger (1895, 1900) described the most iconic anura and lizard species from Pantepui—*Oreophrynella quelchii* and *Riolama leucosticta*—based on the specimens collected by Quelch and McConnell. Large-scale scientific explorations of Pantepui began 30 years later, when the American Museum of Natural History (AMNH) explored the Roraima-tepui traveling from the north of Brazil via Rio Branco (1927–1928), the Cerro Duida in the Amazonas State in Venezuela (1928–1929), and finally the Auyán-tepui in 1937–1938 (Myers, 2000). The Chimantá massif was visited in February 1955 by the Chicago Natural History Museum, and in 1956 the Central University of Venezuela explored the Auyán-tepui (Lindorf, 2006). Although these expeditions have mainly botanical and ornithological objectives, some amphibians and reptiles were collected and described as new species, especially in the case of reptiles (Roze, 1958a,b) (Fig. 12.1). In the 1950s Juan Rivero visited several Venezuelan locations from the Amazonas State—lowlands and uplands adjacent to the Duida–Marahuaka massif—making important collections of amphibians that included several new taxa (Rivero, 1961). Despite the outstanding results from the scarce samples obtained in these pioneering explorations,

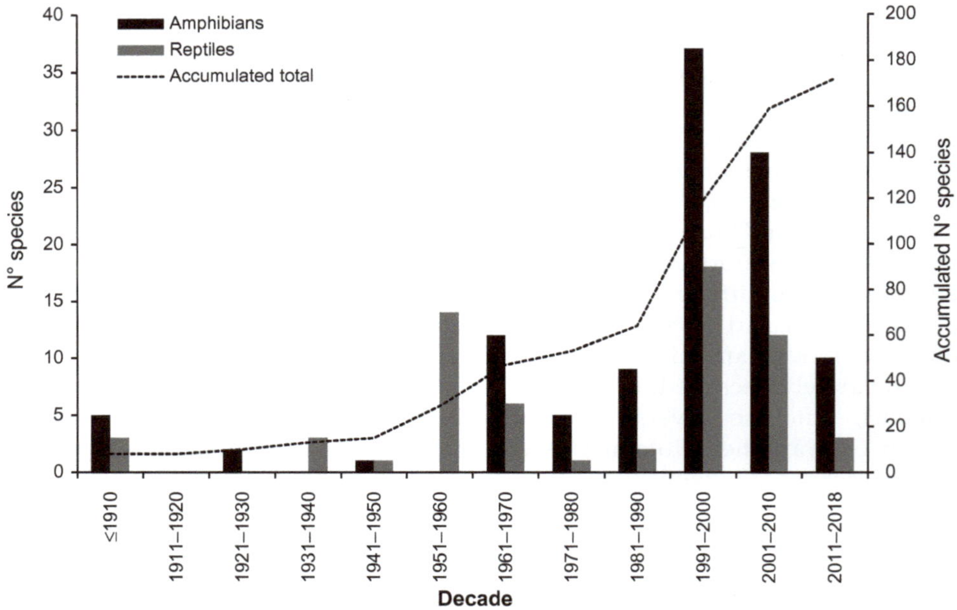

FIGURE 12.1 Discovery rate of amphibians and reptiles from the Guiana Highlands by decade.

the studies on the Pantepui herpetofauna did not have further substantial development until almost two decades later.

In the 1970s, owing to the increasing use of helicopters, biological surveys to remote Guiana areas and tepui summits experienced a remarkable development and included researchers dedicated exclusively to herpetology. During this period the most relevant explorations were to cerros Jaua and Guanacoco in 1974 by the National Museum of Natural History (NMNH) of the Smithsonian Institution and the Venezuelan Society of Natural Sciences of Caracas (Orejas-Miranda and Quesada, 1976); to Cerro Yapacana in 1978 by the AMNH and the Venezuelan government (Myers, 2000); to cerros Marahuaka, Duida, and Huachamakari between 1983 and 1988 by the Terramar Foundation (Michelangeli, 1989, 2005); to the Chimantá massif by the multidisciplinary Chimantá Scientific Group between 1983 and 1986 (Huber, 1992); to the Sierra de la Neblina and adjacent lowlands from 1983 to 1987 (Brewer-Carías, 1988; McDiarmid and Paolillo, 1988); and to the Duida–Marahuaka massif between 1992 and 1994 by the Spanish Agency for International Cooperation (AECI) and the Venezuelan Institute of National Parks (INPARQUES), headed by José Ayarzagüena (Ayarzagüena and Señaris, 1994). Also, the AMNH carried out a series of explorations to Cerro Tamacuari and Serranía Tapirapecó in 1988–1989 (Myers and Donnelly, 1997), to Cerro Guaiquinima in 1990 (Donnelly and Myers, 1991; Mägdefrau et al., 1991), to Auyán-tepui in 1994 (Señaris, 1995; Myers, 1997; Myers and Donnelly, 2008), and a year later to the Yaví, Yutajé, and Guanai tepuis (Myers and Donnelly, 1996, 2001). In the same period, the La Salle Museum of Natural History (MHNLS) visited the Gran Sabana and the slopes and summit of the Roraima-tepui in 1990 and 1991, the Cerro Jaua in 1994 (Señaris and Ayarzagüena, 1996, 2002), and the Auyán-tepui in 1992. Finally, between 1978 and 1993, Stefan Gorzula made the most important and complete herpetological collections in the Venezuelan Guayana, including 55 localities in 29 tepui summits (Gorzula and Señaris, 1999).

McDiarmid and Donnelly (2005) presented an extensive and very detailed update of the herpetological explorations to the Guiana Highlands (mostly in Venezuela), listing 97 species of amphibians (35 not fully identified) and 62 species of reptiles (16 species not totally identified). Later, MacCulloch et al. (2007) synthesized the knowledge on the herpetofauna of Roraima-tepui, and Barrio-Amorós and Brewer-Carías (2008) presented the results of a herpetological expedition to Cerro Sarisariñama in March 2002 (six localities, between 406 and 1375 m asl) and also provided an accounting of the amphibians and reptiles from the slopes and summits of several tepui massifs, based on McDiarmid and Donnelly's (2005) data. Finally, Aubrecht et al. (2012) summarized the herpetofauna found in the caves of five tepuis—Auyán-tepui, Roraima-tepui, Chimantá massif, Cerro Autana, and Cerro Sarisariñama—and Señaris et al. (2014) offered an illustrated guide of the amphibians of Canaima National Park, including the eastern Venezuelan tepuis.

The expeditions of the 1980s provided the specimens from which were described a substantial number of new amphibian and reptile species in successive decades (Fig. 12.1). Sixty-three percent of the amphibians and 44% of reptile species currently known in the Guiana Highlands have been named since 1990. In the last two decades, the exploration of the Venezuelan Guayana highlands has decreased dramatically because of the difficulty in obtaining the permits for fieldwork, sampling, and genetic studies (Rull and Vegas-Vilarrúbia, 2008; Rull et al., 2016), but also due to the lack of scientific and technical personnel and supplies and the insecurity resulting from the political and economic crisis that has afflicted the country in the last 15 years (Bifano, 2016; Requena and Caputo, 2016; Chapter 16: Conservation of Pantepui). In contrast (and fortunately), at the same time, the herpetological studies in Guyana intensified, beginning with the explorations of Bruce Means to Mount Wokomung in 2003 and 2006 (Means and Savage, 2007; but see other field trips to the Venezuelan Guayana in 2006 and 2007 (http://www.brucemeans.com/journal.htm), and Ross MacCulloch and Amy Lathrop of the Royal Ontario Museum to Mount Ayanganna in 2000 and to Mount Wokomung in 2004 (MacCulloch and Lathrop, 2001, 2002, 2005, 2009; Lathrop and MacCulloch, 2007; MacCulloch et al., 2006, 2008a,b, 2009). Cole et al. (2013) synthesized for the first time the herpetofauna of Guyana based on a review of scientific collections and the literature, but also on their 20 years of field trips in that country. Philippe Kok of the Vrije Universiteit Brussels has made intensive explorations to 21 tepui summits/massifs in Guyana and Venezuela (Kok, 2013). The greater sampling and publishing efforts made recently on Guiana Highlands amphibians and reptiles have been performed by this author and his collaborators (Kok, 2008, 2009, 2010, 2013, 2015; Kok and Barrio-Amorós, 2013; Kok et al., 2012, 2013, 2015, 2016, 2017, 2018a,b). Another important contribution is the report of Moraes et al. (2017) about the herpetofauna of the uplands of the Brazilian Serra da Mocidade (600–1365 m asl) collected in a 22-day expedition in 2016. More recently, in November 2017, Miguel Trefaut Rodrigues and his team from the University of São Paulo explored the Pico da Neblina (https://www.bbc.co.uk/news/resources/idt-sh/amazon_discoveries); the study of these interesting herpetological samples will allow a better understanding of the richness and the distribution patterns of the poorly known taxa from this part of the Guiana Highlands.

The herpetological knowledge of the Guiana Highlands is geographically unbalanced. The sampling efforts have been higher in the Eastern Pantepui District (*sensu* Berry et al., 1995; see Chapter 1), which includes the upper part of the Sierra de Lema (La Escalera); the summits of Auyán-tepui; Chimantá massif; the Eastern Tepuis Chain in Venezuela (Roraima-tepui, Kukenán-tepui, Yuruaní-tepui, Wadakapiapué-tepui, Karaurín-tepui, Ilú-tepui, and Tramen-tepui); and Maringma-tepui, Wei-Assipu-tepui, Mount Ayanganna, and Mount Wokomung in Guyana. Also, the Duida–Marahuaka Subdistrict in Venezuela is considered relatively well sampled (Señaris and Ávila-Pires, 2003; McDiarmid and Donnelly, 2005). Part of the Sierra de la Neblina, Jaua-Sarisariñama, and the Yutajé Subdistrict can be considered with a medium degree of herpetological knowledge, but only part of the herpetological collections obtained decades ago from these massifs have been studied in detail (McDiarmid and Paolillo, 1988; Barrio-Amorós and Brewer-Carías, 2008). Most of the other tepui summits and Guiana mountains have only received very low exploration efforts or few opportunistic collections; the herpetofauna of the extensive and remote granitic mountains of Sierra de Maigualida, Serranía de Uasadi, Cerro Parú, Cuao-Sipapo massif, Sierra Marutaní, Sierra de Parima, and Serranía Tapirapecó is unknown.

Therefore, current knowledge of the diversity of amphibians and reptiles that inhabit the Guiana Highlands is vastly underestimated, and the distribution patterns of species from these taxonomic groups is still incipient. Likewise, almost nothing is known about the biology and the ecological requirements of these species, except for scattered data (e.g., McDiarmid and Gorzula, 1989; Solano, 1989; Mägdefrau, 1991; Salerno and Pauly, 2012); neither do we know of any work on population dynamics of Pantepui amphibians and reptiles. For this reason, analyses carried out in this chapter should be considered preliminary and biased by the current taxonomic and biogeographical understanding.

Richness

To date, the herpetofauna of the Guiana Highlands ≥1200 m where the Pantepui biogeographical province is located (Chapter 1), includes 172 described species (Appendix 12.1). This richness is remarkable, especially in amphibians, which comprise about 3% of the megadiverse herpetofauna of South America in only an area of *c*. 7200 km^2 (0.03% of the continental area). Amphibians are represented by 108 anuran species and one caecilian, and reptiles by 34 species of lizards and 29 species of snakes. We are not counting around 20 undescribed taxa indicated by recent molecular analyses as candidate species (e.g., Castroviejo-Fischer et al., 2011; Jungfer et al., 2013; Kok et al., 2016, 2017; Moraes et al., 2017; Vacher et al., 2017) or taxa not fully identified. So the herpetofauna of the Guiana Highlands can easily exceed 200 species.

The most recent report of the Guiana Highlands herpetofauna (McDiarmid and Donnelly, 2005) indicated a total of 104 species (without taking into account undescribed taxa or species registered below 1200 m). Therefore, our update shows an increase of *c*. 40% in species richness in the last 13 years. This remarkable increase is mainly influenced by numerous additions from tepuis in Guyana—Mount Ayanganna, Mount Maringma, and the Wokomung massif—and an important taxonomic effort in the description of new taxa (some of them collected decades ago), corrections of misidentifications, and species splitting based on molecular analyses. Also, we note a better understanding of the distribution patterns of some species that inhabit the Guiana Highlands (especially the Eastern Pantepui District) and that have a wider range than was previously known (e.g., Passos et al., 2013; Fraga et al., 2017).

In the Guiana Highlands, the richest families of anurans are Hylidae, Strabomantidae, and Hemiphractidae. These three families account for about 57% of the amphibian fauna, and together with Aromobatidae and Bufonidae, cover 83% of the total richness. The taxonomic proportion of anuran families in the highlands differs notably from those of adjacent lowlands (Señaris and MacCulloch, 2005; Cole et al., 2013) due to the greater representation of hemipractids, strabomantids, and bufonids and a substantially lesser preponderance of hylid and leptodactylid frogs. Also, some families are absent in the Guiana Highlands (i.e., Allophrynidae, Ceratophryidae, Phyllomedusidae, Pipidae, and Ranidae). The three anuran genera in Pantepui with the most species—*Pristimantis, Stefania*, and *Anomaloglossus*—account for 46% of the total number of species (Table 12.1).

Reptile communities in the Guiana Highlands are considerably less diverse than those at lower altitudes, both in richness and taxonomic composition. Turtles and crocodilians are absent above 500 m elevation, and only the smooth-fronted caiman *Paleosuchus trigonatus* has been recorded up to 1000 m asl in the La Escalera region and in the slopes of the Auyán-tepui (Gorzula and Señaris, 1999; pers. observ.). Amphisbaenids, the lizard families Iguanidae and Gekkonidae, and the snake families Aniliidae and Anomalepidae are absent over 1000 m asl. Other Squamata taxa common in the lowlands are either poorly or not represented above 1200 m [e.g., Phyllodactylidae and Sphaerodactylidae geckos, whiptails and tegus (Teiidae), boids (Boidae), coral snakes (Elapidae), and slender blind snakes of the Leptotyphlopidae and Typhlopidae families]. In Pantepui, the reptile fauna is composed almost exclusively of lizards—especially the Gymnophthalmidae family with 65% of the total richness—and Colubridae snakes (86% of Serpentes) (Table 12.2).

In the tropical highlands, amphibians usually outnumber reptiles. McDiarmid and Donnelly (2005) estimated an amphibian-to-reptile ratio of 1.56:1 in the Guiana Highlands. This update shows an even higher ratio of 1.73:1, a similar value as in the Tsaratanana massif in northern Madagascar (Raxworthy et al., 2008) and some tropical parts of the Andes. Several ecophysiological and evolutionary factors, due to differences in natural history and strategies of thermal adaptation in amphibians and reptiles, might be affecting these diversity patterns (Navas, 2003).

Endemism

Fifty-five percent of the amphibians in the Guiana Highlands are endemic to this region (Table 12.1), but no anuran family is considered to be endemic above 1500 m asl. Ceuthomantidae, recently reestablished at the family level, is the only family restricted to the Guiana Shield. It contains only the genus *Ceuthomantis* (Heinicke et al., 2018), which includes four described species, all found between 930/1200 and 1540 m in the southern and eastern parts of the region; the lowest elevation record is for *Ceuthomantis cavernibardus* at the Brazilian part of Serranía Tapirapecó (Serra do Tapirapecó); however, this record

TABLE 12.1 Number of Amphibia by major taxonomic group in Guiana Highlands (≥1200 m).

Family	No. of species/ family	Genera	No. of species ≥1200 m	No. of total endemics
Order Anura				
Aromobatidae	15	*Allobates*	1	1
		Anomaloglossus	14	7
Bufonidae	13	*Metaphryniscus*	1	1
		Oreophrynella	9	8
		Rhaebo	1	–
		Rhinella	2	–
Centrolenidae	5	*"Cochranella"*	2	2
		Vitreorana	1	–
		Hyalinobatrachium	2	–
Ceuthomantidae	4	*Ceuthomantis*	4	3
Dendrobatidae	1	*Minyobates*	1	1
Eleutherodactylidae	2	*Adelophryne*	2	–
Hemiphractidae	16	*Stefania*	16	12
Hylidae	25	*Boana*	8	1
		Dendropsophus	1	–
		Myersiohyla	5	4
		Nesorohyla	1	–
		Osteocephalus	1	–
		Tepuihyla	7	5
		Scinax	2	–
Leptodactylidae	3	*Adenomera*	1	–
		Leptodactylus	2	–
Microhylidae	2	*Otophryne*	2	1
Strabomantidae	22	*Dischidodactylus*	2	1
		Pristimantis	20	13
Order Gymnophiona				
Rhinatrematidae	1	*Epicrionops*	1	–
		Total	109	60

TABLE 12.2 Number of Reptilia by major taxonomic group in the Guiana Highlands (≥1200 m).

Family	No. of species/ family	Genera	No. of species ≥ 1200 m	No. of endemics ≥ 1200 m
Dactyloidae	5	*Anolis*	5	3
Gymnophthalmidae	22	*Adercosaurus*	1	1
		Anadia	1	1
		Arthrosaura	4	3
		Cercosaura	2	2
		Euspondylus	1	1
		Neusticurus	5	–
		Oreosaurus	1	1
		Pantepuisaurus	1	1
		Riolama	4	4
		Yanomamia	2	1
Scincidae	1	*Copeoglossum*	1	–
Tropiduridae	6	*Plica*	4	–
		Tropidurus	2	1
Colubridae	25	*Atractus*	4	1
		Chironius	3	1
		Dipsas	2	1
		Erythrolamprus	4	1
		Leptophis	2	–
		Leptodeira	1	–
		Mastigodryas	1	–
		Philodryas	1	–
		Tantilla	1	–
		Thamnodynastes	6	5
Elapididae	1	*Micrurus*	1	–
Viperidae	3	*Bothrops*	2	–
		Crotalus	1	–
		Total	63	28

is imprecise in terms of elevation (930/1200 m, Caramaschi and de Niemeyer, 2005). The other lower record of *Ceuthomantis* is from Cerro Sarisariñama (between 1108 and 1375 m), but this was based on specimens uncertainly assigned to this genus (Barrio-Amorós and Brewer-Carías, 2008).

Until now the only Pantepui-endemic genus is the monotypic *Metaphryniscus* (Bufonidae) from the Marahuaka summit at 2600 m. *Oreophrynella* is the other almost exclusive bufonid group in the region, but one of its nine members—*Oreophrynella macconnelli*—extends down to 700 m in the Eastern Guiana. This is also the case for the genus *Myersiohyla*, with two of its six species found below 1000 m: the little-known *Myersiohyla loveridgei* (Rivero, 1961), at 914 m at Cerro Duida, and *Myersiohyla liliae* (Kok, 2006), a species recently transferred to this genus by Pinheiro et al. (2018) found between 400 and 1200 m asl in the Pakaraima Mountains (Fig. 12.2). Other five genera are mostly associated with just the uplands and highlands of the Guiana Shield—the monotypic dendrobatid *Minyobates* from the slopes of Cerro Yapacana; the strabomantid *Dischidodactylus* from the Duida–Marahuaka massif; and the hylid genera *Nesorohyla* (monotypic), *Stefania*, and *Tepuihyla*—with most of these species showing elevational ranges above 1000 m and few or no records below 500 m (Figs. 12.2 and 12.3) (Señaris et al., 1996; MacCulloch and Lathrop, 2002; MacCulloch et al., 2006; Lathrop and

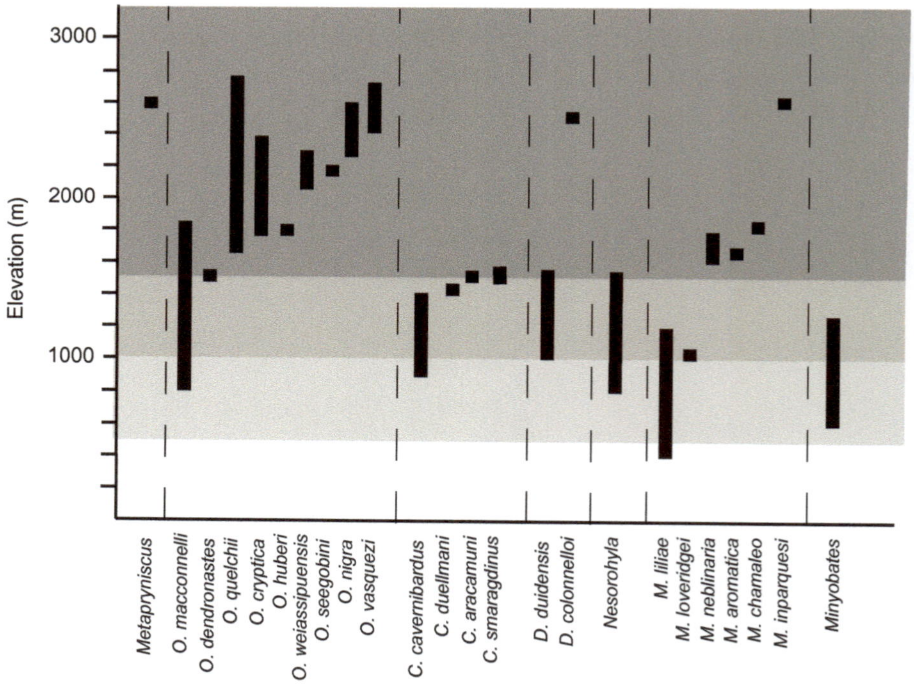

FIGURE 12.2 Altitudinal range distribution of the anuran species of the genera *Metaphryniscus*, *Oreophrynella*, *Ceuthomantis*, *Dischidodactylus*, *Nesorohyla*, *Myersiohyla*, and *Minyobates* in the Guiana Highlands.

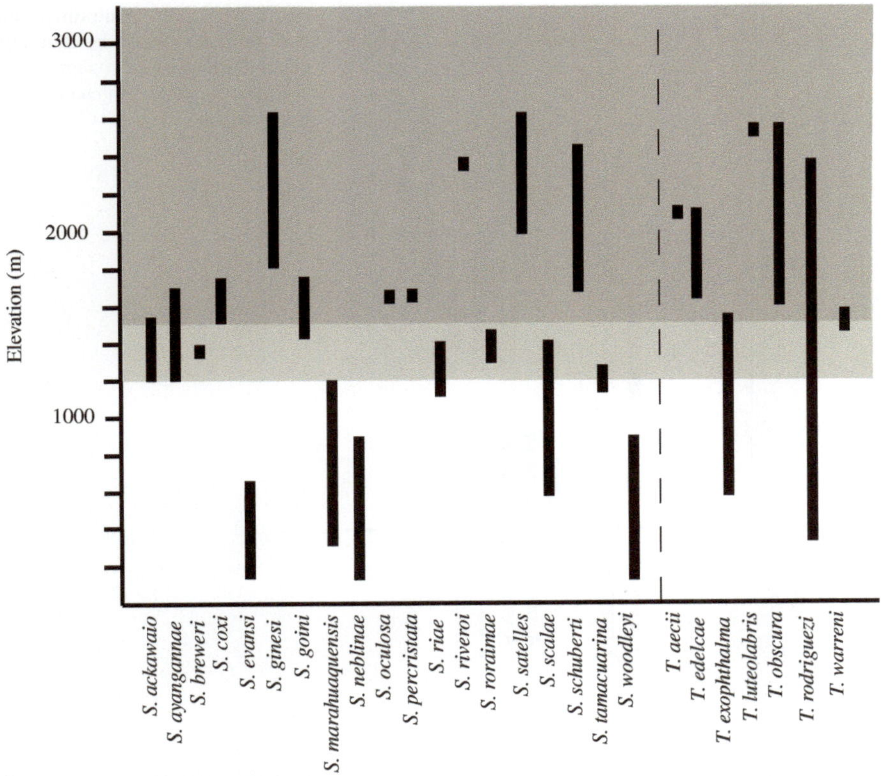

FIGURE 12.3 Altitudinal range distribution of the species of the anuran genera *Stefania* and *Tepuihyla* in the Guiana Highlands.

MacCulloch, 2007; Carvalho et al., 2010; Cole et al., 2013; Faivovich et al., 2013; Kok et al., 2012, 2015, 2016; Pinheiro et al., 2018).

After the genus *Ectopoglossus* was redescribed to accommodate the trans-Andean aromobatid clade of frogs with a median lingual process (Grant et al., 2017), the genus *Anomaloglossus*—with 29 species—has been redefined as a taxon restricted to the Guiana Shield (Fouquet et al., 2015, 2018; Vacher et al., 2017); 38% of its species inhabit the Guiana Highlands, and almost half of them are restricted to elevational belts above 1500 m. The widespread Neotropical genus *Pristimantis* is also the richest in the Guiana Highlands, and 61% of its species are exclusive to Pantepui (Table 12.1). Finally, two glass frogs, *"Cochranella" duidaeana* and *"Cochranella" riveroi*, are considered *incertae sedis* within Centroleninae because molecular data are not yet available and their morphological and behavioral characteristics do not provide unambiguous evidence for their generic placement (Guayasamin et al., 2009). Further studies will be necessary to elucidate the generic placement of these taxa, which could represent other Pantepui endemism.

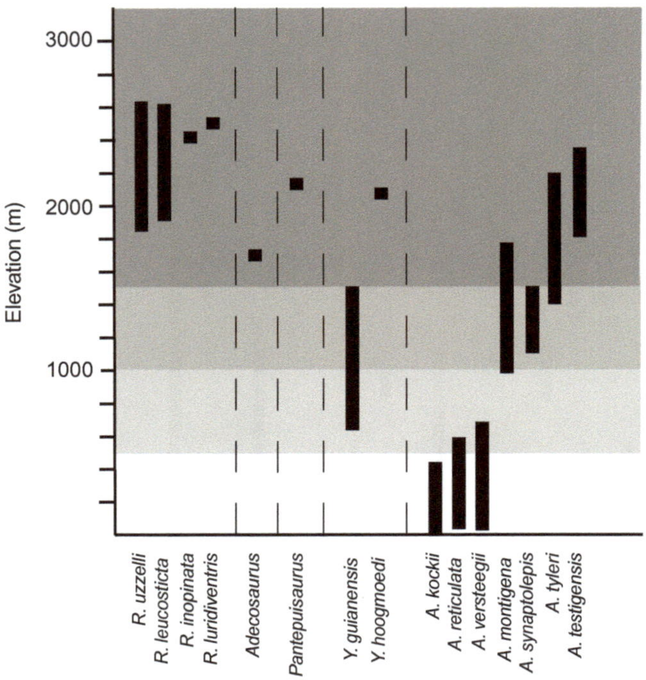

FIGURE 12.4 Altitudinal range distribution of the species of the Gymnophthalmidae genera *Riolama, Adercosaurus, Pantepuisaurus, Yanomamia,* and *Arthrosaura* in the Guiana Highlands.

Forty-four percent of the total reptile species, mainly lizards, are endemic to the Guiana Highlands (Table 12.2). The gymnophthalmid subfamily Riolaminae, with four described species of *Riolama* and two undescribed from Sierra de la Neblina, is restricted to Pantepui (Molina and Señaris, 2003; Kok, 2015; Goicoechea et al., 2016). Two more monospecific gymnophthalmid genera—*Adercosaurus* from Cerro Yutajé at 1700 m and *Pantepuisaurus* at 2080 m in Mount Maringma—are also exclusive to the Guiana Highlands (Myers and Donnelly, 2001; Kok, 2009) (Fig. 12.4). The Guianan genus *Arthrosaura* has four of its seven species restricted to the highlands (Table 12.2). Based on molecular and morphological evidence, Pellegrino et al. (2018) named the genus *Yanomamia* for two species formerly allocated to *Arthrosaura*; *Yanomamia hoogmoedi* (Kok, 2008) is a Pantepuian species known only from Mount Maringma at 2112 m, and *Y. guianensis* (MacCulloch and Lathrop, 2001) is from the Ayanganna and Roraima uplands (MacCulloch and Lathrop, 2001; Cole et al., 2013). An undescribed lizard from the Churí-tepui in the Chimantá massif (Kok et al., 2012) and *Arthrosaura testigensis* (Gorzula and Señaris, 1999) from the Terekyurén-tepui and Murisipán-tepui summits could belong to this new genus, but corroboration based on additional molecular and morphological data is pending (Pellegrino et al., 2018).

No snake family or genus is restricted to the Guiana Shield or to Pantepui. Among Serpentes, half of the endemism corresponds to *Thamnodynastes* (Colubridae) (Table 12.2), many species of which are restricted to the summits of the western tepuis. Nevertheless, the level of endemism in reptiles is probably overestimated, and some of the current

Pantepui "endemisms" actually have a wider distribution, as has been recently documented for some taxa (e.g., Passos et al., 2013; Moraes et al., 2017; Fraga et al., 2017; Kok et al., 2018a; Ávila-Pires et al., 2018).

Patterns of species richness and geographical distribution

Along the altitudinal gradient of the Guiana Shield mountains—here considered those between 1200 and 3014 m—amphibian and reptiles richness is unevenly distributed. On a regional scale, species diversity for both groups decreases from low to high elevations (Fig. 12.5A); in higher altitudes above 2500 m only 13 species of amphibians (12% of the

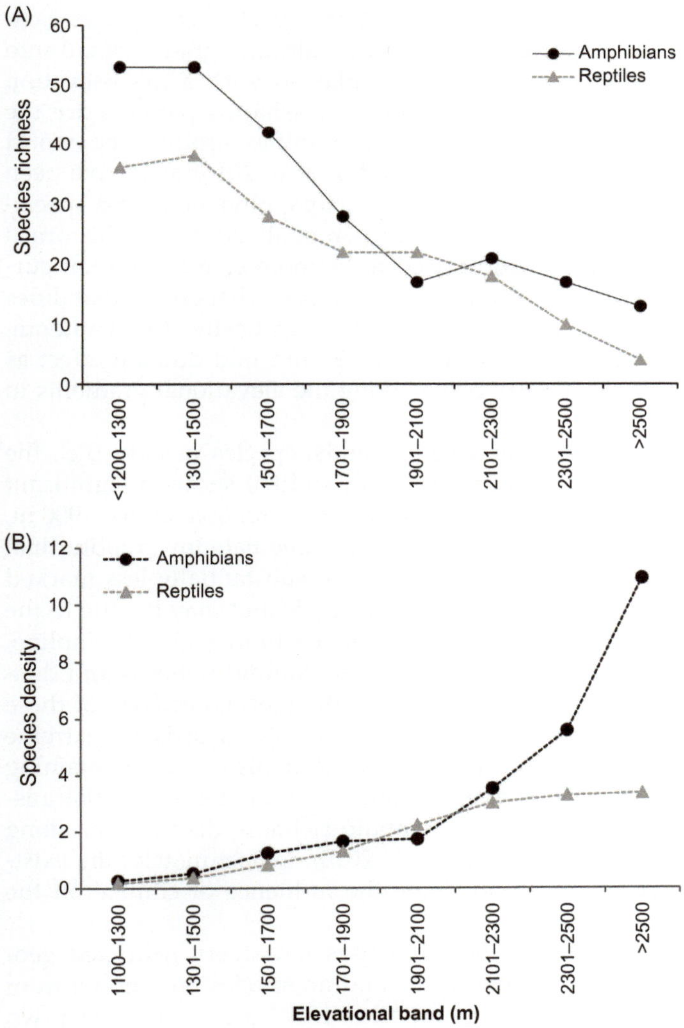

FIGURE 12.5 Elevational patterns of species richness (A) and species density (B) along altitudinal gradient in the Guiana Highlands.

total anuran richness) and four reptiles (6% of reptile fauna) have been recorded. At more local scales, similar patterns have been documented in the Roraima-tepui (MacCulloch et al., 2007), Cerro Sarisariñama (Barrio-Amorós and Brewer-Carías, 2008), and Auyán-tepui (Señaris et al., 2009). This pattern has also been documented in the Sierra de Lema (La Escalera, 200–1400 m), where the herpetofaunal diversity decreases from 37 taxa at lower altitudes to 25 at the highest point, with a turnover in species composition toward a predominance of restricted-range taxa (Duellman, 1997; Barrio-Amorós and Duellman, 2009).

Myers and Donnelly (1997, 2001) stated that "tepuis have relatively depauperate herpetofaunas ... perhaps a general reflection of biotic impoverishment on poor soils" and that the "sizes of tepui faunas are positively correlated with area and negatively correlated with elevation." The decreasing diversity from low to high elevations is a common feature along mountain systems around the world (Lomolino, 2001), and rarely does species richness increase with elevation. Changes in species richness along altitude gradients fall into four patterns: decreasing monotonically, low plateau, low plateau with a mid-elevation peak, and mid-elevation peak (McCain, 2010). Linear and hump-shaped patterns are the most commonly observed in herpetological mountain communities around the world (e.g., Bernal and Lynch, 2008; McCain, 2010; Meza-Joya and Torres, 2016) and have been attributed to species-area effect, mid-domain effect, climatic factors, phylogeny and speciation, and historic events, or a combination of them (e.g., Brown et al., 2014). The historical habitat stability effect on species richness is hypothesized to be more significant than current environmental parameters for taxonomic groups that have poor dispersal capabilities such as amphibians and reptiles (Graham et al., 2006). Specifically for Pantepui, Nogué et al. (2013) suggested a combination of elevation, area, and mid-domain effect as the basic explanation for the diversity of vascular plants along the elevational gradients in these Neotropical table mountains.

Although species richness is lower in the Guiana Highlands, species density (i.e., the number of species adjusted for area in 200-m altitudinal bandwidths) shows a significant increase with elevation (Fig. 12.5B); in amphibians this is especially notable above 1900 m, where species density is ca. three times greater with respect to immediate lower altitudinal belts. This high density of amphibian species on tepui summits (substantially less marked in reptiles) has also been observed in plants (Berry and Riina, 2005) and may be due to the persistence and accumulation of endemic taxa on evolutionary time scales (Lomolino, 2001). The notable difference between amphibians' and reptiles' summit richness and densities might be due to variations in ecological requirements or the thermal ecology of these groups (Navas, 2003), but also the evolutionary and historical factors. Lizards are extreme solar ectotherms with complex physiological and behavioral mechanisms for maintaining their body temperatures (Heatwole, 1982). Temperature, rainfall, and potential evapotranspiration have been reported as the best predictors of reptile richness distribution along some elevation gradients (e.g., Fu et al., 2007; Chettri et al., 2010). Unfortunately, the existing climatic data (Chapter 1) are not enough to evaluate the influence of climate on the reptile diversity patterns in Pantepui.

Most of the amphibians that inhabit the Guiana Highlands have very restricted geographical and altitudinal ranges. Forty-two percent of the anuran species are known from a single tepui summit or slope (i.e., microendemics), and other 18.4% are found in just two

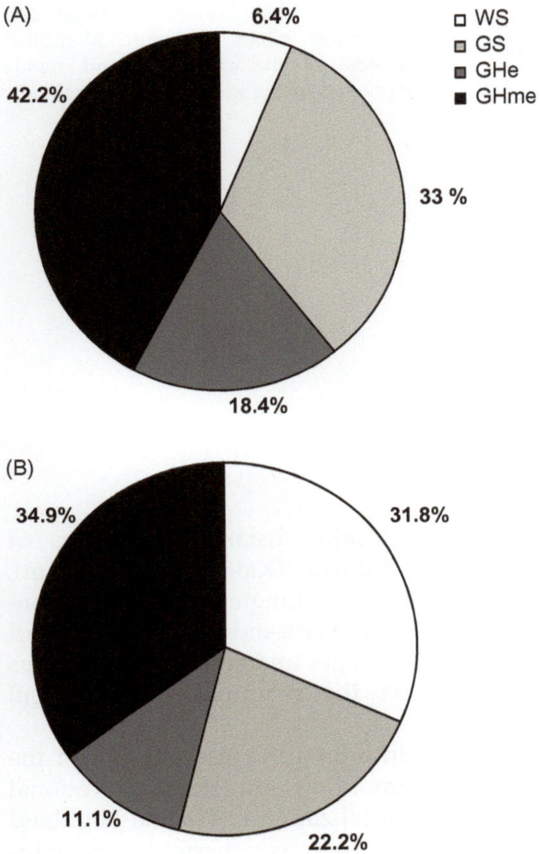

FIGURE 12.6 Composition of distributional patterns for (A) amphibians and (B) reptiles in the Guiana Highlands (GH). *GHe*, known from two or more tepuis or massifs; *GHme*, known only from a single tepui summit; *GS*, restricted to the Guiana Shield; *WS*, widespread, taxa known from areas of South America outside the Guiana Shield.

or three nearby tepuis (Fig. 12.6). There are no reports of any "widespread" amphibian species exclusive to the Guiana Highlands. Around 18% of anuran taxa have a geographical distribution restricted to the Guiana Shield uplands (1000–1500 m asl; Chapter 1) and are recorded on tepui summits. Few taxa with broad geographical and elevational distributions inhabit Pantepui, and most of them are considered complexes of morphologically cryptic species that need detailed studies to define the taxonomic status of lowland and highland populations (Castroviejo-Fischer et al., 2011; Rojas-Runjaic et al., 2017). For example, *Tepuihyla edelcae* was considered a relatively widespread Eastern Pantepui species (Auyán-tepui, Chimantá massif, and Los Testigos massif summits) (Ayarzagüena et al., 1993; Gorzula and Señaris, 1999; McDiarmid and Donnelly, 2005), but further phylogenetic studies showed the existence of two allopatric lineages (Salerno et al., 2012, 2015; Kok et al., 2012; Jungfer et al., 2013). Based mainly on molecular data Kok et al. (2015) described the Chimantá populations as a new species, *Tepuihyla obscura*. In the same paper, *Tepuihyla rimarum*, another Pantepui microemdemic from Ptari-tepui summit at 2400 m asl (Ayarzagüena et al., 1993), was considered a junior synonym of *T. rodriguezi*, a more widely distributed frog in the uplands and highlands of Eastern Guiana (Kok et al., 2015).

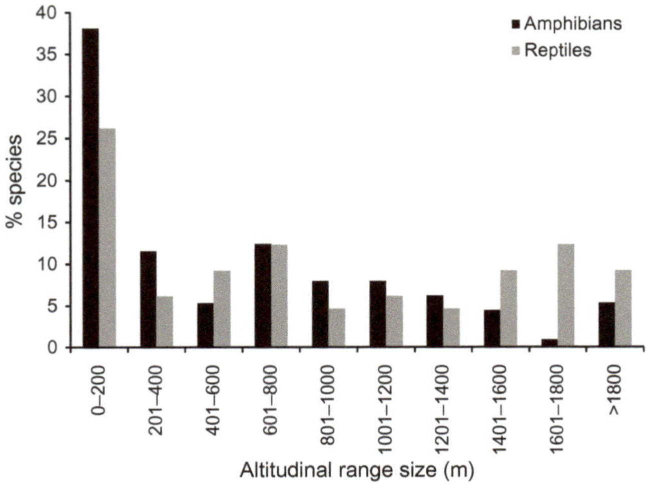

FIGURE 12.7 Altitudinal range size (maximum minus minimum altitudinal records) of the amphibian and reptile species occurring in the Guiana Highlands.

Additionally, genetic divergence between the geographically distant populations of *Tepuihyla galani* (Wadakapiapué-tepui slopes, 1250 m), *T. talbergae* (Kaieteur Falls, 366 m), and *T. rodriguezi* was extremely low (Kok et al., 2012), and then Jungfer et al. (2013) considered the first two species junior synonyms of the last one. With more extended use of phylogenetic analysis, we can anticipate more taxonomic changes and species descriptions (e.g,. genus *Stefania*, Kok et al., 2017) that will affect the percentage of Pantepui microendemisms.

Restricted distributions are accompanied by narrow altitudinal ranges, and half of the anurans have altitudinal range sizes $\leq$400 m (maximum minus minimum elevational records) (Fig. 12.7). This feature, together with thermal specialization in Guiana Highland endemics, are key traits in the impact of climate change on these ectothermic vertebrates (Blaustein et al., 2010; Huey et al., 2012; Foden et al., 2013). Generally, reptiles have broader geographical and altitudinal distributions than amphibians. One third of the species that occur in Pantepui have a widespread distribution and broad altitudinal ranges (Figs. 12.6 and 12.7). Another 35% are microendemics (especially lizards) or inhabit two or more adjacent massifs, but even in these cases their elevational ranges do not exceed 200 m.

Because of the high degree of microendemics, the dissimilarity between herpetofauna assemblages on summits—even in adjacent massifs—is evident, as has been mentioned earlier by Myers and Donnelly (2001) and more recently by Rödder et al. (2010). A preliminary cluster analysis of similarity among amphibian and reptile assemblages using 200-m altitudinal bands revealed distinct groupings along the altitudinal gradient in the Guiana Highlands, apparently strongly influenced by the stepped geomorphology of the region. The amphibians ($>$1700 m) and reptile assemblages ($>$2100 m) are largely distinct from those of lower elevations; very few taxa are shared between these first two groups, and their similarity is less than 10%. In amphibian assemblages, another pronounced differentiation occurs at *c.* 2300 m; reptile assemblages between 1000 and 1500 m show the greatest

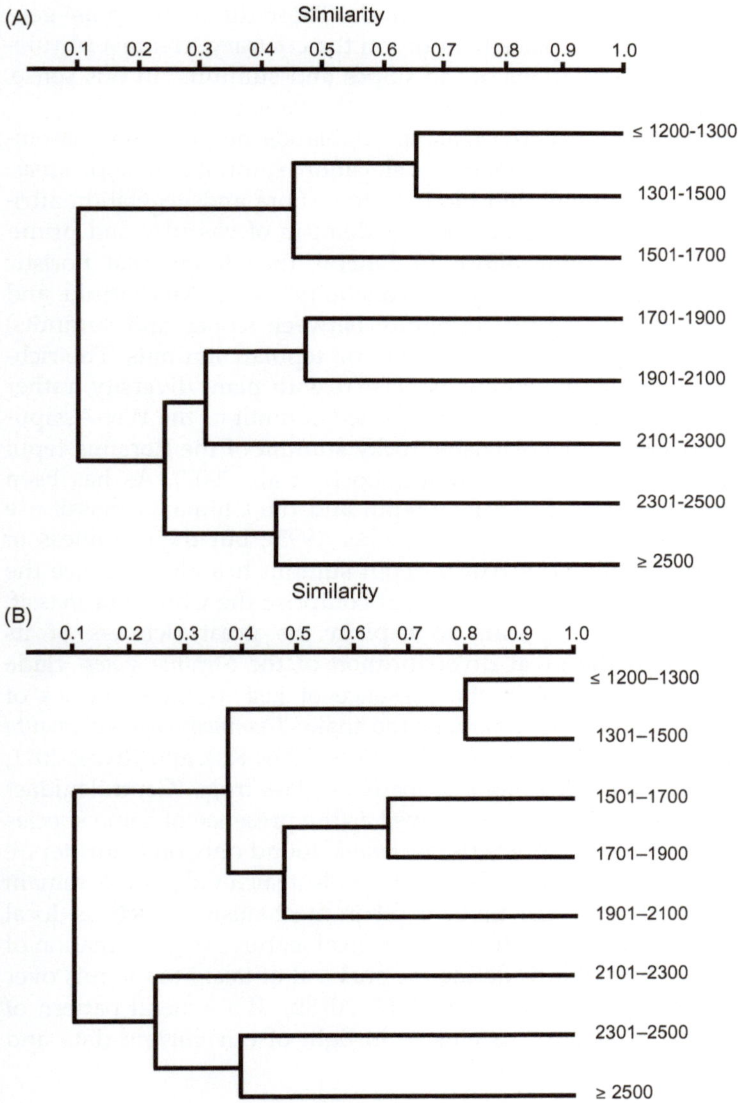

FIGURE 12.8 Similarity dendrogram (Jaccard's coefficient) for 200-m altitudinal belts based on the composition of their amphibian (A) and reptile (B) assemblages.

similarity and are concatenated with the successively higher altitudinal assemblages with which they share less than 40% of the species (Fig. 12.8A and B).

The topography of the Guiana Highlands is extremely variable (Chapter 1). A few tepuis have vertical escarpments that can be 1000 m high, which makes their summits totally isolated horizontally from one another, but also vertically from their surrounding slopes (e.g., Roraima-tepui, Kukenán-tepui, Ptari-tepui). In contrast, the majority of the tepuis and mountains of the Guiana Highlands have more or less continuous slopes without a clear delimitation between them and the summit and, hence, a connection exists between the uplands/lowlands, the slopes, and the highlands. These different tepuis' geomorphologies can allow a gradual habitat replacement or, on the contrary, create a physiographic isolation between species or populations of the slopes and summits. In this sense, isolation alone seems insufficient to explain the endemism of the Pantepui biota.

McDiarmid and Donnelly (2005) examined the Guiana Highlands herpetofauna assemblages, taking into account some abiotic factors (maximum elevation, summit and slope areas, and distance to the nearest neighboring tepui) and biotic factors (flora and vegetation attributes, including number of forest types and richness and endemism of vascular and primitive plants), as well as an index of sampling effort. In general, they found that floristic diversity, slope and summit area, and connectivity ("accessibility" *sensu* McDiarmid and Donnelly, 2005, indicating the geomorphological continuity between slopes and summits) positively influenced amphibian and reptile species' richness on tepuian summits. The richness of similar elevations appears to be more closely associated with plant diversity, rather than summit area. For example, the herpetofauna of the forested summit of the Wei-Assipu-tepui (3 km^2 in area) is twice as diverse as the extensive rocky summit of the Roraima-tepui (34 km^2) (Villareal et al., 2003; Señaris et al., 2005; MacCulloch et al., 2007). As has been pointed out by Myers and Donnelly (2008), the Auyán-tepui and the Chimantá massif are comparable in size (667 km^2 and 615 km^2, respectively, Huber 1995), but their richness in terms of herpetofauna differs substantially. The Auyán-tepui summit has almost twice the number of herpetofauna species than in the *c*. 10 tepuis that comprise the Chimantá massif. The complexity of the Chimantá massif appears to explain the great richness of its flora (Huber 1992, 1995), and perhaps the great diversification of the *Stefania ginesi* clade (Kok et al., 2017) as well, but does not explain the presence of just a single species of *Tepuihyla* (*T. obscura*), the lizard *Oreosaurus mcdiarmidi*, or the snake *Thamnodynastes chimanta* on most of its tepui summits (Gorzula, 1992; Gorzula and Señaris, 1999; Kok and Rivas, 2011; Kok et al., 2015). Likewise, the absence of *Oreophrynella* toads or glass frogs (Centrolenidae) on Chimantá and its high diversity on nearby tepuis, as well as the presence of some species on single summits of the same massif (e.g., *Pristimantis muchimuk*, found only on Churí-tepui, and *Pristimantis abakapa*, recorded only on Abakapá-tepui Rojas-Runjaic et al., 2013), remain unexplained. These differences must therefore involve other mechanisms, such as local extinctions, the occurrence of microrefugia, or particular ecological features (e.g., formation of biophysical habitats, species interactions), which define the survival of taxa atop tepuis over time (Myers and Donnelly, 2008; Rull, 2009; Kok et al., 2017, 2018b). If a general pattern of species composition exists in Pantepui, it remains elusive in light of our current data and knowledge.

Insights on the origin and evolution of the Pantepui herpetofauna

Several hypotheses have been proposed to explain the origin of the Pantepui biota (Chapman, 1931; Tate, 1939; Mayr and Phelps, 1967; Maguire, 1970; Cook, 1974; Haffer, 1974; Chapter 4: Origin and evolution of the Pantepui biota). All of them were reviewed by Hoogmoed (1979), Gorzula (1992), and McDiarmid and Donnelly (2005) from a herpetological perspective using distributional data of amphibians and reptiles. In some studies involving Pantepui taxa, these hypotheses have been suggested as an explanation of the distribution pattern of the species [e.g., the Plateau Theory for the genus *Stefania* by Duellman and Hoogmoed (1984), or for *Oreophrynella, Metaphryniscus,* and *Riolama* by McDiarmid and Donnelly (2005)]. Other theories—such as the Distance Dispersal Theory, the Specialized Habitat Theory, and the Mountain Bridge hypotheses—were discarded as possible explanations for the origin of the herpetofauna due to the impossibility of aerial dispersion, since there are no flying amphibians or reptiles, or due to the lack of geological evidence of a physical connection between the Andes and Pantepui (Hoogmoed, 1979; McDiarmid and Donnelly, 2005). Gorzula (1992) expressed the difference between long-distance dispersal and migration for the Pantepuian amphibians and reptiles by pointing out that "[b]irds can fly and seeds can float/and biogeographers can talk/but to get from one tepui to another/the herpetofauna has to walk." With paleoecological and some molecular phylogenetic evidence, Rull (2005, 2010) reviewed extensively these seminal hypotheses and concluded that both vicariance and biotic interchange (migration and dispersal) have could have contributed to the shaping of present-day biotic Pantepui assemblages.

Recent molecular studies continue to discover an increasing number of Pantepui and Guiana amphibian and reptile species, but the samples on which they are based are mostly limited to the Eastern Guiana Province, highlighting the almost total absence of collections/taxa from the Western Guiana tepui summits. Pantepui-endemic amphibian genera such as *Metaphryniscus* (Bufonidae) and *Dischidodactylus* (Strabomantidae s.f. *incertae sedis*) from the Duida–Marahuaka massif, or the lizard genus *Adercosaurus* (Gymnophthalmidae) from Cerro Yutajé, have never been included in any molecular study. This is also the case of most western Pantepui species of the frog genera *"Cochranella," Myersohyla, Stefania,* and *Tepuihyla* and lizards of *Riolama* and *Cercosaura*.

Contrasting origins of the Pantepui herpetofauna emerge from phylogenetic studies. Heinicke et al. (2018) supported an earliest (Paleocene?) divergence of an ancestral lineage of the genus *Ceuthomantis* from that of all other terraranan—a frog clade comprising nearly 15% of all amphibians—reinforcing a South American (in the case of the Guiana Highlands) origin for New World's direct-developing frogs. Previously, Heinecke et al. (2009) referred to *Ceuthomantis* and other early-branching Guiana Highlands lineages such as *Myersohyla, Stefania,* and *Minyobates*—following the phylogenetic results of Grant et al. (2006), Faivovich et al. (2005), and Wiens et al. (2007)—as "living fossils" and the older Guiana Shield as an evolutionary refuge for them. Kok et al. (2017) reviewed the phylogenetic relationships of *Stefania* species from Eastern Guiana based on a taxon/population sampling that was broader than all other previous studies (e.g., Kok et al., 2012, 2016; Salerno et al., 2012; Castroviejo-Fisher et al., 2015; Duellman, 2015). These studies pointed to a Guiana Highland origin for the genus, with an onset of diversification at *c.* 26 Ma (Oligocene), followed by vicariant events—but also dispersal—between *c.* 22.3 and 14.5 Ma (Miocene) and, finally, recent up-and-down dispersal events during the Pleistocene following glacial–interglacial cycles. This complex diversification scenario would support the Plateau Theory for the early origin of some Pantepui amphibians, but also partially the Vertical Displacement Hypothesis (Rull, 2004) for more recent Pleistocene migrations.

As has been illustrated by several phylogenetic analyses (e.g., Faivovich et al., 2005, 2013; Wiens et al., 2010; Duellman et al., 2016; Pinheiro et al., 2018) *Myersohyla* appears as a basal clade of Cophomantini (Cophomantinae *sensu* Duellman et al., 2016), which may imply a Guiana Highlands origin of earlier divergence for these hylid frog groups.

To resolve the paraphyly of *Myersiohyla* Pinheiro et al. (2018) described the genus *Nesorohyla* for *Hyla kanaima* (Goin and Woodley, 1969) and transferred *Hypsiboas liliae* (Kok, 2006) to *Myersiohyla*. Thus one or two independent origins of the common ancestor of Cophomantini are possible, but always centered on Pantepui. To resolve this puzzle, molecular data of *Myersiohyla aromatica*, *M. inparquesi*, and *M. loveridgei* from Western tepui summits—where this genus is especially diverse—are required.

The recent molecular phylogeny of *Oreophrynella* by Kok et al. (2018b) is exceptional since it includes all (nine) known species in the genus and shows, for first time, a likely Andean origin for a Pantepui-endemic anuran genus. This work also reinforces an Andean origin for the Guiana lowland species of *Atelopus*, as formerly suggested by Noonan and Gaucher (2005). Kok et al. (2018b) biogeographical analyses proposed a colonization of Pantepui by long-distance dispersal of a proto-Andean bufonid ancestor during the Eocene/Oligocene (pre-Miocene), followed by five vicariant diversification events of the main lineages during the Miocene (between 21.73 and 6.84 Ma), probably caused by erosional dissection of the Roraima plateau. Finally, and based on the low genetic diversity among extant summit *Oreophrynella* populations, the same authors suggested a drastic reorganization during the Pleistocene glacial cycles. The remarkable absence of *Oreophrynella* in the huge Chimantá massif is interpreted as a massive local extinction caused by the Pleistocene climatic instability, paralleling the concept of "glacial ghosts" (i.e.,unknown extinct species) of Rull (2005). Except for their origin, the early historical biogeography of *Oreophrynella* follows a similar pattern to *Stefania* (Kok et al., 2017); another remarkable difference between these taxa is the great radiation of *Stefania* in the Eastern Guiana Highlands but especially in the Chimantá massif and peripheral tepuis, where at least seven microendemisms occur (Kok et al., 2017). Most Western tepui summits are inhabited by the endemic genus *Metaphryniscus*: *Metaphryniscus sosae* from Cerro Marahuaca and may be the unidentified bufonids from Neblina cited in McDiarmid and Paolillo 1988; this taxon may represent a current "relict" of an early diversification of *Oreophrynella*, based on its intermediate osteological characteristics with the Andean genus *Osornophryne*, as had been suggested by Señaris et al. (1994).

In contrast to the previously mentioned Pantepui-endemic anuran genera, *Tepuihyla* species are less divergent genetically than previously thought, suggesting that summit populations have evolved in isolation only very recently (3.2–5.3 Ma) from a common ancestor widespread in the Guiana lowlands (Kok et al., 2012; Salerno et al., 2012, 2015; Jungfer et al., 2013). Also, the genetic results could be explained by genetic introgression or incomplete lineage sorting (Salerno et al., 2015) driven by climate change and shifting habitats as in the diversification scenario proposed by Rull (2005).

Neusticurus is an aquatic lizard group with a distribution centered on the Guiana Shield, but with one species distributed in southeastern Amazonia. Based on molecular and morphological data, Marques-Souza et al. (2018) suggested that the genus started to diversify on "Pantepui" and dispersed later to the lowlands towards the west (*Neusticurus medemi*) and east (*N. surinamensis* and *N. bicarinatus*). However, the area referred to as "Pantepui" by Marques-Souza et al. (2018) emcompassed both lowlands and uplands, and Kok et al. (2018a) invoked the Habitat Shift Theory to explain the presence (due to colonization) of *Neusticurus* on tepui summits. None of these lizards is endemic to the Guiana Highlands, and most of them have a broad elevational range.

Although our understanding of the origin and biogeography of Pantepui fauna is growing rapidly, we are still far from having a complete view and suspect that many other taxonomic changes will occur and affect the biogeographic inference as more species are incorporated into these molecular analyses (e.g., Jungfer et al., 2013; Pinheiro et al., 2018; Pellegrino et al., 2018). Sometimes these studies have had unexpected results and have brought more confusion than clarity. For example, Ron et al. (2016) reassigned the enigmatic *Hyla tuberculosa* Boulenger, 1882 to *Tepuihyla* and described *T. shushupe* from Ecuador and Peru, two frogs morphologically and ecologically strikingly different from the rest of the species of the genus, and extended their range far outside the Guiana Shield.

There is little doubt that the current distribution of amphibian and reptile diversity in the Guiana Highlands resulted from a complex interplay of different diversification mechanisms across time (Rull, 2005, 2008, 2011a,b; Kok, 2013; see also Chapter 4), as occurs in other areas of the South American tropics (e.g., Bush, 1994; Noonan and Wray, 2006; Santos et al., 2009; Arteaga et al., 2016; Vacher et al., 2017). No single hypothesis explains adequately the diversification of broad taxonomic groups, because each individual clade has a particular life history and evolution. A "one-size-fits-all" model seems not to exist (Brown et al., 2014; Rull, 2018).

Final remarks

The diversity of amphibians and reptiles of the Guiana Highlands is still underestimated, and knowledge on life-history traits, ecology, and distribution patterns of most taxa inhabiting the tepui summits is almost inexistent. However, it seems that this region has been a "melting pot" for the diversification of the northern South American herpetofauna, including pre-Quaternary vicariance, reorganization of species diversity constrained by climatic instability, and recent invasions (post-Pleistocene) of widespread upland taxa (Rull, 2005; Kok 2013). The increasing amount of phylogenetic studies on Amazonian and Guianan amphibians and reptiles reveals a huge cryptic richness in northern South America, resulting in an increase in new species (i.e., richness) and in new data about already known taxa (e.g., distribution), but also leading to frequent, sometimes ephemeral, taxonomic changes or implying an artificial diversity increase (i.e., "taxonomic inflation" s.l. Padial and De la Riva, 2006; Dubois, 2008). In this sense a collective and multidisciplinary effort from researchers studying the Guiana Highlands biota is needed, starting with the revision and identification of amphibian and reptile samples collected decades ago in the western Venezuelan tepui summits.

Currently, we have plenty of hypotheses about the origin and mechanisms of diversification, but we need more samples to test them. Therefore, field exploration is a priority. Almost 20 years ago, Myers and Donnelly (2001) pointed out that "...additional faunal sampling is needed much more than additional hypothesizing." Unfortunately, in the last two decades Venezuela (where the vast majority of the Guiana Highlands are) has faced a political and economic crisis, and fieldwork has notably decreased (and virtually stopped in some cases) (see Chapter 16). However, for the time being, field explorations can be intensified in the Guiana Highlands of Guyana and Brazil until the appropriate conditions in Venezuela return.

The assemblages of herpetological communities in tepui summits are unique in terms of richness, composition, and endemism, and therefore each one is an essential unit of conservation. The protection of the Guiana Highlands through legal declaration of national parks or natural monuments is not enough to confront the current threats and future scenarios of climate change (Chapter 17: Pantepui and global warmnig). New and creative approaches must be developed soon (e.g., development of models to predict the impact of climate change on herpetofauna communities under different scenarios, identify areas at higher risk and areas of climate refugia for pantepuian amphibians and reptiles, and formulate specific and active management actions for the most threatened species/assemblages *ex situ* conservation plans) (see Chapter 17).

Acknowledgments

We are very grateful to Stefan Gorzula and Donald C. Taphorn for the critical review of the first draft of this manuscript. They provided valuable comments and suggestions that improved and enriched the original text. Grisel Velázquez (UniSIG, IVIC) estimated the area of the different elevational belts and helped us with cartographic issues.

References

Arteaga, A., Pyron, R.A., Peñafiel, N., Romero-Barreto, P., Culebras, J., Bustamante, L., et al., 2016. Comparative phylogeography reveals cryptic diversity and repeated patterns of cladogenesis for amphibians and reptiles in Northwestern Ecuador. PLoS One 11 (4), e0151746.

Aubrecht, R., Barrio-Amorós, C.L., Breure, A.S.H., Brewer-Carías, C., Derka, T., Fuentes-Ramos, O.A., et al., 2012. Venezuelan tepuis: their caves and biota. Acta Geologica Slovaca. Comenius University, Bratislava.

Ávila-Pires, T.C.S., 2005. Reptiles. In: Hollowell T., Reynolds, R.P. (Eds.), Checklist of the Terrestrial Vertebrates of the Guiana Shield. Bulletin of the Biological Society of Washington, No. 13, pp. 25–40.

Ávila-Pires, T.C.S., Hoogmoed, M.S., Silva, M.B., 2018. First records in Brazil for *Arthrosaura versteegii* van Lidth de Jeude, 1904 (Reptilia, Gymnophthalmidae), a possible species complex, and distribution extension for *Arthrosaura montigena* Myers & Donnelly, 2008, in Venezuela. Check List 14 (4), 569–577.

Ayarzagüena, J., Señaris, J.C., 1994. Dos nuevas especies de Hyla (Anura; Hylidae) para las cumbres tepuyanas del Estado Amazonas, Venezuela. Mem. Soc Cienc. Nat. La Salle 53 (139), 127–146.

Ayarzagüena, J., Señaris, J.C., Gorzula, S., 1993. El grupo *Osteocephalus rodriguezi* de las tierras altas de la Guayana venezolana: descripción de cinco nuevas especies. Mem. Soc. Cienc. Nat. La Salle 52, 213–221.

Barrio-Amorós, C.L., Brewer-Carías, C., 2008. Herpetological results of the 2002 expedition to Sarisariñama, a tepui in Venezuelan Guayana, with the description of five new species. Zootaxa 1942, 1–68.

Barrio-Amorós, C.L., Duellman, W.E., 2009. Herpetofauna de la Sierra de Lema, Estado Bolívar, Venezuela. In: Lasso, C.A., Señaris, J.C., Flores, A.L., et al.,Evaluación Rápida de la Biodiversidad de los Ecosistemas Acuáticos del Alto Río Cuyuní, Guayana Venezolana. RAP Bull. Biol. Assessment, vol. 55. Conservation Internacional, Arligton, pp. 137–155.

Bernal, M.H., Lynch, J.D., 2008. Review and analysis of altitudinal distribution of the Andean anurans in Colombia. Zootaxa 1826, 1–25.

Berry, P.E., Riina, R., 2005. Insights into the diversity of the Pantepui flora and the biogeographic complexity of the Guayana Shield. Biol. Skr. 55, 145–167.

Berry, P., Huber, O., Holst, B.K., 1995. Floristic analysis and phytogeography. In: Berry, P.E., Holst, B.K., Yatskievych, K. (Eds.), Flora of the Venezuelan Guayana, vol. 1, Introduction. Missouri Botanical Garden Press, St. Louis, pp. 161–191.

Bifano, C., 2016. Venezuelan science under siege. Nature 535, 336–337.

Blaustein, A.R., Walls, S.C., Bancroft, B.A., Lawler, J.J., Searle, C.L., Gervasi, S.S., 2010. Direct and indirect effects of climate change on amphibian populations. Diversity 2, 281–313.

Boulenger, G.A., 1895. Description of a new batrachian (Oreophryne Quelchii) discovered by Messrs. J. J. Quelch and F. McConnell on the summit of Mount Roraima. Annals Mag. Nat. Hist. Ser. 6 (15), 521–522.

Boulenger, G.A., 1900. Reptiles and batrachians. In: Lankester, E.R. (Ed.), Report on a Collection Bade by Messrs. F. V. McConnell and J. J. Quelch at Mount Roraima in British Guiana. Reptiles. Transactions of the Linnean Society of London, Zoology, London, ser. 2, 8(2), pp. 53–56, pl. 5.

Brewer-Carías, C. (Ed.), 1988. Cerro La Neblina: resultados de la expedición 1983–1987. Fudeci y Editorial Sucre, Caracas.

Brown, J.L., Cameron, A., Yoder, A.D., Vences, M., 2014. A necessarily complex model to explain the biogeography of the amphibians and reptiles of Madagascar. Nat. Commun. 5, 5046.

Bush, M.B., 1994. Amazonian speciation: a necessarily complex model. J. Biogeogr. 21, 5–17.

Caramaschi, U., de Niemeyer, H., 2005. Geographic distribution: *Eleutherodactylus cavernibardus*. Herpetol. Rev. 36, 199.

Carvalho, V.T., MacCulloch, R.D., Bonora, L., Vogt, R.C., 2010. New species of *Stefania* (Anura: Cryptobatrachidae) from Northern Amazonas, Brazil. J. Herpetol. 44, 229–235.

Castroviejo-Fischer, S., Vilá, C., Ayarzagüena, J., Blanc, M., Ernst, R., 2011. Species diversity of *Hyalinobatrachium* glassfrogs (Amphibia: Centrolenidae) from the Guiana Shield, with the description of two new species. Zootaxa 3132, 1–55.

Castroviejo-Fisher, S., Padial, J.M., De la Riva, I., Pombal, J.P., Da Silva, H.R., Rojas-Runjaic, F.J.M., et al., 2015. Phylogenetic systematics of egg-brooding frogs (Anura: Hemiphractidae) and the evolution of direct development. Zootaxa 4004, 1–75.

Chapman, F.M., 1931. The upper zonal bird-life of Mts. Roraima and Duida. Bull. Am. Mus. Nat. Hist. 63, 1–135.

Chettri, B., Bhupathy, S., Acharya, B.K., 2010. Distribution pattern of reptiles along an eastern Himalayan elevation gradient, India. Acta Oecol. 36, 16–22.

Cole, C.J., Townsend, C.R., Reynolds, R.P., MacCulloch, R.D., Lathrop, A., 2013. Amphibians and reptiles of Guyana, South America: illustrated keys, annotated species accounts, and a biogeographic synopsis. Proc. Biol. Soc. Wash. 125, 317–578.

Conservation International, 2003. Conservation Priorities for the Guayana Shield: 2002 Consensus. Center for Applied Biodiversity Science, CI, Washington, DC.

Cook, R.E., 1974. Origin of the highland avifauna of southern Venezuela. Syst. Zool. 23, 257–264.

Donnelly, M.A., Myers, C.W., 1991. Herpetological results of the 1990 Venezuelan Expedition to the summit of Cerro Guaiquinima, with new tepui reptiles. Am. Mus. Novit. 3017, 1–54.

Dubois, A., 2008. A partial but radical solution to the problem of nomenclatural taxonomic inflation. Biol. J. Linn. Soc. 93, 857–863.

Duellman, W.E., 1997. Amphibians of La Escalera region, southeastern Venezuela: taxonomy, ecology, and biogeography. Sci. Pap. Nat. Hist. Mus. Univ. Kansas 2, 1–52.

Duellman, W.E., 1999. Distribution patterns of amphibians in South America. In: Duellman, W.E. (Ed.), Patterns of Distribution of Amphibians. The Johns Hopkins University Press, Baltimore and London, pp. 255–327.

Duellman, W.E., 2015. Marsupial Frogs, *Gastrotheca* and Allied Genera. Johns Hopkins University Press, Baltimore, Maryland, pp. xv–407.

Duellman, W.E., Hoogmoed, M.S., 1984. The taxonomy and phylogenetic relationships of the hylid frog genus *Stefania*. Misc. Publ. Univ. Kans. Mus. Nat. Hist. 75, 1–39.

Duellman, W.E., Marion, A.B., Hedges, S.B., 2016. Phylogenetics, classification, and biogeography of the treefrogs (Amphibia: Anura: Arboranae). Zootaxa 4104, 1–109.

Faivovich, J., Haddad, C.F.B., García, P.C.A., Frost, D.R., Campbell, J.A., Wheeler, W.C., 2005. Systematic review of the frog family Hylidae, with special reference to Hylinae: phylogenetic analysis and taxonomic revision. Bull. Am. Mus. Nat. Hist. 294, 1–240.

Faivovich, J., McDiarmid, R.W., Myers, C.W., 2013. Two new species of *Myersiohyla* (Anura: Hylidae) from Cerro de la Neblina, Venezuela, with comments on other species of the genus. Am. Mus. Novit. 3792, 1–63.

Foden, W.B., Butchart, S.H.M., Stuart, S.N., Vié, J.C., Akcakaya, H.R., Angulo, A., et al., 2013. Identifying the World's most climate change vulnerable species: a systematic trait-based assessment of all birds, amphibians and corals. PLoS ONE 8, e65427.

Fouquet, A., Cassini, C.C., Haddad, C.F.B., Pech, N., Rodrigues, M.T., 2014. Species delimitation, patterns of diversification and historical biogeography of the Neotropical frog genus *Adenomera* (Anura, Leptodactylidae). J. Biogeogr. 41, 855–870.

Fouquet, A., Souza, S.M., Nunes, P.M.S., Kok, P.J.R., Curcio, F.F., Carvalho, C.M., et al., 2015. Two new endangered species of *Anomaloglossus* (Anura: Aromobatidae) from Roraima State, Northern Brazil. Zootaxa 3926, 191–210.

Fouquet, A., Martinez, Q., Zeidler, L., Courtois, E.A., Gaucher, P., Blanc, M., et al., 2016. Cryptic diversity in the *Hypsiboas semilineatus* species group (Amphibia, Anura) with the description of a new species from the eastern Guiana Shield. Zootaxa 4084, 079–104.

Fouquet, A., Vacher, J.P., Courtois, E.A., Villette, H.R., Gaucher, P., Jairam, R., 2018. On the brink of extinction: two new species of *Anomaloglossus* from French Guiana and amended definitions of *Anomaloglossus degranvillei* and *A. surinamensis* (Anura: Aromobatidae). Zootaxa 4379, 1–23.

Fraga, R., Almeida, A.P., Moraes, L.J.C.L., Gordo, M., Pirani, R., Zamora, R.R., et al., 2017. Narrow endemism or insufficient sampling? Geographical range extension and morphological variation of the poorly known *Atractus riveroi* Roze, 1961 (Serpentes: Dipsadidae). Herpetol. Rev. 48, 281–284.

Fu, C., Wang, J., Pu, Z., Zhang, S., Chen, H., Zhao, B., et al., 2007. Elevational gradients of diversity for lizards and snakes in the Hengduan Mountains, China. Biodivers. Conserv. 16, 707–726.

Gehara, M.C.M., Crawford, A.J., Orrico, V.G.D., Rodríguez, A., Lötters, S., Fouquet, A., et al., 2014. High levels of diversity uncovered in a widespread nominal taxon: continental phylogeography of the Neotropical tree frog *Dendropsophus minutus*. PloS ONE 9, e103958.

Goicoechea, N., Frost, D.R., De la Riva, I., Pellegrino, K.C.M., Sites, J.J., Rodrigues, M.T., et al., 2016. Molecular systematics of teioid lizards (Teioidea/Gymnophthalmoidea: Squamata) based on the analysis of 48 loci under tree-alignment and similarity alignment. Cladistics 32, 1–48.

Gorzula, S., 1992. La herpetofauna del macizo del Chimantá. In: Huber, O., (Ed.), El macizo del Chimantá Escudo de Guayana, Venezuela. Un ensayo ecológico tepuyano. Oscar Todtmann Editores, Caracas, pp. 267–280 and 304–310, photos 152–171.

Gorzula, S., Señaris, J.C., 1999. Contribution to the herpetofauna of the Venezuelan Guayana I. A data base. Scientia Guaianae 8, xviii + 270 + 32.

Graham, C.H., Moritz, C., Williams, S.E., 2006. Habitat history improves prediction of biodiversity in rainforest fauna. Proc. Natl. Acad. Sci. 103 (3), 632–636.

Grant, T., Frost, D.R., Caldwell, J.P., Gagliardo, R., Haddad, C.F., Kok, P.J.R., et al., 2006. Phylogenetic systematics of dart-poison frogs and their relatives (Amphibia: Athesphatanura: Dendrobatidae). Bull. Am. Mus. Nat. Hist. 299, 1–262.

Grant, T., Rada, M., Anganoy-Criollo, M., Batista, A., Dias, P.E., Jeckel, A.M., et al., 2017. Phylogenetic systematics of dart-poison frogs and their relatives revisited (Anura: Dendrobatoidea). South Am. J. Herpetol. 12 (Special Issue 1), S1–S90.

Guayasamin, J.M., Castroviejo-Fisher, S., Trueb, L., Ayarzagüena, J., Rada, M., Vilá, C., et al., 2009. Phylogenetic systematics of Glassfrogs (Amphibia: Centrolenidae) and their sister taxon *Allophryne ruthveni*. Zootaxa 2100, 1–97.

Haffer, J., 1974. Avian Speciation in Tropical South America. Publ. Nuttall Ornithol. Club No. 14, Cambridge.

Heatwole, H., 1982. A review of structuring in herpetofaunal assemblages. In: Scott, N.J. (Ed.), Herpetological Communities. U.S. Department of the Interior Fish and Wildlife Service.

Heinicke, M.P., Duellman, W.E., Trueb, L., Means, B., MacCulloch, R.D., Hedges, S.B., 2009. A new frog family (Anura: Terrarana) from South America and an expanded direct developing clade revealed by molecular phylogeny. Zootaxa 2211, 1–35.

Heinicke, M.P., Lemmon, A.R., Lemmon, E.M., McGrath, K., Hedges, S.B., 2018. Phylogenomic support for evolutionary relationships of New World direct-developing frogs (Anura: Terraranae). Mol. Phylogenet. Evol. 118, 145–155.

Hollowell, R., Reynolds, R.P., 2005. Introduction. In: Hollowell, T., Reynolds, R.P. (Eds.), Checklist of the terrestrial vertebrates of the Guiana Shield. Bull. Biol. Soc. Wash. 13, pp. 1–6.

Hoogmoed, M.S., 1979. The Herpetofauna of the Guianan Region. In: Duellman, W.E. (Ed.), The South American Herpetofauna: Its Origin, Evolution and Dispersal. Monograph of the Museum of Natural History, University of Kansas, Kansas, pp. 241–279.

Huber, O., 1987. Consideraciones sobre el concepto de Pantepui. Pantepui 2, 2–10.

Huber, O., (Ed.), 1992. El macizo del Chimantá, un ensayo ecológico tepuyano. Oscar Todtmann Eds., Caracas.

Huber, O., 1995. Geographical and physical features. In: Berry, P., Holst, B.K., Yaskievych, K. (Eds.), Flora of the Venezuelan Guayana. Introduction, vol. 1. Missouri Botanical Garden Press, St. Louis, pp. 1–61.

Huey, R.B., Kearney, M.R., Krockenberger, A., Holtum, J.A.M., Jess, M., Williams, S.E., 2012. Predicting organismal vulnerability to climate warming: roles of behaviour, physiology and adaptation. Philos. Trans. R. Soc. B 367, 1665–1679.

Jungfer, K.H., Faivovich, J., Padial, J.M., Castroviejo-Fisher, S., Lyra, M.M., Berneck, B.V.M., et al., 2013. Systematics of spiny-backed treefrogs (Hylidae: *Osteocephalus*): an Amazonian puzzle. Zool. Scr. 42, 351–380.

Kok, P.J.R., 2008. A new highland species of *Arthrosaura* Boulenger, 1885 (Squamata: Gymnophthalmidae) from Maringma tepui on the border of Guyana and Brazil. Zootaxa 1909, 1–15.

Kok, P.J.R., 2009. Lizard in the clouds: a new highland genus and species of Gymnophthalmidae (Reptilia: *Squamata*) from Maringma tepui, western Guyana. Zootaxa 1992, 53–67.

Kok, P.J.R., 2010. A new species of *Chironius* Fitzinger, 1826 (Squamata: Colubridae) from the Pantepui region, northeastern South America. Zootaxa 2611, 31–44.

Kok, P.J.R., 2013. Islands in the sky: species diversity, evolutionary history, and patterns of en-demism of the Pantepui Herpetofauna (PhD thesis). Leiden University, The Netherlands.

Kok, P.J.R., 2015. A new species of the Pantepui endemic genus *Riolama* (Squamata: Gymnophthalmidae) from the summit of Murisipán-tepui, with the erection of a new gymnophthalmid subfamily. Zool. J. Linn. Soc. 174, 500–518.

Kok, P.J.R., Rivas, G.A., 2011. A new species of *Anadia* (Reptilia, *Squamata*) from the Venezuelan 'Lost World', northern South America. Eur. J. Taxon. 3, 1–18.

Kok, P.J.R., Barrio-Amorós, C.L., 2013. On the taxonomic validity of *Pristimantis tepuiensis* (Schlüter & Rödder, 2007) and *P. stegolepis* (Schlüter & Rödder, 2007), with remarks on the type series of *P. guaiquinimensis* (Schlüter & Rödder, 2007). Zootaxa 3694, 75–80.

Kok, P.J.R., MacCulloch, R.D., Means, D.B., Roelants, K., Van Bocxlaer, I., Bossuyt, F., 2012. Low genetic diversity in tepui summit vertebrates. Curr. Biol. 22, R589–R590.

Kok, P.J.R., Ratz, S., Tagelaar, M., Aubret, F., Means, D.B., 2015. Out of taxonomic limbo: a name for the species of *Tepuihyla* (Anura: Hylidae) from the Chimantá Massif, Pantepui region, northern South America. Salamandra 51, 283–314.

Kok, P.J.R., Russo, V.G., Ratz, S., Aubret, F., 2016. On the distribution and conservation of two "Lost World" tepui summit endemic frogs, *Stefania ginesi* Rivero, 1968 and *S. satelles* Señaris, Ayarzagüena and Gorzula, 1997. Amphib. Reptile Conserv. 10 (1), 5–12.

Kok, P.J.R., Russo, V.G., Ratz, S., Means, D.B., MacCulloch, R.D., Lathrop, A., et al., 2017. Evolution in the South American 'Lost World': insights from multilocus phylogeography of *Stefanias* (Anura, Hemiphractidae, *Stefania*). J. Biogeogr. 44, 170–181.

Kok, P.J.R., Bittenbinder, M.A., van den Berg, J.K., Marques-Souza, S., Nunes, P.M.S., et al., 2018a. Integrative taxonomy of the gymnophthalmid lizard *Neusticurus rudis* Boulenger, 1900 identifies a new species in the eastern Pantepui region, north-eastern South America. J. Nat. Hist. 52. Available from: https://doi.org/10.1080/00222933.2018.1439541.

Kok, P.J.R., Ratz, S., MacCulloch, R.D., Lathrop, A., Dezfoulian, R., Aubret, F., et al., 2018b. Historical biogeography of the palaeoendemic toad genus *Oreophrynella* (Amphibia: Bufonidae) sheds a new light on the origin of the Pantepui endemic terrestrial biota. J. Biogeogr. 45, 26–36.

Lathrop, A., MacCulloch, R.D., 2007. A new species of *Oreophrynella* (Anura: Bufonidae) from Mount Ayanganna, Guyana. Herpetologica 63, 87–93.

Lindorf, H., 2006. La expedición universitaria a la meseta Auyan-tepui, abril 1956. Act. Bot. Venezuelica 29 (1), 177–188.

Lomolino, M.V., 2001. Elevation gradients of species—density: historical and prospective views. Glob. Ecol. Biogeogr. 10, 3–13.

MacCulloch, R.D., Lathrop, A., 2001. A new species of *Arthrosaura* (Sauria: Teiidae) from the highlands of Guyana. Caribb. J. Sci. 37, 174–181.

MacCulloch, R.D., Lathrop, A., 2002. Exceptional diversity of Stefania (Anura: Hylidae) on Mount Ayanganna, Guyana: three new species and new distribution records. Herpetologica 58, 327–346.

MacCulloch, R.D., Lathrop, A., 2005. Hylid frogs from Mount Ayanganna, Guyana: new species, redescriptions, and distributional records. Phyllomedusa 4, 17–37.

MacCulloch, R.D., Lathrop, A., 2009. Herpetofauna of Mount Ayanganna, Guyana: results of the Royal Ontario Museum Ayanganna Expedition 2000. Royal Ontario Mus. Cont. Sci. 4, 1–35.

MacCulloch, R.D., Lathrop, A., Khan, S.Z., 2006. Exceptional diversity of *Stefania* (Cryptobatrachidae) II: six species from Mount Wokomung, Guyana. Phyllomedusa 5, 31–41.

MacCulloch, R.D., Lathrop, A., Reynolds, R.P., Señaris, J.C., Schneider, G.E., 2007. Herpetofauna of Mount Roraima, Guiana Shield Region, northeastern South America. Herpetol. Rev. 38, 24–30.

MacCulloch, R.D., Lathrop, A., Kok, P.J.R., Minter, L.R., Khan, S.Z., Barrio-Amorós, C.L., 2008a. A new species of *Adelophryne* (Anura: Eleutherodactylidae) from Guyana, with additional data on *A. gutturosa*. Zootaxa 1884, 36–50.

MacCulloch, R.D., Lathrop, A., Minter, L.R., Khan, S.Z., 2008b. *Otophryne* (Anura: Microhylidae) from the highlands of Guyana: redescriptions, vocalisations, tadpoles and new distributions. Pap. Avulsos Zool. (São Paulo) 48 (22), 247–261.

MacCulloch, R.D., Lathrop, A., Kok, P.J.R., Ernst, R., Kalamandeen, M., 2009. The genus *Oxyrhopus* (Serpentes: Dipsadidae: Xenodontinae) in Guyana: morphology, distributions and comments on taxonomy. Pap. Avulsos Zool. (São Paulo) 49 (36), 487–495.

Mägdefrau, H., 1991. Rufe, Paarung und Eiablage bei einer Froschart der Gattung *Stefania* (Rivero, 1966). Herpetofauna 13, 7–13.

Mägdefrau, H., Mägdefrau, K., Schlüter, A., 1991. Herpetologische Daten vom Guaiquinima-Tepui, Venezuela. Herpetofauna 13 (70), 13–26.

Maguire, B., 1970. On the flora of the Guayana Highland. Biotropica 2, 85–100.

Marques-Souza, S., Prates, I., Fouquet, A., Camacho, A., Kok, P.J.R., Nunes, P.M.S., et al., 2018. Reconquering the water: evolution and systematics of South and Central American aquatic lizards (Gymnophthalmidae). Zool. Scr. 47, 255–265.

Mayr, E., Phelps, W.H., 1967. The origin of the bird fauna of the South Venezuelan Highlands. Bull. Am. Mus. Nat. Hist. 136, 269–328.

McCain, C.M., 2010. Global analysis of reptile elevational diversity. Glob. Ecol. Biogeogr. 19, 541–553.

McDiarmid, R., Gorzula, S., 1989. Aspects of the reproductive ecology and behavior of the tepui toads, genus *Oreophrynella* (Anura, Bufonidae). Copeia 1989 (2), 445–451.

McDiarmid, R.W., Donnelly, M.A., 2005. The Herpetofauna of the Guayana Highlands: amphibians and reptiles of the Lost World. In: Donnelly, M.A., Crother, B.I., Guyer, C., Wake, M.H., White, M.E. (Eds.), Ecology and Evolution in the Tropics: A Herpetological Perspective. University of Chicago Press, Chicago, pp. 461–560.

McDiarmid, R.W., Paolillo, A., 1988. Herpetological collections—Cerro de la Neblina, updated January 1988. In: Brewer-Carías, C. (Ed.), Cerro de la Neblina. Resultados de la expedición 1983–1987. Fundación para el Desarrollo de las Ciencias Físicas, Matemáticas y Naturales, Caracas, pp. 667–670.

Means, D.B., Savage, J.M., 2007. Three New Malodorous Rainfrogs of the Genus *Pristimantis* (Anura: Brachycephalidae) from the Wokomung Massif in west-central Guyana, South America. Zootaxa 1658, 39–55.

Meza-Joya, F.L., Torres, M., 2016. Spatial diversity patterns of *Pristimantis* frogs in the Tropical Andes. Ecol. Evol. 6 (7), 1901–1913.

Michelangeli, A., 1989. Descripción del cerro Marahuaca. Introducción. Acta Terramaris 1, 1–3.

Michelangeli, A., 2005 (Ed.). Tepuy, Colosos de la tierra. Fundación Terramar, Altolitho C.A., Caracas.

Molina, C., Señaris, J.C., 2003. Una nueva especie del género *Riolama* (Reptilia; Squamata; Gymnophthalmidae) de las tierras altas de la Guayana Venezolana. Mem. Fund. La Salle Cienc. Nat. 155, 5–19.

Moraes, L.J.C.L., Almeida, A.P., Fraga, R., Rojas, R.R., Pirani, R.M., Silva, A.A.A., et al., 2017. Integrative overview of the herpetofauna from Serra da Mocidade, a granitic mountain range in northern Brazil. ZooKeys 715, 103–159.

Myers, C.W., 1997. Preliminary remarks on the summit herpetofauna of Auyantepui, eastern Venezuela. Acta Terramaris 10, 1–5.

Myers, C.W., 2000. A History of Herpetology at the American Museum of Natural History. Bull. Am. Mus. Nat. Hist. 252, 1–232.

Myers, C.W., Donnelly, M.A., 1996. A new herpetofauna from Cerro Yaví, Venezuela: first results of the Robert G. Goelet American Museum-Terramar Expedition to the northwestern tepuis. Am. Mus. Novit. 3171, 1–56.

Myers, C.W., Donnelly, M.A., 1997. A tepui herpetofauna on a granitic mountain (Tamacuri) in the borderland between Venezuela and Brazil: report from the Phipps Tapirapecó Expedition. Am. Mus. Novit. 3213, 1–71.

Myers, C.W., Donnelly, M.A., 2001. Herpetofauna of the Yutajé-Corocoro Massif, Venezuela: second report from the Robert G. Goelet American Museum-Terramar Expedition to the northwestern tepuis. Bull. Am. Mus. Nat. Hist. 261, 1–85.

Myers, C.W., Donnelly, M.A., 2008. The summit herpetofauna of Auyantepui, Venezuela: report from the Robert G. Goelet American Museum-Terramar Expedition, Bull. Am. Mus. Nat. Hist., 308 pp. 1–147.

Naka, L.N., 2011. Avian distribution patterns in the Guiana Shield: implications for the delimitation of Amazonian areas of endemism. J. Biogeogr. 38, 681–696.

Navas, C.A., 2003. Herpetological diversity along Andean elevational gradients: links with physiological ecology and evolutionary physiology. Comp. Biochem. Physiol. 133, 469–485.

Nogué, S., Rull, V., Vegas-Vilarrúbia, T., 2013. Elevational gradients in the neotropical table mountains: patterns of endemism and implications for conservation. Divers. Distrib. 19, 676–687.

Noonan, B.P., Gaucher, P., 2005. Phylogeography and demography of Guayanan harlequin toads (*Atelopus*): diversification within a refuge. Mol. Ecol. 14, 3017–3031.

Noonan, B.P., Wray, K.P., 2006. Neotropical diversification: the effects of a complex history on diversity within the poison frog genus *Dendrobates*. J. Biogeogr. 33, 1007–1020.

Orejas-Miranda, B., Quesada, A., 1976. Ecosistemas frágiles. Cienc. Interamericana 17 (1), 9–15.

Padial, J.M., De la Riva, I., 2006. Taxonomic inflation and the stability of species lists: the Perils of Ostrich's behavior. Syst. Biol. 55 (5), 859–867.

Passos, P., Kok, P.J.R., de Albuquerque, N.R., Rivas, G.A., 2013. Groundsnakes of the Lost World: a review of *Atractus* (Serpentes: Dipsadidae) from the Pantepui region, northern South America. Herpetol. Monogr. 27 (1), 52–86.

Pellegrino, K.C.M., Brunes, T.O., Souza, S.M., Laguna, M.M., Ávila-Pires, T.C.S., Hoogmoed, M.S., et al., 2018. On the distinctiveness of *Amapasaurus*, its relationship with *Loxopholis* Cope 1869, and description of a new genus for *L. guianensis* and *L. hoogmoedi* (Gymnophthalmoidea/Ecpleopodini: Squamata). Zootaxa 4441 (2), 332–346.

Pinheiro, P.D.P., Kok, P.J.R., Noonan, B.P., Means, D.B., Haddad, B.F.B., Faivovich, J., 2018. A new genus of Cophomantini, with comments on the taxonomic status of *Boana liliae* (Anura: Hylidae). Zool. J. Linn. Soc. Available from: https://doi.org/10.1093/zoolinnean/zly030.

Raxworthy, C.J., Pearson, R.G., Rabibisoa, N., Rakotondrazafy, A.M., Ramanamanjato, J., Raselimanana, A.P., et al., 2008. Extinction vulnerability of tropical montane endemism from warming and upslope displacement: a preliminary appraisal for the highest massif in Madagascar. Glob. Change Biol. 14, 1703–1720.

Requena, J., Caputo, C., 2016. Pérdida de talento en Venezuela: migración de sus investigadores. Interciencia 41, 444–453.

Rivas, G.A., Molina, C.R., Ugueto, G.N., Barros, T.R., Barrio-Amorós, C.L., Kok, P.J.R., 2012. Reptiles of Venezuela: an updated and commented checklist. Zootaxa 3211, 1–64.

Rivero, J.A., 1961. Salientia of Venezuela. Bull. Mus. Comp. Zool. 126, 1–207.

Rödder, D., Schlüter, A., Lötters, S., 2010. Is the 'Lost World' lost? High endemism of amphibians and reptiles on South American tepuis in a changing climate. In: Habel, J.C., Assmann, T. (Eds.), Relict Species: Phylogeography and Conservation Biology. Springer, Berlin, pp. 401–416.

Rojas, R.R., Fouquet, A., Ron, S.R., Hernández-Ruz, E.J., Melo-Sampaio, P.R., Chaparro, J.C., et al., 2018. A Pan-Amazonian species delimitation: high species diversity within the genus *Amazophrynella* (Anura: Bufonidae). Peer J. 6, e4941.

Rojas-Runjaic, F.J.M., Camargo, E., de Carvalho, V.T., La Marca, E., 2017. New record and range extension of the Horned Toad, *Rhinella ceratophrys* (Boulenger, 1882) (Anura: Bufonidae), in Venezuela, and confirmation of its presence in Brazil. Check List 13 (1), 2035.

Rojas-Runjaic, F.J.M., Salerno, P.E., Señaris, J.C., Pauly, G.B., 2013. Terraranans of the Lost World: a new species of *Pristimantis* (Amphibia, Strabomantidae) from Abakapá-tepui in the Chimantá massif, Venezuelan Guayana, and additions to the knowledge of *P. muchimuk*. Zootaxa 3686 (3), 335–355.

Ron, S.R., Venegas, P.J., Ortega-Andrade, H.M., Gagliardi-Urrutia, L.A.G., Salerno, P.E., 2016. Systematics of *Ecnomiohyla tuberculosa* with the description of a new species and comments on the taxonomy of *Trachycephalus typhonius* (Anura, Hylidae). ZooKeys 630, 115–154.

Roze, J.A., 1958a. Resultados zoológicos de la expedición de la Universidad Central de Venezuela a la región del Auyan-tepui, Venezuela, basándose en las colecciones de las expediciones de Phelps-Tate, del American Museum of Natural History, 1937–1938, y de la Universidad Central de Venezuela, 1956. Acta Biol. Venez. 2, 243–270.

Roze, J.A., 1958b. Los reptiles del Chimantá Tepui (Estado Bolívar, Venezuela) colectados por la expedición botánica del Chicago Natural History Museum. Acta Biol. Venez. 2, 299–314.

Rull, V., 2004. Biogeography of the 'Lost World': a palaeoecological perspective. Earth-Sci. Rev. 67, 125–137.

Rull, V., 2005. Biotic diversification in the Guayana Highlands: a proposal. J. Biogeogr. 32, 921–927.

Rull, V., 2008. Speciation timing and neotropical biodiversity: the Tertiary–Quaternary debate in the light of molecular phylogenetic evidence. Mol. Ecol. 17, 2722–2729.

Rull, V., 2009. Microrefugia. J. Biogeogr. 36, 481–484.

Rull, V., 2010. The Guayana Highlands: a natural laboratory for the biogeographical and evolutionary study of the neotropical flora. In: Sámchez-Villagra, M.R., Aguilera, O.A., Carlini, A.A. (Eds.), Urumaco and Venezuelan Paleontology. The Fossil Record of the Northern Neotropics. Indiana University Press, Bloomington, pp. 84–102.

Rull, V., 2011a. Neotropical biodiversity: timing and potential drivers. Trends Ecol. Evol. 26, 508–513.

Rull, V., 2011b. Origins of biodiversity. Science 331, 398–399.

Rull, V., 2018. Neotropical diversification: historical overview and conceptual insights. Peer J. Preprints 6, e27294v1. Available from: https://doi.org/10.7287/peerj.preprints.27294v1.

Rull, V., Vegas-Vilarrúbia, T., 2008. Biopiracy rules hinder conservation efforts. Nature 453, 26.

Rull, V., Vegas-Vilarrúbia, T., Safont, E., 2016. The lost World's pristinity at risk. Divers. Distrib. 22, 995–999. Available from: https://doi.org/10.1111/ddi.12469.

Salerno, P.E., Pauly, G.B., 2012. Clutch size variation in egg-brooding *Stefania*. South Am. J. Herpetol. 7 (1), 47–54.

Salerno, P.E., Ron, S.R., Señaris, J.C., Rojas-Runjaic, F.J.M., Noonan, B.P., Cannatella, D.C., 2012. Ancient Tepui summits harbor young rather than old lineages of endemic frogs. Evolution 66, 3000–3013.

Salerno, P.E., Señaris, J.C., Rojas-Runjaic, F.J.M., Cannatella, D.C., 2015. Recent evolutionary history of Lost World endemics: population genetics, species delimitation, and phylogeography of sky island treefrogs. Mol. Phylogenet. Evol. 82, 314–323.

Santos, J.C., Coloma, L.A., Summers, K., Caldwell, J.P., Ree, R., Cannatella, D.C., 2009. Amazonian amphibian diversity is primarily derived from late Miocene Andean lineages. PLoS Biol. 7 (3), e1000056.

Señaris, J.C., 1995. Una nueva especie de *Oreophrynella* (Anura; Hylidae) de la cima del Auyán-tepui, Estado Bolívar, Venezuela. Mem. Soc. Cienc. Nat. La Salle 53 (140), 177–183.

Señaris, J.C., Ávila-Pires, T.C., 2003. Anfibios and reptiles. In: Conservation International (Ed.), Conservation priorities for the Guayana Shield: 2002 Consensus. Washington, DC, pp. 11–13.

Señaris, J.C., Ayarzagüena, J., 1996. Fauna: anfibios y reptiles. In: Rosales, J., Huber, O., (Eds.), Ecología de la Cuenca del Río Caura, Venezuela I. Caracterización general. Scientia Guaianae 6, pp. 80–82.

Señaris, J.C., Ayarzagüena, J., 2002. A new species of *Hyla* (Anura: Hylidae) from the highlands of Venezuelan Guayana. J. Herpetol. 36 (4), 634–640.

Señaris, J.C., Ayarzagüena, J., Gorzula, S., 1994. Los sapos de la Familia Bufonidae (Amphibia; Anura) de las tierras altas de la Guayana Venezolana: descripción de un nuevo género y tres especies. Publ. Aso. Am. Doñana 3, 1–37.

Señaris, J.C., Ayarzagüena, J., Gorzula, S., 1996. Revisión taxonómica del género *Stefania* (Anura; Hylidae) en Venezuela con la descripción de cinco nuevas especies. Publ. Aso. Am. Doñana 7, 1–56.

Señaris, J.C., MacCulloch, R., 2005. Amphibians. In: Hollowell, T., Reynolds, R.P., (Eds.), Checklist of the Terrestrial Vertebrates of the Guiana Shield. Bulletin of the Biological Society of Washington, No. 13, pp. 9–23.

Señaris, J.C., Rojas-Runjaic, F.J.M., (in press). Amphibians and reptiles of Venezuelan Guayana: diversity, biogeography and conservation. In: Rull, V., Carnaval, A. (Eds.), Neotropical Diversification. Springer, Berlin.

Señaris, J.C., DoNascimiento, C., Villareal, O., 2005. A new species of the genus *Oreophrynella* (Anura: Bufonidae) from the Guiana Highlands. Pap. Avulsos Zool. 45 (4), 61–67.

Señaris, J.C., Rivas, G., Molina, C.M., 2009. Capítulo 4. Anfibios y reptiles del Parque Nacional Canaima. In: Señaris, J.C., Lew, D., Lasso, C.A. (Eds.), Biodiversidad del Parque Nacional Canaima: Bases Científicas para la Conservación de la Guayana Venezolana. Fundación La Salle de Ciencias Naturales & The Nature Conservancy, Caracas, pp. 70–92.

Señaris, J.C., Lampo, M., Rojas-Runjaic, F.J.M., Barrio-Amorós, C.L., 2014. Guía ilustrada de los anfibios del Parque Nacional Canaima, Venezuela. Ediciones IVIC, Caracas.
Solano, H., 1989. Aspectos de la biología de *Oreophrynella quelchii* (Boulenger) en los tepuyes venezolanos. Acta Biol. Venez. 12 (3–4), 55–63.
Steyermark, J.A., 1982. Relationships of some Venezuelan forest refuges with lowland tropical floras. In: Prance, G.T. (Ed.), Biological Diversification in the Tropics. Columbia University Press, New York, pp. 182–220.
Tate, G.H., 1939. The mammals of the Guiana region. Bull. Am. Mus. Nat. Hist. 76, 151–229.
Vacher, J.P., Kok, P.J.R., Rodrigues, M.T., Lima, J.D., Lorenzini, A., Martinez, Q., et al., 2017. Cryptic diversity in Amazonian frogs: integrative taxonomy of the genus *Anomaloglossus* (Amphibia: Anura: Aromobatidae) reveals a unique case of diversification within the Guiana Shield. Mol. Phylogenet. Evol. 112, 158–173.
Villareal, O., Señaris, J.C., DoNascimiento, C., 2003. Contribución al conocimiento de la fauna del Wei-Assipu-tepui, Macizo del Roraima, con énfasis en la anurofauna y opiliofauna. Bol. Soc. Venez. Espeleol. 36, 46–50.
Wiens, J.J., Kuczynski, C., Duellman, W.E., Reeder, T.W., 2007. Loss and re-evolution of complex life cycles in marsupial frogs: can ancestral trait reconstruction mislead? Evolution 61, 1886–1899.
Wiens, J.J., Kuczynski, C.A., Hua, X., Moen, D.S., 2010. An expanded phylogeny of treefrogs (Hylidae) based on nuclear and mitochondrial sequence data. Mol. Phylogenet. Evol. 55, 871–882.

Appendix 12.1 Amphibians and reptiles of the Guiana Highlands (≥ 1200 m asl)

WS: widespread, taxa known from areas of South America outside the Guiana Shield; **GS:** restricted to the Guiana Shield; **GHe:** known from two or more tepuis or massifs; **GHme:** known only from a single tepui summit

CLASS AMPHIBIA	Elevational range (m)	Distribution pattern
ORDER ANURA		
Family Aromobatidae		
Subfamily Allobatinae		
Allobates undulatus (Myers et Donnelly, 2001)	1750	GHme
Subfamily Anomaloglossinae		
Anomaloglossus apiau Fouquet, Souza, Nunes, Kok, Curcio, Carvalho, Grant, et Rodrigues, 2015	500–1400	GSwest
Anomaloglossus ayarzaguenai (La Marca, 1997 "1996")	1600	GHme
Anomaloglossus beebei (Noble, 1923)	450–1490	GSeast
Anomaloglossus guanayensis (La Marca, 1997 "1996")	1650–1800	GHme
Anomaloglossus meansi Kok, Nicolai, Lathrop, et McCulloch, 2018	1234–1490	GHe
Anomaloglossus megacephalus Kok, MacCulloch, Lathrop, Willaert, et Bossuyt, 2010	1060–1490	GHe
Anomaloglossus murisipanensis (La Marca, 1997 "1996")	2350	GHme
Anomaloglossus parkerae (Meinhardt et Parmelee, 1996)	860–1300	GSeast
Anomaloglossus praderioi (La Marca, 1997 "1996")	1310–1950	GHe
Anomaloglossus roraima (La Marca, 1997 "1996")	1860–2700	GHe

Anomaloglossus rufulus (Gorzula, 1990 "1988")	2100–2600	GHme
Anomaloglossus shrevei (Rivero, 1961)	350–1829	GSwest
Anomaloglossus tamacuarensis (Myers et Donelly, 1997)	350–1200	GSwest
Anomaloglossus tepuyensis (La Marca, 1997 "1996")	390–1850	GSeast
Family Bufonidae		
Metaphryniscus sosae Señaris, Ayarzagüena et Gorzula, 1994	2600	GHme
Oreophrynella cryptica Señaris, 1995 "1993"	1750–2400	GHme
Oreophrynella dendronastes Lathrop et MacCulloch, 2007	1490	GHme
Oreophrynella huberi Diego-Aransay et Gorzula, 1990 "1987"	1700–1755	GHme
Oreophrynella macconnelli Boulenger, 1900	700–1830	GSeast
Oreophrynella nigra Señaris, Ayarzagüena et Gorzula, 1994	2300–2600	GHe
Oreophrynella quelchii (Boulenger, 1895)	1700–2750	GHe
Oreophrynella seegobini Kok, 2009	2088	GHme
Oreophrynella vasquezi Señaris, Ayarzagüena et Gorzula, 1994	2371–2680	GHe
Oreophrynella weiassipuensis Señaris, DoNascimiento et Villarreal, 2005	2080–2280	GHme
Rhaebo nasicus (Werner, 1903)	100–1410	GSeast
Rhinella ceratophrys (Boulenger, 1882)	412–2713	WS
Rhinella martyi Fouquet, Gaucher, Blanc et Vélez-Rodriguez, 2007	600–1365	GSwest
Family Centrolenidae		
Subfamily Centroleninae		
"*Cochranella*" *duidaeana* (Ayarzagüena, 1992)	2140	GHme
"*Cochranella*" *riveroi* (Ayarzagüena, 1992)	1600	GHme
Vitreorana gorzulae (Ayarzagüena, 1992)	430–1850	GSeast
Subfamily Hyalinobatrachinae		
Hyalinobatrachium cappellei van Lidth de Jeude, 1904	45–2000	WS
Hyalinobatrachium taylori (Goin, 1968 "1967")	30–2000	WS
Family Ceuthomantidae		
Ceuthomantis aracamuni (Barrio-Amorós et Molina, 2006)	1493	GHme
Ceuthomantis cavernibardus (Myers et Donnelly, 1997)	930–1200	GSwest
Ceuthomantis duellmani Barrio-Amorós, 2010	1350	GHme
Ceuthomantis smaragdinus Heinicke, Duellman, Trueb, Means, MacCulloch et Hedges, 2009	1490–1540	GHe

Family Dendrobatidae		
Subfamily Dendrobatinae		
Minyobates steyermarki (Rivero, 1971)	600–1300	GSwest
Family Eleutherodactylidae		
Adelophryne gutturosa Hoogmoed et Lescure, 1984	110–1400	GS
Adelophryne patamona MacCulloch, Lathrop, Kok, Minter, Khan et Barrio-Amoros, 2008	1060–1411	GSeast
Family Hemiphractidae		
Subfamily Hemiphractinae		
Stefania ackawaio MacCulloch et Lathrop, 2002	1234–1559	GHe
Stefania ayangannae MacCulloch et Lathrop, 2002	1234–1700	GHe
Stefania breweri Barrio-Amorós et Fuentes-Ramos, 2003	1250	GHme
Stefania coxi MacCulloch et Lathrop, 2002	1490–1700	GHe
Stefania ginesi Rivero, 1968 "1966"	1850–2600	GHme
Stefania goini Rivero, 1968 "1966"	1400–1700	GHe
Stefania marahuaquensis (Rivero, 1961)	340–1200	GSwest
Stefania oculosa Señaris, Ayarzagüena et Gorzula, 1997 "1996"	1600	GHme
Stefania percristata Señaris, Ayarzagüena et Gorzula, 1997 "1996"	1600	GHme
Stefania riae Duellman et Hoogmoed, 1984	1100–1400	GHe
Stefania riveroi Señaris, Ayarzagüena et Gorzula, 1997 "1996"	2300–2303	GHme
Stefania roraimae Duellman et Hoogmoed, 1984	1376–1410	GHme
Stefania satelles Señaris, Ayarzagüena et Gorzula, 1997 "1996"	2000–2576	GHme
Stefania scalae Rivero, 1970	600–1360	GSeast
Stefania schuberti Señaris, Ayarzagüena et Gorzula, 1997 "1996"	1750–2400	GHme
Stefania tamacuarina Myers et Donnelly, 1997	1160–1270	GHe
Family Hylidae		
Subfamily Cophomantinae		
Boana benitezi (Rivero, 1961)	400–1600	GS
Boana jimenezi (Señaris et Ayarzagüena, 2006)	970–1850	GSeast
Boana rhythmica (Señaris et Ayarzagüena, 2002)	1600	GHme
Boana lemai (Rivero, 1972 "1971")	600–1400	GSeast
Boana roraima (Duellman et Hoogmoed, 1992)	490–1700	GSeast
Boana sibleszi (Rivero, 1972 "1971")	500–1850	GSeast
Boana tepuiana (Barrio-Amorós et Brewer-Carías, 2008)	420–1800	GSeast

Boana xerophylla (Duméril et Bibron, 1841)	0–1700	WS
Myersiohyla aromatica (Ayarzagüena et Señaris, 1994 "1993")	1700	GHme
Myersiohyla chamaleo Faivovich, McDiarmid et Myers, 2013	1820–1880	GHme
Myersiohyla inparquesi (Ayarzagüena et Señaris, 1994 "1993")	2600	GHme
Myersiohyla liliae (Kok, 2006)	400–1200	GSeast
Myersiohyla neblinaria Faivovich, McDiarmid et Myers, 2013	1730–1880	GHme
Nesorohyla kanaima (Goin et Woodley, 1969)	800–1550	GSeast
Osteocephalus taurinus Steindachner, 1862	0–1365	WS
Subfamily Dendropsophinae		
Dendropsophus minutus (Peters, 1872)	0–1600	WS
Subfamily Hylinae		
Tepuihyla aecii (Ayarzagüena, Señaris et Gorzula, 1993 "1992")	2150	GHme
Tepuihyla edelcae (Ayarzagüena, Señaris et Gorzula, 1993 "1992")	1700–2100	GHme
Tepuihyla exophthalma (Smith et Noonan, 2001)	585–10,550	GSeast
Tepuihyla luteolabris (Ayarzagüena, Señaris et Gorzula, 1993 "1992")	2550	GHme
Tepuihyla obscura Kok, Ratz, Tagelaar, Aubret et Means, 2015	1800–2600	GHme
Tepuihyla rodriguezi (Rivero, 1968)	366–2440	GSeast
Tepuihyla warreni (Duellman and Hoogmoed, 1992)	1480–1550	GHe
Scinax danae (Duellman, 1986)	180–1250	GSeast
Scinax exiguus (Duellman, 1986)	70–1230	GSeast
Family Leptodactylidae		
Subfamily Leptodactylinae		
Adenomera lutzi Heyer, 1975	430–1490	GSeast
Leptodactylus rugosus Noble, 1923	230–2100	GS
Leptodactylus longirostris Boulenger, 1882	± 100–1520	GS
Family Microhylidae		
Subfamily Otophryninae		
Otophryne robusta Boulenger in Lankester, 1900	600–1411	GS
Otophryne steyermarki Rivero, 1968 "1967"	1234–2140	GHe
Family Strabomantidae		
Dischidodactylus colonnelloi Ayarzagüena, 1985	2550	GHme
Dischidodactylus duidensis (Rivero, 1968)	1000–1530	GHme
Pristimantis abakapa Rojas-Runjaic, Salerno, Señaris et Pauly, 2013	2245	GHme

Pristimantis aureoventris Kok, Means et Bossuyt, 2011	2210–2260	GHe
Pristimantis auricarens (Myers et Donnelly, 2008)	1600–1750	GHme
Pristimantis avius (Myers et Donelly, 1997)	1160–1460	GHme
Pristimantis cantitans (Myers et Donnelly, 1996)	2150	GHme
Pristimantis dendrobatoides Means et Savage, 2007	1385–1570	GHe
Pristimantis imthurni Kok, 2013	2471	GHme
Pristimantis jamescameroni Kok, 2013	2557–2571	GHme
Pristimantis jester Means et Savage, 2007	1411–1650	GHe
Pristimantis marahuaka (Fuentes-Ramos et Barrio-Amorós, 2004)	2450	GHme
Pristimantis marmoratus (Boulenger, 1900)	600–1800	GSeast
Pristimantis memorans (Myers et Donnelly, 1997)	1160–1270	GHme
Pristimantis muchimuk Barrio-Amorós, Mesa, Brewer-Carías et McDiarmid, 2010	2325	GHme
Pristimantis pruinatus (Myers et Donnelly, 1996)	2150	GHme
Pristimantis pulvinatus (Rivero, 1968)	950–1700	GSeast
Pristimantis saltissimus Means et Savage, 2007	698–1560	GSeast
Pristimantis sarisarinama Barrio-Amorós et Brewer-Carías, 2008	1100–1400	GHme
Pristimantis vilarsi (Melin, 1941)	100–1230	WS
Pristimantis yaviensis (Myers et Donnelly, 1996)	2150	GHme
Pristimantis yuruaniensis Rödder et Jungfer, 2008	2300	GHme
ORDEN GYMNOPHIONA		
Family Rhinatrematidae		
Epicrionops niger (Dunn, 1942)	100–1490	GSeast

CLASS REPTILIA	Elevational range (m)	Distribution pattern
ORDER SQUAMATA		
Family Dactyloidae		
Anolis bellipeniculus (Myers et Donnelly, 1995)	2150	GHme
Anolis carlostoddi (Williams, Praderio et Gorzula, 1996)	2200	GHme
Anolis fuscoauratus D'Orbigny in Duméril et Bibron, 1837	±100–1500	WS
Anolis neblininus (Myers, Williams et McDiarmid, 1993)	1690–2100	GHme
Anolis planiceps Troeschel, 1848	±0–1800	WS
Family Gymnophthalmidae		
Adercosaurus vixadnexus Myers et Donnelly, 2001	1700	GHme
Anadia escalerae Myers, Rivas et Jadin, 2009	1385–1430	Ghe

Arthrosaura montigena Myers et Donnelly, 2008	940–1700	GSeast
Arthrosaura synaptolepis Donnelly, McDiarmid et Myers, 1992	1160–1450	GHe
Arthrosaura testigensis Gorzula et Señaris, 1998	1800–2350	GHe
Arthrosaura tyleri (Burt et Burt, 1931)	1402–2164	GHme
Cercosaura nigroventris (Gorzula et Señaris, 1998)	1650	GHme
Cercosaura phelpsorum (Lancini, 1968)	1380–2150	GHe
Euspondylus auyanensis Myers, Rivas et Jadin, 2009	2325	GHme
Neusticurus arekuna Kok, Bittenbinder, van den Berg, Marques-Souza, Nunes, Laking, et al., 2018	900–2203	GSeast
Neusticurus bicarinatus (Linnaeus, 1758)	±1000–1300	GS
Neusticurus racenisi Roze, 1958	100–1215	GSwest
Neusticurus rudis Boulenger, 1900	678–1500	GSeast
Neusticurus tatei (Burt et Burt, 1931)	±400–1402	GS
Oreosaurus mcdiarmidi Kok et Rivas, 2011	2100–2600	GHme
Pantepuisaurus rodriguesi Kok, 2009	2080	GHme
Riolama inopinata Kok, 2015	2400–2413	GHme
Riolama leucosticta (Boulenger, 1900)	1940–2600	GHe
Riolama luridiventris Esqueda, La Marca et Praderio, 2004	2470–2480	GHme
Riolama uzzelli Molina et Señaris, 2003	1850–2600	GHme
Yanomamia hoogmoedi (Kok, 2008)	2112	GHme
Yanomamia guianensis (MacCulloch et Lathrop, 2001)	700–1490	GSeast
Family Scincidae		
Copeoglossum nigropunctatum Spix, 1825	0–1800	WS
Family Tropiduridae		
Plica plica (Linnaeus, 1758)	±30–1500	WS
Plica lumaria Donnelly et Myers, 1991	780–1380	GScentral
Plica pansticta (Myers et Donnelly, 2001)	180–1220	GSwest
Plica umbra (Linnaeus, 1758)	±0–1500	WS
Tropidurus bogerti Roze, 1958	1600–2080	GHme
Tropidurus hispidus (Spix, 1825)	20–1420	WS
Family Colubridae		
Atractus duidensis Roze, 1961	2050–2358	GHme
Atractus riveroi Roze, 1961	944–1800	GS
Atractus steyermarki Roze, 1958	500–2244	GSeast

Atractus tamessari Kok, 2006	±500–2200	GSeast
Chironius challenger Kok, 2010	1400–2088	GHe
Chironius exoletus (Linnaeus, 1758)	0–1553	WS
Chironius fuscus (Linnaeus, 1758)	±50–1520	WS
Dipsas pakaraima MacCulloch et Lathrop, 2004	1490	GHme
Dipsas catesbyi (Sentzen, 1796)	300–1490	WS
Erythrolamprus breviceps (Cope, 1861 "1860")	±15–1950	WS
Erythrolamprus ingeri (Roze, 1958)	1900–2100	GHme
Erythrolamprus reginae (Linnaeus, 1758)	0–1490	WS
Erythrolamprus trebbaui (Roze, 1958)	1000–1938	GHe
Leptophis ahaetulla (Linnaeus, 1758)	0–1850	WS
Leptophis cupreus (Cope, 1868)	±100–1850	WS
Leptodeira annulata (Linnaeus, 1758)	0–2150	WS
Mastigodryas boddaerti (Sentzen, 1796)	±500–2300	WS
Philodryas cordata Donnelly et Myers, 1991	1030–1520	GHe
Tantilla melanocephala (Linnaeus, 1758)	0–1800	WS
Thamnodynastes chimanta Roze, 1958	1900–2600	GHme
Thamnodynastes corocorensis Gorzula et Ayarzagüena, 1995	2150	GHme
Thamnodynastes duida Myers et Donnelly, 1995	2015	GHme
Thamnodynastes marahuaquensis Gorzula et Aryarzagüena, 1995	2500	GHme
Thamnodynastes pallidus (Linnaeus, 1758)	974–1370	WS
Thamnodynastes yavi Myers et Donnelly, 1995	2150	GHme
Family Elapidae		
Micrurus remotus Roze, 1987	90–1700	GS
Family Viperidae		
Bothrops taeniatus Wagler in Spix, 1824	0–2000	WS
Bothrops atrox (Linnaeus, 1758)	±50–1584	WS
Crotalus durissus (Linnaeus, 1758)	±0–1920	WS

CHAPTER

13

Birds

Jorge L. Pérez-Emán[1,2], Miguel Lentino[2] and Elisa Bonaccorso[3]

[1]Institute of Tropical Zoology and Ecology, Central University of Venezuela, Caracas, Venezuela [2]Phelps Ornithological Collection, Caracas, Venezuela [3]Institute BIOSFERA and Laboratory of Evolutionary Biology, University San Francisco de Quito, Quito, Ecuador

Introduction

The Pantepui biogeographical province lies on the Guiana Shield and represents a biogeographic province defined by the characteristic biota associated with the Guiana Highlands. These highlands are the result of a process of continuous erosion of ancient sandstone sediments (Precambrian age) that are hypothesized to have been initially uplifted during the opening of the tropical Atlantic in the Late Cretaceous (Ghosh, 1985; Sidder and Mendoza, 1991; Briceño and Schubert, 1990). Such a history of erosion has resulted in a unique landscape composed of a series of montane isolates (tepuis or cerros, hereafter included in the word tepui) that differ in elevation (from 800 to 3000 m a.s.l.) and degree of isolation and that are characterized by a localized biota with moderate-to-high endemism in different taxonomic groups (Mayr and Phelps, 1967; Steyermark, 1986; Berry et al., 1995; McDiarmid and Donnelly, 2005).

The ornithological exploration of the Guiana Highlands began almost 200 years ago. The Schomburgk brothers took the first steps to uncover the natural history of this region. Robert Schomburgk discovered Roraima-tepui on his first expedition (1838), and the first specimens were taken from that tepui by Richard Schomburgk on a follow-up expedition (1842). Additional exploration of this tepui and the neighboring area by Henry Whitely and others increased our knowledge of its avifauna, but it was only after the American Natural History Museum expeditions to Roraima-tepui and Cerro Duida, led by G. Tate in 1927 and 1928, that a first synthesis of the avifauna from this region was completed (i.e., Chapman, 1931). Accounts of exploration of the region have been given by Huber and Wurdack (1984), McDiarmid and Donnelly (2005), and others, but it is important to recognize the major contributions of William H. Phelps and W. H. Phelps Jr., who organized,

Biodiversity of Pantepui
DOI: https://doi.org/10.1016/B978-0-12-815591-2.00013-6

funded, and participated in a major regional exploration from 1937–1965, with some additional expeditions in later years. More recently, institutions such as the Foundation for the Development of Physical, Mathematical and Natural Sciences (FUDECI), Terramar Foundation, the American Museum of Natural History, and the Phelps Ornithological Collection organized further explorations in the 1980s and 1990s (Lentino and Bosque, 1989; Willard et al., 1991; Barrowclough and Escalante-Pliego, 1990; Barrowclough et al., 1995, 1997) and in 2000 (Pérez-Emán et al., 2003). Lately, an important number of expeditions have been conducted on different isolated mountains in Colombia, Guyana, and Suriname (Stiles et al., 1995; Barnett et al., 2002; Braun et al., 2003; Ridgely et al., 2005; Robbins et al., 2007; O'Shea et al., 2007; Zyskowski et al., 2011, among others), providing a better insight into the geographical area of the region characterized by the particular avifauna first described by Chapman (1931).

A definition of Pantepui

The history of the term Pantepui began by recognizing that the birds inhabiting the isolated mountains of the Guiana Highlands were unique to the region. The first bird descriptions from Schomburgk's collections (Cabanis, 1848) showed the distinctness of the avifauna from Roraima-tepui. However, major collections by Tate on both Roraima-tepui and Cerro Duida, large tepuis located almost 650 km away from each other, allowed ornithologists to realize that species described after Roraima collections were more widely distributed (Chapman, 1931). The American Museum of Natural History expedition to the Auyán-tepui, organized by Chapman and Phelps (Gilliard, 1941), and later expeditions, set the stage of knowledge needed to provide a thorough synthesis of the regional avifauna.

The term "Pantepui" (from the Greek *pan* = all; all tepuis) was first introduced by Mayr and Phelps (1955, 1967) to encompass the geographical location of the particular avifauna of the region. Their description was basically both geographical and geological ("the sandstone tabletop mountains in the Venezuelan Territorio Amazonas and Estado Bolivar and in the adjacent border regions of Brasil and Guyana") but did not incorporate the biotic component required for Pantepui to be recognized as a biogeographic region. However, apart from the definition itself, Mayr and Phelps included mountains of the Guiana Highlands based on faunistic grounds, particularly the endemic avifauna associated with the submontane and montane habitats found in the slopes and summits of the tepuis ("subtropical avifauna"), and in doing so, they included mountains not supported by their own definition (e.g., granitic mountains, non-tabletop mountains). The reference to a subtropical avifauna was rooted in Merriam's life zones concept, consistently applied by Chapman in his studies on bird distribution of the Neotropical montane regions (Merriam, 1892; Chapman, 1917; Vuilleumier, 2005), in which elevational distribution is thought to be primarily determined by climatic factors, mainly temperature. Consequently, establishing an elevational limit was included as an operational criteria for the definition of the Pantepui province and its biota. Mayr and Phelps (1967) used elevations of 1000 and 1500 m a.s.l. to estimate the area of Pantepui. Huber (1987) provided a currently accepted definition based on both geographical and biological terms (mountain ecosystems in summits and tepui slopes) and restricted the region to elevations from 1200/1500 to

3000 m a.s.l. Huber (1995) subsequently restricted it to above 1500 m (see Chapter 1: Definition and characterization of the Pantepui biogeographical province).

Establishing an elevational limit for the definition of the Pantepui province and the Pantepui biota are two different things. Species presence depends on a broad group of abiotic and biotic variables (e.g., temperature, precipitation, slope and slope orientation, soil type, resources, predators, competitors), as well as species' physiological characteristics, making areas of different elevations as environmentally suitable for the same species or assemblage of species. As such, the elevational distribution of Pantepui species is expected to fall above or below such elevation thresholds (see Chapman, 1931; Borges et al., 2018). However, use of such cutoff values allows the assembly of lists of birds occupying the particular elevational belt that includes the endemic Pantepui biota. On the other hand, using an elevational threshold to define a biogeographical region is not appropriate because such areas should be defined in terms of their unique biota (Cracraft, 1985). Thus rather than excluding lower-elevation tepuis (uplands *sensu* Huber, 1995; <1500 m a.s.l.), we include all regional tepuis if their known avifaunas are characterized by diagnostic elements of the Pantepui biogeographical province (e.g., Acary Mountains, Yapacana, Cerro El Negro, Tafelberg).

The geographical limits of the Pantepui province have changed as new information from bird distributions has become available. Major differences in the literature are related to establishing the eastern and westernmost tepuis included in biogeographical analyses. Early in geological history, the Roraima Formation extended from Suriname to the Sierra de la Macarena in Colombia (Gansser, 1954). However, the limits of the Pantepui province should be based on faunistic grounds. To the east, Mayr and Phelps (1967) included some mountains in Guyana close to the Venezuelan border and named them the "British Mountains." More recently, field work has shown that the avifauna of mountains such as Mount Ayanganna, Potaro Plateau, and Iwokrama and the Acary Mountains, among others in Guyana (Barnett et al., 2002; Braun et al., 2003; Ridgely et al., 2005; Robbins et al., 2007), and Tafelberg in Suriname (Zyskowski et al., 2011), have clear biogeographical affinities with Pantepui and, consequently, should be included in the Pantepui province, with Tafelberg representing its easternmost locality. The western limits, however, are not that clear. Borges et al. (2018) included both the Sierra La Macarena and Chiribiquete Mountains in Colombia, extending the westernmost limits of this biogeographical region substantially when compared to the previously considered Cerro Sipapo in Venezuela. Borges et al. (2018) identified only three endemic taxa associated with Pantepui for each of these localities. Other species included in their analyses (and present in these localities) are widespread in other Neotropical mountains (e.g., *Myioborus miniatus*, *Tangara gyrola*). A review of the three endemics used by Borges et al. (2018) to extend their western limit of Pantepui suggests that their affinities lie elsewhere. For Chiribiquete, *Chlorostilbon olivaresi* and *Hemitriccus margaritaceiventer chiribiquitensis* are likely more related to Magdalena Valley (Colombia) taxa than to Pantepui birds (Stiles et al., 1995; Stiles, 1996). Similarly, for Macarena, *Chlorospingus flavopectus* is clearly an Andean taxon, and *Xiphocolaptes promeropirhynchus macarenae* is also present in the Colombian Andes (Marantz et al., 2018). *Zonotrichia capensis roraimae,* recorded at both western localities and also present in the Guiana Highlands (Rising and Jaramillo, 2019), has a complex geographical variation. A phylogeographic molecular study of this species showed that the few samples analyzed from Roraima and the Colombian Andes are not closely related (Lougheed et al., 2013). Thus biogeographical affinities of these taxa, together with broader avifauna and flora studies of the Chiribiquete and Macarena (Estrada and Fuertes, 1993; Stiles et al., 1995; Cortés-B et al., 1998; Giraldo-Cañas, 2001), indicate that these mountains share affinities with Andean, Amazonian, and Guianan regions rather than with Pantepui. Consequently, we do not include Macarena and Chiribiquete in our analysis of the Pantepui avifauna, and we consider Cerro Sipapo the western limit of Pantepui.

Patterns of diversity and endemism

The avifauna of Pantepui is characterized by a moderate diversity (richness) compared to the Andean region. We identified from the literature a total of at least 141 species occurring in the montane habitats of this region, including 45 tepuis from which some ornithological exploration has been conducted (Fig. 13.1). This number differs from previous publications (e.g., 98 spp. in Mayr and Phelps, 1967; 104 in Willard et al., 1991; 103 in Zyskowski et al., 2011; and 138 in Borges et al., 2018; Appendix I), mostly as a result of differences in species selection criteria. Mayr and Phelps (1967) included endemic species, adding taxa either from montane or lowland habitats with differentiated populations (e.g., *Glaucidium brasilianum duidae*). These authors also included montane species even in the absence of geographical differentiation (e.g., *Colibri delphinae*) or species occurring in lowland habitats but only recorded in montane habitats in the tepuis (e.g., *Phaethornis bourcieri*, *Elaenia ruficeps*, *T. gyrola*). Subsequent changes in bird lists are mostly associated with the addition of not previously recorded montane species and with the exclusion (or not) of species nonexclusive of montane habitats or without differentiated populations associated with highland habitats (Appendix I). Most recently, Borges et al. (2018) produced the most inclusive list, adding species with broad elevational distributions that occur in Pantepui habitats. We believe a more inclusive list is a better approach when assessing diversity of these montane assemblages, characterized by a mixture of lowland species with broad elevational distributions and highland species. In fact, a limited number of these broad elevational species are very characteristic of both summits and slopes of tepuis (e.g., *Myiarchus swainsoni phaeonotus*, *Coereba flaveola roraimae*). Additionally, a better review of the species present in these communities is required because some common elements in these habitats have not been included so far (e.g., *Tachyphonus phoenicius*, *Elaenia chiriquensis*, *Elanoides forficatus*).

Endemism in Pantepui and its geographical distribution

Endemism is high in Pantepui. If we consider endemic taxa, including both species and subspecies, the percentage of endemism is 75% (33% for species; 41 species; Appendix I). These high values (at the subspecific level) are comparable to the endemism levels shown by plants (Berry and Riina, 2005) and the herpetofauna (McDiarmid and Donnelly, 2005), groups in which the subspecies level was not considered. Endemism at the subspecific level includes 164 taxa, slightly differing from Borges et al. (2018) as a result of excluding

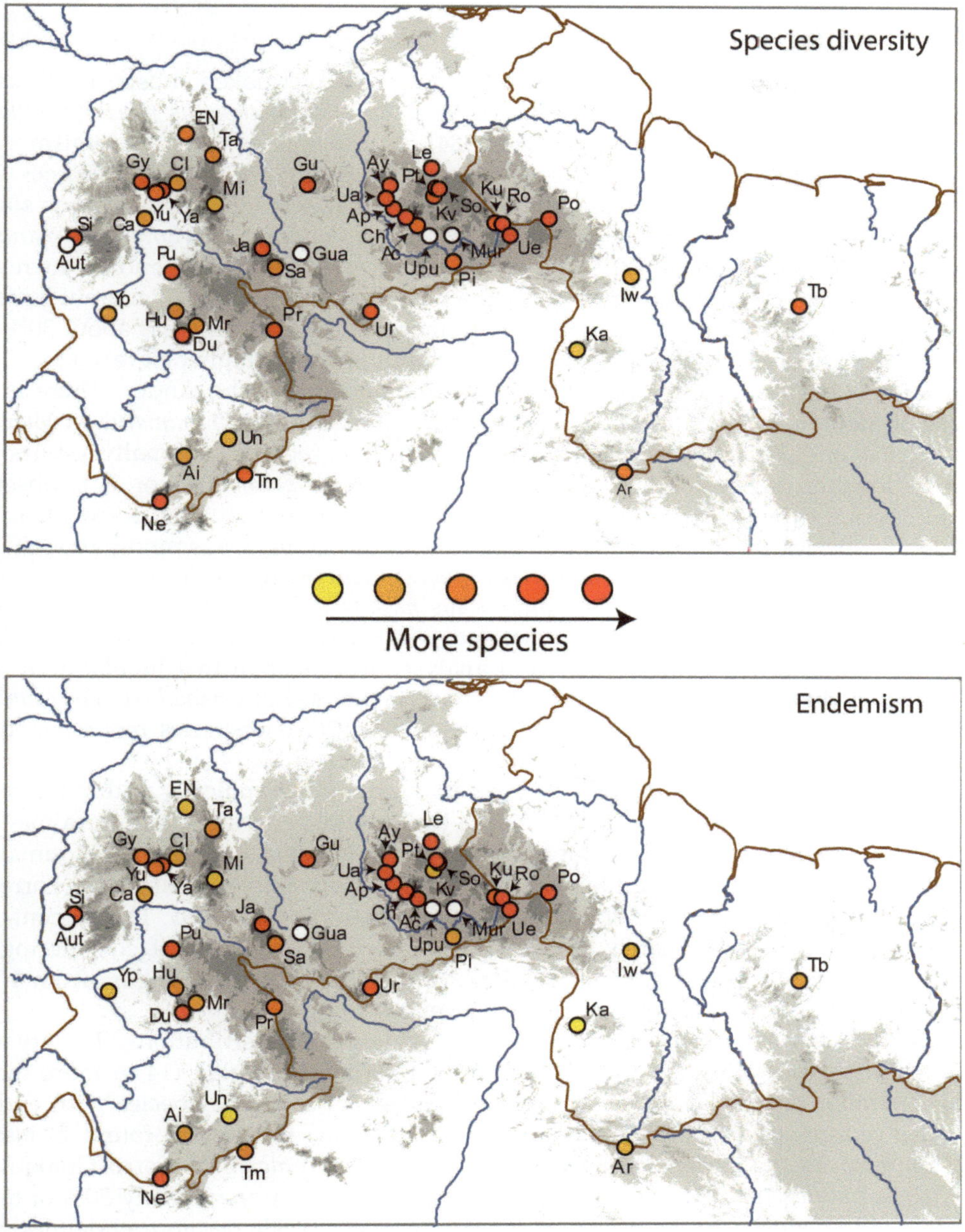

FIGURE 13.1 Species diversity and endemism by tepui in the Pantepui province. Tepui names are as follows (see Fig. 1.2 and Table 1.1 from Chapter 1): *Ar*, Acary Mountains; *Ac*, Akopán-tepui; *Ap*, Aprada-tepui; *Ai*, Cerro Arakamuni; *Ay*, Auyán-tepui; *Cl*, Cerro Calentura; *Ca*, Cerro Camani: *Ch*, Chimantá massif; *Du*, Cerro Duida; *EN*, Cerro El Negro; *Gu*, Cerro Guaiquinima; *Gy*, Cerro Guanay; *Hu*, Cerro Huachamakari; *Iw*, Iwokrama Mountains; *Ja*, Cerro Jaua; *Ka*, Kanuku Mountains; *Kv*, Kavanayén; *Ku*, Kukenán-tepui; *Le*, Sierra de Lema; *Mi*, Sierra de Maigualida; *Mr*, Cerro Marahuaka; *Ne*, Sierra de la Neblina; *Pr*, Sierra de Parima; *Pu*, Parú massif; *Pi*, Paurai-tepui; *Po*, Potaro Plateau; *Pt*, Ptari-tepui; *Ro*, Roraima-tepui; *Sa*, Cerro Sarisariñama; *Si*, Cerro Sipapo (Paraque); *So*, Sororopán-tepui; *Ta*, Cerro Tabaro; *Tb*, Tafelberg Mountains; *Tm*, Cerro Tamacuari; *Ua*, Uaipán-tepui; *Ue*, Uei-tepui; *Un*, Sierra Unturán; *Ur*, Cerro Urutani (Arutani, Marutaní); *Yp*, Cerro Yapacana; *Ya*, Cerro Yaví; *Yu*, Cerro Yutajé. Tepuis with very few information on bird species are represented by white circles and were not included in any analyses.

taxa from La Macarena and Chiribiquete and adding two overlooked subspecies described for Marahuaka (*Troglodytes rufulus marahuacae* and *Zonotrichia capensis perezchinchillae* (as opposed to *roraimae*) (Phelps and Aveledo, 1984). At the species level, endemism values have remained nearly unchanged since Mayr and Phelps' (1967) synthesis (Borges et al., 2018), a likely result of the combination of taxonomic changes resulting in elevating subspecies to the species category (e.g., *Schistocichla leucostigma saturata* vs *Myrmelastes saturatus*, Braun et al., 2005; Remsen et al., 2019; for more changes see Appendix II) and the increase in the number of taxa considered elements of the bird assemblage of the Pantepui avifauna. Most of this endemism (70%) is associated with montane species, whereas about 30% have affinities with lowland species with differentiated populations at higher elevations.

Elevational segregation in Pantepui is less common than in the Andes. There is only one case of montane-endemic species that segregate altitudinally (*M. miniatus* vs *Myioborus castaneocapilla* and *Myioborus albifacies*). In Pantepui, segregation normally occurs with regard to lowland species (e.g., *Trogon collaris* vs *Trogon personatus*) or it is absent as lowland species expand their range into higher elevations without replacement. In fact, nonendemic species present in Pantepui have elevational ranges averaging about 1700 m (Borges et al., 2018). However, some of these species include subspecies replacing lowland populations (e.g., *Xiphocolaptes promeropirhynchus neblinae, Glyphorynchus spirurus coronobscurus, Colaptes rubiginosus viridissimus, Ixothraupis xanthogastra phelpsi*), and studies based on morphological, distributional, and vocal analyses have shown that highland elevation populations likely represent different species. Such is the case of *Lepidothrix suavissima* and *Lepidothrix serena* (Prum, 1994) and of *Mionectes macconnelli* (subspecies *roraimae*), considered a different species (*Mionectes roraimae*; Hilty and Ascanio, 2014).

The number of endemics and species in general varies geographically within the region. Bird richness ranges from 7 to 97 species/tepui (in Kanuku Mountains and Roraima-tepui, respectively), whereas endemics range from 3 in Cerro Yapacana and Acary Mountains to 35 species in Roraima-tepui. Endemism and species diversity are strongly correlated ($R^2 = 0.923$) and show congruent spatial distribution patterns (Fig. 13.1). The percentage of endemism/tepui is approximately 40% or greater (~80% or greater if considering also endemic subspecies) in most but the lowest-elevation tepuis (values ranging from 10% to 20% for Tafelberg, Acary Mountains, and Yapacana).

Most of the avian species in Pantepui have widespread distributions. Twenty-seven endemic species (out of 41) are widely distributed across the region (14 in more than 20 tepuis), a similar pattern found for a large percentage of montane species with endemic subspecies (Borges et al., 2018). In contrast, some endemic species are found in just one tepui, *Emberizoides duidae* in Cerro Duida and *Myioborus cardonai* in Cerro Guaiquinima (Fig. 13.2), a pattern that is stronger at the subspecific level (approximately 30% of taxa in the region; 49 single-tepui endemics). The larger number of single-tepui endemics at the subspecies level are found in Sierra de la Neblina (13), Cerro Sipapo (5), Auyán-tepui (5), and Cerro Duida (4).

Major differences in diversity and species composition, and particularly in the distribution of endemics, have been found within the Pantepui province. A biogeographic subdivision was proposed by Mayr and Phelps (1967) regarding tepui placement in relation to the Caroní River. Cracraft (1985) also recognized two "subcenters" for his Pantepui center of endemism, the Gran Sabana and the Duida subcenters, suggesting that they

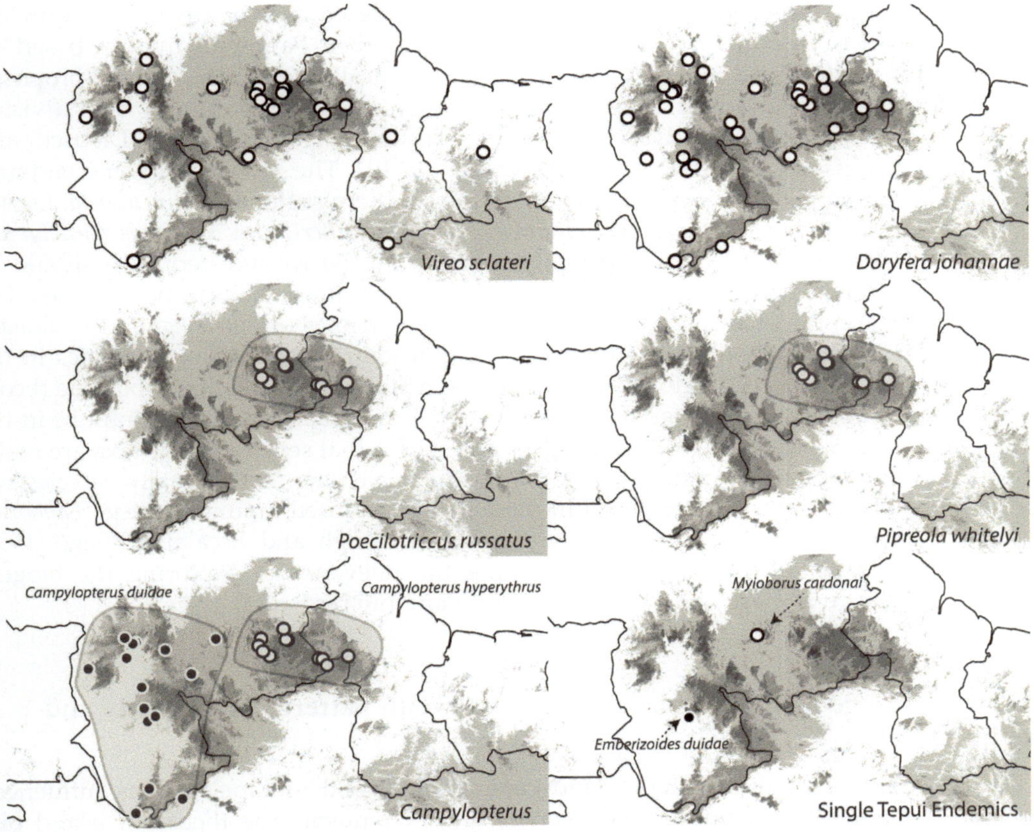

FIGURE. 13.2 Diversity of species geographic distributions in the Pantepui province. Some species are widely distributed, others are restricted to the eastern region, some sister taxa have complementary distributions in the region (*Campylopterus* spp.), and others are highly geographically restricted (i.e., *Myioborus cardonai* and *Emberizoides duidae*).

could be further subdivided with more information, particularly the Duida subcenter. Eastern tepuis are not only more diverse but also include most of the species known from the region (more than 85%; Appendix III). Exclusive taxa of the eastern tepuis are *Crypturellus ptaritepui*, *Myrmelastes saturatus*, *Grallaricula nana kukenamensis*, *Poecilotriccus russatus*, *Pipreola whitelyi*, *Lipaugus streptophorus*, *Cichlopsis leucogenys gularis*, and *Mitrospingus oleagineus* (Fig. 13.2). *Campylopterus hyperythrus* and *Diglossa major* are also exclusive of the eastern tepuis, but have their closest relative in the western tepuis (Mauck and Burns, 2009; McGuire et al., 2014). Some nonendemic species also occur exclusively in the eastern section of Pantepui, such as *Spinus magellanicus*, *Cistothorus platensis*, and *Piranga leucoptera*, some of them potentially associated with habitats available in these tepuis. In contrast, exclusive species for the western tepuis are few and, with the exception of *Emberizoides duidae* (Cerro Duida), are the closest relatives to species present in the eastern tepuis (*Diglossa duidae*, *Campylopterus duidae*, *M. albifacies*, and *M. cardonai*) (Barker et al., 2015; McGuire et al., 2014; Pérez-Emán, 2005).

Biogeographical research in other groups has considered a finer division within the Pantepui province. In particular, Huber (1987) considered a Pantepui division based on both geographical (major drainages) and phytogeographical criteria. This author proposed a regionalization in five sectors, and posteriorly, Berry et al. (1995) refined such division considering just four sectors: Eastern District, Western District, Central District, and Southern District (see Chapter 1) (Fig. 13.3, Appendix III). The absence of some endemic species has a strong geographical component. In the southern sector *Roraimia adusta, Elaenia dayi, Xenopipo uniformis, Macroagelaius imthurni, Polytmus milleri,* and *Setopagis whitelyi* are missing. *P. milleri* and *S. whitelyi* are equally absent from the western tepuis. It should be assessed in the future if such absences are related to factors other than sampling biases. The Central District, on the other hand, has two exclusive endemics (*M. cardonai* and *E. duidae*), and Neblina (Southern District) has the major proportion of single-tepui endemics in the region (13, Borges et al., 2018). However, endemism in Sierra de la Neblina should be reconsidered as exploration of close tepuis such as Cerro Tamacuari and Sierra Unturán in the Tapirapicó massif, as well as Cerro Aracamuni, has shown that seven of these taxa are really endemics to this subregion rather than to Sierra de la Neblina. Additionally, species not recorded in Neblina have been collected in these newly explored mountains (e.g., *Lophornis pavoninus, Chlorophonia cyanea,* and *E. ruficeps*, Barrowclough and Escalante-Pliego, 1990; Barrowclough et al., 1995; Fig. 13.3), suggesting the relevance of exploring the biogeographic significance of these regional divisions for Pantepui birds.

Ecological and geographical factors correlated with patterns of diversity and endemism

The resemblance of the Pantepui landscape to an island archipelago has influenced analytical approaches to studying its diversity and endemism. The theory of island biogeography (MacArthur and Wilson, 1967) has been a major approach used to explain diversity differences among both tepuis and subregions in Pantepui (eastern vs western tepuis). Island biogeography, in its original form, evaluates the role of area and isolation on the diversity of species on an island habitat. Analyses of the Pantepui avifauna, using simple or multiple regression, have included variables such as tepui area (summit or a sum of the summit and slope areas), degree of isolation (distance to closest tepui or average distance to closest and/or largest tepuis), distance to a source of colonization, and elevation (Cook, 1974; Borges et al., 2018). A major weakness of this approach is to assume a unique source of colonists (Andes or Paria Peninsula) and a long-distance dispersal colonization mechanism, aspects that could be addressed more effectively through the study of historical and evolutionary patterns of the avifauna. In fact, distance to the colonization source showed an unexpected positive correlation (the greater the distance to the source, the higher the richness), contrary to island biogeography expectations (Cook, 1974; Borges et al., 2018). This last trend was even more pronounced in the most recent study (Borges et al., 2018), but excluding Sierra La Macarena and Chiribiquete from the analysis shows a clear lack of correlation with distance (Fig. 13.4).

Elevation has been the major factor explaining variation in tepui diversity and endemism. Higher tepuis are characterized by avifaunas with greater diversity and endemism,

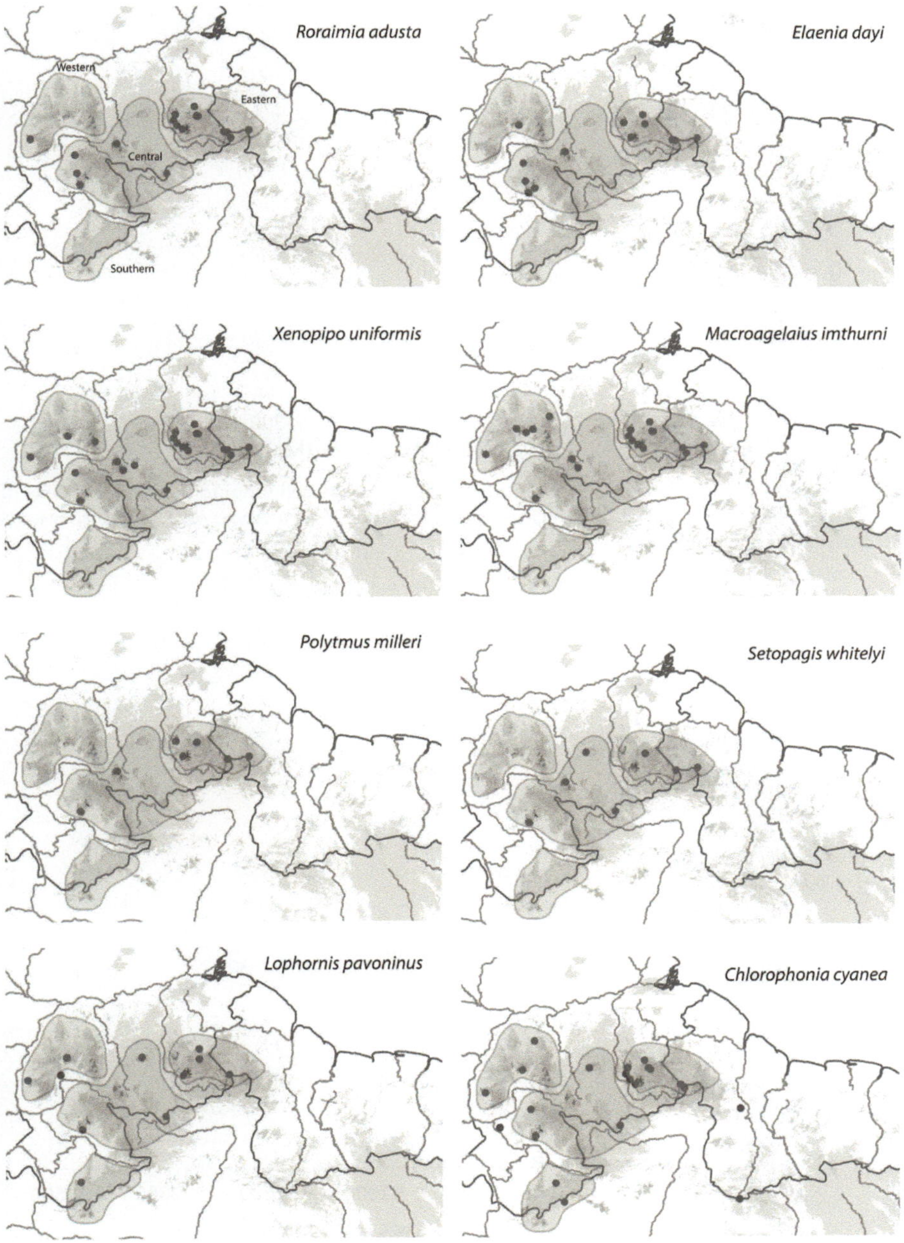

FIGURE 13.3 Distribution of some endemic species in Pantepui according to the finer biogeographical subdivision proposed by Berry et al. (1995) (see text for details). Some distributional gaps suggest potential biogeographic patterns requiring further studies or just sampling biases. For example, the last two species, *Lophornis pavoninus* and *Chlorophonia cyanea*, have not been recorded for Sierra de la Neblina but do occur in the Southern District.

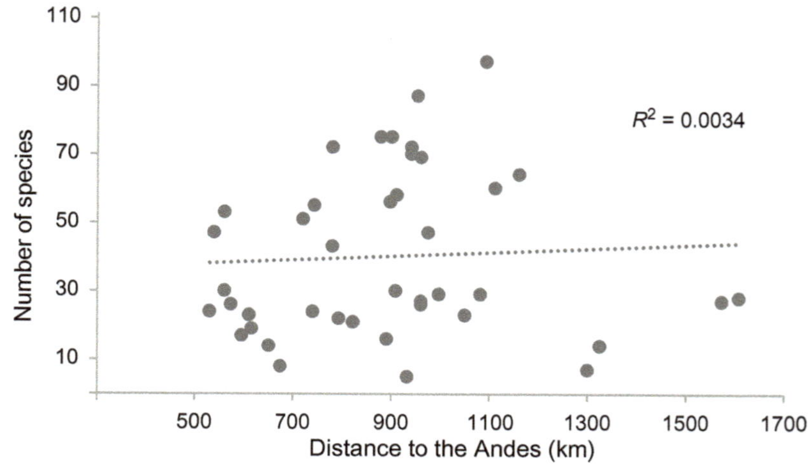

FIGURE 13.4 Linear regression between the minimum distance of each tepui to the Andes and its diversity in terms of number of species. Serrania Los Picachos, located in the Eastern Andes of Colombia, was used as the source of colonists based on Borges et al. (2018).

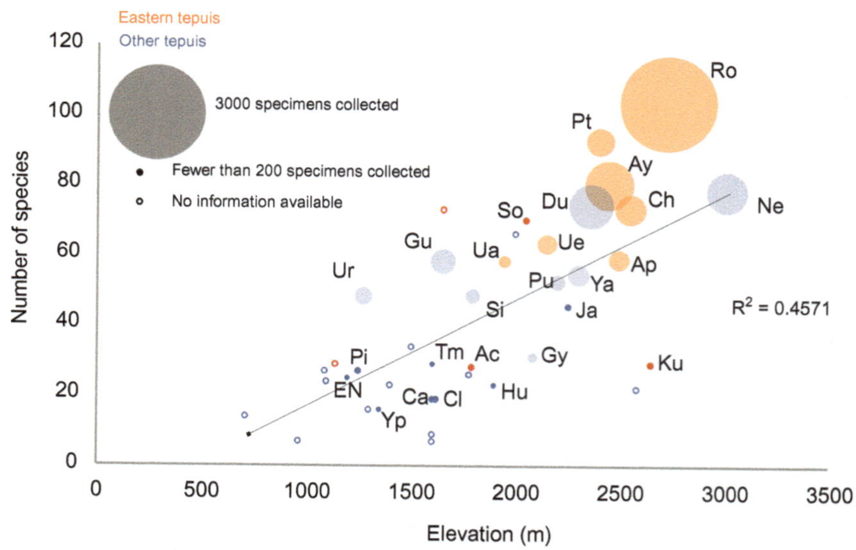

FIGURE 13.5 Linear regression between the elevation of each tepui and its diversity in terms of number of species. The size of symbols indicates the number of specimens collected to date on each mountain. Names for the most sampled mountains follows Fig. 13.1. Notice how some of the highest mountains, which also have the highest diversity, are among those with the best collection efforts; four of them (Roraima-tepui, Ptari-tepui, Auyán-tepui, and Chimantá massif) are tepuis from the Eastern District of Pantepui.

whereas area and isolation explained little variation in these variables (Cook, 1974; Borges et al., 2018). However, tepuis of similar elevation greatly differ in diversity and endemism; in the same way, tepuis of different elevations might be characterized by a similar avian diversity, which could be partially explained by sampling biases (see later, Fig. 13.5). Given these patterns, Cook (1974) realized that environmental heterogeneity or habitat diversity, as potentially correlated to elevation, does not satisfactorily explain the result and proposed that this pattern is a result of the greater degree of connectivity among the eastern tepuis, which are lying on a plateau of about 1000 m a.s.l. (Gran Sabana). In an ecological dynamics of colonization/extinction, as proposed by the island biogeography theory, such connectivity increases the probability of recolonization of locally extinct populations maintaining a higher diversity, a process that is less likely to occur in the more isolated western tepuis (Mayr and Phelps, 1967; Cook, 1974; Willard et al., 1991). Cook (1974) went further and proposed that this dynamic of extinction/colonization was associated with historical factors, in particular, habitat displacements resulting from climate changes during the Pleistocene Epoch (see also Haffer, 1974; Rull, 2004a,b).

The role of sampling biases in studying patterns of diversity and endemism

Understanding sampling biases helps to evaluate our knowledge on bird composition and distribution. Small tepuis (<1500 m a.s.l.) have a smaller number of species and endemics (fewer than 5 spp. in each tepui), a likely result of lower habitat availability (both area and heterogeneity). However, the patchy distribution of most species, especially on tepuis located west of the Caroní River, have resulted in a series of hypotheses concerning the dynamics of extinction/colonization that should be evaluated in light of potential collecting biases.

Collecting effort should provide a way to evaluate how complete our knowledge is from the tepui region. Mayr and Phelps (1967) assumed good sampling for most of the tepuis they included in their analyses, though the effort was quite variable among those tepuis. The number of expeditions, collected individuals, and collecting days and seasons differ among tepuis. Based on literature and collectors field notes, we chose the number of collected individuals (highly correlated to number of expeditions) to explore the completeness of avifauna knowledge on four of these tepuis: Sierra de la Neblina, Cerro Guaiquinima, Auyán-tepui, and Cerro Guanay. For Neblina, 19 new montane species, 14 of which are known to occur in the western tepuis (Willard et al., 1991), were recorded in three new expeditions. Two new Guaiquinima expeditions, one in 1990 and another in 2000, added 27 and 20 extra species, respectively, for a total of 47 new species (Pérez-Emán et al., 2003). Auyán-tepui, with two new expeditions, increased its known summit avifauna by 21 species (Barrowclough et al., 1997) and, for Cerro Guanay, an additional expedition of just 10 days increased its known avifauna by 16 new species, 9 of which were Pantepui endemics. Thus although by the late 1960s more than 10,000 specimens had been collected from Pantepui (Dickerman and Phelps, 1982), knowledge of its avifauna is far from complete. Perhaps the best-known avifaunas are those from Roraima-tepui, Cerro Duida, and Ptari-tepui. Based on the distribution of Pantepui endemics (considering the eastern and western division and their exclusive species), the number of expected endemics is 36 for Eastern tepuis and 30 for western tepuis (Appendix III). Roraima-tepui and

Ptari-tepui are represented by 35 and 34 endemics, respectively. Roraima is missing only *C. ptaritepui*, which is absent from the eastern-chain tepuis (Roraima, Kukenan, and Uei) and Potaro Mountains in Guyana, whereas *Megascops roraimae* and *Streptoprocne phelpsi* are missing from Ptaritepui but are likely to be recorded with new visits to this tepui. Similarly, all potential endemic species for the western tepuis have been recorded in Cerro Duida. Such knowledge is likely a result of the number of expeditions (eight for Roraima), long field work (an American Natural History Museum expedition spent 3–4 months in Cerro Duida), and time of the year of the expedition or seasonality effect (Ptari-tepui was visited three times in the same year in 1944). Additionally, collecting locality (sampled habitats and percentage of area surveyed; see O'Shea et al. (2007) for a comparison of species found in Kopinang Mountains and Roraima in Guyana) and type of collecting/recording (shotguns, mistnets, visual and aural records) should explain a large part of the variation observed in species composition in different tepuis. Consequently, among other biogeographical inferences, the role of extinction to explain the lower diversity in the western tepuis (such as Neblina, Cook, 1974) should be reevaluated in light of these results.

Historical explanations for diversity and endemism in Pantepui

The role of potential barriers, as well as the impact of climatic factors and their association with the landscape, are major historical factors shaping bird distributions. Understanding the role of these factors requires knowledge of the evolutionary history of the taxa included in the analysis (e.g., Smith et al., 2014). Lack of phylogenetic hypotheses requires the assumption that taxonomic categories convey information on the distinctness and time of differentiation among populations of a particular species or sister taxon relationships. Mayr and Phelps (1967), for example, claimed that the presence of endemic genera, species, and subspecies in Pantepui testified to the continuous and long period of avian colonization of the region. Fortunately, a burst of phylogenetic information has contributed to an increased knowledge of genealogical relationships among taxa. For example, a pattern of paraphyletic or polyphyletic relationships has resulted in recognizing previous Pantepui differentiated populations of widespread species as endemic to the region [e.g., *Aulacorhynchus whitelianus* (Bonaccorso et al., 2011), *Elaenia olivina* (Rheindt et al., 2009), *Myioborus castaneocapilla* (Pérez-Emán 2005), *Megascops roraimae* (Dantas et al., 2016), and *Hydropsalis roraimae* (Sigurðsson and Cracraft 2014; but see Remsen et al., 2019)] or recognizing unexpected phylogenetic relationships [e.g., *Myrmelastes saturatus* (Braun et al., 2005), *Vireo sclateri* (Slager et al., 2014)]. Such knowledge of evolutionary relationships provides us with better insights into the historical scenario associated with the building or maintenance of the Pantepui avifauna.

Geographical differentiation and diversification within Pantepui

Has the Pantepui province, with its "island archipelago" landscape, promoted speciation within the region? An exploratory initial approach to assess geographical differentiation within Pantepui is through taxonomy. Distribution of endemic species shows very idiosyncratic patterns, but one that is found in many species is the break associated with the potential barrier represented by the valley carved by the Caroní River (Figs. 13.2 and 13.6),

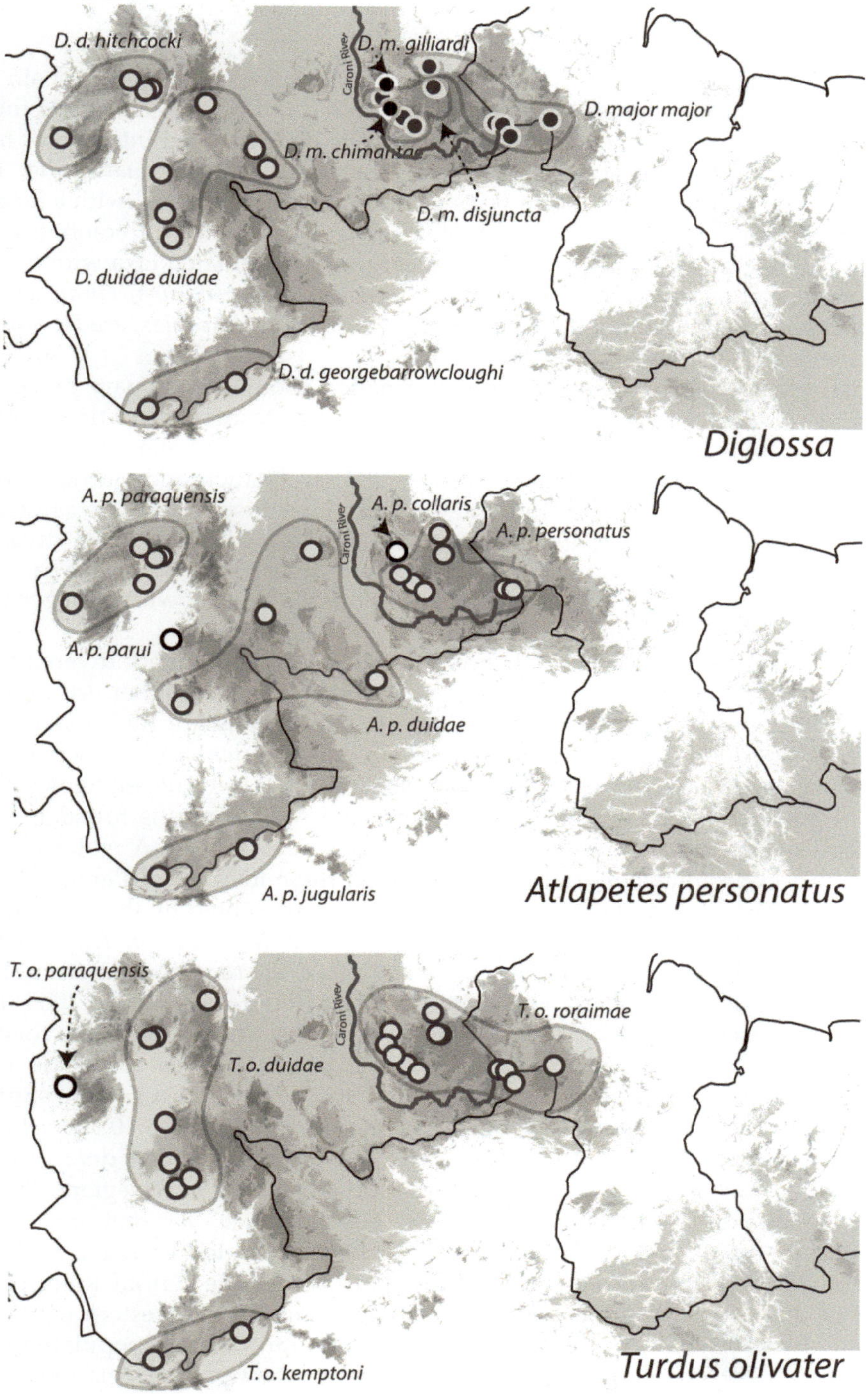

FIGURE 13.6 Some extreme patterns of differentiation in the Pantepui province. Some genera, such as *Diglossa*, have two endemic species, each of them with multiple sub species. *Atlapetes personatus*, an endemic species, and *Turdus olivater*, a nonendemic species, also show large geographic differentiation within Pantepui. Notice the taxonomic break at both sides of the Caroní River and the pattern of larger differentiation (more subspecies) in the western section compared to the eastern group of tepuis for the last two species.

a pattern also documented in plants (Maguire, 1979; Steyermark, 1986; Berry et al., 1995) but not as strongly in amphibians and reptiles (McDiarmid and Donnelly, 2005). At the subspecific level, 24 out of the 41 endemic species (59%) have differentiated into two or more subspecies (up to 6 in the case of *Troglodytes rufulus* and *Atlapetes personatus*, Fig. 13.6). For nonendemic taxa, more than one subspecies is found in 33% of species, with a larger differentiation in montane nonendemic taxa than in lowland taxa with differentiated highland populations (39% vs. 24%). In the first case, the largest differentiated taxa with four subspecies are *Amazilia viridigaster*, *Colaptes rubiginosus*, *Chamaeza campanisona*, *Turdus olivater*, and *Zonotrichia capensis* (Fig. 13.6). In the second case, with three subspecies, are *Hemitriccus margaritaceiventer* and *Platyrinchus mystaceus*.

Molecular phylogenetic studies provide information about evolutionary processes. Pairs or groups of closely related species in Pantepui conform to monophyletic groups for the genera *Campylopterus* and *Myioborus* (McGuire et al., 2014; Pérez-Emán, 2005). This pattern is suggested but not as strongly in the genus *Diglossa* (Mauck and Burns, 2009; Barker et al., 2015), in which both species in Pantepui were long suspected to be associated with different lineages (Vuilleumier, 1969; Graves, 1982). These results suggest diversification within Pantepui and identify the valley of the Caroní River as an important barrier associated with taxonomic breaks and limits of distributions, as also shown by *Trogon personatus* eastern and western populations (Cuervo, 2013).

Molecular studies on *Myioborus* redstarts provide information about diversification and the role of historical factors in geographic differentiation in the region. *M. castaneocapilla* occurs on both sides of the Caroní River, and its populations are traditionally classified into three subspecies (Phelps and Phelps, 1963): *M. c. castaneocapilla*, occurring in the eastern tepuis; *M. c. duidae*, of the Central tepuis; and *M. c. maguirei*, endemic to Cerro La Neblina. *M. cardonai* is endemic to Guaiquinima, and *M. albifacies* is found in the northwestern tepuis (Cerros Sipapo, Camani, Guanay, Yavi, and Yutajé). A molecular phylogenetic study based on mitochondrial DNA recovered three lineages of Pantepui *Myioborus*: (1) *M. albifacies*, distributed in the western region and represented by the populations from Yutajé and Guanay; (2) *M. castaneocapilla maguirei* and *M. cardonai*, from Sierra de la Neblina and Cerro Guaiquinima, respectively; and (3) *M. c. castaneocapilla*, from the eastern region: Auyán-tepui, La Escalera, and Roraima-tepui (Pérez-Emán 2002, 2005, Fig. 13.7). The close relationship between *M. castaneocapilla maguirei* and *M. cardonai*, rendering *M. castaneocapilla* a paraphyletic species, suggests that current species limits do not reflect phylogenetic relationships among populations and species, a pattern equally supported by a study using both mitochondrial and nuclear genes (Lovette et al., 2010). *Myioborus* populations at either side of the Caroní River showed a large nucleotide divergence (>3%) compared with shallow divergences among haplotypes within each region. This result is surprising because populations as far away as 600 km (*M. c. maguirei* and *M. cardonai*) were genetically more closely related than populations separated by less than 200 km (*M. cardonai* and *M. c. castaneocapilla* from Auyán-tepui) across the Caroní River. This study also showed that the extent of geographical differentiation in the western tepuis (as compared to the eastern tepuis) is clearly larger than that found among populations from the eastern section. Given the larger geographical extent of the western region and the larger geographical distances among populations (Fig. 13.7), greater genetic structure is not

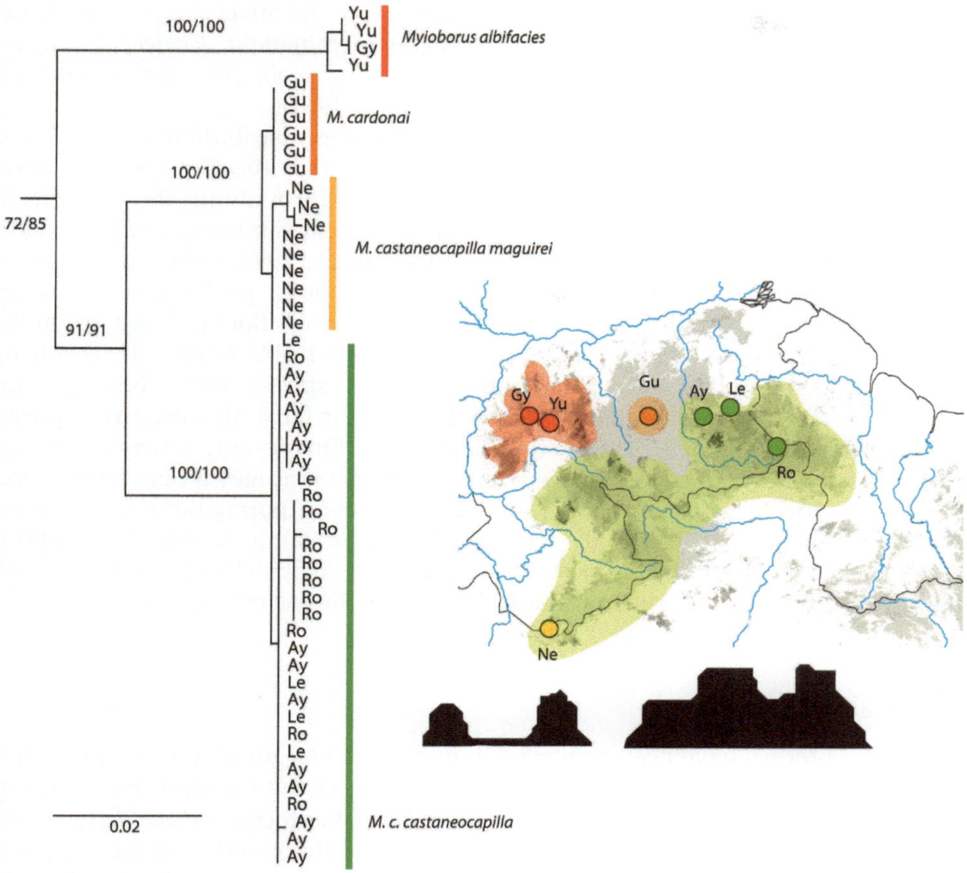

FIGURE 13.7 Molecular phylogeny of Pantepui *Myioborus* showing their geographic distribution. Notice that *Myioborus castaneocapilla* is not monophyletic, with *Myioborus c. maguirei*, from Serranía de la Neblina (yellow), being more related to *M. cardonai*, from Cerro Guaiquinima (orange). The largest genetic break is between *Myioborus albifacies* and the other taxa, whereas populations from the eastern region show a shallow genetic divergence. Silhouettes represent the general topography of the western and eastern section of the Pantepui province. Tepui names follow Fig. 13.1. Numbers refer to node support values: Maximum Likelihood bootstrap and Bayesian Inference posterior probabilities based on Pérez-Emán (2002).

unexpected. In fact, at the subspecies level, the number of endemics is larger in the western tepuis (as a region) than the eastern tepuis (116 vs 102 taxa, Fig. 13.6), a similar pattern found for amphibians and reptiles (McDiarmid and Donnelly, 2005). Moreover, the deep molecular divergence shown by *M. albifacies* compared to the rest of *Myioborus* populations (greater than 5%) is indicative of a long history of independent evolution. In fact, cerros Guanay and Yutajé are similar in species composition to other western tepuis such as cerros Camani, Yaví, and Sipapo (consistent with the idea of considering biogeographic divisions), and we can find taxonomic differences (endemism at the subspecific level) in many species found in these tepuis (*Synallaxis cabanisi yavii, Roraimia adusta obscurodorsalis, Syndactyla roraimae paraquensis, Thamnophilus insignis nigrofrontalis, Chamaeza campanisona yavii, Knipolegus poecilurus paraquensis, Troglodytes rufulus yavii, Atlapetes personatus paraquensis*, and *Diglossa duidae hitchcocki*; Phelps and Phelps, 1963).

The lack of a strong geographic structure in haplotype distribution among *Myioborus* populations from the eastern part of Pantepui suggests that the Gran Sabana plays (or played) an important role in connecting populations throughout the eastern range. Paleoecological data (lack of deposition of organic material, peats) suggest drier conditions in the region before 8000–10,000 years ago (Schubert et al., 1992, 1994; Zink et al., 2011), potentially rejecting hypotheses of climate stability (although peat removal during wet conditions is another possibility, Rull et al., 2013). Moreover, floristics and palynological records indicate the presence of high-tepui flora elements in the Gran Sabana during the Late Pleistocene, suggesting elevational movements of species with changing climatic phases (Rull et al., 2013; Huber 1988, 1995), which might have also occurred during the Holocene (Rull and Montoya, 2017) (see Chapter 2: Climatic and ecological history of Pantepui and surrounding areas). This evidence indicates that elevational belts of vegetation descended during glacial periods, possibly connecting populations from peripheral tepuis to the Gran Sabana (Rull, 2005; Rull and Nogué, 2007), a pattern that supports the scenario of greater connectivity proposed for Mayr and Phelps (1967) and Cook (1974) to explain the greater diversity (origin and maintenance) of the eastern tepuis.

Origin of Pantepui birds

Different hypotheses have been proposed to explain the origin and diversification of the Pantepui avifauna (see Chapter 4: Origin and evolution of the Pantepui biota). Chapman (1931) and Mayr and Phelps (1967) provided up to five hypotheses, which differ in the relative role of vicariance and dispersal on the origin and differentiation of the Pantepui avifauna. From these, the habitat specialization hypothesis refers to available habitat in the region. As such, it is not an historical hypothesis but an ecological one. Habitat specialists include cliff dwellers, such as the swifts, swallows, and *Hirundinea ferruginea* (Appendix I), which have high dispersal abilities and use habitats not restricted to montane habitats (Robbins et al., 2007). Similarly, *Steatornis caripensis* could be assigned to this group, as it is not restricted to the region, but their specific habitat requirements for living and reproduction is provided by the tepuis. The other four hypotheses are the plateau, the distance dispersal, the cool climate, and the habitat shift hypothesis. The first one is a vicariance hypothesis and the other three are hypotheses emphasizing the role of bird movement (dispersal or displacement) to explain the current composition of the Pantepui avifauna, suggesting the origin of Pantepui avifauna was in other areas, either montane regions or areas at lower elevations.

The plateau hypothesis, or more recently named "Lost World" hypothesis (Rull 2004a, 2004b), claimed that differentiation occurred as a result of vicariance due to erosion of a formerly continuous area and isolation of population of birds widely distributed in the region (Chapman, 1931; Tate, 1938; Croizat-Chaley, 1976). Available phylogenetic information (and divergence times) indicate that geographic differentiation is younger than the origin of the current landscape of the region. For example, in *Aulacorhynchus whitelianus*, modest plumage and body size differentiation among populations within Pantepui is not reflected by any degree of mitochondrial differentiation (Bonaccorso et al., 2011), suggesting that such morphological differentiation, if indicative of isolation, is relatively recent. Similar results were found in an analysis of *Mecocerculus leucophrys*, where no genetic differentiation was found among recognized Pantepui subspecies, and even between these and the Andean populations (Cuervo, 2013). Moreover, recent studies on tepui amphibian populations, some of them "summit specialists" or highly divergent in morphology among tepuis, have revealed astonishingly low levels of genetic differentiation, suggesting their relatively young age, as well as a pattern of recent (or even active) dispersal among summits (Kok et al., 2012; Salerno et al., 2012, 2015). Similar patterns were found in phylogenetic studies of characteristic Pantepui plant genera in the families Rapateaceae and Bromeliaceae, such as *Stegolepis* and *Brocchinia* (Givnish et al., 1997, 2000, 2011).

The dispersal hypothesis focuses on colonization of Pantepui by birds from other montane regions either by long-distance or stepping-stone dispersal mechanisms. It was supported by Mayr and Phelps (1967), who suggested that about 50% of the Pantepui avifauna had its origins in the Andes or northern coastal ranges. Alternative scenarios suggest dispersal from Pantepui to the Andes (Prum, 1988) or using the northern Cordilleras of South America as a corridor (Marin, 2010; Marín-Espinoza et al., 2014). This last author claims that presence of Guianan species in the Paria Peninsula and Turimiquire mountains is positive evidence for this hypothesis. However, most of this species are from the lowlands or records are doubtful [one captured but not collected individual of *Campylopterus duidae* in a mangrove habitat in northeastern Venezuela (Lefebvre et al., 1994), an unlikely finding for a montane species distributed at high elevations in Pantepui]. There is no evidence clearly supporting the dispersal hypothesis or its mechanism; however, there are few tepuian species with great dispersal capacity, among them *Streptoprocne phelpsi*, a species that has been recorded in the Cordillera de la Costa (one specimen collected at Rancho Grande, Aragua; Hilty, 2003), and *Nannopsittaca panychlora*, which is able to fly over lowlands and away from the tepuian walls and is a normal resident at the Turimiquire region and the Paria Peninsula (Hilty, 2003).

The climate hypothesis proposes a similar dispersal scenario as the previous hypothesis but associated with changes in climate conditions. Climate change is proposed to connect and fragment bird distributions, causing the closest related taxa to show disjunct distributions (Chapman, 1931; Tate, 1938). Haffer (1974) added that low-elevation mountains or hills, present between the Andean region and Pantepui, acted as resources aiding in a stepping-stone dispersal process. Most of these climate changes are associated with Pleistocene glacial and interglacial periods, and Mayr and Phelps (1967) criticized the hypothesis on the grounds that it did not explain the continuous colonization of Pantepui. However, these authors thought of a unique recent period without considering that such climate changes have occurred at multiple times throughout Earth's history (Haffer, 2008). More recently, a more regional perspective of this hypothesis was based on floristic and palynological grounds. Both Steyermark and Dunsterville (1980) and Huber (1988) argued that floristic similarities between highland and lowland vegetation in Pantepui is a consequence of a lack of effective isolation resulting in current vertical migration of species through valleys and gentle slopes and, additionally, shifts in vegetation associated with cold–warm cycles during the Pleistocene. Rull (2004a,b, 2005) proposed a diversification hypothesis based on connectivity phases (glacials), promoting gene flow between isolated populations of similar species or hybridization between species, and isolation phases (interglacials), particularly in the highlands, promoting diversification through vicariance. Displacement of environmentally sensitive species causing mixing of biotas (transient communities) and adaptation to new environments are important elements of this hypothesis, as proposed in the disturbance-vicariance hypothesis (Bush and Colinvaux, 1990; Bush, 1994).

Recent molecular studies allow for evaluation of both the dispersal and climate hypotheses. These studies have shown that lineages present in Pantepui have their closest relatives in the Andes (e.g., Pantepui subspecies of *Trogon personatus*, Da Costa and Klicka, 2008; Cuervo, 2013; *Myiothlypis bivittata roraimae*, Lovette et al., 2010; *Aulacorhynchus whitelianus*, Bonaccorso and Guayasamin, 2013; *Megascops roraimae*, Dantas et al., 2016) or low-lying hills adjacent to the Andes (e.g., *Thamnophilus insignis*, Brumfield and Edwards, 2007). Just one study has shown that Pantepui species are closely related to Paria species (e.g., *Myioborus*, Pérez-Emán, 2005), but it found no clear phylogenetic relationship of this monophyletic group with any other *Myioborus* species. Moreover, timing of divergence varies among species, but in general, results suggest nonsynchronous colonization dynamics, as shown in Amazonian ecosystems (Smith et al., 2014; Naka and Brumfield, 2018). Indeed, a hummingbird study showed that of the three species of mangoes (Trochilidae) present in Pantepui, each diverged from its closest Andean relative at different times, two of them (*Colibri delphinae, C. coruscans*) less than half a million years ago and the other (*Doryfera johannae*) at almost 3 million years ago (Quintero and Perktas, 2018). In summary, molecular evidence mainly supports a hypothesis of colonization from the Andes in an asynchronous temporal pattern. Such results, however, do not rule out dispersal hypotheses based on individual species mechanisms or dispersal mediated by climate changes.

The final hypothesis, the habitat shift, focuses on the importance of elevational speciation in the origin of the Pantepui avifauna. Mayr and Phelps (1967) proposed that populations differentiate through time at higher elevations due to benign environmental conditions compared to adjacent lowlands. However, contrary to this view, Bush and Colinvaux (1990) and Bush (1994) argued that mountain slopes have greater opportunities to hold transient communities (largest disturbance) and, as such, increase the likelihood of isolation and speciation promoted by abiotic and biotic factors. Regardless of the mechanisms associated with differentiation, some of the assumed parapatric or elevationally segregated sister taxon relationships given by Mayr and Phelps (1967) to support this hypothesis have proved to be incorrect (*S. whitelyi* is not closely related to *Nyctipolus nigrescens*, Sigurðsson and Cracraft, 2014; *Hylophilus sclateri* is no longer an *Hylophilus* but a *Vireo*, Slager et al., 2014; *Herpsilochmus roraimae* is not the closest relative of *Herpsilochmus dorsimaculatus*, G. Bravo, pers. comm.; *Thamnophilus insignis* is closer to *Thamnophilus divisorius* than to *Thamnophilus amazonicus*, Brumfield and Edwards, 2007, and *Myrmelastes saturatus* was found to be a different species and potentially unrelated to its parapatric neighbor in the Pantepui province *Myrmelastes leucostigma*, Braun et al., 2005). On the other hand, recent phylogenetic data have also shown that a number of Pantepui endemics (species or subspecies) are sisters to widely distributed lowland taxa (e.g., *Megascops choliba duidae*, Dantas et al., 2016; *Lipaugus streptophorus* and *L. vociferans*, Berv and Prum, 2014; *Myrmothera campanisona* vs *Myrmothera simplex*, Carneiro et al., 2018; *Mionectes macconnelli* vs *M. roraimae*, Hilty and Ascanio, 2014). Thus, how lowland taxa, which undoubtedly have broad elevational ranges (Borges et al., 2018), differentiate into highland endemics is a fertile ground in the understanding of the processes that generate Pantepui diversity and endemism.

Future prospects and conservation

Current knowledge of the Pantepui avifauna has come a long way since the first biogeographical analyses were published. As our knowledge of the current distribution of its avifauna improves (as more basic surveys are done) and their phylogenetic relationships become available, more thorough and informative analyses could be done. Global evaluation of the origin of the Pantepui avifauna, including an assessment of potential mechanisms of dispersal, the description of patterns of elevational segregation in the Pantepui avifauna and further testing of different evolutionary hypotheses about its origin are just some of the questions that need to be addressed in the future. The last question has the potential to evaluate the likelihood of parapatric speciation or secondary contact in explaining such population replacements, a process associated with the buildup of biodiversity in montane ecosystems (Cadena et al., 2019).

Ecological studies in Pantepui have largely been neglected. These studies should range from ecosystem-level studies, assessing ecological services provided by birds (e.g., dispersal, pollination, prey regulation), to community-level studies that focus on the spatial and temporal dynamics of avian communities or assemblages. Migration studies are lacking and could include the likelihood of species altitudinal migration, the importance of these habitats/ecosystems to latitudinal migrants, and short-range migratory behavior (from northern Cordilleras to the tepuis). Nearctic–Neotropical migratory birds occur in low abundance on Pantepui, but the high number of observed species suggests that the region is an important area for migrants that has yet to be evaluated.

Population-level studies that provide details on the demography and natural history of the species are of vital importance to understand the vulnerability (or not) of these ecosystems. There is sparse information on the phenology of molt and reproduction published as a result of specimen collection, but only Willard et al. (1991) provided some analyses. Moreover, available information is biased temporally, as most tepui expeditions have been conducted during the dry season. An interesting pattern has been observed in lowland species distributed at different elevations on the tepui slopes. Individuals of *Xiphorhynchus pardalotus* that inhabit the lowlands start molting a month later than individuals at higher elevations, potentially in response to regional rainfall patterns. Questions such as how general these patterns are, what the association could be with differential responses of species to climate change, and what factors are associated with differentiation along elevation ranges are only some of the topics that could be addressed with these studies. Even in the face of all potential limitations to studying these ecosystems, a clear key to success is the logistics necessary to conduct ecological studies in the area, particularly in the eastern region, due to the availability of roads, their easy accessibility, and the large regional area crossed by them.

The conservation status of the Pantepui avifauna needs to be evaluated. The avifauna of the region is distributed in areas that are not currently threatened by human influence (though tourism activity might require some evaluation). However, the impact of mining and habitat degradation in the lowlands (and lower slopes) adjacent to the tepuis might have an effect on the ecological dynamics of these birds. Additionally, an important conservation aspect to consider is the impact of climate change on Pantepui bird populations. Careful evaluation and descriptions of avian distribution along elevational gradients, spatial connectivity in the region, and the potential impact of climate changes could identify threatened populations and/or species and highlight critical aspects to consider for the conservation and management of these ecosystems (Rull and Nogué, 2007; Nogué et al., 2013).

Acknowledgments

The authors thank both V. Rull and J.V. Remsen Jr. for their comments and helpful editorial remarks that greatly improved this manuscript. We also want to recognize the work done by all people, in particular museum collectors, that has greatly contributed to our knowledge of the Pantepui avifauna, making it possible to write the kind of synthetic work we have done here.

References

Barker, F.K., Burns, K.J., Klicka, J., Lanyon, S.M., Lovette, I.J., 2015. New insights into New World biogeography: an integrated view from the phylogeny of blackbirds, cardinals, sparrows, tanagers, warblers, and allies. Auk 132, 333–348.

Barnett, A., Shapley, R., Benjamin, P., Henry, E., Garrell, M., 2002. Birds of the Potaro Plateau, with eight new species for Guyana. Cotinga 18, 19–36.

Barrowclough, G.F., Escalante-Pliego, P., 1990. Notes on the birds of the Sierra de Unturán, southern Venezuela. Bull. Br. Ornithol. Club 110, 167–169.

Barrowclough, G.F., Escalante-Pliego, P., Aveledo-Hostos, R., Pérez-Chinchilla, L., 1995. An annotated list of the birds of the Cerro Tamacuarí region, Serranía de Tapirapecó, Federal Territory of Amazonas, Venezuela. Bull. Br. Ornithol. Club 115, 211–219.

Barrowclough, G.F., Lentino, M.R., Sweet, P.R., 1997. New records of birds from Auyán-tepui, Estado Bolívar, Venezuela. Bull. Br. Ornithol. Club 117, 194–198.

Berry, P.E., Huber, O., Holst, B.K., 1995. Floristic analysis and phytogeography. In: Steyermark, J.A., Berry, P.E., Holst, B.K. (Eds.), Flora of the Venezuelan Guayana. Introduction, vol. 1. Timber Press, Oregon, pp. 161–191.

Berry, P.E., Riina, R., 2005. Insights into the diversity of the Pantepui flora and the biogeographic complexity of the Guayana Shield. Biol. Skr. 55, 145–167.

Berv, J.S., Prum, R.O., 2014. A comprehensive multilocus phylogeny of the Neotropical cotingas (Cotingidae, Aves) with a comparative evolutionary analysis of breeding system and plumage dimorphism and a revised phylogenetic classification. Mol. Phyl. Evol. 81, 120–136.

Bonaccorso, E., Guayasamin, J.M., Peterson, A.T., Navarro-Sigüenza, A.G., 2011. Molecular phylogeny and systematics of Neotropical toucanets in the genus *Aulacorhynchus* (Aves, Ramphastidae). Zool. Scr. 40, 336–349.

Bonaccorso, E., Guayasamin, J.M., 2013. On the origin of Pantepui montane biotas: a perspective based on the phylogeny of *Aulacorhynchus* toucanets. PLoS One 8, e67321. Available from: https://doi.org/10.1371/journal.pone.0067321.

Borges, S.H., Dantas Santos, M.P., Moreira, M., Baccaro, F., Capurucho, J.M.G., Ribas, C., 2018. Dissecting bird diversity in the Pantepui area of endemism, northern South America. J. Ornithol. 159, 1073–1086.

Braun, M.J., Robbins, M.B., Milensky, C.M., O'Shea, B.J., Barber, B.R., Hinds, W., et al., 2003. New birds for Guyana from Mts Roraima and Ayanganna. Bull. Br. Ornithol. Club 123, 24–33.

Braun, M.J., Isler, M.L., Isler, P.R., Bates, J.M., Robbins, M.B., 2005. Avian speciation in the Pantepui: the case of the Roraiman Antbird (*Percnostola* [*Schistocichla*] "*leucostigma*" *saturata*). Condor 107, 327–341.

Briceño, H.O., Schubert, C., 1990. Geomorphology of the Gran Sabana, Guayana Shield, southeastern Venezuela. Geomorphology 3, 125–141.

Brumfield, R.T., Edwards, S.V., 2007. Evolution into and out of the Andes: a Bayesian analysis of historical diversification in *Thamnophilus* antshrikes. Evolution 61, 346–367.

Bush, M.B., 1994. Amazonian speciation: a necessarily complex model. J. Biogeogr. 21, 5–17.

Bush, M.B., Colinvaux, P.A., 1990. A long climatic and vegetation record from lowland Panama. J. Veg. Sci. 1, 105–118.

Cabanis, J.V., 1848. In: Schomburg, R. (Ed.), Reisen in Britisch-Guiana in den Jahren 1840–1844, vol. 3. J. J. Weber, Leipzig, pp. 662–765.

Cadena, C.D., Pérez-Emán, J.L., Cuervo, A.M., Céspedes, L.N., Epperly, K.L., Klicka, J.T., 2019. Extreme genetic structure and dynamic range evolution in a montane passerine bird: implications for tropical diversification. Biol. J. Linn. Soc. 126, 487–506.

Carneiro, L., Bravo, G.A., Aristizábal, N., Cuervo, A.M., Aleixo, A., 2018. Molecular systematics and biogeography of lowland antpittas (Aves, Grallariidae): the role of vicariance and dispersal in the diversification of a widespread Neotropical lineage. Mol. Phyl. Evol. 120, 375–389.

Chapman, R.M., 1917. The distribution of bird-life in Colombia. A contribution to a biological survey in South America. Bull. Am. Mus. Nat. Hist. 36, 1–729.

Chapman, R.M., 1931. The upper zonal bird-life of Mts. Roraima and Duida. Bull. Am. Mus. Nat. Hist. 63, 1–135.

Cook, R.E., 1974. Origin of the highland avifauna of southern Venezuela. Syst. Zool. 23, 257–264.

Cortés-B, R., Franco-R, P., Rangel-C, J.O., 1998. La flora vascular de la Sierra de Chiribiquete, Colombia. Caldasia 20, 103–141.

Cracraft, J., 1985. Historical biogeography and patterns of differentiation within the South American avifauna: areas of endemism. Ornithol. Monogr. 36, 49–84.

Croizat-Chaley, L., 1976. Biogeografía analítica y sintética ("Panbiogeografía") de las Américas. Academia de Ciencias Físicas. Matemáticas y Naturales, Caracas.

Cuervo, A.M., 2013. Evolutionary Assembly of the Neotropical Montane Avifauna (Ph.D. dissertation). Louisiana State University, 125 pp.

Da Costa, J.M., Klicka, J., 2008. The Great American Interchange in birds: a phylogenetic perspective with the genus *Trogon*. Mol. Ecol. 17, 1328–1343.

Dantas, S.M., Weckstein, J.D., Bates, J.M., Krabbe, N.K., Cadena, C.D., Robbins, M.B., et al., 2016. Molecular systematics of the new world screech-owls (*Megascops*: Aves, Strigidae): biogeographic and taxonomic implications. Mol. Phyl. Evol. 94, 626–634.

Dickerman, R.W., Phelps Jr., W.H., 1982. An annotated list of the birds of Cerro Urutaní on the border of Estado Bolívar, Venezuela, and Territorio Roraima, Brazil. Am. Mus. Novit. 2732, 1–20.

Estrada, J., Fuertes, J., 1993. Estudios botánicos en la Guayana colombiana: 4. Notas sobre la vegetación y la flora de la Sierra de Chiribiquete. Revista de la Academia Colombiana de Ciencias Exactas Físicas y Naturales 18, 483–497.

Gansser, A., 1954. The Guiana Shield (South America). Eclogae Geol. Helv. 47, 77–117.

Ghosh, S., 1985. Geology of the Roraima Group and its implications. Boletín de Geología (Venezuela), Publicación Especial 10, 33–50.

Gilliard, E.T., 1941. The birds of Mt. Auyan-tepui, Venezuela. Bull. Am. Mus. Nat. Hist. 67, 439–508.

Giraldo-Cañas, D., 2001. Relaciones fitogeográficas de las sierras y afloramientos rocosos de la Guayana colombiana: un estudio preliminar. Revista Chilena de Historia Natural 74, 353–364.

Givnish, T.J., Evans, T.M., Pires, J.C., Sytsma, K.J., 1997. Molecular evolution and adaptive radiation in *Brocchinia* (Bromeliaceae: Pitcairnioideae) atop the tepuis of the Guayana Shield. In: Givnish, T.J., Sytsma, K. J. (Eds.), Molecular Evolution and Adaptive Radiation. Cambridge University Press, Cambridge, pp. 259–311.

Givnish, T.J., Evans, T.M., Zjhra, M.L., Patterson, T.B., Berry, P.E., Sytsma, K.J., 2000. Molecular evolution, adaptive radiation, and geographic diversification in the amphiatlantic family Rapateaceae: evidence from *ndh* F sequences and morphology. Evolution 54, 1915–1937.
Givnish, T.J., Barfuss, M.H.J., van Ee, B., Riina, R., Schulte, K., Horres, R., et al., 2011. Phylogeny, adaptive radiation, and historical biogeography in Bromeliaceae: insights from an eight-locus plastid phylogeny. Am. J. Bot. 98, 872–895.
Graves, G., 1982. Speciation in the Carbonated Flower-Piercer (*Diglossa carbonaria*) complex of the Andes. Condor 84, 1–14.
Haffer, J., 1974. Avian speciation in tropical South America. Pub. Nuttall Ornithol. Club 14, 1–390.
Haffer, J., 2008. Hypotheses to explain the origin of species in Amazonia. Braz. J. Biol. 68 (Suppl), 917–947.
Hilty, S.L., 2003. Birds of Venezuela., second ed. Princeton University Press, Princeton, USA.
Hilty, S.L., Ascanio, D., 2014. McConnell's Flycatcher *Mionectes macconnelli* is more than one species. Bull. Br. Ornithol. Club 134, 270–279.
Huber, O., 1987. Consideraciones sobre el concepto de Pantepui. Pantepui 2, 2–10.
Huber, O., 1988. Guayana highlands versus Guayana lowlands, a reappraisal. Taxon 37, 595–614.
Huber, O., 1995. Geographical and physical features. In: Steyermark, J.A., Berry, P.E., Holst, B.K. (Eds.), Flora of the Venezuelan Guayana. Introduction., vol. 1. Timber Press, Oregon, pp. 1–61.
Huber, O., Wurdack, J.J., 1984. History of botanical exploration in Territorio Federal Amazonas, Venezuela. Smithsonian Contrib. Bot. 56, 1–83.
Kok, P.J.R., MacCulloch, R.D., Means, D.B., Roelants, K., van Bocxlaer, I., Bossuyt, F., 2012. Curr. Biol. 22, R587–R589.
Lefebvre, G., Poulin, B., McNeill, R., 1994. Temporal dynamics of mangrove bird communities in Venezuela with special reference to migrant warblers. Auk 111, 405–415.
Lentino, M., Bosque, C., 1989. Avifauna preliminar de Marahuaca, T.F. Amazonas, Venezuela. Acta Terramaris 1, 65–70.
Lougheed, S.C., Campagna, L., Dávila, J.A., Tubaro, P.L., Lijtmaer, D.A., Handford, P., 2013. Continental phylogeography of an ecologically and morphologically diverse Neotropical songbird, *Zonotrichia capensis*. BMC Evol. Biol. 13, 58.
Lovette, I.J., Pérez-Emán, J.L., Sullivan, J.P., Banks, R.C., Fiorentino, I., Córdoba-Córdoba, S., et al., 2010. A comprehensive multilocus phylogeny for the wood-warblers and a revised classification of the Parulidae (Aves). Mol. Phyl. Evol. 57, 753–770.
MacArthur, R.H., Wilson, E.O., 1967. The Theory of Island Biogeography. Princeton University Press, Princeton.
Maguire, B., 1979. Guayana, Region of the Roraima Sandstone Formation. In: Larsen, K., Holm-Nielsen, L.B. (Eds.), Tropical Botany. Academic Press, London, pp. 185–221.
Marantz, C.A., Aleixo, A., Bevier, L., Patten, M.A., 2018. Strong-billed Woodcreeper (Xiphocolaptes promeropirhynchus. In: del Hoyo, J., Elliott, A., Sargatal, J.J., Christie, D.A., de Juana, E. (Eds.), Handbook of the Birds of the World Alive. Lynx Edicions, Barcelona, Retrieved from: https://www.hbw.com/node/56609 (10.09.18).
Marin, G., 2010. Acerca del origen y biogeografía de la avifauna del Pantepuy o Croizat versus Mayr: una revisión crítica. Biologist (Perú) 8, 79–108.
Marín-Espinoza, G., González-Bruzual, L.G., Navarro-Rodríguez, R., 2014. *Elaenia ruficeps* (Aves: Tyrannidae, Elaeniinae): nuevo indicio de la interconexión biogeográfica de las avifaunas del Macizo Guayanés y la Cordillera Nororiental de Venezuela. Biologist 12, 1–7.
Mauck III, W.M., Burns, K.J., 2009. Phylogeny, biogeography, and recurrent evolution of divergent bill types in the nectar-stealing flowerpiercers (Thraupini: *Diglossa* and *Diglossopis*). Biol. J. Linn. Soc. 98, 14–28.
Mayr, E., Phelps Jr., W.H., 1955. Origin of the bird fauna of Pantepui. In: Portman, A., Sutter, E. (Eds.), Acta Xl Congressus Internationalis Omithologici. XI International Ornithological Congress, Basel, pp. 399–400.
Mayr, E., Phelps Jr., W.H., 1967. The origin of the bird fauna of the South Venezuelan Highlands. Bull. Am. Mus. Nat. Hist. 136, 269–328.
McDiarmid, R.W., Donnelly, M.A., 2005. The herpetofauna of the Guayana Highlands: amphibians and reptiles of the Lost World. In: Donnelly, M.A., Crother, B.I., Guyer, C., Wake, M.H., White, M.E. (Eds.), Ecology and Evolution in the Tropics: A Herpetological Perspective. University of Chicago Press, Chicago, Illinois, pp. 461–560.

McGuire, J.A., Witt, C.C., Remsen Jr., J.V., Corl, A., Rabosky, D.L., et al., 2014. Molecular phylogenetics and the diversification of hummingbirds. Curr. Biol. 24, 910–916.

Merriam, C.H., 1892. The geographic distribution of life in North America with special reference to the Mammalia. Proc. Biol. Soc. Wash. 7, 1–64.

Naka, L.N., Brumfield, R.T., 2018. The dual role of Amazonian rivers in the generation and maintenance of avian diversity. Sci. Adv. 4, eaar8575.

Nogué, S., Rull, V., Vegas-Vilarrúbia, T., 2013. Elevational gradients in the neotropical table mountains: patterns of endemism and implications for conservation. Divers. Distrib. 19, 676–687.

O'Shea, B.J., Milensky, C.M., Claramunt, S., Schmidt, B.K., Gebhard, C.A., Schmitt, C.G., et al., 2007. New records for Guyana, with description of the voice of Roraiman Nightjar *Caprimulgus whitelyi*. Bull. Br. Ornithol. Club 127, 118–128.

Pérez-Emán, J., 2002. Molecular systematics, biogeography, and population history of the genus *Myioborus* (Aves, Parulinae) (Ph.D. dissertation). University of Missouri-St. Louis, 132 pp.

Pérez-Emán, J.L., 2005. Molecular phylogenetics and biogeography of the Neotropical redstarts (*Myioborus*; Aves, Parulinae). Mol. Phyl. Evol. 37, 511–528.

Pérez-Emán, J., Sharpe, C.J., Lentino, M., Prum, R.O., Carreño F, I.J., 2003. New records of birds from the summit of Cerro Guaiquinima, Estado Bolívar, Venezuela. Bull. Br. Ornithol. Club 123, 78–90.

Phelps Jr., W.H., Aveledo, R.H., 1984. Dos nuevas subespecies de aves (Troglodytidae, Fringillidae) del Cerro Marahuaka, Territorio Amazonas, Venezuela. Boletín de la Sociedad Venezolana de Ciencias Naturales 39, 5–10.

Phelps, W.H., Phelps Jr., W.H., 1963. Lista de las aves de Venezuela con su distribución. Segunda edición. Volumen 1, Parte 2, Passeriformes. Boletín de la Sociedad Venezolana de Ciencias Naturales 24 (104, 106), 1–479 pp.

Quintero, E., Perktas, U., 2018. Phylogeny and biogeography of a subclade of mangoes (Aves, Trochilidae). J. Ornithol. 159, 29–46.

Prum, R.O., 1994. Species status of the White-fronted Manakin, *Lepidothrix serena* (Pipridae), with comments on conservation biology. Condor 96, 692–702.

Prum, R.O., 1988. Historical relationships among avian forest areas of endemism in the Neotropics. In: Quellet, H. (Ed.), Acta XIX Congressus Internationalis Ornithologici. XIX International Ornithological Congress, Ottawa, pp. 2562–2572.

Remsen Jr., J.V., Areta, J.I., Cadena, C.D., Claramunt, S., Jaramillo, A., Pacheco, J.F., et al., 2019. A classification of the bird species of South America. American Ornithologists' Union. Available from: http://www.museum.lsu.edu/~Remsen/SACCBaseline.html.

Rheindt, F.E., Christidis, L., Norman, J.A., 2009. Genetic introgression, incomplete lineage sorting and faulty taxonomy create multiple polyphyly in a montane clade of *Elaenia* flycatchers. Zool. Scr. 38, 143–153.

Ridgely, R.S., Agro, D., Joseph, L., 2005. Birds of Iwokrama Forest. Proc. Acad. Nat. Sci. Philadelphia 154, 109–121.

Rising, J., Jaramillo, A., 2019. Rufous-collared Sparrow (*Zonotrichia capensis*). In: del Hoyo, J., Elliott, A., Sargatal, J., Christie, D.A., de Juana, E. (Eds.), Handbook of the Birds of the World Alive. Lynx Edicions, Barcelona, Retrieved from: https://www.hbw.com/node/61910 (04.02.19).

Robbins, M.B., Braun, M.B., Milensky, C.M., Schmidt, B.K., Prince, W., Rice, N.H., et al., 2007. Avifauna of the Upper Essequibo River and Acary Mountains, southern Guyana. Ornitol. Neotrop. 18, 339–368.

Rull, V., 2004a. Biogeography of the "Lost World": a palaeoecological perspective. Earth-Sci. Rev. 67, 125–137.

Rull, V., 2004b. Is the Lost World really lost? Palaeoecological insights into the origin of the peculiar flora of the Guayana Highlands. Naturwissenschaften 91, 139–142.

Rull, V., 2005. Biotic diversification in the Guayana Highlands: a proposal. J. Biogeogr. 32, 921–927.

Rull, V., Montoya, E., 2017. Holocene vegetation dynamics on the Apakará summit of the neotropical Guayana Highlands and potential environmental drivers. Rev. Palaeobot. Palynol. 240, 22–32.

Rull, V., Nogué, S., 2007. Potential migration routes and barriers for vascular plants of the Neotropical Guyana Highlands during the Quaternary. J. Biogeogr. 34, 1327–1341.

Rull, V., Montoya, E., Nogué, S., Vegas-Vilarrúbia, T., Safont, E., 2013. Ecological palaeoecology in the Neotropical Gran Sabana region: long-term records of vegetation dynamics as a basis for ecological hypothesis testing. Perspect. Plant Ecol. Evol. Syst. 15, 338–359.

Schubert, C., Fritz, P., Aravena, R., 1992. Investigaciones paleoambientales: resultados preliminares. In: Huber, O. (Ed.), Chimantá. Escudo de Guayana, Venezuela. Un ensayo ecológico tepuyano. Oscar Todtmann Editores, C. A., Caracas, Venezuela, pp. 97–110.
Schubert, C., Fritz, P., Aravena, R., 1994. Late Quaternary paleoenvironmental studies in the Gran Sabana (Venezuelan Guayana Shield). Quat. Int. 21, 81–90.
Salerno, P.E., Ron, S.R., Señaris, C., Rojas-Runjaic, F.J.M., Noonan, B.P., Cannatella, D.C., 2012. Ancient tepui summits harbor young rather than old lineages of endemic frogs. Evolution 66, 3000–3013.
Salerno, P.E., Señaris, J.C., Rojas-Runjaic, F.J.M., Cannatella, D.C., 2015. Recent evolutionary history of Lost World endemics: population genetics, species delimitation, and phylogeography of sky-island treefrogs. Mol. Phyl. Evol. 82, 314–323.
Sidder, G.B. and S.V. Mendoza. 1991. Geology of the Venezuelan Guayana Shield and its Relation to the entire Guayana Shield. U.S. Geological Survey Open File Report, 91–141.
Sigurðsson, S., Cracraft, J., 2014. Deciphering the diversity and history of New World nightjars (Aves: Caprimulgidae) using molecular phylogenetics. Zool. J. Linn. Soc. 170, 506–545.
Slager, D.L., Battey, C.J., Bryson Jr., R.W., Voelker, G., Klicka, J., 2014. A multilocus phylogeny of a major New World avian radiation: the Vireonidae. Mol. Phyl. Evol. 80, 95–104.
Smith, B.T., McCormack, J.E., Cuervo, A.M., Hickerson, M.J., Aleixo, A., Cadena, C.D., et al., 2014. The drivers of tropical speciation. Nature 515, 406–409.
Steyermark, J.A., 1986. Speciation and endemism in the flora of the Venezuelan tepuis. In: Vuilleumier, F., Monasterio, M. (Eds.), High Altitude Tropical Biogeography. Oxford University Press, New York, pp. 317–373.
Steyermark, J.A., Dunsterville, G.C.K., 1980. The lowland floral element on the summit of Cerro Guaiquinima and other cerros of the Guayana highland of Venezuela. J. Biogeogr. 7, 283–303.
Stiles, F.G., Tellería, J.L., Díaz, M., 1995. Observaciones sobre la composición, ecología y zoogeografía de la avifauna de la Sierra de Chiribiquete, Caquetá, Colombia. Caldasia 17, 481–500.
Stiles, F.G., 1996. A new species of Emerald hummingbird (Trochilidae, *Chlorostibon*) from the Sierra de Chiribiquete, southeastern Colombia, with a review of the *C. mellisugus* complex. Wilson Bull. 108, 1–27.
Tate, G.H.H., 1938. Notes on the Phelps Venezuelan Expedition. Geograph. Rev. 28, 452–474.
Vuilleumier, F., 1969. Pleistocene speciation in birds living in the high Andes. Nature 223, 1179–1180.
Vuilleumier, F., 2005. Dean of American ornithologists: the multiple legacies of Frank M. Chapman of the American Museum of Natural History. Auk 122, 389–402.
Willard, D.E., Foster, M.S., Barrowclough, G.F., Dickerman, R.W., Cannell, P.F., Coats, S.L., et al., 1991. The birds of Cerro de la Neblina, Territorio Federal Amazonas, Venezuela. Fieldiana. Zoology 65, 1–80.
Zink, J.A., García, P., van de Plicht, J., 2011. Tepui peatlands: age record and environmental changes. In: Zink, J. A., Huber, O. (Eds.), Peatlands of the Western Guayana Highlands, Venezuela. Springer, Heidelberg, pp. 189–236.
Zyskowski, K., Mittermeier, J.C., Ottema, O., Rakovic, M., O'Shea, B.J., Lai, J.E., et al., 2011. Avifauna of the easternmost tepui, Tafelberg in Central Suriname. Bull. Peabody Mus. Nat. Hist. 52, 153–180.

Appendix I List of the 141 species included in previous analyses of Pantepui avifauna and this study. Category: PE, Pantepui endemic; MSUB, species with montane subspecies; LSUB, lowland species with highland differentiated populations (subspecies); M, montane species; L, lowland species; H, habitat specialist; B, species with broad distribution and high dispersal abilities

Species	Category	Mayr and Phelps (1967)	Willard et al. (1991)	Zyskowski et al. (2011)	Borges et al. (2018)	This study
Crypturellus ptaritepui	PE	X	X	X	X	X
Tigrisoma fasciatum	M		X	X	X	X
Buteogallus solitarius	B		X	X	X	X
Geranoaetus melanoleucus	B		X	X	X	X
Patagioenas fasciata	MSUB	X	X	X	X	X
Megascops choliba	LSUB	X	X	X	X	X
Megascops roraimae	PE	X	X	X	X	X
Glaucidium brasilianum	LSUB	X	X	X	X	X
Aegolius harrisii	M		X	X	X	X
Steatornis caripensis	H		X		X	X
Hydropsalis roraimae	PE	X	X	X	X	X
Setopagis whitelyi	PE	X	X	X	X	X
Streptoprocne phelpsi	PE	X	X	X	X	X
Streptoprocne zonaris	H				X	X
Chaetura cinereiventris	H	X	X	X		X
Aeronautes montivagus	MSUB	X	X	X	X	X
Cypseloides cryptus	H				X	X
Phaethornis augusti	LSUB	X	X	X	X	X
Phaethornis bourcieri	L	X	X			X
Phaethornis griseogularis	M		X	X	X	X
Doryfera johannae	MSUB	X	X	X	X	X
	PE	X	X	X	X	X

(*Continued*)

(Continued)

Species	Category	Mayr and Phelps (1967)	Willard et al. (1991)	Zyskowski et al. (2011)	Borges et al. (2018)	This study
Campylopterus hyperythrus						
Campylopterus duidae	PE	X	X	X	X	X
Colibri coruscans	MSUB	X	X	X	X	X
Colibri delphinae	M		X	X	X	X
Lophornis pavoninus	PE	X	X	X	X	X
Lophornis ornatus	L				X	X
Polytmus milleri	PE	X	X	X	X	X
Heliodoxa xanthogonys	PE	X	X	X	X	X
Amazilia viridigaster	MSUB	X	X	X	X	X
Amazilia lactea	LSUB				X	X
Amazilia versicolor	L				X	X
Amazilia tobaci	L				X	X
Chlorostilbon mellisugus	LSUB				X	X
Trogon personatus	MSUB	X	X	X	X	X
Trogon collaris	L				X	X
Aulacorhynchus whitelianus	PE	X	X	X	X	X
Veniliornis kirkii	MSUB	X	X	X	X	X
Colaptes rubiginosus	MSUB	X	X	X	X	X
Falco deiroleucus	L				X	X
Pyrrhura egregia	PE	X	X	X	X	X
Pyrrhura picta	L				X	X
Nannopsittaca panychlora	M	X	X	X	X	X
Amazona dufresniana	L				X	X
Taraba major	LSUB	X	X	X	X	X
Thamnophilus insignis	PE	X	X	X	X	X

(*Continued*)

(Continued)

Species	Category	Mayr and Phelps (1967)	Willard et al. (1991)	Zyskowski et al. (2011)	Borges et al. (2018)	This study
Dysithamnus mentalis	MSUB	X	X	X	X	X
Dysithamnus leucostictus	M			X	X	X
Myrmotherula behni	MSUB	X	X	X	X	X
Herpsilochmus roraimae	PE	X	X	X	X	X
Euchrepomis callinota	MSUB				X	X
Myrmelastes saturatus	PE	X	X	X	X	X
Myrmelastes caurensis	L			X	X	X
Grallaria guatimalensis	MSUB	X	X	X	X	X
Grallaricula nana	MSUB	X	X	X	X	X
Myrmothera simplex	PE	X	X	X	X	X
Chamaeza campanisona	MSUB	X	X	X	X	X
Synallaxis macconnelli	MSUB	X	X	X	X	X
Cranioleuca demissa	PE	X	X	X	X	X
Philydor rufum	MSUB				X	X
Roraimia adusta	PE	X	X	X	X	X
Automolus subulatus	LSUB				X	X
Clibanornis rubiginosus	MSUB				X	X
Syndactyla roraimae	PE	X	X	X	X	X
Sclerurus mexicanus	L				X	X
Lochmias nematura	MSUB	X	X	X	X	X
Xiphocolaptes promeropirhynchus	LSUB	X	X	X	X	X
Glyphorynchus spirurus	LSUB				X	X
Xiphorhynchus pardalotus	L				X	X

(*Continued*)

(Continued)

Species	Category	Mayr and Phelps (1967)	Willard et al. (1991)	Zyskowski et al. (2011)	Borges et al. (2018)	This study
Phyllomyias burmeisteri	MSUB	X		X	X	X
Leptopogon amaurocephalus	LSUB		X		X	X
Elaenia cristata	LSUB				X	X
Elaenia dayi	PE	X	X	X	X	X
Elaenia olivina	PE	X	X	X	X	X
Elaenia ruficeps	L	X				X
Mecocerculus leucophrys	MSUB	X	X	X	X	X
Pogonotriccus chapmani	PE	X	X	X	X	X
Phylloscartes nigrifrons	PE	X	X	X	X	X
Mionectes macconnelli	MSUB	X	X	X	X	X
Mionectes oleagineus	LSUB				X	X
Myiophobus roraimae	MSUB	X	X	X	X	X
Hemitriccus margaritaceiventer	LSUB	X	X	X	X	X
Poecilotriccus russatus	PE	X	X	X	X	X
Platyrinchus mystaceus	LSUB	X	X	X	X	X
Hirundinea ferruginea	H	X	X	X	X	X
Contopus fumigatus	MSUB	X	X	X	X	X
Contopus nigrescens	M				X	X
Knipolegus poecilurus	MSUB	X	X	X	X	X
Myiarchus swainsoni	L				X	X
Oxyruncus cristatus	MSUB	X	X	X	X	X
Pipreola whitelyi	PE	X	X	X	X	X
Lipaugus streptophorus	PE	X	X	X	X	X
Rupicola rupicola	L		X		X	X

(*Continued*)

(Continued)

Species	Category	Mayr and Phelps (1967)	Willard et al. (1991)	Zyskowski et al. (2011)	Borges et al. (2018)	This study
Procnias albus	M				X	X
Procnias averano	M				X	X
Lepidothrix suavissima	PE	X	X	X	X	X
Xenopipo uniformis	PE	X	X	X	X	X
Ceratopipra cornuta	PE	X	X	X	X	X
Pachyramphus castaneus	LSUB	X	X	X	X	X
Vireo sclateri	PE	X	X	X	X	X
Pygochelidon cyanoleuca	H	X	X		X	X
Alopochelidon fucata	H				X	X
Cistothorus platensis	M	X	X	X	X	X
Henicorhina leucosticta	L				X	X
Coereba flaveola	L					X
Pheugopedius coraya	LSUB	X		X	X	X
Troglodytes rufulus	PE	X	X	X	X	X
Microcerculus ustulatus	PE	X	X	X	X	X
Cichlopsis leucogenys	MSUB	X	X	X	X	X
Turdus flavipes	MSUB	X	X	X	X	X
Turdus leucops	M	X	X	X	X	X
Turdus olivater	MSUB	X	X	X	X	X
Turdus ignobilis	LSUB	X	X	X	X	X
Pipraeidea melanonota	M	X	X	X	X	X
Ixothraupis xanthogastra	LSUB	X	X	X	X	X
Ixothraupis punctata	L		X			X
Ixothraupis guttata	MSUB	X	X	X	X	X
Tangara gyrola	L	X			X	X
Stilpnia cyanoptera	MSUB	X	X	X	X	X

(*Continued*)

(Continued)

Species	Category	Mayr and Phelps (1967)	Willard et al. (1991)	Zyskowski et al. (2011)	Borges et al. (2018)	This study
Cyanerpes caeruleus	LSUB				X	X
Hemithraupis guira	LSUB				X	X
Diglossa duidae	PE	X	X	X	X	X
Diglossa major	PE	X	X	X	X	X
Haplospiza rustica	MSUB	X	X	X	X	X
Emberizoides duidae	PE	X	X	X	X	X
Catamenia homochroa	MSUB	X	X	X	X	X
Mitrospingus oleagineus	PE	X	X	X	X	X
Zonotrichia capensis	MSUB	X	X	X	X	X
Atlapetes personatus	PE	X	X	X	X	X
Piranga flava	MSUB	X	X	X	X	X
Piranga leucoptera	M		X	X	X	X
Setophaga pitiayumi	L	X	X	X	X	X
Myioborus miniatus	M	X	X	X	X	X
Myioborus castaneocapilla	PE	X	X	X	X	X
Myioborus cardonai	PE	X	X	X	X	X
Myioborus albifacies	PE	X	X	X	X	X
Myiothlypis bivittata	MSUB	X	X	X	X	X
Basileuterus culicivorus	LSUB				X	X
Macroagelaius imthurni	PE	X	X	X	X	X
Chlorophonia cyanea	MSUB	X	X	X	X	X
Spinus magellanicus	M	X		X	X	X
Total number of species		98	104	103	136	141

Appendix II Changes in the taxonomy of Pantepui birds since Mayr and Phelps (1967). Current names are based on the South American Classification Committee (Remsen et al., 2019)

Previous name(s)	Current name
Harpyhaliaetus solitarius	*Buteogallus solitarius*
Otus guatemalae	*Megascops roraimae*
Megascops guatemalae	
Megascops vermiculatus roraimae	
Caprimulgus longirostris	*Systellura longirostris*
Hydropsalis roraimae	
Caprimulgus whitelyi	*Setopagis whitelyi*
Hydropsalis whitelyi	
Cypseloides phelpsi	*Streptoprocne phelpsi*
Aulacorhynchus derbianus	*Aulacorhynchus whitelianus*
Piculus rubiginosus	*Colaptes rubiginosus*
Schistocichla caurensis	*Myrmelastes caurensis*
Percnostola leucostigma	*Myrmelastes leucostigma*
	Myrmelastes saturatus
Margarornis adusta	*Roraimia adusta*
Synallaxis moesta	*S. macconnelli*
Cranioleuca curtata demissa	*C. demissa*
Philydor hylobius	Not valid (described from juvenile males of *Syndactyla roraimae*)
Automolus roraimae	*S. roraimae*
Acrochordopus zeledoni	*Phyllomyias burmeisteri*
Pipromorpha macconnelli	*Mionectes macconnelli*
Todirostrum mystaceus	*Platyrinchus mystaceus*
Idioptilon margaritaceiventer	*Hemitriccus margaritaceiventer*
Todirostrum russatum	*Poecilotriccus russatus*
Chloropipo uniformis	*Xenopipo uniformis*
Pipra cornuta	*Ceratopipra cornuta*
Hylophilus sclateri	*Vireo sclateri*
Notiochelidon cyanoleuca	*Pygochelidon cyanoleuca*
Atticora cyanoleuca	

(*Continued*)

(Continued)

Previous name(s)	Current name
Thryothorus coraya	*Pheugopedius coraya*
Platycichla flavipes	*Turdus flavipes*
Tangara xanthogastra	*Ixothraupis xanthogastra*
Tangara chrysophrys	*Ixothraupis guttata*
Tangara guttata	
Tangara punctata	*Ixothraupis punctata*
Tangara cyanoptera	*Stilpnia cyanoptera*
Spodiornis rusticus	*Haplospiza rustica*
Emberizoides herbicola duidae	*Emberizoides duidae*
Parula pitiayumi	*Setophaga pitiayumi*
Basileuterus bivittatus	*Myiothlypis bivittata*
Carduelis magellanica	*S. magellanicus*
Sporagra magellanica	

Appendix III Endemic species from Pantepui and its distribution based on biogeographic sectors as defined by Berry et al. (1995)

Species	Eastern district	Western district	Central district	Southern district
Crypturellus ptaritepui	1	0	0	0
Megascops roraimae	1	1	1	1
Hydropsalis roraimae	1	1	1	1
Setopagis whitelyi	1	0	1	0
Myrmelastes saturatus	1	0	0	0
Campylopterus hyperythrus	1	0	0	0
Campylopterus duidae	0	1	1	1
Lophornis pavoninus	1	1	1	1
Poecilotriccus russatus	1	0	0	0
Heliodoxa xanthogonys	1	1	1	1
Aulacorhynchus whitelianus	1	1	1	1
Polytmus milleri	1	0	1	0
Syndactyla roraimae	1	1	1	1
Herpsilochmus roraimae	1	0	1	1
R.oraimia adusta	1	1	1	0
Myrmothera simplex	1	1	1	1
Cranioleuca demissa	1	1	1	1
Pyrrhura egregia	1	0	1	0
Streptoprocne phelpsi	1	1	1	1
Elaenia dayi	1	1	1	0
Elaenia olivina	1	1	1	1
Pipreola whitelyi	1	0	0	0
Phylloscartes nigrifrons	1	1	1	1
Phylloscartes chapmani	1	1	1	1
Lipaugus streptophorus	1	0	0	0
Lepidothrix suavissima	1	1	1	1
Xenopipo uniformis	1	1	1	0
Vireo sclateri	1	1	1	1
Ceratopipra cornuta	1	1	1	1

(*Continued*)

(Continued)

Species	Eastern district	Western district	Central district	Southern district
Troglodytes rufulus	1	1	1	1
Thamnophilus insignis	1	1	1	1
Microcerculus ustulatus	1	1	1	1
Diglossa duidae	0	1	1	1
Diglossa major	1	0	0	0
Emberizoides duidae	0	0	1	0
Mitrospingus oleagineus	1	0	0	0
Atlapetes personatus	1	1	1	1
Myioborus castaneocapilla	1	0	1	1
Myioborus cardonai	0	0	1	0
Myioborus albifacies	0	1	0	0
Macroagelaius imthurni	1	1	1	0
Total per district	36	26	32	23

CHAPTER

14

Mammals

Daniel Lew[1] *and Burton K. Lim*[2]

[1]Biodiversity Unit, Venezuelan Institute of Scientific Research (IVIC), Caracas, Venezuela

[2]Department of Natural History, Royal Ontario Museum (ROM), Toronto, ON, Canada

Introduction

The Guiana Shield covers an approximate area of 2.5 million km^2 located roughly between the main channels of the Orinoco and Amazon Rivers (Huber and Foster, 2002). The geographic limits are the Amazon River (Brazil) to the south, the Japurá-Caquetá River to the southwest, the Serranía de Chiribiquete to the extreme west (Colombia), the Guaviare River to the northwest, the Orinoco to the north (Venezuela), and the Atlantic Ocean to the east along the coastal drainage area of Guyana, French Guiana, Suriname, and Brazil. The Guiana Shield is defined by a set of geological, climatic, physiographic, and biogeographic characteristics. The topographical landscape is dominated by a distinctive feature, including a mountainous complex of tablelands recognized by Huber (2006: 470) as the phytogeographic Pantepui province, which "includes all high mountain ecosystems of the Guayana Shield mountains above an elevation of c.1500 m a.s.l. The province is discontinuous and occurs mainly in the meso (12°C–18°C) and submicrothermic (8°C–12°C) belts of southern Venezuela, but also in northwest Guyana and along the Brazilian–Venezuelan watershed." From a biological point of view Pantepui stands out for its great variety of ecosystems and a high proportion of endemic species of fauna and flora.

With the purpose of characterizing the mammalian fauna of Pantepui, the scope of our analysis has been primarily delimited to the Guianas, including southeastern Venezuela (restricted to the Amazonas, Bolívar, and Delta Amacuro states), Guyana, French Guiana, and Suriname. This region includes 81.4% of the territories with elevations between 1000 and 1500 m a.s.l. in the Guiana Shield and 94% of the elevations above 1500 m a.s.l. The remaining 6% of this highland region is located in the contiguous territory of the Brazilian border with Venezuela and Guyana (Fig. 14.1), which is an area that is poorly known and requires more biological exploration.

Biodiversity of Pantepui
DOI: https://doi.org/10.1016/B978-0-12-815591-2.00014-8

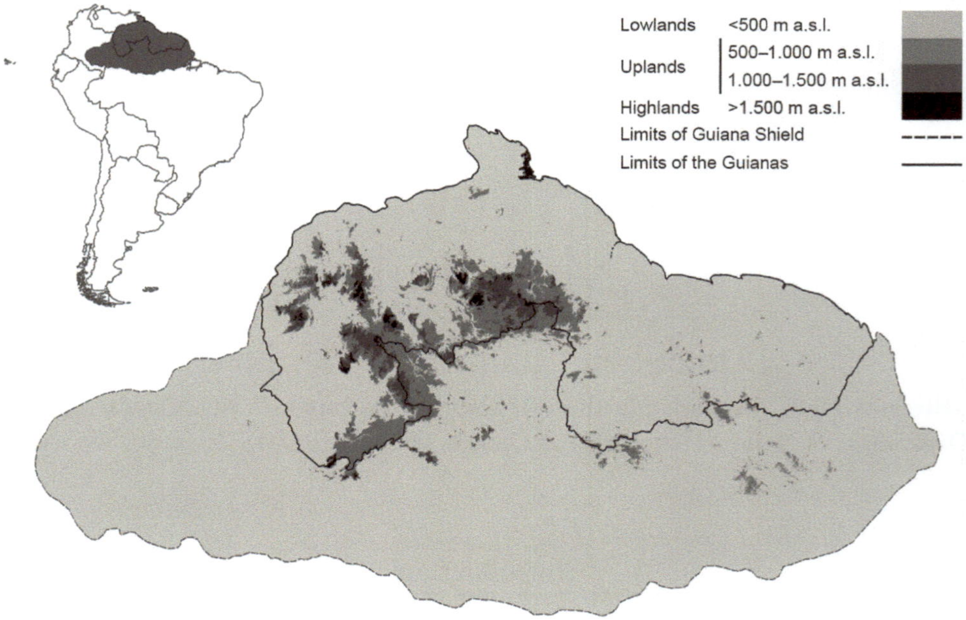

FIGURE 14.1 Elevation map of the Guiana Shield with the Guianas demarcated by the black line.

To analyze the elevational distribution of the mammals of the Guianas and understand the composition of species across the extreme topography from the lowlands to the summit of the mountains (tepuis), we subdivide the vertical space into three main categories (Fig. 14.2) following Huber (1995, 2006). The area under analysis covers a total of 865,139 km^2 (approximately 35% of the shield). The lowlands are the peripheral plains extending from sea level to 500 m a.s.l. and cover 712,783 km^2 dominated by a relief of plains and undulating hills. The uplands have an altitudinal range from 500 to 1500 m a.s.l. and cover 145,916 km^2 of an elevated plateau that is topographically more irregular, including some low mountains. Only 0.7% are the highlands, with elevations above 1500 m and having a characteristic landscape of shear-cliff tabletop mountains, which account for an area of 6440 km^2. However, the elevational demarcations of the categories are not rigid because the physiography and landscape across the Guiana Shield are highly variable (Huber, 1995). For example, the eastern edge of the Pakaraima upland plateau in Guyana dips close to 400 m a.s.l. at Kaieteur Falls, and Tafelberg mountain in Suriname is the eastern outlier of the Roraima geological formation, but its sloping tepui tabletop summit ranges from only 600 to 1000 m a.s.l. Likewise, several species of bats and rats were collected at the base of the Mount Wokomung summit on a plateau at 1420 m a.s.l. in Guyana, which is undeniably highland tepui habitat. Another categorization of the Guiana Shield is by phytogeography and floristic regions (Berry et al., 1995). The Western and Eastern Guiana provinces roughly correspond to the lowlands, the Central Guiana province corresponds to the uplands, and Pantepui corresponds to the highlands (Huber, 2006).

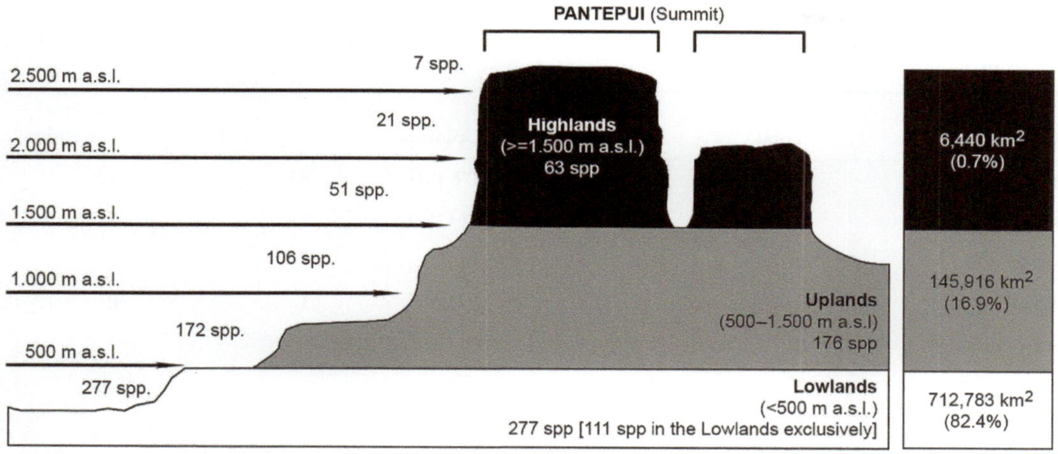

FIGURE 14.2 Schematic profile of Guiana Shield topography at 500 m elevational intervals with the three main categories of altitudinal distribution of mammal species and area occupied by each category.

Mammalian richness of the Guianas

We have updated the list of known species for the Guianas, which was last compiled by Lim (2012), to 294 taxa composed of 12 orders (Table 14.1; Appendix 14.1) with a clear dominance of bats (Chiroptera; 51.4%) followed by rodents (Rodentia; 19.7%) and opossums (Didelphimorphia; 9.2%). The nine remaining orders account for <20% of the species: 17 primates, 17 carnivores, 6 sloths and anteaters (Pilosa), 6 armadillos (Cingulata), 5 deer and peccaries (Artiodactyla), 2 dolphins (Cetacea), 2 manatees (Sirenia), 2 rabbits (Lagomorpha), and 1 tapir (Perissodactyla). Although whales (Cetacea) are now considered closely related to and included within the even-toed ungulates (Artiodactyla) and referred to as the superorder Cetartiodactyla (Montgelard et al., 1997), the taxonomic ranks are not well established (Spaulding et al., 2009), so we retain the old classification of orders for convenience and familiarity.

As a result of the addition of 15 previously unregistered (2), elevated (5), or recently described species (8) and the deletion of five species because of synonymy (3) or distributional restriction to outside of the Guiana Shield (2), the current checklist for the Guianas contains 10 more species than earlier reported by Lim (2012). The species added to the list include four opossums: *Cryptonanus* sp., recently recorded from French Guiana, far from its previously known Amazonian distribution (species designation pending; Baglan and Catzeflis, 2016); *Marmosops pakaraimae*, recently described as a new species from five localities in the highlands of western Guyana and eastern Venezuela (Voss et al., 2013); *Monodelphis arlindoi*, recently described from central-south Guyana and Brazil in southeastern Roraima to eastern Amazonas and northern Pará (north of the Amazon River; Pavan et al., 2012); and *Monodelphis touan*, recently resurrected from synonymy (with *Monodelphis brevicaudata*) by Pavan et al. (2012) based on molecular and

TABLE 14.1 Number of species of mammalian families in the Guianas by altitudinal categories and endemics. Number of highland (Pantepui) species in the lowlands and uplands are in parentheses for orders and all mammals in the Guianas.

Taxonomy	Total	Lowlands	Uplands	Highlands	Guiana endemics
Mammalia	**294**	**277 (54)**	**176 (61)**	**63**	**40**
Didelphimorphia	**27**	**24 (4)**	**17 (7)**	**7**	**7**
Didelphidae	27	24	17	7	7
Pilosa	**6**	**6 (1)**	**2 (1)**	**1**	**0**
Bradypodidae	2	2			
Megalonychidae	1	1			
Myrmecophagidae	3	3	2	1	
Cingulata	**6**	**6 (0)**	**2 (0)**	**0**	**0**
Dasypodidae	6	6	2		
Chiroptera	**151**	**147 (27)**	**90 (27)**	**28**	**7**
Emballonuridae	16	16	7		
Noctilionidae	2	2	2		
Mormoopidae	7	6	6		1
Phyllostomidae	77	77	60	22	5
Natalidae	1	1			
Furipteridae	1	1			
Thyropteridae	3	3	1		
Vespertilionidae	18	16	10	4	1
Molossidae	26	25	4	2	
Primates	**17**	**16 (3)**	**6 (3)**	**3**	**7**
Cebidae	6	6	2	1	1
Atelidae	4	4	2	1	1
Aotidae	1	1	1		1
Pitheciidae	6	5	1	1	4
Carnivora	**17**	**16 (2)**	**15 (3)**	**3**	**0**
Canidae	2	2	2		
Felidae	6	6	6	1	
Mustelidae	5	4	3	1	
Procyonidae	4	4	4	1	

(*Continued*)

TABLE 14.1 (Continued)

Taxonomy	Total	Lowlands	Uplands	Highlands	Guiana endemics
Cetacea	**2**	**2 (0)**	**0 (0)**	**0**	**0**
Delphinidae	1	1			
Platanistidae	1	1			
Sirenia	**2**	**2 (0)**	**0 (0)**	**0**	**0**
Trichechidae	2	2			
Perissodactyla	**1**	**1 (1)**	**1 (1)**	**1**	**0**
Tapiridae	1	1	1	1	
Artiodactyla	**5**	**5 (1)**	**5 (1)**	**1**	**0**
Tayassuidae	2	2	2		
Cervidae	3	3	3	1	
Rodentia	**58**	**53 (15)**	**38 (18)**	**19**	**19**
Sciuridae	5	5	4	1	1
Cricetidae	32	28	23	15	13
Erethizontidae	2	2	1	1	1
Caviidae	1	1	1		
Hydrochaeridae	1	1	1		
Dasyproctidae	5	5	2		1
Cuniculidae	1	1	1	1	
Echimyidae	11	10	5	1	3
Lagomorpha	**2**	**2 (0)**	**0 (0)**	**0**	**0**
Leporidae	2	2			

morphological analyses. Additionally, there are three taxonomic name changes since Lim (2012), including one at the level of genus, with *Micoureus* designated a subgenus of *Marmosa* (Voss et al., 2014), and two at the species level, with *Philander canus* replacing *P. mondolfi* (a junior synonym; Voss et al., 2018) and *Marmosops caucae* replacing *M. neblina* (also a junior synonym; Díaz-Nieto et al., 2016).

For mormoopid bats, there are several changes resulting from taxonomic revisions of the *Pteronotus parnellii* complex. Among them, *P. parnellii* (formerly listed as part of the Guianan fauna) is now recognized as a West Indian-endemic species. Primarily based on molecular and acoustic data, *Pteronotus rubiginosus* (formerly ranked as a subspecies of *P. parnellii*) was raised to species level (Dávalos, 2006; de Thoisy et al., 2014) and was shown to be broadly sympatric with *Pteronotus alitonus*, newly described from specimens collected in Brazil, Guyana, French Guiana, and Suriname (Pavan et al., 2018). These two species are allopatric from *Platyrrhinus fuscus* (another former subspecies of *P. parnellii*), which was recognized as a distinct species by Pavan and Marroig (2016) based on samples from Trinidad, Saint Vincent and Grenadines, Venezuela, and, more recently, from northwestern Guyana (Pavan et al., 2018).

Five Phyllostomidae are added to the list of bats and one is deleted, including Guianan specimens of *Lonchorhina* previously identified as *Lonchorhina aurita* reidentified as *Lonchorhina inusitata* by Williams and Genoways (2008). Recently Mantilla-Meluk and Montenegro (2016) referred one specimen from the Amazonas state in Venezuela to *L. aurita*. The taxonomic status of the widely distributed *Platyrrhinus helleri* is more complicated, but this species is now restricted to Central America and northwestern South America (Velazco et al., 2010). In the Guianas, four cryptic species are now recognized, including the newly described *Platyrrhinus angustirostris*, with records in Venezuela from the lowlands of Bolívar and Amazonas to the Gran Sabana plateau, and *Platyrrhinus fusciventris* from the lowlands and highlands of eastern Venezuela, Guyana, Suriname, and French Guiana (Velazco et al., 2010). Velazco and Lim (2014) subsequently described *Platyrrhinus guianensis* as endemic to the lowlands of Guyana and Suriname. Lastly, *Platyrrhinus incarum* (formerly treated as a subspecies of *P. helleri*) was elevated to species status and occurs in the Guiana lowlands (Velazco and Patterson, 2008; Velazco and Lim, 2014).

Other taxonomic changes in the phyllostomid bats include the synonymizing of *Micronycteris microtis* under *Micronycteris megalotis* based on the molecular and morphological review by Porter et al. (2007), who did not find any data to support its distinction, although they never made a recommendation for taxonomic change. However, Moras et al. (2014) still recognized *M. microtis* based solely on morphological differences. There are three changes at the genus level, including *Mimon crenulatum* being placed in the new genus *Gardnerycteris* based on the phylogenetic study of Hurtado and Pacheco (2014), *Lonchophylla thomasi* placed in the new genus *Hsunycteris* (Parlos et al., 2014), and *Vampyressa bidens* placed in the genus *Vampyriscus* (Porter and Baker, 2004). In addition, *Lichonycteris obscura* is restricted to northwestern South America and Central America, whereas *Lichonycteris degener* is recognized as morphologically distinct and allopatrically distributed east of the Andes, where its range includes the Guianas (Griffiths and Gardner, 2008).

Another addition is a new species of molossid bat, *Molossus fentoni*, that was previously confused with *Molossus molossus* (see Loureiro et al., 2018). However, *Molossus barnesi* is removed because it is considered a junior synonym of *Molossus coibensis* (see Catzeflis et al., 2016). In the bat family Vespertilionidae, *Lasiurus* was recently split into several genera (Baird et al., 2015) based on molecular and morphological analyses that resulted in the following relevant taxonomic changes: *Aeorestes cinereus*, *Aeorestes egregius*, and *Dasypterus ega*. However, Zeigler et al. (2016) recommended recognition at the subgenus level, which Baird et al. (2017: 24) did not agree with because their "revised lasiurine taxonomy more accurately reflects deep morphological and genetic diversity within the tribe at the generic level."

Three species of primates are added to the list, including *Saimiri cassiquiarensis* in southern Venezuela, eastern Colombia, and northwestern Brazil, which is now considered distinct from *Saimiri sciureus* (see Lynch Alfaro et al., 2015); the new species *Cacajao ayresi* in northern Brazil and southern Venezuela (Boubli et al., 2008; Urbani and Portillo-Quintero, 2018); and *Alouatta seniculus*, whose presence in the area is recognized by Urbani and Portillo-Quintero (2018) as bounded by the Guainía, Casiquiare, Orinoco, and Atapapo Rivers in the southwestern state of Amazonas in Venezuela. Other taxonomic changes in primates include the deletion of *Chiropotes israelita*, which is synonymized under *Chiropotes chiropotes* (see Veiga et al., 2008), and *Cebus apella*, which is transferred to the genus *Sapajus* (Lynch Alfaro et al., 2012).

Based on molecular data, the small gray brocket deer *Mazama nemorivaga* has been recognized as distinct from *M. gouazoubira* (Duarte et al., 2008). Similarly, the coastal dolphin *Sotalia guianensis* is considered a separate species from the riverine dolphin *Sotalia fluviatilis* (Caballero et al., 2007). A taxonomic revision of olingos synonymized *Bassaricyon beddardi* under *Bassaricyon alleni* (see Helgen et al., 2013).

In rodents, *Nectomys rattus* is now considered a senior synonym for the Guianan taxon formerly known either as *N. squamipes melanius* or as *N. melanius* (see Musser and Carleton, 2005). The species formerly known as *Echimys semivillosus* has been transferred to the new genus *Pattonomys* (Emmons, 2005), which now also includes *Pattonomys punctatus* (formerly a junior synonym of *E. semivillosus*); the latter is restricted to the northwestern limits of the Guianas along the Orinoco River in Venezuela (Emmons et al., 2015). We follow the taxonomic ruling of the International Commission of Zoological Nomenclature (1998) for the correct spelling of the generic name for capybara so that the species is now considered to be *Hydrochoerus hydrochaeris* and not *Hydrochoeris hydrochaeris*.

It is also noted that an editorial formatting error in the appendix of the checklist in Lim (2012) omitted seven species (*Mormoops megalophylla*, *Anoura latidens*, *Vampyriscus brocki*, *Eumops maurus*, *Speothos venaticus*, *Sciurus flammifer*, and *Coendou melanurus*) from the last line of each page of the table. However, these omissions did not affect the species counts in the text. And finally, there was a misspelling for the manatees (*Trichechidae* and *Trichechus*) in the table.

Elevational occurrence

Of the 294 species of mammals reported from the Guianas, most (94%) are found in the lowlands, more than half (60%) occur in the uplands, and 21% are documented in the highlands (Table 14.1; Appendix 14.1). Overall, Guianan-endemic species account for 14% of the mammalian biodiversity. The lowlands have 111 species that occur only below 500 m a.s.l., whereas 12 species have not been registered in the lowlands: *Marmosa tyleriana*, *M. pakaraimae*, *Monodelphis reigi*, *Pteronotus davyi*, *Histiotus humboldti*, *A. cinereus*, *Myotis oxyotus*, *Mustela frenata*, *Oecomys* sp. 1, *Podoxymys roraimae*, *Rhipidomys macconnelli*, and *Rhipidomys wetzeli*. The number of highland species was relatively high in the two other elevational categories, with the lowlands having 84% (54 of 63) of these Pantepui species. Although bats have the highest species diversity in all elevational categories, their proportion decreases as elevation increases, and rodents represent a higher proportion of the highland fauna than in the Guianas in general.

None of the six species in the three orders of whales, manatees, and rabbits occur in the uplands or higher elevations of the Guiana Shield, and no armadillos have been reported in the highlands (Table 14.1). Opossums and tapir have been reported from all three elevational categories. At the family level, unlike anteaters (Myrmecophagidae), no sloths are found in the uplands or above. Similarly within bats, two of the nine families (Natalidae and Furipteridae) occur only in the lowlands. By contrast, three families (Phyllostomidae, Vespertilionidae, and Molossidae) are documented in the highlands and lower elevations. For rodents, five of the eight families also occur in all elevational categories, whereas the other three families are found in the uplands and below. Within primates, carnivores, and even-toed ungulates (Artiodactyla), only one family has not been reported in the highlands.

Forty species of mammals are currently recognized as endemic to the Guianas (Table 14.1), but only in 4 of the 12 orders: Didelphimorphia, Chiroptera, Primates, and Rodentia. Except for the latter, each order has seven endemic species, which represent 26.9%, 4.6%, and 41.2% of their totals, respectively, and 17.5% each of all endemics. The remaining 19 species are rodents and equivalent to 32.8% of this order in the Guianas and 47.5% of all endemics. Although bats have nearly three times more species than rodents, the percentage of endemic species is inversely proportional.

Pantepui mammals

The most recent summary of mammals from the Pantepui highlands region, approximately >1500 m a.s.l., documented 40 species (Lim, 2012). Our update increases the total by 23 species, including the addition of two opossums, 11 bats, 3 primates, 1 carnivore, 1 deer, and 5 rodents (Table 14.2). Most of these new distributional records are primarily based on a review of Venezuelan specimens, but one addition is represented by the description of *M. pakaraimae* as a new species of mouse opossum (Voss et al., 2013).

Of the 63 species of mammals (21% of the Guianas) known from the Pantepui highlands, the bats, rodents, and opossums comprise 45%, 30%, and 11%, respectively (Table 14.2). Most of these species (95%) have distributions in the uplands, whereas 83% are found in the lowlands. Only 20% are endemic to the Guianas, whereas nine of these species are further endemics to the Guiana upland plateau >500 m a.s.l. (with a couple of exceptions), including three opossums, five rodents, and one bat. Only one of these species, the Roraima mouse, is endemic to the summit region of Pantepui. *P. roraimae* is known only from the top of two tepuis between 2216 and 2628 m a.s.l. by six specimens collected exclusively in the Venezuelan and Guyanese territories of Roraima-tepui (summit area of 34 km^2) (Anthony, 1929; Pérez-Zapata et al., 1992) and a seventh specimen recently captured on the nearby Wei-Assipu-tepui (summit area of 2.5 km^2) (Leite et al., 2015). *M. tyleriana* is a mouse opossum collected on the summit, slopes, and plateaus of Auyán-tepui, Cerro Jaua, and Cerro Duida in humid forests between 1300 and 2100 m a.s. l. (Ochoa, 1985; Creighton and Gardner, 2008). *M. reigi* is a short-tailed opossum described by Lew and Pérez-Hernández (2004) from the northern slopes of the Gran Sabana uplands at 1440 m a.s.l. on Sierra de Lema. The distribution was extended to the slopes and summit of Mount Ayanganna in the Potaro-Siparuni region between 1100 and 2050 m a.s.l., which

TABLE 14.2 Species of mammals documented from the Guiana Highlands (Pantepui) and their occurrence across the Guiana Shield. New additions to this fauna since Lim (2012) are indicated by an asterisk (*).

Taxonomy	Lowlands (<500 m a.s.l.)	Uplands (500–1500 m a.s.l.)	Guiana Shield endemic	Pantepui endemic	Elevation range (m a.s.l.)	Altitude interval (m)	Distribution (Tate, 1939)
Mammalia 63	**53**	**60**	**13**	**9**			
Didelphimorphia 7	4	7	4	3			
Didelphidae 7	4	7	4	3			
Didelphis imperfecta	X	X	X		50–2550	2500	8
*Didelphis marsupialis**	X	X			5–1800	1795	1
Marmosa tyleriana		X	X	X	1300–2010	710	8
Marmosops caucae	X	X			140–2000	1860	4
*Marmosops pakaraimae**		X	X	X	800–1500	700	8
Marmosa demerarae	X	X			25–2100	2075	2
Monodelphis reigi		X	X	X	1100–2050	950	8
Pilosa 1	1	1					
Myrmecophagidae 3	1	1					
Tamandua tetradactyla	X	X			100–2150	2050	2
Chiroptera 28	28	28	2	1			
Phyllostomidae 22	22	22	2	1			
Phyllostominae 3	3	3	1				
Lophostoma schulzi	X	X	X		20–1420	1400	8
Micronycteris megalotis	X	X			25–1420	1395	1
Trachops cirrhosus	X	X			5–1420	1415	1
Glossophaginae 4	4	4					
Anoura caudifer	X	X			30–2150	2120	2
Anoura geoffroyi	X	X			80–2550	2470	1
Anoura latidens	X	X			50–2000	1950	4
*Glossophaga soricina**	X	X			0–1500	1500	1

(Continued)

TABLE 14.2 (Continued)

Taxonomy	Lowlands (<500 m a.s.l.)	Uplands (500–1500 m a.s.l.)	Guiana Shield endemic	Pantepui endemic	Elevation range (m a.s.l.)	Altitude interval (m)	Distribution (Tate, 1939)
Carolliinae 3	3	3					
Carollia brevicauda	X	X			15–1800	1785	2
*Carollia perspicillata**	X	X			10–1500	1490	1
*Rhinophylla pumilio**	X	X			5–1500	1495	2
Stenodermatinae 12	12	12	1	1			
*Ametrida centurio**	X	X			0–1600	1600	5
*Artibeus amplus**	X	X			50–2100	2050	7
Artibeus bogotensis	X	X			30–1500	1470	4
*Artibeus concolor**	X	X			10–1500	1490	5
*Artibeus lituratus**	X	X			5–1500	1495	1
*Artibeus planirostris**	X	X			0–1600	1600	2
*Mesophylla macconnelli**	X	X			40–1515	1475	1
Platyrrhinus aurarius	X	X	X	X	340–2200	1860	8
*Sphaeronycteris toxophyllum**	X	X			100–1500	1400	4
Sturnira lilium	X	X			15–1800	1785	1
Sturnira tildae	X	X			25–1800	1775	2
*Uroderma bilobatum**	X	X			15–1500	1485	1
Vespertilionidae 4	2	2					
Histiotus humboldti					1820–1820	0	4
Aeorestes cinereus					2085–2100	15	4
Myotis oxyotus	X	X			851–1800	949	4
Myotis riparius	X	X			45–1600	1555	1

Molossidae 2	2	2					
Molossus molossus	X	X			0–1600	1200	1
Nyctinomops macrotis	X	X			100–2100	2000	1
Primates 3	3	3	1				
Cebidae 1	1	1					
*Cebus olivaceus**	X	X			10–2000	1990	5
Atelidae 1	1	1					
*Alouatta macconnelli**	X	X			10–1960	1950	5
Pitheciidae 1	1	1	1				
*Cacajao hosomi**	X	X	X		70–1500	1430	8
Carnivora 3	2	3					
Felidae 1	1	1					
*Puma concolor**	X	X			50–1750	1700	1
Mustelidae 1		1					
Mustela frenata		X			1050–2500	1450	4
Procyonidae 1	1	1					
Nasua nasua	X	X			35–2550	2515	2
Perissodactyla 1	1	1					
Tapiridae 1	1	1					
Tapirus terrestris	X	X			50–2150	2100	2
Artiodactyla 1	1	1					
Cervidae 1	1	1					
*Odocoileus cariacou**	X	X			100–1500	1400	6
Rodentia 19	15	18	6	5			
Sciuridae 1	1	1					
*Sciurus aestuans**	X	X			15–1800	1785	5

(Continued)

TABLE 14.2 (Continued)

Taxonomy	Lowlands (<500 m a.s.l.)	Uplands (500–1500 m a.s.l.)	Guiana Shield endemic	Pantepui endemic	Elevation range (m a.s.l.)	Altitude interval (m)	Distribution (Tate, 1939)
Cricetidae 15	11	14	4	4			
Necromys urichi	X	X			100–2550	2450	4
*Nectomys rattus**	X	X			30–1500	1470	2
Neusticomys venezuelae	X	X			300–400	1100	4
Oecomys sp. 1		X	X	X	1100–1550	450	8
Oecomys trinitatis	X	X			306–2000	1694	1
*Oligoryzomys fulvescens**	X	X			15–1900	1885	1
Euryoryzomys macconnelli	X	X			107–1420	1313	4
Hylaeamys megacephalus	X	X			10–1420	1410	5
Hylaeamys yunganus	X	X			10–1420	1410	3
Podoxymys roraimae			X	X	2216–2628	412	8
Rhipidomys leucodactylus	X	X			138–1800	1662	3
Rhipidomys macconnelli		X	X	X	550–2600	2050	8
Rhipidomys nitela	X	X			10–1480	1470	3
Rhipidomys wetzeli		X	X	X	587–1800	1213	8
*Zygodontomys brevicauda**	X	X			10–1900	1890	6
Erethizontidae 1	1	1	1				
Coendou melanurus	X	X	X		10–1829	1819	8
Cuniculidae 1	1	1					
*Cuniculus paca**	X	X			10–1800	1790	1
Echimyidae 1	1	1	1	1			
Proechimys hoplomyoides	X	X	X	X	135–1550	1415	8

is the highest tepui wholly within Guyana (Lim et al., 2010). *M. pakaraimae* is another mouse opossum that occurs on the talus slopes of six tepuis (Roraima-tepui, Mount Ayanganna, Mount Wokomung, Sierra de Lema, Auyán-tepui, and Churí-tepui) from 800 to 1500 m a.s.l. (Voss et al., 2013; García et al., 2014) but has not been recorded from the summits. *R. macconnelli* is a common and widely distributed rat found on many tepui slopes and summits of the Guiana Shield (Tribe et al., 2015), including the first reports from Mount Ayanganna, Roraima-tepui, and Mount Wokomung in Guyana. *R. wetzeli* is not as common as its sympatric congener but "is found at middle elevations on the tepui table mountains" (Tribe et al., 2015: 617). In addition to reporting it from the slopes of Mount Ayanganna, Roraima-tepui, and Mount Wokomung in Guyana, we document the occurrence of this species at 587 m elevation on the Tafelberg summit in Suriname. An undescribed species of arboreal rice rat (*Oecomys*) has been collected on the slopes of mounts Ayanganna and Wokumong from 1100 to 1550 m a.s.l. It is genetically most similar to *Oecomys rex* in the lowlands of Guyana, Suriname, and French Guiana, but the mitochondrial DNA sequence divergence is about 12% (Gomes Júnior et al., 2016), which is at the level of nonsister species relationship for the genus (Baker and Bradley, 2006). *Proechimys hoplomyoides* is a terrestrial spiny rat "limited to the tepui area of southeastern and southern Venezuela" (Patton et al., 2015: 987). However, one specimen was collected at 135 m a.s.l. from Tama on the Rio Orinoco near Cerro Duida in Amazonas state (Handley, 1976) and others from 200 to 580 m a.s.l. at Serranía de Los Pijiguaos in Bolivar state (Ochoa et al., 1988). In addition, it has been documented in Guyana from 600 to 1550 m a.s.l. on the slopes of Mount Ayanganna, Roraima-tepui, and Mount Wokomung. *Platyrrhinus aurarius* is the only bat found endemic to the Guiana upland region. It occurs at 600–700 m a.s.l. on the summit of Tafelberg in Suriname (Williams et al., 1983). In Guyana it has been documented from 1500 m a.s.l. on the slopes of Mount Ayanganna, Roraima-tepui, and Mount Wokomung to Kaieteur Falls on the edge of the Pakaraima plateau at 414 m a.s.l. In Venezuela this species is found from 700 m a.s.l. on Cerro Duida (Handley, 1976) to 2100 m a.s.l. on Serranía de la Neblina (Gardner, 1988). However, in Brazil it has been reported from 340 m a.s.l. at Missaõ Marari on the Rio Katana near Serra do Aracá in Amazonas state (Velazco and Gardner, 2009). Tafelberg and Aracá were designated isolated low-mountain outliers of the Central Guiana floristic province ranging from 300 to 1500 m a.s.l. (Berry et al., 1995).

Altitudinal distribution

Considering successively increasing intervals of 500 m of elevation, the number of species decreases progressively, with >2500 m a.s.l. having only seven species (2.4%) in four orders of mammals: *Didelphis imperfecta*, *Anoura geoffroyi*, *M. frenata*, *Nasua nasua*, *Necromys urichi*, *P. roraimae*, and *R. macconnelli* (Table 14.3). Only four species are found in all elevational intervals: *D. imperfecta*, *A. geoffroyi*, *N. nasua*, and *N. urichi*. The number of highland species in each 500 m interval remains relatively stable until 2000 m a.s.l. in terms of both total species in the Guianas and only highland species. However, their numbers fall drastically by more than half between 2000 and 2500 m a.s.l.

Most species of mammals in the Guianas (around 45%) have an elevational range of 500 m or less between the altitudinal extremes of their documented occurrence (Fig. 14.3).

TABLE 14.3 Species of mammals in the Guianas by 500 m intervals of altitude (the number of highland species is in parentheses).

Taxonomy	Total	≤500	500–1000	1000–1500	1500–2000	2000–2500	>2500
Didelphimorphia	27 (7)	24 (4)	15 (5)	12 (7)	7 (7)	5 (5)	1 (1)
Pilosa	6 (1)	6 (1)	2 (1)	2 (1)	1 (1)	1 (1)	0
Cingulata	6 (0)	6 (0)	2 (0)	1 (0)	0	0	0
Chiroptera[a]	151 (29)	145 (27)	90 (27)	48 (27)	23 (23)	7 (7)	1 (1)
Primates	17 (3)	17 (3)	6 (3)	5 (3)	3 (3)	1 (1)	0
Carnivora	17 (3)	16 (2)	14 (2)	7 (3)	3 (3)	2 (2)	2 (2)
Cetacea	2 (0)	2 (0)	0	0	0	0	0
Sirenia	2 (0)	2 (0)	0	0	0	0	0
Perissodactyla	1 (1)	1 (1)	1 (1)	1 (1)	1 (1)	1 (1)	0
Cetartiodactyla	5 (1)	5 (1)	5 (1)	5 (1)	1 (1)	0	0
Rodentia[a]	58 (19)	51 (15)	37 (16)	25 (16)	12 (12)	4 (4)	3 (3)
Lagomorpha	2 (0)	2 (0)	0	0	0	0	0
Total	**294 (64)**	**277 (54)**	**172 (56)**	**106 (59)**	**51 (51)**	**21 (21)**	**7 (7)**
% of total species	**21.8**	**18.4**	**19.0**	**20.1**	**17.3**	**7.1**	**2.4**
% of Highland species	**100**	**84.4**	**87.5**	**92.2**	**79.7**	**32.8**	**10.9**

[a]*Two species of rodents and one species of bat without elevation data.*

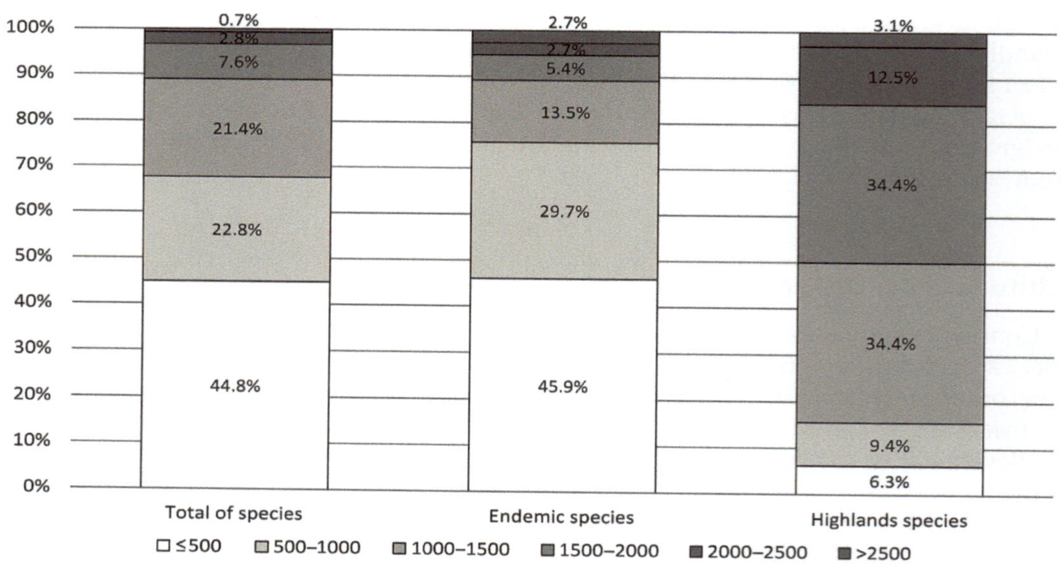

FIGURE 14.3 Proportion of species in the Guianas, endemic species, and highland species by altitudinal range of occurrence (altitudinal extremes of their documented occurrence).

A similar number of species (128) have an elevational range between 500 and 1500 m. Only 11% of the species have an elevational range >1500 m. These proportions are similar to that observed for endemic species, with a constant decrease in the number of species with increasing altitude range. But this pattern changes dramatically for the highland species, with nearly 70% of the species with altitudinal extremes between 1000 and 2000 m. The remaining 30% are evenly distributed below 1000 m (14.7%) and above 2000 m (15.6%). The greatest amplitude of altitudinal range >2000 m was recorded for 11 species, of which two are endemic to the Guianas and are identified with an asterik: *Marmosa demerarae, D. imperfecta* (*), *Tamandua tetradactyla, Anoura caudifer, A. geoffroyi, Artibeus amplus, Nyctinomops macrotis, N. nasua, Tapirus terrestris, N. urichi*, and *R. macconnelli* (*). Only two species exceeded an elevational range of 2500 m: *D. imperfecta* (2500 m, 50–2550 m a.s.l.) and *N. nasua* (2515 m, 35–2550) (Fig. 14.3; Appendix 14.1).

Geographic distribution

The distributional patterns for 92 species of mammals recognized by Tate (1939) to understand the affinities and relationships of the Guianan fauna to their geographic surroundings were updated by Lim (2012) to 284 species. Our current list of 294 species shows that one-third of the diversity in the Guianas has a wide distribution from South America to Central America (Category 1; Fig. 14.4). This broad Neotropical distributional category is more than twice as diverse as any other category across elevational ranges, except in the highlands where it is more similar to three other categories. There is a general decrease in diversity in each distributional category of lesser inclusive area, except the total number of endemic species in the Guiana Shield (Category 8), which is relatively high across each of the elevational categories. Another exception to this pattern is the number of species with affinities to the western Amazon and the foothills of the Andes (Category 4), which is proportionally more diverse in the highlands. No species in the Guianas is restricted to northern South America north of the Amazon River (Category 9).

Species composition of massifs and tepuis

Eleven of the 22 massifs in the Guianas (Huber, 1995) include 10 or more species with records above approximately 1000 m of altitude (Table 14.4). The better-known massifs are located in southern Bolívar state in Venezuela and the vicinity of western Guyana, where the Eastern tepui chain and the Gran Sabana uplands are the most diverse. However, this may be a sampling bias of greater surveying effort in this area compared to the western tepuis. The mammals associated with six massifs are virtually unknown: Los Testigos massif, Aprada-tepui, Cerro Sipapo uplands, Parú massif, Cuao-Sipapo massif, and Yapacana uplands.

Among the species recorded above 1000 m a.s.l., *R. macconnelli* is the most widely occurring in Pantepui, with records in or proximal to 12 different massifs. The next most broadly distributed species are *A. caudifer* (10); *D. imperfecta* and *A. geoffroyi* (9); *Carollia brevicauda* and *P. aurarius* (8); *M. demerarae, Lionycteris spurrelli, Artibeus bogotensis*, and *N. nasua* (7); *Sturnira tildae* and *Sciurus aestuans* (6); and *Marmosa murina, Artibeus obscurus*,

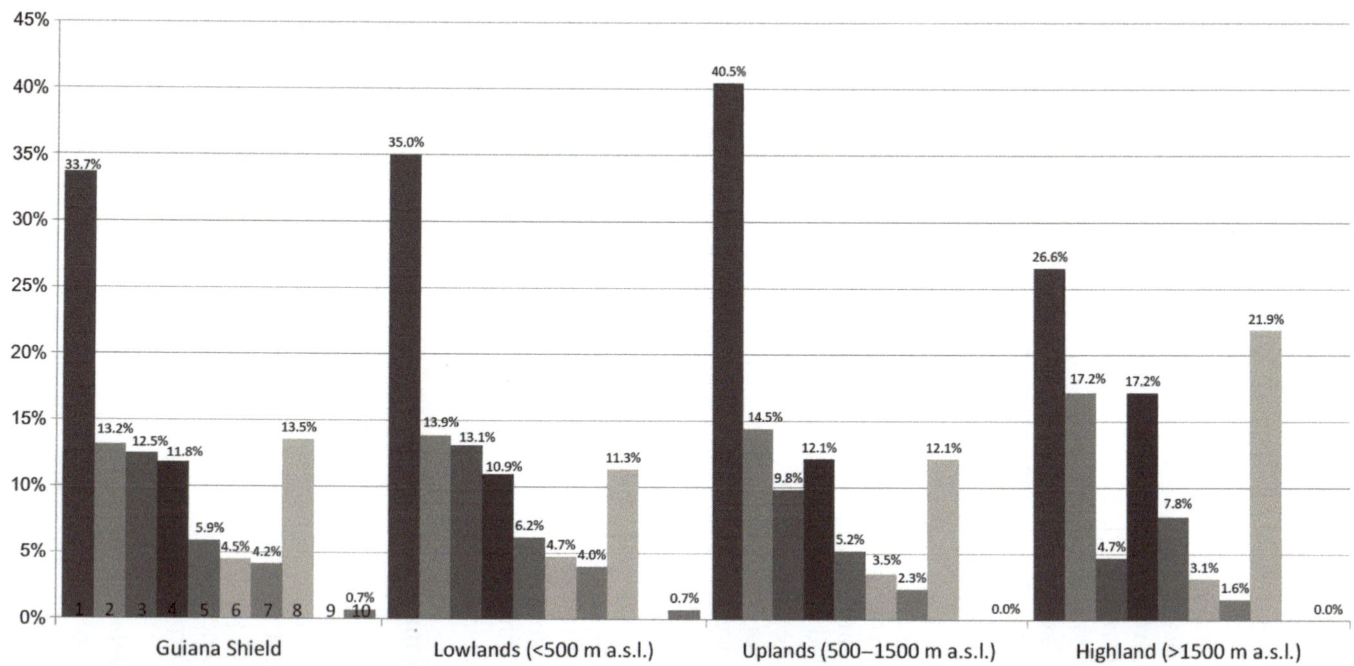

FIGURE 14.4 Proportion of mammalian species in the Guianas based on the distributional categories of Tate (1939), including (1) widely distributed in Central and South America; (2) widely distributed in South America but does not extend west of Panama; (3) distributed in the Amazon basin, including the Guianas; (4) distributed from the Guianas west into upper Amazonia and the Andean foothills; (5) distributed from the Guianas east into Brazil; (6) occurring primarily in savanna or dry open areas; (7) distributed from the Guianas into Colombia and Panama (and into Central America for several species of bats); (8) endemic to the Guiana Shield; (9) distributed in northern South America north of the Amazon River (without records); and (10) distributed in the northern edge of the Guianas in Venezuela and Colombia.

TABLE 14.4 Number of mammal species recorded above 1000 m a.s.l. at each massif and tepui (associated massif number in parentheses).

Massif	Number of species >1000 m a.s.l.	Tepui	Number of species >1000 m a.s.l.
(1) Eastern tepui Chain	61	(2) Sierra de Lema	57
(2) Gran Sabana uplands	60	(3) Ptari-tepui	39
(3) Ptari massif	39	(5) Auyán-tepui	34
(5) Auyan massif	34	(1) Roraima-tepui	32
(17) Duida-Marahuaka massif	26	(1) Mount Ayanganna	29
(22) Imeri massif	24	(1) Mount Wokomung	27
(12) Yutajé massif	15	(22) Sierra de la Neblina	24
(8) Chimantá massif	11	(12) Cerro Guanay	15
(18) Parima uplands	11	(17) Cerro Marahuaka	14
(20) Tapirapecó massif	10	(17) Cerro Duida	13
(10) Jaua massif-Maigualida massif	9	(8) Chimantá	11
(11) Yaví massif	7	(18) Sierra Parima	11
(19) Unturán uplands	3	(20) Cerro Tamacuari	9
(6) Canaima uplands	1	(10) Cerro Jaua	8
(9) Paragua uplands	1	(11) Cerro Yaví	6
(21) Arakamuni-Avispa uplands	1	(10) Sierra de Maigualida	3
(4) Los Testigos massif	0	(12) Serranía Yutajé	3
(7) Aprada massif	0	(19) Sierra Unturán	3
(13) Sipapo uplands	0	(10) Cerro Sarisariñama	2
(14) Parú massif	0	(1) Uei-tepui	1
(15) Cuao-Sipapo massif	0	(1) Kukenán (Matauí)-tepui	1
(16) Yapacana uplands	0	(9) Cerro Ichún	1
		(12) Coro Coro	1
		(21) Cerro Arakamuni	1

N. urichi, and *Oligoryzomys fulvescens* (5). The other 95 species have been recorded from four to one massif (Table 14.5). Of 183 species without an evident association to any massif, only five reach localities equal to or higher than 1000 m a.s.l. (*Myotis keaysi*, *Nyctinomops laticaudatus*, *Aotus trivirgatus*, *Cacajao hosomi*, and *Sciurus gilvigularis*), and 112 are strictly below 500 m a.s.l.

TABLE 14.5 Number of mammal species known by number of massifs or tepuis where recorded (for 312 records associated with massifs and 372 records associated with tepuis above 1000 m a.s.l.).

Number of massifs	Number of species (>1000 m a.s.l.)	Number of tepuis	Number of species (>1000 m a.s.l.)
12	1	17	1
11	0	16	1
10	1	15	0
9	2	14	1
8	2	13	0
7	4	12	0
6	2	11	3
5	4	10	0
4	15	9	2
3	17	8	2
2	20	7	5
1	43	6	7
0	183	5	6
	294	4	10
		3	12
		2	18
		1	39
		0	187
			294

A. geoffroyi is associated with the most tepuis (17), followed by *R. macconnelli* (16), *A. caudifer* (14), and *D. imperfecta*, *C. brevicauda*, and *P. aurarius*, with 11 tepuis each. Another 101 species are present on 9–1 tepui. Of 187 remaining species 180 have been recorded below 1000 m a.s.l. in the Guiana Shield (112 strictly below 500 m a.s.l.) and 24 of them are endemic to the Guianas.

Similarity analysis of upland species composition in massifs and tepuis gives insights into the biogeographic affinities between different montane regions in the Guianas. We used presence (with records above 1000 m a.s.l.) or absence of mammal species among the better-surveyed massifs and tepuis to calculate Jaccard's coefficient and cluster by the unweighted pair-group method using arithmetic averages (UPGMA).

The species composition between the six massifs with 20 or more species recorded above 1000 m a.s.l. has geographic structuring (Fig. 14.5) that groups (1) the Eastern tepui

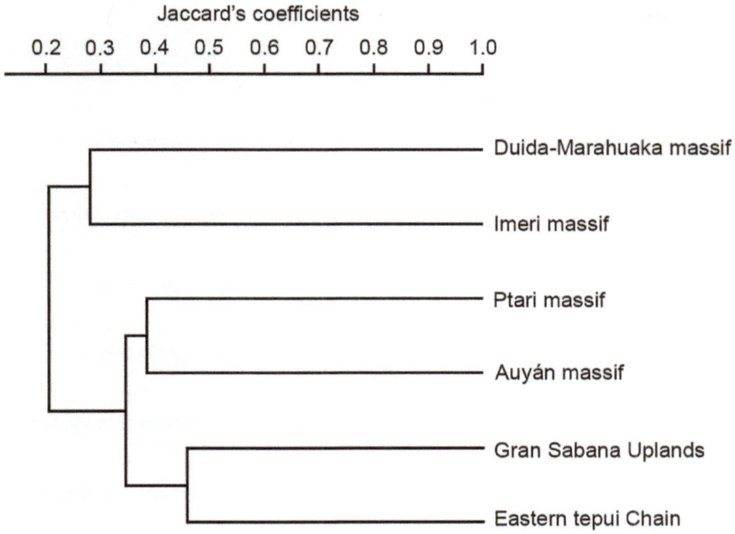

FIGURE 14.5 Similarity analysis among six massifs in the Guianas with 20 or more species of mammals recorded above 1000 m a.s.l. from an unweighted pair-group method with arithmetic mean (UPGMA) cluster analysis of Jaccard's coefficient.

chain and the Gran Sabana uplands tableland formations in eastern Bolívar state, Venezuela, and western Guyana; (2) Auyán-tepui and Ptari massifs in southern Bolívar state, Venezuela; and (3) Imeri and Duida-Marahuaka massifs in southwestern Amazonas state, Venezuela.

The grouping of seven tepuis with 25 or more species recorded above 1000 m a.s.l. shows geographic structuring (Fig. 14.6): (1) Eastern tepui chain of Mount Ayanganna and Mount Wokomung in Guyana and Roraima-tepui; (2) three tepuis belonging to different massifs but all located proximally in the central uplands of the Guianas, including Auyán-tepui, Ptarí-tepui, and Sierra de Lema on the northern slope of the Gran Sabana uplands plateau; and (3) Sierra de la Neblina (Imeri massif) in southwestern Amazonas state in Venezuela (border with Brazil).

Imeri massif in southwestern Amazonas state has one-third of its species from the Guianas west into the upper Amazon and the Andean foothills (Tate category 4), which is twice as much as any of the other massifs analyzed. Likewise, a similar distributional composition of species was observed for tepuis, with Sierra de la Neblina having a high proportion of species ranging to the Andes.

Biogeography

Based on the distributional categories of Tate (1939), species that are widely occurring from South America into Central America are the most numerous (27%) of the Pantepui mammals; however, the majority of these (12 species) are bats (Table 14.2). By contrast,

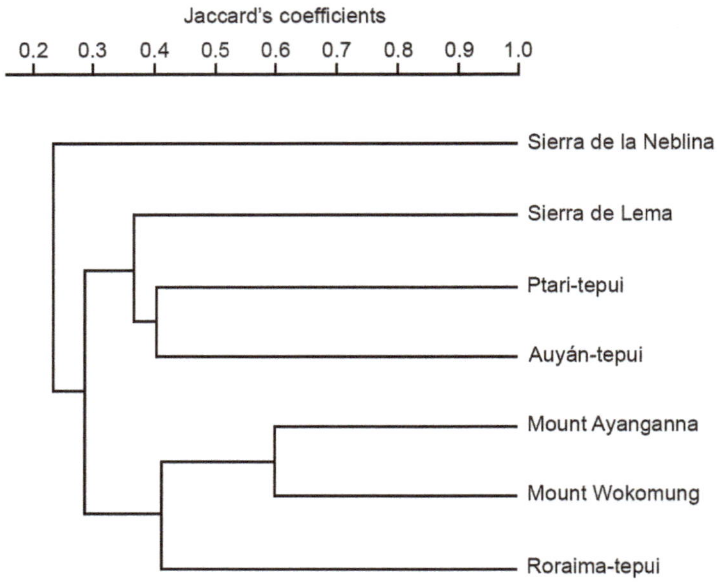

FIGURE 14.6 Similarity analysis among seven tepuis in the Guianas with 25 or more species of mammals recorded above 1000 m a.s.l. from an unweighted pair-group method with arithmetic mean (UPGMA) cluster analysis of Jaccard's coefficient.

more than half of the opossums and almost one-third of the rodents in the highlands are endemic to the Guiana Shield, indicating a greater vagility in bats. But in general, about half of the mammal species known from the Guianas also range into northwestern South America and Central America.

Several biogeographic origins have been suggested for the nine species of mammals endemic to Pantepui, including the Amazon, Andes, and Cerrado of the Brazilian Shield, but hypotheses of ancestral areas are poorly understood and equivocal for many species level phylogenies (Voss et al., 2013). Nonetheless, recent studies are increasing our knowledge of the evolution of mammals in the Guiana Highlands. *P. roraimae* was included in a molecular phylogenetic analysis of akodontine rodents, and its closest relatives (*Thalpomys* and *Necromys*) are primarily found to the southeast in the open Cerrado habitats of the Brazilian Shield with divergence estimated in the Late Pliocene about 3 million years ago (Leite et al., 2015). *M. tyleriana* is the sister taxon to a clade of three species from the surrounding Amazonian lowlands (Voss et al., 2014). *M. reigi* diverged during the Early Pleistocene (2.4 million years ago) from its sister species *Monodelphis adusta* and other closely related taxa in the western Amazon lowlands, although one derived lineage occurs in the Andes (Pavan et al., 2016). *M. pakaraimae* is the sister species to *Marmosops parvidens* from the immediately adjacent eastern Amazon lowlands (Voss et al., 2013; Díaz-Nieto et al., 2016). *R. macconnelli* and *R. wetzeli* were successive basal lineages in a recent phylogenetic analysis of the genus (Brito et al., 2017), which indicates that the initial diversification of *Rhipidomys* occurred in the Guiana Highlands. Although times of divergence were not calculated, the phylogeny suggests that the most recent common ancestors were cold-adapted taxa that speciated in isolation on tepui summits during subsequent warmer periods in the Late Pliocene or Early Pleistocene, similar to the Cool Climate Theory of Mayr and Phelps (1967). Furthermore, the next two basal lineages are Andean species, which suggests long-distance dispersal across the lowlands during cooler temperatures. *O. rex* "L" of Gomes Júnior et al. (2016) has a mitochondrial sequence divergence of 12% from its sister species *O. rex* "K" in the Eastern Guiana lowlands of Huber (2006). This biogeographic pattern is similar to that of *M. pakaraimae* and *M. parvidens*. *P. hoplomyoides* has not been included in a phylogenetic analysis but has also been placed in the sister genus *Hoplomys* (Moojen, 1948; Cabrera, 1961), which occurs in northwestern South America and Central America. However, speculation on its biogeographic affinities at this time is premature. *P. aurarius* has its sister species (*P. infuscus*) distributed in the western Amazonian lowlands, but this relationship is poorly supported and other closely related taxa have primarily an Andean occurrence (Velazco and Patterson, 2008). Resolving these deeper splits will give better predictions of modes of speciation for this speciose genus.

The level of mammalian endemism in Pantepui and more generally in the Guianas is low compared to most other organismal groups primarily because of the dispersal abilities of bats, which comprise over half of the species diversity. Furthermore, the ancient Guiana Shield has acted as a stable core area to more recent palaeoenvironmental changes in neighboring regions such as the Amazon basin and Andean Mountains (e.g., Lim, 2008). Recent molecular studies give a better understanding of the relationships between some species of Pantepui and their closest relatives in the biogeographical regions surrounding the Guiana Shield. However, different affinities in the analysis of the mammalian community occupying the highlands of the Guianas do not suggest one biogeographic pattern that characterizes Pantepui, nor do they support a single hypothesis that has been previously postulated for the origin of the Guiana Highlands fauna (Chapman, 1917, 1931; Mayr and Phelps, 1967; Haffer, 1974; Hoogmoed, 1979). With the exception of the evidence of speciation in the Guiana Highlands (*Rhipidomys*) and close relationships with the lowlands of eastern Guiana (*Oecomys* and *Marmosops*), the other Pantepui endemics show a greater affinity with taxa from the surrounding Amazonian lowlands or from more distant regions such as the Brazilian Shield or the Andes. This supports the proposal that several mechanisms, operating individually or jointly, would be necessary to explain the origin of the various faunal elements that make up the mammalian communities of the Guianas and Pantepui (Pérez-H and Lew, 1998).

References

Anthony, H.E., 1929. Two new genera of rodents from South America. Am. Mus. Novit. 383, 1–6.

Baglan, A., Catzeflis, F., 2016. Barn owl pellets collected in coastal savannas yield two additional species of small mammals for French Guiana. Mammalia 80 (1), 91–95.

Baird, A.B., Braun, J.K., Mares, M.A., Morales, J.C., Patton, J.C., Tran, C.Q., et al., 2015. Molecular systematic revision of tree bats (Lasiurini): doubling the native mammals of the Hawaiian Islands. J. Mammal. 96 (6), 1255–1274.

Baird, A.B., Braun, J.K., Engstrom, M.D., Holbert, A.C., Huerta, M.G., Lim, B.K., et al., 2017. Nuclear and mtDNA phylogenetic analyses clarify the evolutionary history of two species of native Hawaiian bats and the taxonomy of Lasiurini (Mammalia: Chiroptera). PLoS One 12 (10), 1–27.

Baker, R.J., Bradley, R.D., 2006. Speciation in mammals and the genetic species concept. J. Mammal. 87 (4), 643–662.

Berry, P.E., Huber, O., Holst, B.K., 1995. Floristic analysis and phytogeography. In: Steymark, J.A., Berry, P.E., Holst, B.K. (Eds.), Flora of the Venezuelan Guayana. Introduction, vol. 1. Missouri Botanical Garden Press, St. Louis, MO, pp. 161–191. , Timber Press, Portland.

Boubli, J.P., da Silva, M.N.F., Amado, M.V., Hrbek, T., Pontual, F.B., Farias, I.P., 2008. A taxonomic reassessment of *Cacajao melanocephalus* Humboldt (1811), with the description of two new species. Int. J. Primatol. 29, 723–741.

Brito, J.M., Tinoco, N., Chávez, D., Moreno-Cárdenas, P., Batallas, P., Ojala-Barbour, R., 2017. New species of arboreal rat of the genus *Rhipidomys* (Cricetidae, Sigmodontinae) from Sangay National Park, Ecuador. Neotrop. Biodivers. 3 (1), 65–79.

Caballero, S., Trujillo, F., Vianna, J.A., Barrios-Garrido, H., Montiel, M.G., Beltrán-Pedreros, S., et al., 2007. Taxonomic status of the genus *Sotalia*: species level ranking for "Tucuxi" (*Sotalia fluviatilis*) and "Costero" (*Sotalia guianensis*) dolphins. Mar. Mammal Sci. 23 (2), 358–386.

Cabrera, A., 1961. Catálogo de los mamíferos de América del Sur. Rev. Mus. Argentino Cien. Nat. "Bernardino Rivadavia". Cien. Zool. 4 (2), 309–732.

Catzeflis, F., Gager, Y., Ruedid, M., de Thoisy, B., 2016. The French Guianan endemic *Molossus barnesi* (Chiroptera: Molossidae) is a junior synonym for *M. coibensis*. Mammal. Biol. 81, 431–438.

Chapman, F.M., 1917. The distribution of bird-life in Colombia: a contribution to a biological survey of South America. Bull. Am. Mus. Nat. Hist. 36, 1–729.

Chapman, F.M., 1931. The upper zonal bird-life of Mts. Roraima and Duida. Bull. Am. Mus. Nat. Hist. 63, 1–135.

Creighton, G.K., Gardner, A.L., 2008. Genus *Marmosa*. In: Gardner, A.L. (Ed.), Mammals of South America. Marsupials, Xenarthrans, Shrews, and Bats, vol. 1. University of Chicago Press, Chicago, IL, pp. 51–61.

Dávalos, L.M., 2006. The geography of diversification in the mormoopids (Chiroptera: Mormoopidae). Biol. J. Linn. Soc. 88 (1), 101–118.

Díaz-Nieto, J.F., Jansa, S.A., Voss, R.S., 2016. D.N.A. sequencing reveals unexpected recent diversity and an ancient dichotomy in the American marsupial genus *Marmosops* (Didelphidae: Thylamyini). Zool. J. Linn. Soc. 176 (4), 914–940.

Duarte, J.M.B., Gonzalez, S., Maldonado, J.E., 2008. The surprising evolutionary history of South American deer. Mol. Phylogenet. Evol. 49 (1), 17–22.

Emmons, L.H., 2005. A revision of the genera of arboreal Echimyidae (Rodentia: Echimyidae, Echimyinae), with descriptions of two new genera. In: Lacey, E.A., Myers, P. (Eds.), Mammalian Diversification: From Chromosomes to Phylogeography (A Celebration of the Career of James L. Patton). University of California Publications in Zoology, vol. 133. University of California Press, Berkeley, CA, pp. 247–310.

Emmons, L.H., Leite, Y.L.R., Patton, J.L., 2015. Family Echimyidae Gray, 1825. In: Patton, J.L., Pardiñas, U.F.J., D'Elía, G. (Eds.), Mammals of South America. Rodents, vol. 2. University of Chicago Press, Chicago, IL, pp. 877–950.

García, F.J., Sánchez-H, J., Semedo, T.B.F., 2014. Descripción de una nueva especie de comadrejita ratona del género *Marmosops* Matschie, 1916 (Didelphimorphia, Didelphidae). Therya 5 (3), 701–723.

Gardner, A.L., 1988. The mammals of Parque Nacional Serranía de La Neblina, Territorio Federal Amazonas, Venezuela. In: Brewer-Carías, C. (Ed.), Cerro de La Neblina: resultados de la expedición 1983–1987. Fundación para el Desarrollo de las Ciencias Físicas Matemáticas y Naturales, Venezuela, pp. 693–754.

Gomes Júnior, R.G., Schneider, C.H., de Lira, Y., Moura Carvalho, N.D., Feldberg, E., Ferreira da Silva, M.N., et al., 2016. Intense genomic reorganization in the genus *Oecomys* (Rodentia, Sigmodontinae): comparison between DNA barcoding and mapping of repetitive elements in three species of the Brazilian Amazon. Comp. Cytogenet. 10 (3), 401–426.

Griffiths, T.A., Gardner, A.L., 2008. Subfamily Glossophaginae. In: Gardner, A.L. (Ed.), Mammals of South America. Marsupials, Xenarthrans, Shrews, and Bats, vol. 1. University of Chicago Press, Chicago, IL, pp. 224–243.

Haffer, J., 1974. Avian Speciation in Tropical South America. With a Systematic Survey of the Toucans (Ramphastidae) and Jacamars (Galbulidae), vol. 14. Publications of the Nuttall Ornithological Club, Cambridge, pp. 1–390.

Handley Jr., C.O., 1976. Mammals of the Smithsonian Venezuelan Project. Brigham Young Univ. Sci. Bull., Biol. Ser. 20 (5), 1–91.

Helgen, K.M., Pinto, C.M., Kays, R., Helgen, L.E., Tsuchiya, M.T.N., Quinn, A., et al., 2013. Taxonomic revision of the olingos (*Bassaricyon*), with description of a new species, the Olinguito. Zookeys 324, 1–83.

Hoogmoed, M.S., 1979. The herpetofauna of the Guianan Region. In: Duellman, W.E. (Ed.), The South American Herpetofauna: Its Origin, Evolution and Dispersal. Monograph Mus. Nat. Hist. Univ. Kansas, vol. 7. pp. 241–279.

Huber, O., 1995. Geographical and physical features. In: Steyermark, J., Berry, P., Holst, B. (Eds.), Flora of the Venezuelan Guayana. Introduction, vol. 1. Missouri Botanical Garden, St. Louis, MO, pp. 1–61.

Huber, O., 2006. Herbaceous ecosystems on the Guayana Shield, a regional overview. J. Biogeogr. 33, 464–475.

Hurtado, N., Pacheco, V., 2014. Análisis filogenético del género *Mimon* Gray, 1847 (Mammalia, Chiroptera, Phyllostomidae) con la descripción de un nuevo género. Therya 5 (3), 751–791.

Huber, O., Foster, M.N. (Eds.), 2002. Prioridades de conservación para el Escudo de Guayana. Conservation International, Washington, DC, p. 99.

International Commission of Zoological Nomenclature, 1998. Opinion 1894. Regnum Animals..., Ed. 2 (M.J. Brisson, 1762): rejected for nomenclatural purposes, with the conservation of the mammalian generic names *Philander* (Marsupialia), *Pteropus* (Chiroptera), *Glis*, *Cuniculus* and *Hydrochoerus* (Rodentia), *Meles*, *Lutra* and *Hyaena* (Carnivora), *Tapirus* (Perissodactyla), *Tragulus* and *Giraffa* (Artiodactyla). Bull. Zool. Nomencl. 55 (1), 64–71.

Leite, Y.L.R., Kok, P.J.R., Weksler, M., 2015. Evolutionary affinities of the 'Lost World' mouse suggest a late Pliocene connection between the Guiana and Brazilian shields. J. Biogeogr. 42, 706–715.

Lew, D., Pérez-Hernández, R., 2004. Una nueva especie del género *Monodelphis* (Didelphimorphia: Didelphidae) de la Sierra de Lema, Venezuela. Mem. Fund. La Salle Cien. Nat. 159–160, 7–25.

Lim, B.K., 2008. Historical biogeography of New World emballonurid bats (tribe Diclidurini): taxon pulse diversification. J. Biogeogr. 35, 1385–1401.

Lim, B.K., 2012. Biogeography of mammals from the Guianas of South America. In: Patterson, B.D., Costa, L.P. (Eds.), Bones, Clones and Biomes: The History and Geography of Recent Neotropical Mammals. University of Chicago Press, Chicago, IL, pp. 230–258.

Lim, B.K., Engstrom, M.D., Patton, J.C., Bickham, J.W., 2010. Molecular phylogenetics of Reig's Short-Tailed Opossum (*Monodelphis reigi*) and its distributional range extension into Guyana. Mammal. Biol. 75, 287–293.

Loureiro, L.O., Lim, B.K., Engstrom, M.D., 2018. A new species of mastiff bat (Chiroptera, Molossidae, *Molossus*) from Guyana and Ecuador. Mammal. Biol. 89, 10–21.

Lynch Alfaro, J.W., Boubli, J.P., Olson, L.E., Di Fiore, A., Wilson, B., Gutiérrez Espeleta, G.A., et al., 2012. Explosive Pleistocene range expansion leads to widespread Amazonian sympatry between robust and gracile capuchin monkeys. J. Biogeogr. 39, 272–288.

Lynch Alfaro, J.W., Boubli, J.P., Paim, F.P., Ribas, C.C., da Silva, M.N.F., Messias, M.R., et al., 2015. Biogeography of squirrel monkeys (genus *Saimiri*): South-central Amazon origin and rapid pan-Amazonian diversification of a lowland primate. Mol. Phylogen. Evol. 82, 436–454.

Mantilla-Meluk, H., Montenegro, O., 2016. A new species of *Lonchorhina* (Chiroptera: Phyllostomidae) from Chiribiquete, Colombian Guayana. Rev. Biodiv. Neot. 6 (2), 171–187.

Mayr, E., Phelps Jr., W.H., 1967. The origin of the bird fauna of the south Venezuelan highlands. Bull. Am. Mus. Nat. Hist. 136, 269–328.

Montgelard, C., Catzetfis, F.M., Douzery, E., 1997. Phylogenetic relationships of artiodactyls and cetaceans as deduced from the comparison of cytochrome b and 12s rRNA mitochondrial sequences. Mol. Biol. Evol. 14 (5), 550–559.

Moojen, J., 1948. Speciation in Brazilian spiny rats (genus *Proechimys*, family Echimyidae). Univ. Kansas Publ. Mus. Nat. Hist. 1 (19), 301–406.

Moras, L., e Gomes, A., da Cunha Tavares, V., 2014. Distribution and taxonomy of the common big-eared bat *Micronycteris microtis* (Chiroptera: Phyllostomidae) in South America. Mammalia 79 (4), 439–447.

Musser, G.G., Carleton, M.D., 2005. Superfamily Muroidea. In: third ed. Wilson, D.E., Reeder, D.M. (Eds.), Mammal Species of the World, a Taxonomic and Geographic Reference, vol. 2. Johns Hopkins University Press, pp. 894–1531.

Ochoa, G.J., 1985. Nueva localidad para *Marmosa tyleriana* (Marsupialia: Didelphidae) en Venezuela. Doñana Acta Vertebrata 12 (1), 183–185.

Ochoa, G.J., Sánchez-H, J., Bevilacqua, M.P., Rivero, R., 1988. Inventario de los mamíferos de la Reserva Forestal de Ticoporo y la Serranía de Los Pijiguaos, Venezuela. Acta Cient. Venez. 39, 269–280.

Parlos, J.A., Timm, R.M., Swier, V.J., Zeballos, H., Baker, R.J., 2014. Evaluation of paraphyletic assemblages within Lonchophyllinae, with description of a new tribe and genus. Occas. Pap. Mus. Texas Tech Univ. 320, 1–23.

Patton, J.L., Leite, R.N., 2015. Genus Proechimys JA Allen, 1899. In: Patton, J.L., Pardiñas, U.F.J., D'Elía, G. (Eds.), Mammals of South America., vol. 2. University of Chicago Press, Chicago, IL, pp. 950–989.

Pavan, A.C., Marroig, G., 2016. Integrating multiples evidences in taxonomy: species diversity and phylogeny of mustached bats (Mormoopidae: *Pteronotus*). Mol. Phylogenet. Evol. 103, 184–198.

Pavan, A.C., Bobrowiec, P.E.D., Percequillo, A.R., 2018. Geographic variation in a South American clade of mormoopid bats, *Pteronotus* (Phyllodia), with description of a new species. J. Mammal. 99 (3), 624–645.

Pavan, S.E., Rossi, R.V., Schneider, H., 2012. Species diversity in the *Monodelphis brevicaudata* complex (Didelphimorphia: Didelphidae) inferred from molecular and morphological data, with the description of a new species. Zool. J. Linn. Soc.: Lond 165, 190–223.

Pavan, S.E., Jansa, S.A., Voss, R.S., 2016. Spatiotemporal diversification of a low-vagility neotropical vertebrate clade (short-tailed opossums, Didelphidae: *Monodelphis*). J. Biogeogr. 43, 1299–1309.

Pérez-H, R., Lew, D., 1998. Las clasificaciones e hipótesis biogeográficas para la Guayana venezolana. Interciencia 26 (9), 373–382.

Pérez-Zapata, A., Lew, D., Aguilera, M., Reig, O.A., 1992. New data on the systematics and karyology of *Podoxymys roraimae* (Rodentia, Cricetidae). Mammal. Biol. 57, 216–224.

Porter, C.A., Baker, R.J., 2004. Systematics of *Vampyressa* and related genera phyllostomid bats as determined by cytochrome-*b* sequences. J. Mammal. 85 (1), 126–132.

Porter, C.A., Hoofer, S.R., Cline, C.A., Hoffmann, F.G., Baker, R.J., 2007. Molecular phylogenetics of the phyllostomid bat genus *Micronycteris* with descriptions of two new subgenera. J. Mammal. 88 (5), 1205–1215.

Spaulding, M., O'Leary, M.A., Gatesy, J., 2009. Relationships of Cetacea (Artiodactyla) among mammals: increased taxon sampling alters interpretations of key fossils and character evolution. PLoS ONE 4 (9), 1–14.

Tate, G.H.H., 1939. Mammals of the Guiana region. Bull. Am. Mus. Nat. Hist. 76, 151–229.

de Thoisy, B., Pavan, A.C., Delaval, M., Lavergne, A., Luglia, T., Pineau, K., et al., 2014. Cryptic diversity in common mustached bats *Pteronotus* cf. *parnellii* (Mormoopidae) in French Guiana and Brazilian Amapa. Acta Chiropterol. 16 (1), 1–13.

Tribe, C.J., 2015. Genus *Rhipidomys* Tschudi, 1845. In: Patton, J.L., Pardiñas, U.F.J., D'Elía, G. (Eds.), Mammals of South America, vol. 2. University of Chicago Press, Chicago, IL, pp. 583–617.

Urbani, B., Portillo-Quintero, C., 2018. Consideraciones sobre la distribución y estado de conservación de los primates de la Guayana venezolana. In: Urbani, B., Kowalewski, M., Cunha, R.G.T., de la Torre, S., Cortés-Ortiz, L. (Eds.), La primatología en Latinoamérica 2—A primatologia na America Latina 2. Ediciones IVIC. Instituto Venezolano de Investigaciones Científicas (IVIC), Venezuela, pp. 677–690.

Veiga, L.M., Silva Jr., J.S., Mittermeier, R.A., Boubli, J.P., 2008. *Chiropotes chiropotes*. In: IUCN 2010. IUCN Red List of Threatened Species. Version 2010.4 <www. iucnredlist.org> (downloaded 14.07.18).

Velazco, P.M., Gardner, A.L., 2009. A new species of *Platyrrhinus* (Chiroptera: Phyllostomidae) from western Colombia and Ecuador, with emended diagnoses of *P. aquilus*, *P. dorsalis*, and *P. umbratus*. Proc. Biol. Soc. Wash. 122 (3), 249–281.

Velazco, P.M., Lim, B.K., 2014. A new species of broad-nosed bat *Platyrrhinus* Saussure, 1860 (Chiroptera: Phyllostomidae) from the Guianan Shield. Zootaxa 3796 (1), 175–193.

Velazco, P.M., Patterson, B.D., 2008. Phylogenetics and biogeography of the broad-nosed bats, Genus *Platyrrhinus* (Chiroptera: Phyllostomidae). Mol. Phylogenet. Evol. 49, 749–759.

Velazco, P.M., Gardner, A.L., Patterson, B.D., 2010. Systematics of the *Platyrrhinus helleri* species complex (Chiroptera: Phyllostomidae), with descriptions of two new species. Zool. J. Linn. Soc.: Lond. 159, 785–812.

Voss, R.S., Lim, B.K., Díaz-Nieto, J.F., Jansa, S.A., 2013. A new species of *Marmosops* (Marsupialia: Didelphidae) from the Pakaraima highlands of Guyana, with remarks on the origin of the endemic Pantepui mammal fauna. Am. Mus. Novit. 3778, 1–27.

Voss, R.S., Gutiérrez, E.E., Solari, S., Rossi, R.V., Jansa, S.A., 2014. Phylogenetic relationships of Mouse Opossums (Didelphidae, *Marmosa*) with a revised subgeneric classification and notes on sympatric diversity. Am. Mus. Novit. 3817, 1–27.

Voss, R.S., Díaz-Nieto, J.F., Jansa, S.A., 2018. A revision of *Philander* (Marsupialia: Didelphidae). Part 1: *P. quica*, *P. canus*, and a new species from Amazonia. Am. Mus. Novit. 3891, 1–70.

Williams, S.L., Genoways, H.H., 2008. Subfamily Phyllostominae. In: Gardner, A.L. (Ed.), Mammals of South America. Marsupials, Xenarthrans, Shrews, and Bats, vol. 1. University of Chicago Press, Chicago, IL, pp. 82–107.

Williams, S.L., Genoways, H.H., Groen, J.A., 1983. Results of the Alcoa Foundation-Suriname Expeditions. VII. Records of mammals from central and southern Suriname. Ann. Carnegie Mus. 52, 329–336.

Ziegler, A.C., Howarth, F.G., Simmons, N.B., 2016. A second endemic land mammal for the Hawaiian Islands: a new genus and species of fossil bat (Chiroptera: Vespertilionidae). Am. Mus. Novit. 3854, 1–52.

Appendix 14.1 Species of mammals documented from the Guianas by altitudinal categories (lowlands, uplands, and highlands), endemic species (Guiana Shield and highlands or Pantepui), elevation occurrence, altitudinal interval, and geographic distribution.

Taxonomy	Lowlands (<500)	Uplands (500–1500)	Highlands (>1500)	Endemic of Guiana Shield	Endemic of Pantepui	Elevation range (m a.s.l.)	Altitude interval (m)	Distribution (Tate, 1939)
Mammalia 294								
Didelphimorphia 27								
Didelphidae 27								
Caluromys lanatus	X	X				100–850	750	4
Caluromys philander	X	X				0–700	700	5
Chironectes minimus	X					15–460	445	1
Cryptonanus sp.	X					6–6	0	6
Didelphis imperfecta	X	X	X	X		50–2550	2500	8
Didelphis marsupialis	X	X	X			5–1800	1795	1
Gracilinanus emiliae	X	X				25–850	825	3
Hyladelphys kalinowskii	X					10–230	220	3
Lutreolina crassicaudata	X					298–440	142	5
Marmosa lepida	X					90–90	0	3
Marmosa murina	X	X				10–1347	1337	2
Marmosa tyleriana		X	X	X	X	1300–2010	710	8
Marmosops caucae	X	X	X			140–2000	1860	4
Marmosops pakaraimae		X	X	X	X	800–1500	700	8
Marmosops parvidens	X	X				10–500	490	3
Marmosops pinheiroi	X	X				50–1374	1324	5

(Continued)

(Continued)

Taxonomy	Lowlands (<500)	Uplands (500–1500)	Highlands (>1500)	Endemic of Guiana Shield	Endemic of Pantepui	Elevation range (m a.s.l.)	Altitude interval (m)	Distribution (Tate, 1939)
Metachirus nudicaudatus	X	X				20–1046	1026	1
Marmosa demerarae	X	X	X			25–2100	2075	2
Monodelphis arlindoi	X	X		X		70–670	600	8
Monodelphis brevicaudata	X	X		X		20–1318	1298	8
Monodelphis reigi		X	X	X	X	1100–2050	950	8
Monodelphis sp.	X					53–160	107	6
Monodelphis touan	X					9–227	218	3
Philander andersoni	X					100–180	80	4
Philander deltae	X			X		10–20	10	7
Philander canus	X	X				100–800	700	2
Philander opossum	X					10–280	270	5
Pilosa 6								
Bradypodidae 2								
Bradypus tridactylus	X					30–120	90	5
Bradypus variegatus	X					10–250	240	1
Megalonychidae 1								
Choloepus didactylus	X					10–325	315	3
Myrmecophagidae 3								
Cyclopes didactylus	X					10–200	190	1
Myrmecophaga tridactyla	X	X				50–1000	950	1
Tamandua tetradactyla	X	X	X			100–2150	2050	2
Cingulata 6								
Dasypodidae 6								
Cabassous unicinctus	X	X				10–880	870	2
Dasypus kappleri	X					50–331	281	3

Dasypus novemcinctus	X	X	112–1420	1308	1
Dasypus sabanicola	X		160–170	10	6
Euphractus sexcinctus	X		275–275	0	6
Priodontes maximus	X		110–250	140	2
Chiroptera 151					
Emballonuridae 16					
Centronycteris maximiliani	X		50–250	200	2
Cormura brevirostris	X	X	10–640	630	1
Cyttarops alecto	X		26–56	30	1
Diclidurus albus	X		140–473	333	1
Diclidurus ingens	X		0–300	300	3
Diclidurus isabellus	X		50–317	267	3
Diclidurus scutatus	X	X	155–900	745	3
Peropteryx kappleri	X		100–150	50	1
Peropteryx leucoptera	X		50–120	70	2
Peropteryx macrotis	X	X	70–900	830	1
Peropteryx trinitatis	X		50–374	324	2
Rhynchonycteris naso	X	X	0–800	800	1
Saccopteryx bilineata	X	X	10–1300	1290	1
Saccopteryx canescens	X	X	10–700	690	3
Saccopteryx gymnura	X		50–250	200	5
Saccopteryx leptura	X	X	4–700	696	1
Noctilionidae 2					
Noctilio albiventris	X	X	10–1100	1090	1
Noctilio leporinus	X	X	0–800	800	1

(*Continued*)

(Continued)

Taxonomy	Lowlands (<500)	Uplands (500–1500)	Highlands (>1500)	Endemic of Guiana Shield	Endemic of Pantepui	Elevation range (m a.s.l.)	Altitude interval (m)	Distribution (Tate, 1939)
Mormoopidae 7								
Mormoops megalophylla	X					400–400	0	1
Pteronotus alitonus	X	X		X		50–708	658	3
Pteronotus davyi		X				550–620	70	7
Pteronotus fuscus	X	X				50–800	750	7
Pteronotus gymnonotus	X	X				50–960	910	1
Pteronotus personatus	X	X				10–700	690	1
Pteronotus rubiginosus	X	X				5–670	665	4
Phyllostomidae 77								
Phyllostominae 29								
Chrotopterus auritus	X	X				0–650	650	1
Glyphonycteris daviesi	X	X				50–500	450	1
Glyphonycteris sylvestris	X	X				40–600	560	1
Lampronycteris brachyotis	X					45–140	95	1
Lonchorhina aurita	X					200–400	200	1
Lonchorhina fernandezi	X			X		90–90	0	8
Lonchorhina inusitata	X	X				90–1030	940	5
Lonchorhina orinocensis	X	X				30–700	670	6
Lophostoma brasiliense	X					0–280	280	1
Lophostoma carrikeri	X	X				70–700	630	3
Lophostoma schulzi	X	X	X	X		20–1420	1400	8
Lophostoma silvicolum	X	X				20–880	860	1
Macrophyllum macrophyllum	X					58–400	342	1
Micronycteris brosseti	X					50–81	31	3

Micronycteris hirsuta	X	X		50–700	650	1
Micronycteris megalotis	X	X	X	25–1420	1395	1
Micronycteris minuta	X	X		50–500	450	1
Micronycteris schmidtorum	X			58–155	97	1
Mimon bennettii	X	X		50–670	620	2
Gardnerycteris crenulatum	X	X		0–1080	1080	1
Phylloderma stenops	X	X		26–700	674	1
Phyllostomus discolor	X	X		0–700	700	1
Phyllostomus elongatus	X	X		10–830	820	2
Phyllostomus hastatus	X	X		45–980	935	1
Phyllostomus latifolius	X	X		100–587	487	3
Tonatia saurophila	X	X		26–620	594	1
Trachops cirrhosus	X	X	X	5–1420	1415	1
Trinycteris nicefori	X	X		10–620	610	1
Vampyrum spectrum	X	X		50–1250	1200	1
Glossophaginae 11						
Anoura caudifer	X	X	X	30–2150	2120	2
Anoura geoffroyi	X	X	X	80–2550	2470	1
Anoura latidens	X	X	X	50–2000	1950	4
Choeroniscus godmani	X	X		160–500	340	7
Choeroniscus minor	X	X		90–920	830	2
Glossophaga longirostris	X	X		0–1080	1080	6,7
Glossophaga soricina	X	X	X	0–1500	1500	1
Lichonycteris degener	X			85–240	155	2
Lionycteris spurrelli	X	X		30–1300	1270	2
Hsunycteris thomasi	X	X		30–1150	1120	1
Scleronycteris ega	X			190–190	0	3

(Continued)

(Continued)

Taxonomy	Lowlands (<500)	Uplands (500–1500)	Highlands (>1500)	Endemic of Guiana Shield	Endemic of Pantepui	Elevation range (m a.s.l.)	Altitude interval (m)	Distribution (Tate, 1939)
Carolliinae 5								
Carollia brevicauda	X	X	X			15–1800	1785	2
Carollia castanea	X					160–300	140	7
Carollia perspicillata	X	X	X			10–1500	1490	1
Rhinophylla fischerae	X					140–140	0	3
Rhinophylla pumilio	X	X	X			5–1500	1495	2
Stenodermatinae 30								
Ametrida centurio	X	X	X			0–1600	1600	5
Artibeus amplus	X	X	X			50–2100	2050	7
Artibeus bogotensis	X	X	X			30–1500	1470	4
Artibeus cinereus	X					0–275	275	2
Artibeus concolor	X	X	X			10–1500	1490	5
Artibeus gnomus	X	X				15–1100	1085	2
Artibeus jamaicensis	X					15–400	385	7
Artibeus lituratus	X	X	X			5–1500	1495	1
Artibeus obscurus	X	X				5–1170	1165	2
Artibeus planirostris	X	X	X			0–1600	1600	2
Artibeus sp.	X			X				8
Chiroderma trinitatum	X	X				15–1170	1155	2
Chiroderma villosum	X	X				48–1080	1032	1
Enchisthenes hartii	X					155–155	0	4
Mesophylla macconnelli	X	X	X			40–1515	1475	1
Platyrrhinus aurarius	X	X	X	X	X	414–2200	1786	8
Platyrrhinus brachycephalus	X					5–170	165	2

Platyrrhinus angustirostris	X	X			50–850	800	4
Platyrrhinus fusciventris	X	X			5–1170	1165	8
Platyrrhinus guianensis	X	X		X	50–500	450	8
Platyrrhinus incarum	X	X			50–500	450	3
Sphaeronycteris toxophyllum	X	X	X		100–1500	1400	4
Sturnira lilium	X	X	X		15–1800	1785	1
Sturnira tildae	X	X	X		25–1800	1775	2
Uroderma bilobatum	X	X	X		15–1500	1485	1
Uroderma magnirostrum	X	X			5–900	895	1
Vampyriscus bidens	X	X			5–900	895	3
Vampyriscus brocki	X				40–280	240	3
Vampyressa thyone	X	X			45–1250	1205	1
Vampyrodes caraccioli	X	X			100–1170	1070	1
Desmodontinae 2							
Desmodus rotundus	X	X			15–900	885	1
Diaemus youngi	X	X			40–670	630	1
Natalidae 1							
Natalus tumidirostris	X				50–120	70	6
Furipteridae 1							
Furipterus horrens	X				10–180	170	1
Thyropteridae 3							
Thyroptera discifera	X				9–227	218	1
Thyroptera devivoi	X				140–140	0	6
Thyroptera tricolor	X	X			15–1000	985	1

(*Continued*)

(Continued)

Taxonomy	Lowlands (<500)	Uplands (500–1500)	Highlands (>1500)	Endemic of Guiana Shield	Endemic of Pantepui	Elevation range (m a.s.l.)	Altitude interval (m)	Distribution (Tate, 1939)
Vespertilionidae 18								
Eptesicus andinus	X					240–240	0	4
Eptesicus brasiliensis	X	X				10–1030	1020	1
Eptesicus chiriquinus	X	X				50–500	450	1
Eptesicus diminutus	X	X				10–1190	1180	4
Eptesicus furinalis	X	X				10–1420	1410	1
Histiotus humboldti			X			1820–1820	0	4
Lasiurus atratus	X			X		75–260	185	8
Lasiurus blossevillii	X					75–250	175	1
Aeorestes cinereus			X			2085–2100	15	4
Aeorestes egregius	X					175–311	136	5
Dasypterus ega	X	X				155–851	696	1
Myotis albescens	X	X				10–1347	1337	1
Myotis keaysi	X	X				140–1200	1060	4
Myotis nigricans	X	X				10–900	890	1
Myotis oxyotus		X	X			851–1800	949	4
Myotis riparius	X	X	X			45–1600	1555	1
Rhogeessa hussoni	X					270–270	0	5
Rhogeessa io	X					10–290	280	1
Molossidae 26								
Cynomops abrasus	X					0–280	280	2
Cynomops greenhalli	X					10–10	0	7
Cynomops paranus	X					50–170	120	3
Cynomops planirostris	X					10–155	145	2

Eumops auripendulus	X			80–280	200	1
Eumops maurus	X			100–100	0	10
Eumops nanus	X			200–200	0	7
Eumops dabbenei	X			300–400	100	4
Eumops glaucinus	X			120–155	35	1
Eumops hansae	X			30–180	150	1
Eumops trumbulli	X			40–200	160	3
Molossops neglectus	X			30–260	230	2
Molossops temminckii	X			10–187	177	1
Molossus aztecus	X			50–185	135	7
Molossus coibensis	X			50–300	250	1
Molossus molossus	X	X	X	0–1600	1200	1
Molossus pretiosus	X			20–100	80	1
Molossus rufus	X	X		30–500	470	1
Molossus sinaloae	X			100–245	145	1
Molossus fentoni	X			60–245	185	9
Neoplatymops mattogrossensis	X			40–290	250	2
Nyctinomops gracilis	X			138–350	212	4
Nyctinomops laticaudatus	X	X		80–1200	1120	1
Nyctinomops macrotis	X	X	X	100–2100	2000	1
Promops centralis	X			105–105	0	1
Promops nasutus	X			105–155	50	2
Primates 17						
Cebidae 6						
Cebus albifrons	X			100–290	190	4
Sapajus apella	X			150–250	100	3

(*Continued*)

(Continued)

Taxonomy	Lowlands (<500)	Uplands (500–1500)	Highlands (>1500)	Endemic of Guiana Shield	Endemic of Pantepui	Elevation range (m a.s.l.)	Altitude interval (m)	Distribution (Tate, 1939)
Cebus olivaceus	X	X	X			10–2000	1990	5
Saguinus midas	X			X		40–120	80	8
Saimiri cassiquiarensis	X					95–270	175	5
Saimiri sciureus	X	X				50–550	500	3
Atelidae 4								
Alouatta macconnelli	X	X	X			10–1960	1950	5
Alouatta seniculus	X					100–250	150	4
Ateles belzebuth	X	X				120–1000	880	4
Ateles paniscus	X			X		120–240	120	8
Aotidae 1								
Aotus trivirgatus	X	X		X		140–1000	860	8
Pitheciidae 6								
Cacajao ayresi	X			X		140–140	0	8
Cacajao hosomi	X	X	X	X		70–1500	1430	8
Cacajao melanocephalus	X					90–90	0	4
Callicebus lugens	X					90–450	360	4
Chiropotes chiropotes	X			X		80–300	220	8
Pithecia pithecia	X			X		15–250	235	8
Carnivora 17								
Canidae 2								
Cerdocyon thous	X	X				60–1100	1040	6
Speothos venaticus	X	X				120–880	760	1
Felidae 6								
Leopardus pardalis	X	X				10–970	960	1
Leopardus tigrinus	X	X				120–1440	1320	1

Leopardus wiedii	X	X		120–870	750	1
Panthera onca	X	X		10–1300	1290	1
Puma concolor	X	X	X	50–1750	1700	1
Puma yagouaroundi	X	X		5–600	595	1
Mustelidae 5						
Eira barbara	X	X		100–980	880	1
Galictis vittata	X	X		200–600	400	1
Mustela frenata		X	X	1050–2500	1450	4
Lontra longicaudis	X			100–266	166	1
Pteronura brasiliensis	X			50–330	280	2
Procyonidae 4						
Bassaricyon alleni	X	X		90–1415	1325	4
Potos flavus	X	X		100–700	600	1
Nasua nasua	X	X	X	35–2550	2515	2
Procyon cancrivorus	X	X		50–600	550	1
Cetacea 2						
Delphinidae 1						
Sotalia guianensis	X			0–50	50	3
Platanistidae 1						
Inia geoffrensis	X			0–0	0	3
Sirenia 2						
Trichechidae 2						
Trichechus inunguis	X			200–200	0	3
Trichechus manatus	X			20–40	20	1
Perissodactyla 1						
Tapiridae 1						
Tapirus terrestris	X	X	X	50–2150	2100	2

(Continued)

(Continued)

Taxonomy	Lowlands (<500)	Uplands (500–1500)	Highlands (>1500)	Endemic of Guiana Shield	Endemic of Pantepui	Elevation range (m a.s.l.)	Altitude interval (m)	Distribution (Tate, 1939)
Cetartiodactyla 5								
Tayassuidae 2								
Pecari tajacu	X	X				25–1083	1058	1
Tayassu pecari	X	X				80–1100	1020	1
Cervidae 3								
Mazama americana	X	X				15–1311	1296	2
Mazama nemorivaga	X	X				70–1365	1295	3
Odocoileus cariacou	X	X	X			100–1500	1400	6
Rodentia 58								
Sciuridae 5								
Sciurillus pusillus	X					40–360	320	3
Sciurus flammifer	X	X		X		150–658	508	8
Sciurus aestuans	X	X	X			15–1800	1785	8
Sciurus gilvigularis	X	X				138–1200	1062	4
Sciurus igniventris	X	X				100–600	500	4
Cricetidae 32								
Calomys hummelincki	X					40–40	0	7
Holochilus sciureus	X	X				15–658	643	4
Neacomys dubosti	X			X		90–275	185	8
Neacomys guianae	X	X		X		180–1046	866	8
Neacomys paracou	X	X		X		50–1318	1268	8
Necromys urichi	X	X	X			100–2550	2450	4
Nectomys rattus	X	X	X			30–1500	1470	2
Nectomys palmipes	X					5–297	292	7

Neusticomys oyapocki	X			X		214–214	0	8
Neusticomys venezuelae	X	X	X			300–1400	1100	4
Oecomys auyantepui	X	X		X		50–800	750	8
Oecomys bicolor	X	X				15–1046	1031	2
Oecomys concolor	X	X				40–1200	1160	4
Oecomys rex	X			X		100–465	365	8
Oecomys roberti	X	X				100–830	730	4
Oecomys rutilus	X	X		X		70–700	630	8
Oecomys sp. 1		X	X	X	X	1100–1550	450	8
Oecomys sp. 2	X			X				8
Oecomys speciosus	X					10–306	296	6
Oecomys trinitatis	X	X	X			306–2000	1694	1
Oligoryzomys fulvescens	X	X	X			15–1900	1885	1
Oligoryzomys sp.*	X			X				8
Euryoryzomys macconnelli	X	X	X			107–1420	1313	4
Hylaeamys megacephalus	X	X	X			10–1420	1410	5
Hylaeamys yunganus	X	X	X			10–1420	1410	3
Podoxymys roraimae			X	X	X	2216–2628	412	8
Rhipidomys leucodactylus	X	X	X			138–1800	1662	3
Rhipidomys macconnelli		X	X	X	X	550–2600	2050	8
Rhipidomys nitela	X	X	X			10–1480	1470	3
Rhipidomys wetzeli		X	X	X	X	587–1800	1213	8
Sigmodon alstoni	X	X				40–923	883	6
Zygodontomys brevicauda	X	X	X			10–1900	1890	6

(Continued)

(Continued)

Taxonomy	Lowlands (<500)	Uplands (500–1500)	Highlands (>1500)	Endemic of Guiana Shield	Endemic of Pantepui	Elevation range (m a.s.l.)	Altitude interval (m)	Distribution (Tate, 1939)
Erethizontidae 2								
Coendou melanurus	X	X	X	X		10–1829	1819	8
Coendou prehensilis	X	X				150–550	400	2
Caviidae 2								
Cavia aperea	X	X				100–1331	1231	6
Hydrochoerus hydrochaeris	X	X				45–1336	1291	2
Dasyproctidae 5								
Dasyprocta fuliginosa	X					93–150	57	4
Dasyprocta guamara	X			X		10–10	0	8
Dasyprocta leporina	X	X				50–900	850	5
Myoprocta acouchy	X					20–160	140	5
Myoprocta pratti	X	X				135–1000	865	4
Cuniculidae 1								
Cuniculus paca	X	X	X			10–1800	1790	1
Echimyidae 11								
Dactylomys dactylinus	X	X				180–570	390	3
Echimys chrysurus	X					100–150	50	5
Pattonomys punctatus	X					10–160	150	10
Isothrix orinoci	X			X		100–138	38	8
Isothrix sinnamariensis	X			X		100–250	150	8
Makalata didelphoides	X					50–400	350	3
Mesomys hispidus	X	X				100–500	400	3
Proechimys cuvieri	X	X				10–506	496	3

Proechimys guyannensis	X	X				37–1350	1313	3
Proechimys hoplomyoides	X	X	X	X	X	135–1550	1415	8
Proechimys quadruplicatus	X					50–270	220	4
Lagomorpha 2								
Leporidae 2								
Sylvilagus brasiliensis	X					30–300	270	1
Sylvilagus floridanus	X					54–100	46	6

CHAPTER

15

Vertebrate parasites

Ricardo Guerrero

Institute of Tropical Zoology, Faculty of Sciences, Central University of Venezuela, Caracas, Venezuela

Introduction

This chapter briefly summarizes the information available to date about the ectoparasites and endoparasites of Pantepui vertebrates. The account is subdivided into two sections. The first section corresponds to the collections made during historical zoological expeditions to Pantepui and the further studies on vertebrate parasites resulting from them. The second section accounts for occasional fieldtrips carried out by the author and other colleagues during recent decades. Some preliminary conclusions are derived that will be refined with further studies.

Historical expeditions

The information on vertebrate parasites from Pantepui is scarce, and most of the organisms known to date are mammal parasites. The first known record was from samples collected by J.J. Quelch on the Roraima summit and identified and described by Arrow (1907) as *Amblyopinus angustus* (Coleoptera: Staphylinidae). The most likely host was the rodent *Rhipidomys macconnelli*, described by De Winton (1900) from Quelch's collections. Between 1965 and 1968, 270 mammal species from >100 localities around Venezuela were collected under the auspices of the Smithsonian Venezuelan Project (Handley, 1976). Several sampled localities were on the Guiana Highlands (Sierra de Lema and the Duida and Marahuaka slopes) (see Fig. 1.2 in Chapter 1: Definition and characterization of the Pantepui biogeographical province)

Biodiversity of Pantepui
DOI: https://doi.org/10.1016/B978-0-12-815591-2.00015-X

where a number of mammal ectoparasites were described (Brennan and van Bronswijk, 1973; Brennan and Goff, 1977; Brennan and Reed, 1976, 1974; Emerson and Price, 1975; Furman, 1972a,b; Herrin and Tipton, 1975; Jones et al., 1972; Johnson, 1972; Machado-Allison and Barrera, 1972; McDaniel, 1972; Saunders, 1975; Tipton and Machado-Allison, 1972; Wenzel, 1976). Further descriptions of mammal ectoparasites and endoparasites from Sierra de Lema, Sierra de la Neblina, Cerro Marahuaka, Cerro Duida, Cerro Guaiquinima, and other tepuis and Guiana Highland mountains were provided by Guerrero (1988, 1989, 1994a,b, 1995a,b, 1996a,b). Guerrero (in Myers and Donelly, 1996) reported the first lizard ectoparasite known to date. All these records are summarized in Table 15.1.

Other reports exist whose Pantepui character is less clear. For example, Díaz-Ungria (1963) reported *Tepuinema verganii* (Nematoda: Subuluridae) on a common rat (*Mus musculus*) from Auyán-tepui, but this species was described in the surrounding lowlands of this tepui; therefore it is not restricted to Pantepui. Several species of feather mites (Acarina: Astigmata) were found on avian skins, sometimes without precise locations. An example is *Toxerodectes biasculatus* (Acarina: Proctophyllodidae) (Chong and Atyeo, 1973) found in the hummingbird *Colibri delphinae* skin collected on the Duida slopes by Chapman (1931). However, according to Mayr and Phelps (1967), this hummingbird is not a Pantepui species. Other parasites collected on non-Pantepui bird species include *Rhytidelasma dilatata* (Acarina: Pterolichidae) and *Aralichus nobilis* (Acarina: Pterolichidae) found on the skin of the parrot *Diopsyttaca nobilis nobilis* from Roraima-tepui and Auyán-tepui (Atyeo, 1988; Atyeo and Pérez, 1988). In this case, the parrot host was collected by Chapman (1931), who identified it as *Diopsittaca nobilis*, on the slopes of these tepuis below the Pantepui elevational range (Fig. 15.1).

Recent collections

This section describes the vertebrate parasites found on specific tepuian summits visited occasionally by the author and other colleagues during recent decades, with the specific aim of finding and describing vertebrate parasites from the Pantepui biogeographical province. Part of this information is still under study, but a report on the state of the art would merit publication in a book on Pantepui biodiversity like this. The report includes 16 tepuis and many new species, still undescribed, from already known genera. Some of these species were found on a single tepui and others on several of them, depending on the distribution of the host species. As a result of these studies, the better-known parasites are batflies (Diptera: Streblidae) living on bats and mites (Acarina) and fleas (Siphonaptera) living on several rodents. Most of the batflies recorded are of wide Neotropical distribution (Table 15.2), whereas mites and fleas are more restricted to Pantepui and nearby areas (Table 15.3).

TABLE 15.1 Species of Pantepui vertebrate parasites (historical expeditions). The location of the tepuis can be seen in Chapter 1 (Fig. 1.2). (A), Aves; (C), Chiroptera; (L), Lacertilia; (M), Marsupialia; (R), Rodentia; (Sl), Slopes; (Su), Summits. Species with an asterisk are depicted in Fig. 15.1.

Ectoparasites

Class: Insecta

Order	Family	Species	Host	Location	Reference
Anoplura	Hoplopleuridae	*Hoplopleura angulata*	*Rhipidomys macconnelli* (R)	Sierra de Lema, Cerro Marahuaka	Johnson (1972)
		Hoplopleura sp.	*Necromys urichi* (R), *R. macconnelli* (R)	Neblina (Sl, Su), Cerro MarahuaKa	Guerrero (1988, 1989)
Coleoptera	Staphylinidae	*Amblyopinus angustus*	*R. macconnelli* (R)	Roraima-tepui	Arrow (1907)
		Amblyopinus proximus	*R. macconnelli* (R)	Not given	Machado-Allison and Barrera (1972)
		Amblyopinus sp.	*N. urichi* (R)	Cerro Marahuaka (Su)	Guerrero (1989)
Diptera	Streblidae	*Anastrebla caudiferae*	*Anoura caudifer* (C)	Auyán-tepui, Cerro Guaiquinima, Neblina (Sl, Su), Sierra de Lema	Wenzel (1976), Guerrero (1996a)
		Anastrebla modestini	*Anoura geoffroyi* (C)	Sierra de Lema, Cerro Marahuaka (Sl)	Wenzel (1976), Guerrero (1996a)
		Anastrebla spurrelli	*Lionycteris spurrelli* (C)	Cerro Guaiquinima, Sierra de Lema, Cerro Marahuaka (Sl)	Wenzel (1976)
		Aspidoptera falcata	*Sturnira lilium* (C), *Sturnira tildae* (C)	Cerro Guaiquinima, Neblina (Sl), Sierra de Lema	Wenzel (1976), Guerrero (1995b)
		Anatrichobius scorzai	*Myotis oxyotus* (C)	Sierra de Lema	Wenzel (1976)
		Exastinion clovisi	*A. geoffroyi* (C)	Auyán-tepui, Cerro Guaiquinima, Sierra de Lema, Cerro Marahuaka (Sl, Su), Neblina (Sl, Su)	Wenzel (1976), Guerrero (1988, 1995b)
		Neotrichobius sp.	*Rhinophylla pumilio* (C)	Sierra de Lema	Wenzel (1976)
		Nycterophilia parnelli	*Pteronotus parnellii* (C)	Sierra de Lema	Wenzel (1976)

(Continued)

TABLE 15.1 (Continued)

Ectoparasites

Class: Insecta

Order	**Family**	**Species**	**Host**	**Location**	**Reference**
		Paratrichobius longicrus	*Artibeus lituratus* (C), *Platyrrhinus aurarius* (C)	Cerro Guaiquinima, Neblina (Sl), Sierra de Lema	Wenzel (1976), Guerrero (1994b)
		Paratrichobius lowei	*Artibeus cinereus* (C), *Artibeus bogotensis* (C)	Sierra de Lema, Cerro Guaiquinima, Cerro Marahuaka (Sl), Neblina (Sl)	Wenzel (1976), Guerrero (1994b)
		Paratrichobius sp.	*P. aurarius* (C)	Sierra de Lema, Cerro Marahuaka (Sl)	Wenzel (1976)
		Speiseria peytoni	*Carollia brevicauda* (C)	Neblina (Sl), Sierra de Lema	Wenzel (1976), Guerrero (1994b)
		Strebla guajiro	*C. brevicauda* (C)	Neblina (Sl), Sierra de Lema	Wenzel (1976), Guerrero (1996a)
		Strebla chrotopteri	*Chrotopterus auritus* (C)	Cerro Duida (Sl)	Guerrero (1996a)
		Strebla harderi	*Anoura latidens* (C)	Sierra de Lema	Wenzel (1976)
		Strebla paramirabilis	*Artibeus jamaicensis* (C), *A. geoffroyi* (C), *P. aurarius* (C)	Cerro Arakamuni, Cerro Marahuaka (Sl), Neblina (Sl), Tamacuari	Wenzel (1976), Guerrero (1996a)
		Trichobius angulatus	*P. aurarius* (C)	Sierra de Lema, Cerro Marahuaka (Sl), Cerro Guaiquinima, Neblina (Sl)	Wenzel (1976), Guerrero (1995a)
		Trichobius caecus	*P. parnellii* (C)	Sierra de Lema	Wenzel (1976)
		Trichobius joblingi	*C. brevicauda* (C), *Carollia perspicillata* (C)	Sierra de Lema, Cerro Guaiquinima, Tamacuari	Wenzel (1976), Guerrero (1988, 1994a)
		Trichobius lionycteridis	*L. spurrelli* (C)	Cerro Marahuaka (Sl)	Wenzel (1976)
		Trichobius propinquus	*A. geoffroyi* (C)	Sierra de Lema	Wenzel (1976)
		Trichobius sparsus	*P. parnellii* (C)	Sierra de Lema	Wenzel (1976)

		Trichobius tiptoni	*A. caudifer* (C)	Sierra de Lema, Cerro Marahuaka (Sl), Neblina (Sl), Tamacuari	Wenzel (1976), Guerrero (1994a)
		Trichobius uniformis	*Glossophaga soricina* (C)	Sierra de Lema	Wenzel (1976)
Phthiraptera	Gyropidae	*Gliricola handleyi*	*Proechimys hoplomyoides* (R)	Sierra de Lema	Emerson and Price (1975)
	Trimenoponidae	*Cummingsia gardneri*	*Marmosa impavida* (M)	Neblina (Sl)	Price and Emerson (1986)
Siphonaptera	Rhopalopsyllidae	*Polygenis k. klagesi*	*N. urichi* (R)	Sierra de Lema	Tipton and Machado-Allison (1972)
		Polygenis roberti beebei	*N. urichi* (R), *R. macconnelli* (R)	Cerro Marahuaka (Sl), Sierra de Lema	Tipton and Machado-Allison (1972), Guerrero (1989)
		Polygenis versuta	*Hylaemys yunganus* (R), *R. macconnelli* (R)	Cerro Marahuaka (Sl)	Guerrero (1989)
	Hystrichopsyllidae	*Adoratopsylla antiquorum discreta*	*N. urichi* (R), *R. macconnelli* (R)	Cerro Duida, Cerro Marahuaka (Sl)	Tipton and Machado-Allison (1972)
		Adoratopsylla antiquorum rara	*Marmosa murina* (M)	Sierra de Lema	Tipton and Machado-Allison (1972)
	Stephanocercidae	*Craneopsylla minerva minerva**	*R. macconnelli* (R)	Cerro Duida, Cerro Marahuaka (Sl)	Tipton and Machado-Allison (1972)
	Ischnopsyllidae	*Hormopsylla trux*	*Nyctinomops macrotis* (R)	Cerro Marahuaka (Sl)	Hastriter and Guerrero (1998)
Class: Acarina					
Ixodida	Ixodidae	*Amblyomma oblongoguttatum*	*Rupicola rupicola* (A)	Cerro Marahuaka (Su)	Guerrero (1996b)
		Amblyomma sp.	*R. macconnelli* (R)	Sierra de Lema	Jones et al. (1972)
		Ixodes lasallei	*Marmosops neblina* (M)	Neblina (Su)	Guerrero (1996b)
		Ixodes luciae	*Philander andersoni* (M)	Tamacuari	Guerrero (1996b)
Mesostigmata	Laelapidae	*Androlaelaps fahrenholzi*	*N. urichi* (R), *H. yunganus* (R), *R. macconnelli* (R)	Cerro Marahuaka (Sl)	Guerrero (1989)

(*Continued*)

TABLE 15.1 (Continued)

Ectoparasites

Class: Insecta

Order	Family	Species	Host	Location	Reference
		Androlaelaps rotundus	*N. urichi* (R)	Neblina (Su)	Guerrero (1988)
		Gigantolaelaps oudemansi	*N. urichi* (R), *Oecomys concolor* (R), *H. yunganus* (R), *R. macconnelli* (R)	Cerro Marahuaka (Sl), Neblina (Sl)	Guerrero (1988, 1989)
		*Laelaps conula**	*H. yunganus* (R), *R. macconnelli* (R)	Cerro Marahuaka (Sl, Su), Sierra de Lema	Furman (1972a), Guerrero (1989)
		Laelaps surcomata	*R. macconnelli* (R)	Sierra de Lema	Furman (1972a)
		Laelaps paulistanensis	*R. macconnelli* (R)	Not given	Furman (1972a)
		Mysolaelaps sp.	*N. urichi* (R), *H. yunganus* (R), *R. macconnelli* (R)	Cerro Marahuaka (Sl, Su)	Guerrero (1989)
	Macronyssidae	*Lepidodorsum* sp.	*R. macconnelli* (R)	Cerro Marahuaka (Sl)	Guerrero (1989)
		Macronyssoides sp.	*A. bogotensis* (R)	Cerro Marahuaka (Sl)	Guerrero (1989)
		Ornithonyssus bacoti	*R. macconnelli* (R)	Cerro Marahuaka (Sl)	Saunders (1975), Guerrero (1989)
		Parichoronyssus sp.	*Anoura geoffoyi* (C), *A. Latidens* (C)	Cerro Marahuaka (Sl, Su)	Guerrero (1989)
	Spinturnicidae	*Periglischrus vargasi*	*A. geoffoyi* (C), *A. latidens* (C)	Cerro Marahuaka (Sl, Su)	Guerrero (1989)
Sarcoptiformes	Chirodiscidae	*Alabidocarpus furmani**	*A. geoffroyi* (C), *A. latidens* (C), *N. macrotis* (C)	Cerro Marahuaka (Sl)	McDaniel (1972), Guerrero (1989)
		Pseudoalabidocarpus sp.	*A. geoffroyi* (C)	Cerro Marahuaka (Sl, Su)	Guerrero (1989)
Trombidiformes	Trombiculidae	*Crotiscus desdentatus desdentatus*	*R. macconnelli* (R)	Cerro Marahuaka (Sl)	Guerrero (1989)
		Eutrombicula tropica	*R. macconnelli* (R)	Cerro Marahuaka (Sl)	Brennan and Reed (1974)

		Hoffmannina dianneae	*Neusticomys venezuelae* (R), *R. macconnelli* (R)	Cerro Marahuaka (Sl), Cerro Duida	Brennan and Goff (1977)
		Kymocta aitkeni	*R. macconnelli* (R)	Cerro Marahuaka (Sl)	Brennan and van Bronswijk (1973)
		Kymocta inca	*R. macconnelli* (R)	Cerro Marahuaka (Sl)	Brennan and van Bronswijk (1973)
		Microtrombicula sp.	*N. macrotis* (C)	Cerro Marahuaka (Sl)	Guerrero (1989)
		Nycterinastes primus	*A. latidens* (C)	Sierra de Lema	Brennan and Reed (1973)
		Parasecia sp.	*Euspondylus goeleti* (L)	Cerro Yaví	Guerrero in Myers and Donelly (1996)
		Quadraseta sp.	*N. urichi* (R), *H. yunganus* (R), *R. macconnelli* (R)	Cerro Marahuaka (Sl)	Guerrero (1989)
		Whartonia nudosetosa	*A. latidens* (C)	Cerro Marahuaka (Sl)	Guerrero (1989)
		Wharonia sp.	*N. macrotis* (C)	Cerro Marahuaka (Sl)	Guerrero (1989)
Endoparasites					
Cestoda					
Cyclophylloidea	Taenidae	*Cysticercus* sp.	*N. urichi* (R)	Cerro Marahuaka (Su)	Guerrero (1989)
Nematoda					
Heterakoidea	Aspidoderidae	*Ansiruptodera* sp.	*N. urichi* (R)	Cerro Marahuaka (Su)	Guerrero (1989)
Rictularioidea	Rictulariidae	*Pterigodermatites* sp.	*N. urichi* (R)	Cerro Marahuaka (Su)	Guerrero (1989)
Trichinelloidea	Trichuridae	*Trichuris* sp.	*N. urichi* (R), *Sciurus aestuans* (R)	Cerro Marahuaka (Sl, Su)	Guerrero (1989)
Trichostrongyloidea	Heligmonellidae	*Sciurodendrium* sp.	*S. aestuans* (R)	Cerro Marahuaka (Sl)	Guerrero (1989)
	Heligmosomida	*Stilestrongylus* sp.	*N. urichi* (R), *R. macconnelli* (R)	Cerro Marahuaka (Sl, Su)	Guerrero (1989)
	Viannaiidae	*Viannaia guyanensis*	*Didelphis imperfecta* (M)	Sierra de Lema	Guerrero (1985)

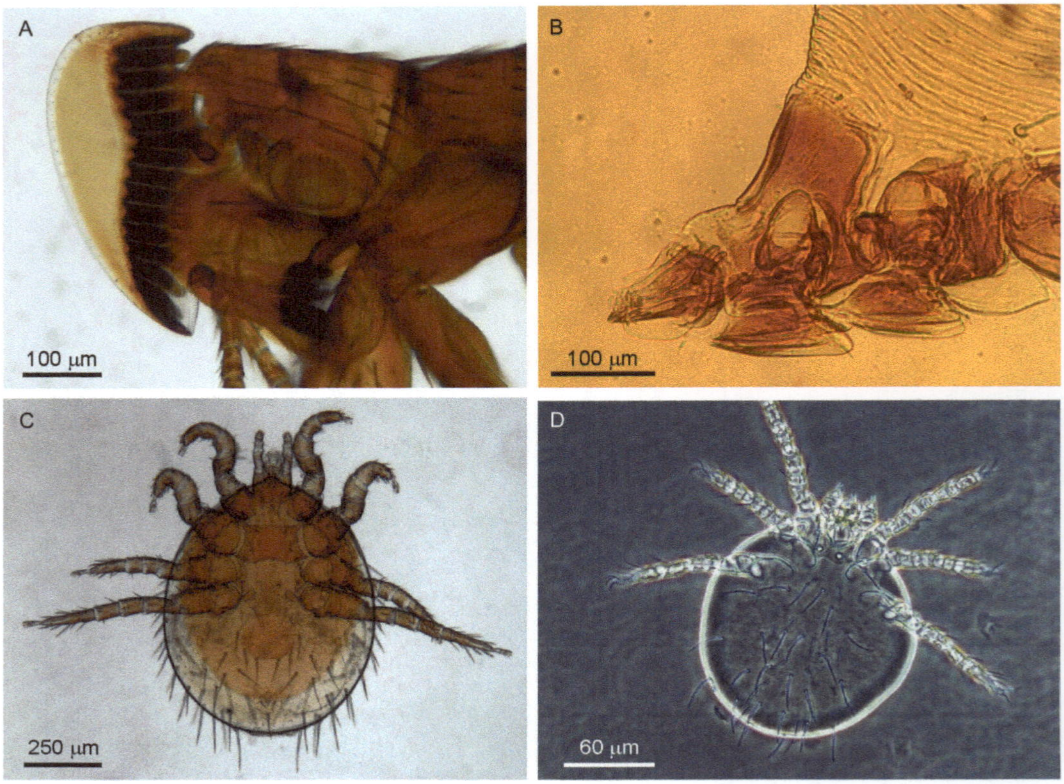

FIGURE 15.1 Selected examples of vertebrate parasites from Pantepui. (A) Head of *Craneopsylla minerva minerva* (Siphonaptera: Stephanocircidae) on *Rhipidomys macconnelli* (Rodentia) from Cerro Duida (Tables 15.1 and 15.3). (B) Head of *Alabidocarpus furmani* (Acarina: Chirodiscidae) on *Anoura geoffroyi* (Chiroptera) from Cerro Marahuaka (Table 15.1). (C) *Laelaps conula* (Acarina: Laelapidae) on *R. macconnelli* (Rodentia) from Auyán-tepui (Tables 15.1 and 15.3). (D) Unidentified Trombiculidae (Acarina) on *R. macconnelli* (Rodentia).

Preliminary conclusions

The parasite fauna of Pantepui is still scarcely known. From the information available to date, most parasites correspond to mammals, except for one lizard and one bird; no anuran (frogs, toads) parasites have reported so far. All ectoparasites are insects, notably Diptera, and acari, mostly Mesostigmata and Trombidiformes. The more frequent endoparasites are Nematoda followed by Cestoda; the absence of Trematoda is remarkable. Among the 67 ectoparasites and seven endoparasites found on the tepuian summits, only one, the acari *Laelaps conula* (a rodent ectoparasite), seems to be endemic to Pantepui, but this may change after the study of new material collected recently. There are no endemic genera or families. It seems too soon still for sound biogeographical inferences, but it is clear that parasites largely depend on the distribution patterns of their respective hosts.

TABLE 15.2 Species of streblid batflies and their geographical distribution (recent collections). 1, Dick et al. (2016); 2, Guerrero (1994a); 3, Guerrero (1995a); 4, Guerrero (1994b); 5, Guerrero (1995b); 6, Guerrero (1996a); 7, Wenzel (1976); n.r., new records.

Locality **Altitude (m.)**	**Arakamuni** **1550**	**Auyán-tepui** **1800–2200**	**Guaiquinima** **1100**	**Guanay** **1220**	**Lema** **1250**	**Marahuaka** **1100–2100**	**Neblina** **1210–2050**	**Tamacuari** **1550**	**Yaví** **2150**	**Yutajé** **1780**	**Distribution**
Species											
Anatrichobius scorzai				n.r.	7						Neotropical
Aspidoptera falcata			5		7, 5		5				Neotropical
Exastinion clovisi		5		n.r.	7, 5	5	5			n.r.	Neotropical
Megistopoda proxima					4		4				Neotropical
Neotrichobius bisetosus	4										Neotropical
Neotrichobius delicatus			4	n.r.	7	4,7					Neotropical
Paratrichobius longicrus		n.r.	4	n.r.	4	7	4		n.r.		Neotropical
Paratrichobius lowei			4	n.r.	7	4	4				Neotropical
Speiseria peytoni					7						Neotropical
Trichobius angulatus		n.r.	3	n.r.	3	7	3		n.r.	n.r.	Colombia, Venezuela
Trichobius assimilis	3				7	7					Peru, Venezuela
Trichobius caecus					2						Neotropical
Trichobius handleyi											Neotropical
Trichobius joblingi			3				3	3			Neotropical
Trichobius jubatus				n.r.							Neotropical
Trichobius lionycteridis	2		2		7	7		2			Neotropical
Trichobius parasparsus					2						Colombia, Venezuela
Trichobius propinquus					7						Brazil, Venezuela
Trichobius sparsus					7						Neotropical

(Continued)

TABLE 15.2 (Continued)

Locality Altitude (m.)	Arakamuni 1550	Auyán-tepui 1800–2200	Guaiquinima 1100	Guanay 1220	Lema 1250	Marahuaka 1100–2100	Neblina 1210–2050	Tamacuari 1550	Yaví 2150	Yutajé 1780	Distribution
Trichobius tiptoni				n.r.	3	3	3	n.r.		n.r.	South America
Trichobius uniformis					7						Neotropical
Strebla guajiro							6				Neotropical
Strebla harderi					7	7					Neotropical
Strebla chrotopteri						6					Neotropical
Strebla mirabilis					7						Neotropical
Strebla paramirabilis		n.r.	6	n.r.	6, 7	6, 7	6	6			South America
Anastrebla modestini		n.r.	6	n.r.	6, 7	6, 7	6			n.r.	South America
Anastrebla caudiferae					7	6, 7				n.r.	Neotropical
Anastrebla spurrelli			6		7	6, 7					Peru, Venezuela
Paraeuctenodes similis					7						Neotropical

TABLE 15.3 Species of laelapid mites and their geographical distribution (recent collections). 1, Furman (1972a); 2, Tipton and Machado-Allison (1972); 3, Guerrero (1989); n.r., new records. Species with an asterisk are depicted in Fig. 15.1.

Location Altitude (m.)	Arakamuni 1550	Auyantepui 1800–2200	Guaiquinima 1100	Guanay 1220	Lema 1250	Marahuaka 1100–2100	Murisipán 2350	Neblina 1210–2050	Roraima 2600	Tamacuari 1550	Yaví 2150	Yutajé 1780	Distribution
Species													
Acarina: Laelapidae													
Androlaelaps fahrenholzi		n.r.				3		n.r.	n.r.				Cosmopolitan
*Laelaps conula**	n.r.	n.r.	n.r.	n.r.	1	1,3	n.r.	n.r.	n.r.	n.r.	n.r.	n.r.	Pantepui
Laelaps surcomata	n.r.				1	n.r.							Brazil, Pantepui
Mysolaelaps heteronychus		n.r.	n.r.			n.r.		n.r.	n.r.		n.r.		Brazil, Venezuela
Insecta: Siphonaptera													
Polygenis roberti beebei	n.r.	n.r.	n.r.		3	2			n.r.	n.r.	n.r.		Panama, Venezuela
Polygenis versuta		n.r.		n.r.		2			n.r.		n.r.	n.r.	Brazil, Pantepui
Adoratopsylla antiquorum discreta		n.r.			2	2							Colombia, Venezuela
*Craneopsylla minerva minerva**					2	2		n.r.			n.r.	n.r.	South America

Acknowledgements

I am grateful to Armando Michelangeli, Charles Brewer-Carías, Eugenio de Bellard-Pietri, and Charles W. Myers, who led several expeditions to the tepuis and made possible parasite sampling on them. My wife, Barbara Szymanska, has been my right hand during many collection trips and made all drawings from my publications. Many friends, colleagues, and coworkers helped me in the field by collecting animals to be analyzed for parasites.

References

Arrow, G.J., 1907. On two new parasitic coleoptera (Fam. Staphylinidae) from South America. Ann. Mag. Nat. Hist. 19, 125–127.

Atyeo, W.T., 1988. Feather mites of the *Aralichus canestrinii* (Trouessart) complex (Acarina, Pterolichidae) from New World Parrots (Psittcidae). I. From the genera *Ara Lecepede* and *Anodorhynchus Spix*. Field. Zool. New Ser. 47, 1–26.

Atyeo, W.T., Pérez, T.M., 1988. Morphotypes of new world *Rhytidelasma* Gaud (Acarina, Pterolichidae), with (re) descriptions of five named and one new species. Acarologia 29, 175–187.

Brennan, J.M., van Bronswijk, J.E.M.H., 1973. Endoparasitic chiggers. VII. A redescription of *Kymocta* Yunker & Brennan, 1962, a new species from Venezuela and other records (Acarina: Trombiculidae). J. Med. Entomol 10, 449–451.

Brennan, J.M., Goff, M.L., 1977. The neotropical genus Hoffmannina: four new species and other records from Mexico, Panama, and Venezuela (Acarina: Trombiculidae). J. Parasitol. 63, 908–914.

Brennan, J.M., Reed, J.T., 1973. More new genera and species of chiggers (Acarina: Trombiculidae) from Venezuela. J. Parasitol. 59, 706–710.

Brennan, J.M., Reed, J.T., 1974. The genus *Eutrombicula* in Venezuela (Acarina: Trombiculidae). J. Parasitol. 60, 699–711.

Chapman, F.M., 1931. The upper zonal bird-life of Mts. Roraima and Duida. Bull. Am. Mus. Nat. Hist. 63, 1–135.

Chong, K.P., Atyeo, W.T., 1973. The Pterodectinae feather mites of hummingbirds; the genus *Toxerodectes* Park and Atyeo (The *hastifolia* group). J. Georgia Entomol. Soc. 8, 21–233.

De Winton, W.E., 1900. Mammalia. In: Lankester E.R., Report on a collection made by Messrs. F.V. McConnell and J.J. Quelch at Mount Roraima in British Guiana. Trans. Linn. Soc. London, Zool. 8, 52.

Dick, W.C., Gracioli, G., Guerrero, R., 2016. Family Streblidae. Zootaxa 4122, 784–802.

Díaz-Ungria, C., 1963. Nématodes parasites, nouveaux ou intéressants, du Venezuela. Ann. Parasitol. Hum. Comp. 31, 893–913.

Emerson, K.C., Price, R.D., 1975. Mallophaga of Venezuelan mammals. Brigham Young Univ. Sci. Bull., Biol. Ser. 20, 1–77.

Furman, D.P., 1972a. New species of *Laelaps* (Acarina: Laelapidae) from Venezuela. J. Med. Entomol 9, 35–46.

Furman, D.P., 1972b. Laelapid mites (Laelapidae: Laelapinae) of Venezuela. Brigham Young Univ. Sci. Bull. Biol. Ser. 17, 1–58.

Guerrero, R., 1985. Nematoda:Trichostrongyloidea Parásitos de Mamíferos Silvestres de Venezuela. II. Revisión del Género *Viannaia* Travassos, 1914. Mem. Soc. Ciencias Nat. La Salle 45, 9–47.

Guerrero, R., 1988. Informe preliminar sobre la parasitofauna de los mamíferos colectados en el Parque Nacional "Serranía de La Neblina" durante los días 7 al 24 de Marzo de 1984. In: Brewer-Carías, C. (Ed.), Cerro de La Neblina. Resultados de la Expedición 1983–1987. FUDECI, Caracas, pp. 581–589.

Guerrero, R., 1989. Lista de los parásitos de mamíferos del Cerro Marahuaka, Territorio Federal Amazonas, Venezuela. Acta Terramaris 1, 85–96.

Guerrero, R., 1994a. Catálogo de los Streblidae (Diptera: Pupipara) parásitos de Murciélagos (Mammalia: Chiroptera) del Nuevo Mundo. II. Los grupos: *pallidus*, *caecus*, *major*, *uniformis* y *longipes* del género *Trichobius* Gervais, 1844. Acta Biol. Venez. 15, 1–18.

Guerrero, R., 1994b. Catálogo de los Streblidae (Diptera: Pupipara) parásitos de Murciélagos (Mammalia: Chiroptera) del Nuevo Mundo. IV. Trichobiinae con alas desarrolladas. Bol. Entomol. Venez. N. S. 9, 161–192.

Guerrero, R., 1995a. Catálogo de los Streblidae (Diptera: Pupipara) parásitos de Murciélagos (Mammalia: Chiroptera) del Nuevo Mundo. III. Los grupos: *dugesii, dunni* y *phyllostomae* del género *Trichobius* Gervais, 1844. Acta Biol. Venez. 15, 1–27.

Guerrero, R., 1995b. Catálogo de los Streblidae (Diptera: Pupipara) parásitos de Murciélagos (Mammalia: Chiroptera) del Nuevo Mundo. V. Trichobinae con alas reducidas o ausentes y miscelaneos. Boletin Entomol. Venez. N. S. 10, 135–160.

Guerrero, R., 1996a. Catálogo de los Streblidae (Diptera: Pupipara) parásitos de Murciélagos (Mammalia: Chiroptera) del Nuevo Mundo. VI. Streblinae. Acta Biol. Venez. 16, 1–25.

Guerrero, R., 1996b. Garrapatas de Venezuela (Acarina: Ixodoidea). Listado de especies y claves para su identifición. Bol. Dir. Malariol. Saneam. Amb. 36, 1–24.

Handley, C.O., 1976. Mammals of the Smithsonian Venezuelan Project. Brigham Young Univ. Sci. Bull., Biol. Ser. 20, 1–89.

Hastriter, M.W., Guerrero, R., 1998. A review of the bat flea genus *Hormopsylla* (Siphonaptera: Ischnopsyllidae). Proc. Entomol. Soc. Wash. 100, 247–251.

Herrin, C.S., Tipton, V.J., 1975. Spinturnicid mites of Venezuela (Acarina: Spinturnicidae). Brigham Young Univ. Sci. Bull. Biol. Ser. 20, 1–72.

Jones, E.K., Clifford, C.M., Keirans, J.E., Kohls, G.M., 1972. The ticks of Venezuela (Acarina: Ixodoidea) with a key to *Amblyomma* in the western hemisphere. Brigham Young Univ. Sci. Bull. Biol. Ser. 17, 1–40.

Johnson, P.T., 1972. Sucking lice of venezuelan rodents, with remarks on related species (Anoplura). Brigham Young Univ. Sci. Bull. Biol. Ser. 17, 1–62.

Machado-Allison, C.E., Barrera, A., 1972. Venezuelan Amblyopinini (Insecta: Coleoptera; Staphylinidae). Brigham Young Univ. Sci. Bull. Biol. Ser. 17, 1–14.

McDaniel, B., 1972. Labidocarpid bat-mites of Venezuela (Listrophoroidea: Labidocarpidae). Brigham Young Univ. Sci. Bull. Biol. Ser. 17, 15–32.

Mayr, E., Phelps Sr, W.H., 1967. The origen of the bird fauna on the South Venezuelan highlands. Bull. Am. Mus. Nat. Hist. 136, 269–328.

Myers, C.W., Donelly, M.A., 1996. A new herpetofauna from Cerro Yaví, Venezuela: first results of the Robert G. Goelet American Museum-Terramar expedition to the Northwestern tepuis. Am. Mus. Novit. 3172, 1–56.

Price, R.D., Emerson, K.C., 1986. New species of *Cummingsia* Ferris (Mallophaga: Trimenoponidae) from Peru and Venezuela. Proc. Biol. Soc. Wash. 99, 748–752.

Saunders, R.C., 1975. Venezuelan Macronyssidae (Acarina: Mesostigmata). Brigham Young Univ. Sci. Bull. Biol. Ser. 20, 75–90.

Tipton, V.J., Machado-Allison, C.E., 1972. Fleas of Venezuela. Brigham Young Univ. Sci. Bull. Biol. Ser. 17, 1–115.

Wenzel, R.L., 1976. The streblid batflies of Venezuela (Diptera: Streblidae). Brigham Young Univ. Sci. Bull. Biol. Ser. 20, 1–177.

PART IV

Biodiversity Conservation

CHAPTER

16

Conservation of Pantepui: between complex emergency and climate change

Mariapia Bevilacqua[1], *Celsa Señaris*[2] *and Otto Huber*[3]

[1]Venezuelan Association for the Conservation of Natural Areas - ACOANA, Caracas, Venezuela [2]Laboratory of Population Genetics, Center of Ecology, Venezuelan Institute for Scientific Research (IVIC), Caracas, Venezuela [3]Trauttmansdorff Botanical Gardens, Merano, Italy

Introduction

The historical and present context of the conservation of the highlands of the Guiana Shield (Pantepui) was reviewed from the commitments and actions of the regional governments towards the conservation practice in the region. It has been concluded that they have been insufficient. This is due, on the one hand, to the social imagination that considers the remote, isolated, and difficult-to-access condition of Pantepui a guarantee for its perpetual conservation and, on the other hand, to the emergence of progressive and populist regional governments that promote the intensive use of natural resources, with the premise of redistributing income for social welfare. In this context, the protected areas of the Guiana Shield become vulnerable spaces and their borders become highly permeable to extractivism, given the failure of the natural resource management institutions. Some issues analyzed were the causes of this institutional failure and the factors that have made conservation priorities invisible and underestimated the importance of Pantepui as the last globally important frontier of tropical areas remaining on the planet (Potapov, 2017; WWF, 2012). Henceforth a question arises: Who is interested in Pantepui? Awareness needs to be raised of the importance of Pantepui's heritage and to promote cross-border collaboration to develop research-action capacities, such as interculturality, the defense of nature, and the construction of peace, with the long-term support of donors

Biodiversity of Pantepui
DOI: https://doi.org/10.1016/B978-0-12-815591-2.00016-1

and cooperators of the international community. Finally, it is declared that the conservation of Pantepui will be sustainable to the extent that the fundamental role of indigenous people and traditional communities and their fundamental rights are recognized in the conservation and management of territories and good management practices of the common resources are promoted that allow their enjoyment for the next generations.

Antecedents in the conservation of Pantepui

The conservation of the high mountains of the Guiana Shield began in the 1960s with the creation of the Talferberg Nature Reserve in Suriname and the Canaima National Park in Venezuela (Fig. 16.1). This provided legal protection to the eastern peak of the sandstone mountains of the Guiana Shield in the Sipaliwini district, Suriname, and the tepuis in the western sector of the Gran Sabana municipality, Venezuela—including the Auyán-tepui with the highest waterfall in the world, Angel Falls, one of the most geographical landmarks in the region. Its creation, based on landscape, biological, and ecological values and to protect the headwaters of the Caroni for the Guri Hydroelectric Project (Castro-Morales and Gorzula, 1986; Huber, 1995; Bevilacqua, 2003; Yerena, 2011), followed the conservation model of that time, which was based on the belief that the protection of biodiversity is best achieved through the creation of protected areas where ecosystems can function isolated from human disturbance.

In the 1970s, several milestones marked a turning point in the development of international policy towards the environment from two complementary approaches: the mitigation

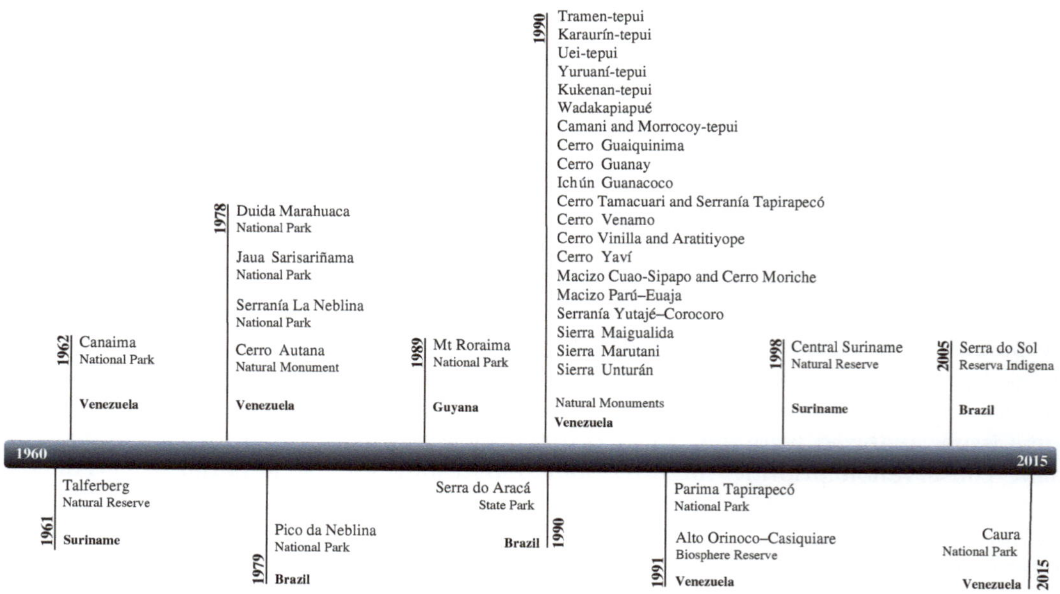

FIGURE 16.1 Timeline showing the sequence of creation of the protected areas of the highlands of the Guiana Shield.

of the environmental impact of industrial development and the protection of nature. The most important milestones were the First Conference of the Human Environment, Economic Development and Environment of the United Nations Organization (Stockholm, 1972), the Second World Congress of National Parks (Yellowstone, 1972), the institutionalization of the "World Heritage" approach by UNESCO, and the Annual Report of the International Union for the Conservation of Nature for the creation of protected areas in 1978. At the Stockholm Conference, the problem of pollution associated with industrial development rose for the first time in the diplomatic sphere, and in the Second World Congress of National Parks, global attention was called for regarding the urgent need to protect geographic spaces of interest to save the planet, promote human development, achieve prosperity, and promote universal peace. As a result of these conferences and their policy guidelines towards the environment, a radical change in the existing paradigms of welfare, growth, and global economic development is demanded of the international industrial community, because these were being carried out at an environmentally high and unsustainable cost over time (Bruntland, 1987). The global commitment to the conservation of natural resources acquires greater relevance, and soon the conservation interest oriented towards the tropics, where the sites with the greatest biological diversity in the planet are concentrated. Venezuela welcomes and implements international environmental policies through (1) the expansion of the Canaima National Park in 1975 to 2,000,000 hectares in order to include the Roraima-tepui, which is shared with Brazil and Guyana; (2) the creation of the Ministry of Environment and Renewable Natural Resources (1977), the first in Latin America; and (3) the creation of three national parks and a natural monument (Fig. 16.1), which together comprise 1,900,030 ha, an unprecedented extension in the conservation of the Guiana Shield mountains (Bevilacqua, 2003). Brazil and Guyana, for their part, created the national parks Pico da Neblina (1979) and Mount Roraima (1989), which protect the highest tepuis of the shield.

In the 1980s international concern about the environment continued to grow, given the evidence of the impact of population growth and the volumes of industrial production documented in the Caring for the Earth report (IUCN, UNEP, WWF, 1991a,b). Here a warning was issued about the threat to the future of humanity if no action is taken to safeguard the planet's vitality or to move towards "sustainable development" as the only rational option to guarantee the survival of man. With the evolution of the concept of "biodiversity" in the 1990s towards a conceptual approach broader than the biological aspect, biodiversity acquired a discursive centrality of global political interest (Núñez et al., 2003). Biodiversity was central on the agenda of issued debated at the United Nations Conference on Environment and Development (Rio, 1992), a meeting among heads of state to reach agreements related to the environment, sustainable development, climate change, and biodiversity. The agenda was also present in the IV World Congress of National Parks and Other Protected Areas (Caracas, 1992), which brought together for the first time more than 2000 delegates to discuss the future of protected area management and technical guidance, issues of growing global interest to all nations. At the United Nations Conference (Rio, 1992), the Convention on Climate Change, the Declaration of Principles Relating to Forests, and the Convention on Biological Diversity were approved, which are universal legal frameworks that promoted the creation of new protected areas throughout the world, including in the countries of the Guiana Shield.

In the context of this political and scientific effervescence associated with biodiversity and sustainable development—but also from the perspective of two new approaches, such as environmental services and ecological viability—Venezuela has taken the unprecedented regional initiative to prohibit all forestry and mining activities in the Amazonas state (Venezuela, 1978, 1989), as well as in the upper Caroní River basin (Venezuela, 1991) and protected all highlands in the Guiana and Amazonas states under a strict form of legal protection. In 1990 an unpublished declaration of an "archipelago" of 25 tepuis under a single natural monument was made. A year later, two new protected areas were decreed encompassing the largest area under legal protection of all the Guiana Shield region and one of the most extensive in the world (approximately 80,000 km^2), designed to protect the biological diversity of the southern Venezuelan Amazon, its tepuis, and the territories of ancestral occupation of the Yanomami people. The areas decreed were the Alto Orinoco-Casiquiare Biosphere Reserve (1991), the first of its kind in the country and in the Guiana Shield, and the Parima-Tapirapecó National Park (Fig. 16.1). The first and most extensive biological corridor of the Amazon basin and the Guiana Shield is thus legally constituted in the group formed by the binational parks La Neblina, Parima-Tapirapecó (Venezuela), and Pico da Neblina (Brazil), the latter with 22,200 km^2. For its part, in 1990 Brazil protected the highlands of Serra do Aracho with a state park. The same year, Suriname extended the protection of the Talferberg Nature Reserve by including it as part of the 16,000-km^2 complex of lowland forests and mountains in the Central Suriname Natural Reserve.

The current set of protected areas of the Guiana Shield demonstrates the commitment of the countries of the region to adopt and implement environmental conservation strategies, primarily promoted during 1970–99. The objectives of creation and management of the protected areas at the time reflect the evolution of the conceptual approaches for the conservation of landscapes, sites with high biodiversity and endemism, but also of the interest in attending to the complex relations between the people and ecological feasibility in response to the growing pressure of society towards a management of "parks for life," sustainable development ,and the mitigation of climate change. In addition to the striking physiographic landscape of large plateaus with slopes and summits above 1000 m a.s.l., broad valleys and low hill systems are incorporated into the design of the protected areas, densely covered by subtropical, moist broadleaf forests on gently rolling plains or peneplains that separate the tepuis, with the objective of protecting a set of ecosystems that have conservation value for the ecological viability and the protection of the territories of native indigenous societies. Relevant forest areas are zoned as buffer areas for the use of natural resources, for indigenous ancestral occupation, and for intangible protection of lands surrounding Pantepui. Accordingly, Brazil has decreed the Serra do Sol Indigenous Reserve (2005) with an area of 17,430 km^2 and Venezuela declared the extensive ecoregion of the Caura River as El Caura National Park (2015) with c.75,000 km^2. Both measures open a new chapter for the conservation of the Guiana Shield, and this is impressive due to the magnitude of the surfaces, the contradictions in conception, and the socioenvironmental conflicts with the territories of existing indigenous peoples.

Today the greater representation of the natural heritage of Pantepui has a legal conservation status under one of the most restrictive management categories II and III of the universal protected areas system: national park and natural monument (IUCN, 1994). For many years it has been considered that the isolated conditions, the remote location, and the difficulty in accessing the tepuis are sine qua non factors for the conservation of Pantepui. However, as will be discussed later, these conditions are no longer a guarantee against the effects of climate change and global changes in the regional economic, social, and political contexts. The Declaration of Paramaribo on conservation priorities in the Guiana Shield (Huber and Foster, 2003; Bernard et al., 2011) marked a milestone in terms of international attention to the relevance of the natural and cultural heritage, traditionally invisible in the geopolitical boundaries of the Amazon region and its regional cooperation treaties.

Failure of natural resource management institutions

With the arrival of the new millennium, the interdependence between social, economic, and environmental agendas was recognized at the highest global level through the approach adopted by the United Nations in the Sustainable Development Goals (UN, 2017). However, the set of nations still fails to translate this awareness and commitment into action and change. The paradigm of human development is still associated with economic growth, and, especially in Latin America, the idea prevails that it will generate well-being in the population (employment, consumption, etc.) through the intense and efficient appropriation of natural resources, to the increase in exports, and to foreign investment (Gudynas, 2011; Svampa, 2013). Thus global policies towards the environment continue to be subordinated to the economic policies that set the course for human development and economic growth, the consequence of which is the ecological mark documented in the annual report of the WWF (2016). This document indicates that human activities and the demand for associated resources have grown so dramatically that the environmental conditions that encouraged this development and growth are beginning to deteriorate to such an extent that the resulting damage to human health and ecosystems threatens to undermine the economic and social advances of industrialization itself.

In the new millennium, a series of progressive and popular governments are emerging in Latin America. Their policies converge on the premise that inequality and poverty are essentially solved by economic means (Gudynas, 2011; Moreira, 2017). The thesis of promoting increases sustained in the extraction of minerals, hydrocarbons, and various extensive crops is booming in the region as a mechanism to finance social programs, reverse inequalities, promote inclusion and social mobility, recover infrastructure, and build a new critical consciousness to capitalism (García-Guadilla, 2014; Gudynas, 2017). In this way, Latin America enters a new economic and political–ideological order sustained by the boom in international prices of raw materials and consumer goods (commodities), increasingly demanded by the most industrialized countries and emerging powers (Svampa, 2013). The governments in the region increase their role as suppliers of raw materials, for their economic growth and to increase their monetary reserves, consolidating a neoextractivist style of development (Gudynas, 2011), based on the overexploitation of natural resources, especially nonrenewable ones. Economic growth expands to other geographic spaces, including those traditionally protected due to their value for the conservation of biodiversity, water, forest, and natural reserves for indigenous populations

(RIGBYY, 2017). A dynamic invasion of territories occurs as a result, destroying regional economies, national institutions, and biodiversity and deepening the process of land appropriation; displacing rural, peasant, and indigenous communities; and violating citizen decision processes, as well as generating new asymmetries and social, economic, environmental, and political–cultural conflicts (Svampa, 2013). These conflicts mark the opening of a new cycle of struggles, which increasingly affect the region, as recorded by the Latin American Observatory of Environmental Conflicts (http://olca.cl/oca/index.htm). The criminalization of human rights also advances, and the socioenvironmental problem is sacrificed or made invisible, and is considered secondary in view of the serious problems of poverty and exclusion in societies (Terán-Mantovani, 2016).

In this context, protected areas emerge as highly vulnerable geographic spaces, and their limits become more flexible and more permeable in the face of the ideological discourse that the social goal justifies the means and economic growth is based on the appropriation of nature. New socioeconomic asymmetries affect local populations and other actors linked to protected areas, mediated by the absence of individual capacities to convert rights into real freedoms and achieve what people value and aspire to, according to the terms described in Sen's approach to capabilities (1985, 1999). The absence of freedom to maintain lifestyle and culture, linked to the territory and access to natural resources, is a permanent source of conflict in environmental management and constitutes a vital force that perpetuates the perverse cycle of poverty–environmental crisis–poverty in the region.

Acheson (2006) suggests that the cause of global environmental degradation is institutional because the rules, mechanisms, and governance structures that control access and regulate the use of natural resources are absent, do not apply, or otherwise are not effective, and then the institution fails in its pursuit of the conservation of common resources. In the Guiana Shield in Venezuela, there is evidence of the institutional failure to manage natural resources. Novo and Díaz (2007), in their environmental assessment of the Canaima National Park, conclude that the conservation of biodiversity in the highlands is at risk due to the lack of adequate government resources in terms of budget, park rangers, and a plan for the management of such a large and remote protected area. On the other hand, Bevilacqua et al. (2009) delve into the factors that contribute to the failure of the natural resource management institutions of the Canaima National Park and suggest that, in addition to clear rules to regulate the use of natural resources for conservation purposes, two groups of other factors must be taken into consideration. The first group refers to the broad participation of stakeholders and their homogeneity, the recognition of rights, the social and community sense, their dependence on the resource, and adaptive management structures and techniques, as well as conservation costs and benefits shared in a fair and equitable manner. The second group refers to knowledge and understanding of the location and distribution of natural resources, especially biodiversity, and their state of conservation, as well as diagnoses of opportunities and threats to their sustainable use. In this regard, the extreme difficulties and limitations to produce knowledge relevant to the practice of conservation as a consequence of the bureaucratic process in Venezuela to obtain research permits in the tepuis of the Canaima National Park and the Guiana region can be observed (Rull and Vegas-Vilarrúbia 2008; Rull et al., 2008).

As the government of Venezuela deepens its progressive and populist model based on the appropriation of nature, the social and solidary community economy permeates the borders of protected areas (García-Guadilla, 2014). This disturbs its regulatory frameworks and invades spaces that are integral to protection and managing the natural environment for conservation purposes, thus affecting the integrity of biodiversity and ecosystem functioning. Environmental impacts permeate the boundaries of the tepuis, due to intensive and uncontrolled tourism use (Rull et al., 2016). Yerena (2011) concludes that the future of the environmental heritage of the Guiana region is uncertain, describing the changes in Venezuela's environmental policy towards protected areas in Bolívar and Amazonas states, which resulted in the withdrawal of conservation efforts in favor of a development model based on geopolitical and economic alliances with neighboring governments (Guyana and Brazil), sustained at the local level by corrupt military, civil, and community alliances that were complacent when it came to extractivism. The invasion of territories with the purpose of extracting mineral resources in natural and protected areas in the Venezuelan Amazon has been a chronic and cyclical problem, which has had governmental impunity, causing severe violations to the human rights of the Yanomami people, including massacres in indigenous communities and environmental impact to their territories (Bello and Tillet, 2015; RIGBYY, 2017). In the Venezuela of the 21st century, mining extractivism is consolidated as a state policy towards the territories of the Guiana region (Venezuela, 2018). At the same time, the vision of the new paradigm of economic development associated with permissive mining, protected by military alliances, paramilitaries, and corrupt criminals from the highest governmental level, has been consolidated in the social imagination, as has been described in similar contexts of extractive policies in Latin America (Gudynas, 2017) and global forest policies (Lloyd, 2008).

Given all this, the failure of institutions to manage natural resources is also due to a third set of factors linked to dynamic forces that link the governments with developmentalist and extractive policies to destroy with premeditated intent, the institutional bases that allowed the state to design, create, and manage the extensive set of protected areas for conservation and human welfare in general.

Complex crisis

In the mid-2000s, the developmental cycle in Latin America came to an end as a consequence of various factors that prevented governments from maintaining economic growth (León, 2015). Various changes were introduced in the political, economic, social, and environmental spheres in the region, but an inability or lack of will to achieve the sustainable development objectives that governments committed to comply with in global forums prevails (UN, 2017). Throughout the region, environmental heritage is still widely threatened by the continuity of extractive policies and, in the case of Venezuela, the complex crisis promises to bring the environmental and cultural impacts on the protected areas of the Guiana Shield to an unprecedented level. Next, the Venezuelan case will be analyzed in detail not only due to personal experience but because the greatest geographic representation of Pantepui is concentrated in this country, and its conservation will depend on the national and international response to this complex emergency.

Over time, the voices warning about the complex emergency in Venezuela, characterized by a partial or total breakdown of the legal, institutional, and administrative structure of the state (ACAPS, 2018; Bermúdez et al., 2018; España and Ponce, 2018; OPS, 2018), have increased. This situation has also profoundly and structurally affected environmental management. This translates into multiple and massive violations of the environmental legal framework with widespread damage to society in general, but particularly important to the indigenous people and the natural heritage of the Guiana Shield. In the 21st century, Venezuela went from the failure of the natural resource management institutions described by Acheson (2006) to an environmental crisis induced by the same factors that underlie the complex humanitarian emergency, as described by human rights defenders (ACNUDH, 2018; Civilis Derechos Humanos, Acción Solidaria and Codevida, 2018; PROVEA, 2018; OEA, 2018). From the environmental point of view, this can be specified in terms of (1) groups with power conflicts around the control of natural resources and territorial domains with strategic value in the economic order; (2) dismantling legitimate institutions and the rule of law with environmental competences; (3) denial of guarantees to a timely and effective protection and to an independent justice in the form of nonconforming occupation of the territory and affecting natural resources; (4) systematic use of violence by official, military, and civil forces, as well as fragmentation of power linked to the use of natural resources; and (5) the appearance of corrupt economic and social forms that impacts natural resources, especially in the lowlands of the Guiana Shield. The combination of these factors destroys the nation's capabilities to protect the environment and assist the population that depends on it. Since the Venezuelan government deepens the neoextractivist model of metallic and nonmetallic minerals as a state policy for export purposes (Venezuela, 2018), it will be a matter of time before the induced socioeconomic asymmetries promote groups of people to look for mineral resources in other territorial spaces with less pressure of current use, such as the tepuis, encapsulated in the historical imagination of the promise of "El Dorado." Local communities report incipient mining exploration on slopes in tepuian formations at low altitude in the Paragua River basin and the Caroní and Caura Rivers, and a binational information exchange network has an extensive record of illegal mining (mechanized and manual) as well as clandestine landing strips in protected areas and indigenous territories of the Yanomami and Ye'kwana peoples in the Amazonian and Guiana region (RIGBYY, 2017). These activities are carried out in the context of an absolute rupture of national and local environmental authority, but also in a breakdown of the traditional indigenous authorities and their customary institutions for the management of common resources. In the absence of environmental authority, other uses may increase in the tepuis, such as illegal biodiversity trade; mass tourism; and expeditions for scientific, extreme sports, documentary, recreational, adventure, and elite purposes, which increase environmental impacts due to the fragility of tepuian ecosystems. This will extend the impact already documented in some tepuis (Gorzula and Huber, 1992; Rull et al., 2016; Fernandez-Delgado, 2016). The remote condition of these areas has not prevented the realization of these activities in the past, so an increase of them is feasible before the breakdown of environmental management institutions and the advance of neoextractivism occur. This situation brings us to the verge of one of the greatest ecological crimes against mankind's natural heritage, which should cause an alert, raise awareness, and mobilize action from the international environmental and academic community.

Climate change

The First Academic Report on Climate Change in Venezuela (Villamizar et al., 2018) reveals the lack of evidence to attribute the effects of climate change on protected areas, and the capacity for adaptation and resilience of each of the protected ecosystems is unknown. Future assessments are necessary to evaluate whether this lack of evidence is a recurrent pattern in the rest of the protected areas in the highlands of the Guiana Shield and a product of geopolitical positions in the context of climate change (Terán-Mantovani, 2016). Given the lack of evidence, the contributions based on predictive models and the traditional ecological knowledge of indigenous peoples become relevant for policies of adaptation and mitigation to climate change. Some of the findings on this topic are hereby presented.

As explained in Chapter 17: Pantepui and global warming, estimates of the effects of global warming in the Guiana highlands based on simulation models predict severe threats to plants and communities of Pantepui due to a substantial habitat loss. Given these predictions, the authors propose conservation strategies based on ex situ techniques to mitigate the loss of genetic and species diversity (Rull et al., 2009; Safont et al., 2012). On the other hand, the main author of this chapter started a participatory action-research group in 2017 focused on the perceptions of climate change, ethnoclimatological knowledge, and its implications in the processes of change and adaptation by the Ye'kwana people. They are an indigenous society of hunters, collectors, and horticulturists, with a population that lives in the upper basin of the Caura River and at the foot of Jaua-Sarisariñama-tepui and Sierra Maigualida in Bolívar state, Venezuela. Their preliminary findings show that more than 89% of Ye'kwana respondents perceive some type of climate change in their territories, including the decrease in rainfall, the general increase in temperature, the modification of floods and droughts, and changes in climate seasonality (Bevilacqua, 2018). The Ye'kwana also perceive an increase in the uncertainty associated with the prediction of climate, the frequency of abnormal or extreme weather events in their traditional territory, climatic seasons occurring outside of time, flood pulses, and descending rivers synchronized with the flowering and fruiting of species of interest, which are all considered alarming. The study also identified environmental and astronomical indicators widely used by Ye'kwana elders for climate prediction, including plant phenology and behavior and movement of birds, animals, and insects (Table 16.1). Taking into account that the contributions of local climate observations are recognized for their contribution to studies on climate change (Reyes-García et al., 2015; Alexander, 2011), research on Guiana Shield highlands can benefit from data collection and indigenous observations in lands surrounding the tepuis to improve climate models, expand records in localities where information is lacking, and inform strategies for effective adaptation of biodiversity that are finely tuned to the specific characteristics of unique environments and contexts such as the tepuis.

TABLE 16.1 Type and quantity of bioindicators and environmental and astronomical indicators used in climate prediction by indigenous households of the Ye'kwana people, in the Caura River basin, Bolívar state, Venezuela.

Generic use	Type of indicator	Quantity
Announces arrival of summer (agriculture cycle)	Zoo-indicator	11
	Phyto-indicator	3
	Astronomic	3
Announces arrival of winter	Zoo-indicator	9
Predicts rain	Zoo-indicator	3
	Astronomic	4
Predicts flood	Phyto-indicator	2
Announce warmer days	Atmospheric	1

Final reflection

Seen in retrospect, the performance and commitments of regional governments towards the practice of conservation in the highlands of the Guiana Shield have been insufficient. This is partly due to the social imagination that its remote, isolated, and difficult-to-access condition is a guarantee for its conservation. On the other, it is due to the urgency of addressing the conservation of biodiversity in the lowlands where human populations are concentrated, the occupation of the territory, and the challenges for sustainable development. These factors have made conservation priorities invisible and underestimated the importance of Pantepui as the last frontier of tropical wilderness.

In the last decade, two initiatives have concentrated invaluable effort on conservation actions for the Guiana Shield. The first, the Guiana Shield Initiative (2000–2008) with UICN-NL funds that later evolved into the multidonor platform Guiana Shield Facility (2011–2014), funded by the European Union, promoted the goal of empowering the six countries that share the great ecoregion of the Guiana Shield, and its local communities, to benefit from the conservation and sustainable development of its natural resources. In practice, the initiative focused efforts on lowland territories with occupation and use pressure in areas of conservation value in Colombia, Guyana, and Suriname. The initiative led to the creation of the International Society of Biodiversity of the Guiana Shield and formed ties of cooperation between conservation institutions, communities, and universities that continue to this day.

The second initiative is the Andes-Amazon-Atlantic Ecological–Cultural Corridor (See https://www.gaiaamazonas.org/), a macroregional effort promoted from Colombia, which aims to safeguard the cultural and biological diversity and encourage sustainable models of development through the effective protection of 2.6 million km^2 of forest that connect the Amazon with the ecosystems of the Andes and the Atlantic Ocean, through nine South American countries. The area of interest is defined, on the one hand, by hydrographic limits of the Amazon River basin and, on the other, by extensive protected areas and interconnected indigenous territories. The approach leaves out the largest concentration of extensive, emblematic, biodiverse, legendary, and culturally relevant mountains of the central-eastern area of the Guiana Shield in Venezuela and Guyana. Thus an opportunity is missed to capitalize on the interconnection of protected areas and indigenous territories already existing in countries that share the Guiana Shield, which would allow the Pantepui province as a whole to be included in the strategic vision of large-scale conservation of the environmental and cultural heritage of the South American continent.

The tepui ecosystems have gone through many climatic fluctuations in the last millennia, and certainly, the evidence of climate change will have its repercussion in Pantepui, and we must prepare ourselves to anticipate actions of protection and management of the highland biodiversity. Nevertheless, we want to draw attention to the fact that the threats and impacts due to anthropogenic use of the protected areas and the tepuian complexes from local sources are very real and already carry environmental and socioeconomic costs that are constantly accelerating, which must mobilize an urgent call and action plan to protect the Guiana Shield highlands and the world heritage as a whole.

The conservation of Pantepui, and the management of protected areas which contain them, requires courage, coherence, consistency, resilience, and, above all, long-term continuity in action, based on universal values of peace, freedom, human rights, and inclusion. In countries where political, social, and economic contexts limit academic freedoms—as evidenced in Venezuela with the decrease of academic production capacity (Requena and Caputo, 2016), the destruction of spaces for the exercise of science at the service of social welfare (See http://factor.prodavinci.com/institutodemedicinatropical/index.html), the progressive deterioration of autonomy and academic freedom in the country's universities (Coalición de Cátedras y Centros de Derechos Humanos, 2017), and the extreme bureaucracy that limits scientific cooperation for environmental studies (Rull and Vegas-Vilarrúbia, 2008; Rull et al., 2008)—cooperation to study, understand, defend, and conserve Pantepui—require creative and innovative forms and mechanisms of association to maintain the research and practice of conservation. The key challenge is to increase awareness of the importance of the heritage of Pantepui and boost cross-border collaboration to develop community, academic, and technical-professional capacities in interculturality, defense of nature, and peace building, from action research and long-term support of the academy and the international community of donors. Finally, the conservation of Pantepui will be sustainable to the extent that the management of protected areas and other territories that contain it recognize indigenous rights, especially the collective ownership of their ancestral lands, and foster integration of traditional ecological knowledge into natural common-pool resource management and successfully address co-management plans.

References

ACAPS, 2018. Venezuela humanitarian crisis. Briefing Note 23 may 2018. The Assessment Capacities Project. Geneva, Switzerland. https://www.acaps.org/search?search_query = venezuela (consulted 08.10.18).

Acheson, J.M., 2006. Institutional failure in resource management. Annu. Rev. Anthropol. 35, 117–134.

ACNUDH, 2018. Violaciones de los derechos humanos en la República Bolivariana de Venezuela: una espiral descendente que no parece tener fin. Oficina del Alto Comisionado de las Naciones Unidas para los Derechos Humanos (ACNUDH). Nueva York, EEUU. https://www.ohchr.org/Documents/Countries/VE/VenezuelaReport2018_SP.pdf (consulted 10.09.18).

Alexander, C., Bynum, N., Johnson, E., et al., 2011. Linking indigenous and scientific knowledge of climate change. BioScience 6, 477–484.

Bello, L.J., Tillet, A., 2015. Minería en la Amazonía Venezolana: Derechos Indígenas y Ambientales. El caso del pueblo Yanomami. Grupo de Trabajo Ambiental de la Amazonía WATANIBA y Red Jurídica para la Defensa de la Amazonas RAMA. Caracas, Venezuela.

Bermúdez, Y., Mazuera-Arias, R., Albornoz-Arias, N., Morffe Peraza, M.A., 2018. Informe sobre la movilidad humana venezolana. Realidades y perspectivas de quienes emigran Servicio Jesuita a Refugiados (SJR). San Cristóbal, Venezuela.

Bernard, C., Upgren, A., Honzák, M., 2011. Review of the Shield Priority Setting Outcomes: Narrative Report. Conservation International, Georgetown, Guyana.

Bevilacqua, M., 2003. In: Aguilera, M., Azocar, A., González, E.J. (Eds.), Áreas Bajo Régimen de Administración Especial. Biodiversidad de Venezuela. M. CONICIT-Fundación Polar, Caracas, pp. 922–945.

Bevilacqua, M., Medina, D.A., Cárdenas, L., 2009. Manejo de recursos naturales en el Parque Nacional Canaima: desafíos institucionales para la conservación. In: Señaris, J.C., Lew, D., Lasso, C. (Eds.), Biodiversidad del Parque Nacional Canaima. Bases Técnicas Para la Conservación de la Guayana Venezolana. Fundación La Salle de Ciencias Naturales, Caracas, Venezuela, pp. 209–221.

Bevilacqua, M., Rodríguez, R., Domínguez J., 2018. Perception of climate change among Ye'kwana people, Venezuela. XVI Congreso de la Sociedad Internacional de Etnobiologia. Belem, Brasil 7 al 10 de agosto.

Bruntland, G., 1987. Our Common Future, Report of the World Commission on Environment and Development. World Commission on Environment and Development, 1987. Published as Annex to General Assembly document A/42/427.

Castro-Morales, L., Gorzula, S., 1986. The interrelations of the Caroní river basin ecosystems and hydroelectric power projects. Interciencia 6, 271–277.

Civilis Derechos Humanos, Acción Solidaria and Codevida, 2018. Derecho a la salud en una emergencia humanitaria compleja Guía Práctica de Grupos Interdisciplinarios. Caracas, Venezuela, pp. 33.

Coalición de Cátedras y Centros de Derechos Humanos, 2017. El pensamiento bajo amenaza Situación de la libertad académica y la autonomía universitaria en Venezuela. Caracas, Venezuela. https://www.derechos.org.ve/web/wp-content/uploads/Informe-LA-y-AU-FIN.pdf (consulted el 18.09.18).

España, L.P., Ponce, M.G., 2018. Encuesta Nacional de Condiciones de Vida 2017. Revisado el 28/04/2018. https://www.ucab.edu.ve/investigacion/centros-e-institutos-de-investigacion/encovi-2017/ (consulted 28.08.18).

Fernández-Delgado, M., Giarrizzo, J.G., García-Amado, M.A., Contreras, M., Salazar, V., Barton, H., et al., 2016. Evidence of *Helicobacter* spp. in freshwaters from Roraima tepui, Guayana Shield, South America. Antonie van Leeuwenhoek 109, 529.

García-Guadilla, M.P., 2014. Neo-extractivism, neo-rentierism and social movements in Venezuela's 21st century ecosocialism: discourses, conflicts and resistance. In: Thimothy Doyle, T., MacGregor, S. (Eds.), Environmental Movements around the World: Shades of Green in Politics and Culture. Praeguer, pp. 77–104.

Gorzula, S., Huber, O., 1992. Consideraciones finales. In: Huber, O. (Ed.), El Macizo del Chimantá, Escudo de Guayana, Venezuela. Todtmann Editores, Caracas, Venezuela, pp. 325–330.

Gudynas, E., 2011. El nuevo extractivismo progresista en América del Sur tesis sobre un viejo problema bajo nuevas expresiones. Colonialismo del siglo XXI. Negocios extractivos y defensa del territorio de América Latina. Icaria Editorial, Barcelona, pp. 75–92.

Gudynas, E., 2017. Extractivismo y corrupción en América del Sur. RevIISE 10, 73–87.

Huber, O., 1995. Conservation of the Venezuelan Guayana. In: Steyermark, J.A., Berry, P.E., Yatskievych, K., Holst, B.K. (Eds.), Flora of the Venezuelan Guayana, 1. Missouri Botanical Garden Press, St. Louis, pp. 193–218.

Huber, O., Foster, M.N., 2003. Conservation Priorities for the Guayana Shield. Conservation International, Washington.

IUCN, 1994. Guidelines for Protected Areas Management Categories. IUCN, Cambridge, UK and Gland, Switzerland.

IUCN, UNEP, WWF, 1991a. Caring for the Earth. A Strategy for Sustainable Living. The World Conservation Union, United Nations Environmental Programe and World Wide Fund for Nature, Gland, Switzerland.

IUCN, UNEP, WWF, 1991b. Caring for the Earth. The World Conservation Union. United Nations Environmental Program and WWF International, Gland, Switzerland.

León, V., 2015. ¿Fin del ciclo progresista? América Latina en Movimiento, Edición Digital. Agencia Latinoamericana de Información, Quito, Ecuador.

Lloyd, I.C., 2008. State failure, corruption, and warfare: challenges for forest policy. J. Sust. For. 3, 189–223.

Moreira, C., 2017. El largo ciclo del progresismo latinoamericano y su freno. Los cambios políticos en América Latina de la última década (2003–2015). Rev. Bras. Ci. Soc. 93, 1–28.

Novo, I., Díaz, D., 2007. Informe final de la evaluación del Parque nacional Canaima. Venezuela, como sitio de Patrimonio Natural de la Humanidad. INPARQUES. Caracas, Venezuela.

Núñez, I., González-Gaudiano, E., Barahona, A., 2003. La biodiversidad: historia y contexto de un concepto. Interciencia 7, 387–393.

OEA, 2018. Informe de la secretaría general de la Organización de los Estados Americanos y del panel de expertos internacionales independientes sobre la posible comisión de crímenes de lesa humanidad en Venezuela. Washington, DC, EEUU. https://reliefweb.int/sites/reliefweb.int/files/resources/Informe-Panel-Independiente-Venezuela-ES.pdf (consulted 10.09.18).

OPS, 2018. Respuesta de la OPS[PAHO] para mantener una agenda eficaz de cooperación técnica en Venezuela y en los estados miembros vecinos. 162.a Sesión del Comité Ejecutivo. Punto 7.22 del orden del día provisional. CE162/INF/22, Rev. 1. Organización Panamericana de la Salud OPS[PAHO] Washington, EEUU.

Potapov, P., Hansen, M.C., Laestadius, L., Turubanova, S., Yaronshenko, A., Thies, C., et al., 2017. The last frontiers of wilderness: tracking loss of intact forest landscapes from 2000 to 2013. Sci. Adv. 3, 1–13.

PROVEA, 2018. Informe Anual sobre la situación de los Derechos Humanos en Venezuela, correspondiente al período enero-diciembre de 2017. Programa Venezolano de Educación Acción en Derechos Humanos. Caracas, Venezuela. https://www.derechos.org.ve/informe-anual/informe-anual-enero-diciembre-2017 (consulted 10.07.18).

Requena, J., Caputo, C., 2016. Pérdida de talento: fuga de investigadores en Venezuela. Interciencia 7, 444–453.

Reyes-García, V., Fernández-Llamazares, A., Guèze, M., Garcés, A., Mallo, M., Vila-Gómez, M., et al., 2015. Local indicators of climate change: the potential contribution of local knowledge to climate research. Wiley Interdiscip. Rev. Clim. Change 1, 109–124.

RIGBYY, 2017. Mapa Garimpo illegal nos Territórios Yanomami e Ye'kwana (Brasil-Venezuela). Rede de Informação Geográfica Binacional Yanomami e Ye'kwana. https://acervo.socioambiental.org/sites/default/files/maps/2018-08/yad00485.pdf (consulted 28.09.18).

Rull, V., Vegas-Vilarrúbia, T., 2008. Biopiracy rules hinder conservation efforts. Nature 453, 26.

Rull, V., Vegas-Vilarrúbia, T., Nogué, S., Montoya, E., 2008. Bureaucratic obstruction of conservation science in the Guayana Highlands. Conserv. Biol. 22, 508–509.

Rull, V., Vegas-Vilarrúbia, T., Nogué, S., Huber, O., 2009. Conservation of the unique neotropical vascular flora of the Guayana Highlands in the face of global warming. Conserv. Biol. 5, 1323–1327.

Rull, V., Vegas-Vilarrúbia, T., Safont, E., 2016. The Lost World's pristinity at risk. Divers. Distrib. 22, 995–999.

Safont, E., Vegas-Vilarrúbia, T., Rull, V., 2012. Use of Environmental Impact Assessment (EIA) tools to set priorities and optimize strategies in biodiversity conservation. Biol. Conserv. 149, 113–121.

Sen, A., 1985. Well-being, agency and freedom: the Dewey Lectures 1984. J. Philos. 82, 169–221.

Sen, A., 1999. Development as Freedom. Random House Inc., NewYork.

Svampa, M., 2013. Consenso de los Commodities y lenguajes de valoración en América Latina. Nueva Sociedad 244, 30–46.

Terán-Mantovani, E., 2016. Extractivismo, desarrollo y cambio climático: ¿hacia dónde marcha la Revolución Bolivariana en plena crisis ambiental global? In: GriffonBriceño, D. (Ed.), Narrativas contrahegemónicas de la crisis climática. Amalivaca Ediciones. Fundación de Educación Ambiental, Caracas, Venezuela, pp. 87–129.

UN, 2017. The Sustainable Development Goals Report 2017. http://sdgactioncampaign.org/wp-content/uploads/2017/07/TheSustainableDevelopmentGoalsReport2017.pdf (consulted 16.09.18).

Venezuela, 1978. Decreto 2552. Gaceta Oficial de la República de Venezuela No. 31.408. Enero 19, 1978. Caracas, Venezuela.

Venezuela, 1989. Decreto 269. Gaceta Oficial de la República de Venezuela No. 4.106 Extraordinario. Junio 9, 1989. Caracas, Venezuela.

Venezuela, 1991. Decreto 1742. Gaceta Oficial de la República de Venezuela, No. 31.783. Agosto 23, 1991. Caracas, Venezuela.

Venezuela, 2018. Minería Responsable en Venezuela. Una oportunidad de inversión. Ministerio del Poder Popular para Desarrollo Minero Ecológico. República Bolivariana de Venezuela. Caracas, Venezuela. http://desarrollominero.gob.ve/wp-content/uploads/2018/04/Catalogo_Minerales_2018_espanol_Baja.pdf (consulted 16.09.18).

Villamizar, A., Buroz, E., Lairet, R., Gómez, J.A. (Eds.), 2018. Primer Reporte Académico de Cambio Climático. Resumen para Responsables de Políticas en Cambio Climático para Venezuela: Contribución de los Grupos de Trabajo I, II y III. Secretaría Académica de Cambio Climático de la Academia de Ciencias Físicas. Matemáticas y Naturales de Venezuela. Ediciones ACFIMAN—CITECI, Caracas, Venezuela.

WWF, 2012. Living Guianas report 2012. State of the Guianas Drivers and pressures Towards Green Economies. World Wildlife Fund, Paramaribo, Suriname.

WWF, 2016. Living Planet report 2016. Risk and Resilience in a New Era. WWF International, Gland, Switzerland.

Yerena, E., 2011. La Guayana venezolana: sostenibilidad ambiental incierta. Mundo Nuevo 6, 339–357. Caracas, Venezuela.

CHAPTER

17

Pantepui and global warming

Valentí Rull[1], *Sandra Nogué*[2], *Elisabet Safont*[3] *and Teresa Vegas-Vilarrúbia*[3]

[1]**Institute of Earth Sciences Jaume Almera (ICTJA-CSIC), Barcelona, Spain** [2]**Department of Geography and Environment, University of Southampton, Southampton, United Kingdom** [3]**Department of Evolutionary Biology, Ecology and Environmental Sciences, Faculty of Biology, University of Barcelona, Barcelona, Spain**

Introduction

Pantepui conservation has traditionally been addressed from the perspective of more or less immediate threats, with an emphasis on direct anthropogenic threats, such as vegetation trampling, biopiracy, the introduction of exotic invader species, water and soil contamination, and fire (e.g., Huber, 1992, 1995c; Safont et al., 2014, 2016; Fernández-Delgado et al., 2016; Rull et al., 2016, and literature therein). Current conservation entities have been defined using criteria based on these types of risks. Roughly a decade ago, the ongoing effects of global warming (GW) began to be considered an indirect threat to Pantepui biota, linked to the risk of extinction by habitat loss. Indeed, the usually flat topography of the tepuian summits and their relatively low elevation compared to other cone-shaped Neotropical mountains such as the Andes will prevent summit species from migrating upwards following the expected warming trend projected by the end of this century (Rull and Vegas-Vilarrúbia, 2006).

GW is considered a significant threat to mountain biota worldwide, as species may respond by shifting their altitudinal distributions, leading to changes in the diversity and composition of their communities and reductions, fragmentation, or loss of their habitat. An increasing number of studies show that this is already occurring and warn of the potential threat of GW to summit species and communities (e.g., Thuiller et al., 2005; Pauli et al., 2007, 2012; Colwell et al., 2008; Kelly and Goulden, 2008; Lenoir et al., 2008; Kreyling et al., 2010; Walther, 2010; Dirnböck et al., 2011; Engler et al., 2011; Feeley et al., 2011; Sheldon et al., 2011; Gottfried et al., 2012; Grabherr et al., 2010; Jump et al., 2012). Tropical mountains are of particular concern because of their high biodiversity and levels of

Biodiversity of Pantepui
DOI: https://doi.org/10.1016/B978-0-12-815591-2.00017-3

endemism (Davis et al., 1997; Myers et al., 2000; Malcom et al., 2006; Laurence et al., 2011; Nogué et al., 2013).

Several studies have been conducted in Pantepui to quantitatively estimate the potential extinction by GW-induced habitat loss and to evaluate the possible conservation actions (Nogué et al., 2009; Safont et al., 2012, 2014; Vegas-Vilarrúbia et al., 2012). Emphasis has been given to vascular plants, the better-known Pantepui organisms in terms of biodiversity and elevational ranges, which is a critical parameter for estimating the risk of habitat loss. Due to the lack of autoecological information on the Pantepui vascular flora, current estimates have assumed homogeneous responses among all species, and the idiosyncratic features of species that eventually facilitate tolerance, acclimation, or adaptation to GW have not been evaluated (Rull and Vegas-Vilarrúbia, 2017). This chapter reviews the results of studies conducted to date in relation to the risk of extinction of the Pantepui vascular flora as a result of habitat loss and discusses the eventual conservation actions that should be performed. It also discusses the main weaknesses of these estimates and provides perspectives on future research into the less developed aspects, such as tolerance, acclimation, adaptation, and the potential genetic and ecological mechanisms that are involved. As current estimates of potential GW-driven biodiversity depletion are based on outdated GW projections (Houghton et al., 2001; Solomon et al., 2007), eventual modifications introduced by the latest predictions (Allen et al., 2018) are also evaluated. Studies of the potential consequences of GW for untouched biomes and ecosystems are rare, and the case of Pantepui could serve as a pilot survey.

Current extinction estimates

Estimates of potential extinction caused by habitat loss in Pantepui have been obtained using three complementary methods, namely, the altitudinal range shift (ARS) method, the species–area relationship (SAR) method, and the climate envelope distribution model (CEDM) method. The ARS method is simple and uses the adiabatic lapse rate, which is −0.6°C/100 m elevation in the Guayana region (Huber, 1995a), to predict the eventual altitudinal displacement of the range of each species under a given temperature increase. This method was first applied in a prospective study (Rull and Vegas-Vilarrúbia, 2006) to a species endemic to Pantepui and other emblematic genera, such as *Bonnetia* and *Stegolepis*, assuming GW of 2°C–4°C by the end of this century (Houghton et al., 2001). Using the current adiabatic lapse rate, this represents an upward displacement of 330–670 m, which would threaten c. 20%–40% of the analyzed taxa as a result of habitat loss (Fig. 17.1).

Some years later, the ARS and SAR methods were combined using geographic information system (GIS) techniques to estimate the potential habitat loss for the entire Pantepui vascular flora and for the Pantepui–endemic species under the same IPCC (Intergovernmental Panel on Climate Change) predictions of 2°C–4°C for the year 2100 (Nogué et al., 2009). The first step was to establish SARs of the type $S = S_0A^z$ (S, number of species; A, area; S_0, intercept; z, slope). These relationships were highly significant for both all ($S = 39{,}08A^{0.354}$; $r^2 = 0.706$; $P < .0001$) and endemic–Pantepui species ($E = 27.41A^{0.354}$; $r^2 = 0.968$; $P > .002$). Then the adiabatic lapse rate was used to estimate the reduction in the Pantepui area using a digital elevation model (DEM). Finally, the previously obtained

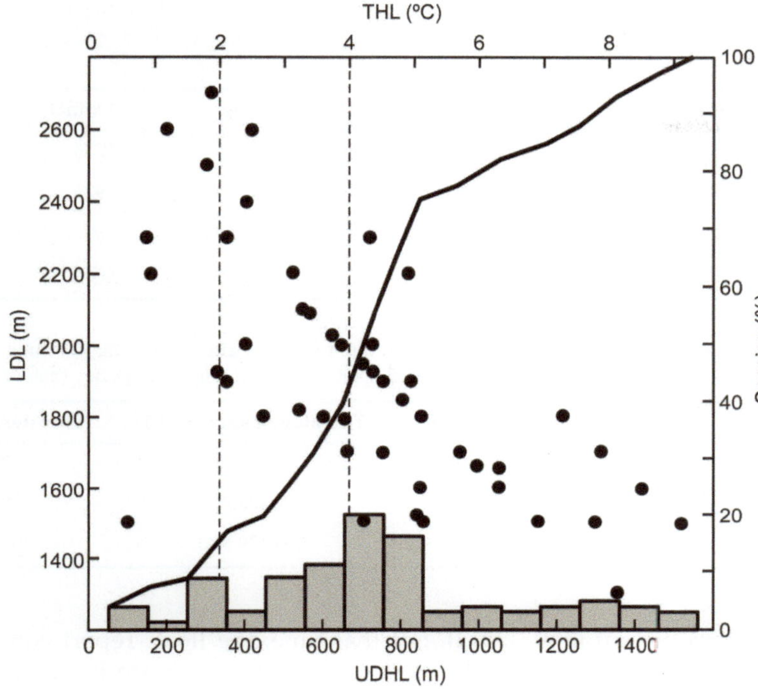

FIGURE 17.1 Graphical display of the altitudinal range shift (ARS) method. Black dots are the species analyzed. UDHL is the upward displacement required for total habitat loss, THL is the temperature required for total habitat loss, and LDL is the lower distribution limit of the corresponding genera. Solid line represents the cumulative risk of habitat loss in species percentage, and gray bars are the rate of change of this parameter. Broken lines indicate the range of the predicted IPCC temperature increase for the end of the present century. *Source: Redrawn from Rull, V., Vegas-Vilarrúbia, T., 2006. Unexpected biodiversity loss under global warming in the neotropical Guayana Highlands: a preliminary appraisal. Glob. Change Biol. 12, 1–6.*

significant SARs were used to estimate the number of species predicted for the GW-reduced Pantepui area (Nogué et al., 2009). According to this analysis, under the more optimistic IPCC scenario (B1), the reduction in the vascular plant diversity may be almost 75%, including 28% of endemics, whereas in the more pessimistic scenario (A2), the biodiversity depletion may reach 83%, with 54% of endemics (Table 17.1). An individual species-by-species analysis based on their specific estimated reductions in habitat confirmed that 166–343 of the endemic species (22%–46%) would be threatened by habitat loss (Nogué et al., 2009). In addition, virtually all of the analyzed species showed a fragmented potential distribution by the end of the century, regardless of the IPCC scenario. Similar general results were obtained by Rödder et al. (2010) using CEDMs, but no specific evaluations at the species level were conducted.

These first estimates were revised after the publication of an updated IPCC report (Solomon et al., 2007), with an emphasis on Pantepui–endemic species. The figures were similar, with ~30% (B1) to ~50% (A2) endemic species endangered by extinction as a

TABLE 17.1 Estimates of the number of species living and extinct due to habitat loss by 2100 using the species–area models for the total (SAR; $r^2 = 0.706$, $P < .001$) and endemic (EAR; $r^2 = 0.969$, $P < .002$) species (Nogué et al., 2009).

Model	Equation	IPCC scenario	Living	Extinct	% Extinct
SAR	$S = 39{,}08A^{0.354}$	B1 ($\Delta T = 2°C$)	589	1733	74.6
SAR	$S = 39{,}08A^{0.354}$	A2 ($\Delta T = 4°C$)	393	1929	83.1
EAR	$E = 27.41A^{0.354}$	B1 ($\Delta T = 2°C$)	545	209	27.7
EAR	$E = 27.41A^{0.354}$	A2 ($\Delta T = 4°C$)	348	406	53.8

TABLE 17.2 Endemic species threatened with extinction due to habitat loss considering the three warming phases of the IPCC (Solomon et al., 2007), using the ARS method applied to each individual species (Safont et al., 2012).

Phase	Years	B1 (ΔT °C)	B2 (ΔT °C)	Threatened species (B1)	Threatened species (A2)
I	2011–2030	1.0	1.0	45 (5.7%)	35 (5.7%)
II	2046–2065	2.0	2.5	140 (22.7%)	184 (29.8%)
III	2080–2099	2.5	4.0	184 (29.8%)	307 (49.7%)

result of habitat loss (Safont et al., 2012), but that time, the IPCC report considered three phases within the 21st century with different rates of temperature increase (Table 17.2). This facilitated the hierarchization of the threatened species into 10 priority conservation categories under the B1 scenario and 13 under the A2 scenario. These categories were obtained by defining environmental impact values of the type $EIV = \Sigma(\lambda_i \cdot C_i)$, where C_i represents assessment criteria (endemism type, degree of endemism, keystone species, geographic distribution, and altitudinal range) and λ_i is a weight factor derived from expert criteria, ranging from 10% to 25%. Safont et al. (2012) provide an exhaustive list of Pantepui–endemic species ranked by their respective EIVs and priority categories. Prioritizing species according to extinction threat may be a more cost-effective way to invest conservation resources than comprehensive plans intended to preserve all species at once or only the most emblematic (Jiménez-Alfaro et al., 2010; Pärtel et al., 2005).

Conservation insights

The possible effects of GW on the Pantepui biota is still a developing field of research, and it is too soon to recommend specific measures based on the available data. However, the threat is already underway, and it is time to analyze potential actions to be discussed on the basis of the existing information and to identify the most urgent research needs to develop preliminary conservation measures. In this section, we briefly review some in situ and ex situ actions that have been considered for Pantepui in the recent literature. The aim is not to propose the straightforward application of these measures, but rather to present them as a basis for discussion with the corresponding authorities and stakeholders.

In situ actions

In situ practices, notably the enhancement of the degree of protection or the creation of protected areas, are considered the best strategy for long-term biodiversity conservation (Frankel et al., 1995; Primack, 2002). In Pantepui, this type of protection is already represented by conservation figures as, for example, a national park, a biosphere reserve, or a World Heritage Site. In addition, the tepuis themselves were declared a natural monument in 1990, which involves additional and specific protections (Huber, 1995c). However, the risk of extinction caused by habitat loss due to GW cannot be addressed with in situ practices alone, and ex situ measures seem to be required to preserve the Pantepui biodiversity. Indeed, GW is a general phenomenon that cannot be circumvented or palliated at a particular site or in a particular region unless this area can be isolated from the general environment by a greenhouse-like structure that is able to maintain a constant climate by artificial means. Given the dimensions, the physiography, and the remoteness of Pantepui, this would be unfeasible, and most likely undesirable, as it would represent not only a regrettable transformation of the Pantepui landscape but also a barrier to dispersal and eventual biotic interactions.

The only natural possibility for in situ conservation relies on the existence of natural refugia for the Pantepui biota. ARS–GIS modeling predicts that, in the worst scenario (A2; 4°C increase), >80% of the present Pantepui surface would disappear by the end of the century (Nogué et al., 2009). Considering only the tepuis with reliable and sufficient floristic information (Huber, 1995b), fragmentation analysis indicates that half of them would completely lose the Pantepui area and therefore all their Pantepui species (Table 17.3). Most of the others would lose 80%–90% of the Pantepui surface, and only two (Marahuaka and Chimantá) would retain ~40% of their present Pantepui surface. Therefore Chimantá and Marahuaka, especially the former (Fig. 17.2), are the better-suited tepuis to act as potential refugia for the Pantepui vascular flora by the end of the century under the worst IPCC scenario of a 4°C increase (Vegas-Vilarrúbia et al., 2012). If special in situ conservation actions should be conducted, these two tepuis, especially the Chimantá massif, seem to be the better suited targets.

Another possibility for in situ survival would be the occurrence of microrefugia, or tiny sites that are able to support small vegetation stands due to the presence of favorable microclimates (Rull, 2009). However, the presence and distribution of such microrefugia are virtually impossible to predict, and hence premeditated conservation actions based on this possibility are difficult to implement. The possibility of building artificial microrefugia atop the tepuis to preserve the most representative flora cannot be ruled out, although the investment needed to create and maintain such structures would be high.

Some ex situ possibilities

Several possibilities for ex situ conservation in the face of GW have been discussed for the Pantepui flora, including the creation of germplasm banks and living plant collections, and managed relocation (Rull and Vegas-Vilarrúbia, 2006; Nogué et al., 2009; Safont et al., 2012). Germplasm preservation, especially the preservation of seeds, is a common activity in biodiversity conservation, and its methods have already been validated and standardized. This option would be the most inexpensive and easy to implement. Given the high

TABLE 17.3 Pantepui area loss predicted for the end of this century on the tepuian summits with reliable and sufficient floristic information (Vegas-Vilarrúbia et al., 2012). Predictions are based on ARS–GIS modeling considering the IPCC scenario of 4°C increase (A2) (Nogué et al., 2009). Abbreviations refer to the general Pantepui map (Fig. 1.2). Some of these tepuis and massifs can be seen in Plate 17.1.

Tepui	Abbreviation	Present area (km^2)	2100 area ($\Delta T = 4°C$)	Area loss (%)
Guaiquinima	Gq	31.4	0.0	100
Neblina	Nb	357.0	40.1	89
Aracamuni	Am	56.0	0.0	100
Cuao	Cu	59.2	0.0	100
Sororopán	Sp	24.3	0.0	100
Angasima	Ts	9.6	0.0	100
Sipapo	Si	96.7	0.0	100
Guanay	Gy	84.8	0.0	100
Huachamacari	Hu	7.0	0.0	100
Camani	Cm	4.3	0.0	100
Jaua	Ja	48.00	12.1	97
Uei	Ui	18.4	0.0	100
Carrao	Ca	38.3	0.92	98
Aparamán	Aa	3.4	0.0	100
Sarisariñama	Sr	295.0	13.6	95
Kamarkawarai–Tereké-Yurén	Ts	36.4	2.39	93
Duida	Du	448.0	10.6	98
Auyán	Ay	509.0	11.0	98
Corocoro–Yutajé	Co–Yt	275.0	7.19	97
Ilú	Iu	82.3	10.9	87
Roraima–Kukenán	Ro–Ku	262.0	61.1	77
Marahuaca	Mk	214.0	78.4	63
Chimantá	Ch	952.0	408.0	57

diversity of the Pantepui flora, the initial selection of the most threatened species according to the priority criteria established in the former environmental impact analysis (Safont et al., 2012) would be a reasonable starting point to establish the best procedures. Growing Pantepui plants under controlled conditions in already existing botanical gardens has also been suggested. In addition to species preservation, this strategy would facilitate the development of autoecological studies, especially those focused on the biotic responses to different climatic and environmental conditions. The same priority criteria proposed for

PLATE 17.1 Some tepuis and tepuian massifs representative of Table 17.1 (see Fig. 1.2 for geographical reference). (A) From left to right: Kukenán-tepui, Roraima-tepui, and Wei-assipu-tepui. (B) Ilú–Tramén massif: Tramén-tepui, Ilú-tepui, and Kararaurín-tepui. (C) Los Testigos massif: Aparamán-tepui, Murisipán-tepui, Tereké-yurén-tepui, and Kamarkawarai-tepui. (D) Chimmantá massif: Tirepón-tepui. (E) Angasima-tepui. (F) Vertical cliffs of the Chimantá massif. (F) Aerial view of the Chimantá summit. *Source: Photos: V. Rull.*

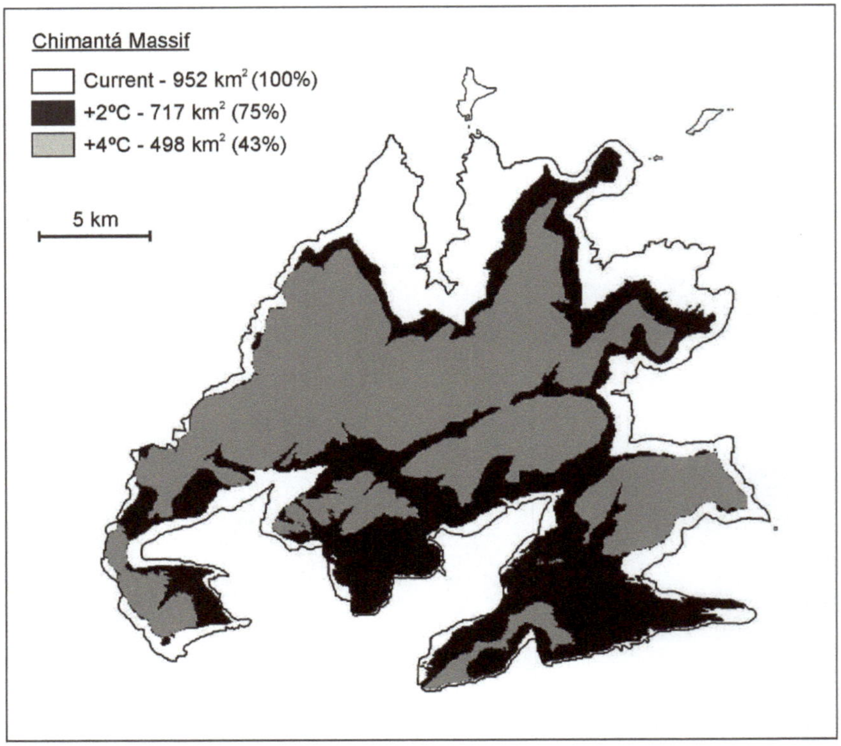

FIGURE 17.2 ARS–GIS simulation of the Pantepui area at the top of the Chimantá massif for the end of the present century under the more optimistic (B1; 2°C increase) and the more pessimistic (A2; 4°C increase) IPCC scenarios (Houghton et al., 2001). *Source: Redrawn from Nogué, S., Rull, V., Vegas-Vilarrúbia, T., 2009. Modeling biodiversity loss by global warming in Pantepui, northern South America: projected upward migration and potential habitat loss. Clim. Change 94, 77–85 and Vegas-Vilarrúbia, T., Nogué, S., Rull, V., 2012. Global warming, habitat shifts and potential refugia for biodiversity conservation in the neotropical Guayana Highlands. Biol. Conserv. 152, 159–168.*

the germplasm banks seem appropriate here for selecting the species to preserve and analyze. In both cases—germplasm banks and botanical gardens—the ultimate objective is to reintroduce the preserved species into their original areas, provided the former threats have disappeared and the original environmental conditions have been reestablished. For a successful reintroduction, the preserved material must have genetic variability representative of the original species to guarantee that the population dynamics and evolutionary trends are as natural as possible (Hardwick et al., 2011; Vitt et al., 2010).

Managed relocation is a highly controversial topic due to the potential occurrence of unexpected ecological consequences for recipient ecosystems (Hunter, 2007; McLachlan et al., 2007; Davidson and Simkanin, 2008; Hoegh-Guldberg et al., 2008; Ricciardi and Simberloff, 2009; Seddon et al., 2009). To minimize these impacts, the relocated species may be restricted to selected areas along elevational transects according to their individual requirements and isolated from autochthonous communities. The northern Andes have been proposed as a possible recipient area for Pantepui species, as these species would be able to migrate upwards, up to 5000 m elevation or more, which could minimize GW-induced extinction caused by habitat loss (Rull et al., 2009). However, potential relocations must be carefully planned on an ecologically sound basis to ensure that the creation of novel ecosystems for stressed populations does not damage the recipient ecosystem (Safont et al., 2012). For now, this option is highly speculative due to the lack of sufficient autoecological and synecological knowledge regarding the translocation candidates and eventual recipient ecosystems. Much more ecological and evolutionary research is still needed to determine whether the managed relocation of Pantepui species is a feasible and convenient option.

Weaknesses and future research

The most urgent research need in order to obtain a true appraisal of the extent and magnitude of the GW threat in Pantepui is to refine the existing habitat loss and extinction estimates, which should be considered preliminary and a first step towards more realistic projections that consider all relevant environmental factors, as well as the ecological features of the species concerned that could affect their individual responses to climate change. Using the available data, this is all that can be done currently. To move beyond this limitation, the following points should be considered: (1) environmental factors other than temperature could act as drivers of species range shifts; (2) the migration capacity of a species could be limited by the lack of suitable substrates or topographical barriers; (3) not all all species will respond to warming by migrating upwards at the same rates; and (4) updated IPCC estimates (Allen et al., 2018) introduce a new and somewhat disturbing perspective with respect to former reports.

Environmental factors driving species range shifts

In terms of climatic factors other than temperature, precipitation intensity and distribution are predicted to undergo significant shifts during this century worldwide, but such changes seem to be of relatively little relevance in the Pantepui area. On the one hand, the 0%–5% reduction in mean annual rainfall forecasted by the IPCC by the end of this century (Solomon et al., 2007; Stocker et al., 2013) is almost negligible considering the perhumid climate of the tepui summits, where precipitation ranges from 2500 to 4000 mm per year with little seasonal variation (Huber, 1995a). On the other hand, temperature has been recognized as the main factor controlling the altitudinal distribution of vascular plant species in Pantepui (Huber, 1995a), as in many other areas (e.g., Guisan and Theurillat, 2000; Thuiller et al., 2005). The projected increase in the concentration of atmospheric CO_2 is another factor that should be considered, and evaluating its effects requires detailed ecophysiological studies of selected species, such as those in the priority categories mentioned earlier. Unfortunately, such studies are not available and are difficult to perform due to logistic and bureaucratic constraints (see later).

Influence of topography and substrate

The influence of local topographical and edaphic factors (e.g., slope orientation, substrate availability, microclimatic conditions, vegetation structure) might be relevant to controlling altitudinal plant migration and could be better understood with data obtained during intensive field work in combination with GIS tools. Again, acquiring detailed knowledge of these aspects has been hindered by the difficulty in conducting ecological studies atop the tepuis, but the need for these types of studies is indisputable, and methods that are applicable globally, such as those developed by the GLORIA (Global Observatory Research Initiative in Alpine Environments) project (Pauli et al. 2005), have been encouraged (Rull et al., 2009). These methods have been especially designed to measure the eventual GW-driven upslope migration of extant plants in mountains all over the world through the installation of permanent observation and meteorological stations on selected mountain slopes. The flat surface of many tepui summits is a handicap for such studies, but some large tepuis and tepuian massifs have elevational gradients on their summits that are suitable for GLORIA-type methods. The best example is, once more, the Chimantá massif, which seems to be a keystone tepuian complex for studying the future of the tepuian biota under the action of GW.

Autoecological features of the involved species

Increasing the autoecological knowledge of the potential responses of threatened plant species is imperative. Current predictive models for the GH assume homogeneous responses, but this should be considered approximate, as different species may respond to warming in different ways. Species with higher phenotypic plasticity might be able to tolerate warming by altering their morphological, physiological, or life history traits to accommodate changing climates. Examples include changing photosynthesis rates and growth in response to increasing atmospheric CO_2 concentrations or modifying phenological features, such as growth and flowering times and duration, in response to changing meteorological seasonality (Ainsworth and Rogers, 2007; Gunderson et al., 2010; Medeiros and Ward, 2013; Liancourt et al., 2015). The combined effects of warming, water stress, and elevated CO_2 should also be taken into account (e.g., Xu and Zhou, 2006; Xu et al., 2014, 2016).

At the ecosystem level, such changes could influence the resilience of species in response to climate change and, consequently, their competitive ability, symbiotic interactions, and fitness, which may affect community composition and ecological functioning (Kimball et al., 2012). According to Becklin et al. (2016), phenotypic plasticity may alleviate the effects of climate change but may not guarantee long-term persistence, especially if environmental change exceeds the variability that species have experienced historically. The species that are unable to accommodate GW in this way should adapt, in an evolutionary sense, to the new environmental conditions or migrate; otherwise, they will become extinct in situ. The potential occurrence of rapid evolutionary adaptation (i.e., genetic change) in response to GW is a controversial issue (Merilä, 2012) that remains difficult to demonstrate (Franks et al., 2007). A necessary condition for a species to undergo rapid evolutionary change in response to climate change is sufficient genetic variation in the traits under selection pressure (Becklin et al., 2016), but such knowledge of the Guiana Highlands does not exist for the same reasons mentioned earlier and should be urgently addressed in light of the priority categories established for the more threatened species (Safont et al., 2012).

Latest IPCC estimates and warming acceleration

The recently approved GW report of the IPCC (Allen et al., 2018) is focused on the differences in the effects, impacts, and required adaptations if GW results in a 1.5°C increase compared to a 2°C increase by 2030. Current measurements indicate that human activities have already raised global temperatures by 1°C above preindustrial levels and that a 1.5°C increase will be reached between 2030 and 2052. This estimate surpasses the 1°C increase projected by the IPCC for the period 2011–2030 (Table 17.2) by 0.5°C (Solomon et al., 2007). Hence, GW seems to be taking place faster than formerly expected, although the magnitude and rate of warming will depend on the geographic location in question. The new IPCC report also emphasizes that the differences in regional climate characteristics between the present day and GW-induced increases in temperature of 1.5°C and between 1.5°C and 2°C will be quite different and include increases in mean temperature and heat extremes. Consequently, impacts on terrestrial ecosystems and biodiversity, including species loss and extinction and biodiversity-related risks (e.g., forest fires, dissemination of invasive species), are expected to be higher at a 2.0°C increase in temperature induced by GW compared to a 1.5°C increase. These projections suggest that the time available for Pantepui species to acclimate or to move upwards has shrunken, that the pace of habitat loss is moving faster, and that ecological threshold situations may be accelerated by the increase in the number of hot days projected for the tropics. Henceforth, an urgent revision of former impact assessments on Pantepui species and habitats is required in order to amend risk assessments and plan conservation actions to be undertaken in a timely manner.

Bureaucratic constraints

In addition to the remoteness and low accessibility of the Guiana Highlands, a recurrent obstacle to a sound assessment of the plant species and communities threatened with extinction by habitat loss is the difficulty in obtaining fieldwork and sampling permits. This problem (i.e., severe sampling restrictions, tedious paperwork, and prohibitively long waiting times to obtain the permits) was first noted several years ago (Rull and Vegas-Vilarrúbia, 2008; Rull et al., 2008), but the situation has not changed. The lack of sufficient environmental management tools (Novo and Díaz, 2007) and the difficulty in establishing effective control mechanisms in a region as vast as the Guiana Highlands are serious challenges to adequate protection policies (Rull et al., 2016). Currently, visits to most of the tepuis are prohibited, and scientific surveys, especially those related to genetic studies, are subject to serious restrictions to prevent biopiracy (Rull and Vegas-Vilarrúbia, 2008). These restrictions should be relaxed for well-defined academic purposes to allow proper appraisal of the potential effects of GW on Guiana Highlands species and their ecosystems. Current estimates based on the available databases and methods, such as the ARS and CEDM methods, are all that can currently be accomplished. Without new and more detailed ecological and physiological studies of selected Guiana Highlands plants and ecosystems, the background information required for optimizing conservation practices will remain unknown. Under these conditions, the use of remote sensing techniques would be useful, especially those involving radiation measurements related to ecophysiological features, such as energy balance or photosynthesis (Jones et al., 2003). Such measurements might be conducted using conventional and high-resolution satellite imagery (up to 30 cm) able to provide detailed information on biomass, moisture, vegetation cover, and photosynthesis, or newly developed techniques, such as unmanned aerial vehicles (UAVs), also known as drones (Salamí et al., 2014).

Conclusions and recommendations

The main conclusion is that the potential effects of GW on the Pantepui flora and vegetation—and likely on the whole Pantepui biota—could be devastating, and it is urgent to increase research aimed at providing reliable data to inform conservation measures. Preliminary studies estimated the potential magnitude of the expected biodiversity loss by the end of 2100 and established a risk classification for species and tepuis, leading to the definition of priority categories for conservation. The Chimantá massif has been identified as the best-suited site for conducting the research needed to inform conservation practices. Further research should focus on the autoecology of the more threatened species, especially in relation to their responses to climate change, and the use of field methods to document and measure the eventual GW-induced upward migration of plant species. Predictions based on these data can be used to identify the best in situ or ex situ conservation actions to undertake. The research needed to protect Pantepui species from extinction caused by GW is almost impossible due to the difficulty in obtaining the corresponding official fieldwork permits. This has dangerously delayed the adoption of pertinent and urgently needed conservation measures, whereas GW is ongoing and will not wait for us to find a solution. The latest IPCC estimates suggest an acceleration of GW, which makes any delay especially worrisome.

Acknowledgments

Results summarized in this chapter were obtained in research projects funded by the BBVA Foundation (projects PANTEPUI; BIOCON 2004-90/05 and PANTEPUI-II; BIOCON 2008-031 to V.R.), the Higher Council for Scientific Research or CSIC (projects PATAM; 200830I258 and PALAB; 201130E075 to V.R.), the Catalan Agency for Management of University and Research Grants (grant GREB; 2009-SGR439 to V.R. and T.V.) and the University of Barcelona (grant APIF to E.S.). Critical review by Suzette Flantua contributed to the improvement of the manuscript.

References

Ainsworth, E.A., Rogers, A., 2007. The response of photosynthesis and stomatal conductance to rising [CO_2]: mechanisms and environmental interactions. Plant Cell. Environ. 30, 258–270.

Allen, M., Babiker, M., Chen, Y., de Coninck, H., Connors, S., van Diemen, R., et al. , 2018. Global warming of 1.5°C. IPCC Spec. Rep. 1.5, Incheon (Korea).

Becklin, K.M., Anderson, J.T., Gerhard, L.M., Wadgymar, S.M., Wessinger, C.A., Ward, J.K., 2016. Examining plant physiological responses to climate change through an evolutionary lens. Plant Physiol. 172, 635–649.

Colwell, R.K., Brehm, G., Cardelús, C.L., Gilman, A.C., Longino, J.T., 2008. Global warming, elevational range shifts, and lowland biotic attrition in the wet tropics. Science 322, 258–261.

Davidson, I., Simkanin, C., 2008. Skeptical of assisted colonization. Science 322, 1048–1049.

Davis, S.D., Heywood, V.H., Herrera-MacBride, O., Villalobos, J., and Hamilton, A., 1997. Centres of Plant Diversity: A Guide and Strategy for their Conservation. The Americas, vol. 3. IUCN Publications Unit, Cambridge.

Dirnböck, T., Essl, F., Rabitsh, W., 2011. Disproportional risk for habitat loss of high-altitude endemic species under climate change. Glob. Change Biol. 17, 990–996.

Engler, R., Randin, C.F., Thuiller, W., Dullinger, S., Zimmermann, N.E., Araújo, M.B., et al., 2011. 21st century climate change threatens mountain flora unequally across Europe. Glob. Change Biol. 17, 2330–2341.

Feeley, K.J., Silman, M.R., Bush, M.B., Farfan, W., García Cabrera, K., Malhi, Y., et al., 2011. Upslope migration of Andean trees. J. Biogeogr. 38, 783–791.

Fernández-Delgado, M., Giarrizzo, J.G., García-Amado, M.A., Contreras, M., Salazar, V., Barton, H., et al., 2016. Evidence of *Helicobacter* spp. in freshwaters from Roraima Tepui, Guayana Shield, South America. Antoine Van Leeuwenhoeck 109, 529–542.

Frankel, O.H., Brown, A.H.D., Burdon, J.J., 1995. The Conservation of Plant Biodiversity. Cambridge University Press, Cambridge.

Franks, S.J., Sim, S., Wels, A.E., 2007. Rapid evolution of flowering time by an annual plant in response to a climate fluctuation. Proc. Natl. Acad. Sci. U.S.A. 104, 1278–1282.

Gottfried, M., Pauli, H., Futschik, A., Akhalkatsi, M., Baran Čok, P., Benito Alonso, J.L., et al., 2012. Continentwide response of mountain vegetation to climate change. Nat. Clim. Change 2, 111–115.

Grabherr, G., Gottfried, M., Pauli, H., 2010. Climate change impacts in alpine environments. Geogr. Compass 4/8, 1133–1153.

Guisan, A., Theurillat, J.P., 2000. Assessing alpine plant vulnerability to climate change: a modeling perspective. Integr. Assess. 1, 307–320.

Gunderson, C.A., O'Hara, K.H., Campion, C.M., Walker, A.V., Edwards, N.T., 2010. Thermal plasticity of photosynthesis: the role of acclimation in forest responses to a warming climate. Glob. Change Biol 16, 2272–2286.

Hardwick, K.A., Fiedler, P., Lee, L.C., Pavlik, B., Hobbs, R.J., Aronson, J., et al., 2011. The role of botanic gardens in the science and practice ofecological restoration. Conserv. Biol. 25, 265–275.

Hoegh-Guldberg, O., Hughes, L., McIntyre, S., Lindenmayer, D.B., Parmesan, C., Possingham, H.P., et al., 2008. Assisted colonization and rapid climate change. Science 321, 345–346.

Houghton, J.T., Ding, Y., Griggs, D.J., Noguer, M., van der Linden, P.J., Dai, X., et al., 2001. Climate Change 2001: The Scientific Basis. Cambridge University Press, Cambridge.

Huber, O., 1992. El Macizo del Chimantá. Un ensayo ecológico tepuyano. O. Todtman, Caracas.

Huber, O., 1995a. Geographical and physical features. In: Berry, P.E., Holst, B.K., Yatskievych, K. (Eds.), Flora of the Venezuelan Guayana. Introduction, vol. 1. Missouri Botanical Graden Press, St. Louis, pp. 1–61.

Huber, O., 1995b. History of botanical exploration. In: Berry, P.E., Holst, B.K., Yatskievych, K. (Eds.), Flora of the Venezuelan Guayana. Introduction, vol. 1. Missouri Botanical Graden Press, St. Louis, pp. 63–95.

Huber, O., 1995c. Conservation of the Venezuelan Guayana. In: Berry, P.E., Holst, B.K., Yatskievych, K. (Eds.), Flora of the Venezuelan Guayana. Introduction, vol. 1. Missouri Botanical Graden Press, St. Louis, pp. 193–218.

Hunter, M.L., 2007. Climate change and moving species: furthering the debate on assisted colonization. Conserv. Biol. 21, 1356–1358.

Jiménez-Alfaro, B., Colubi, A., González-Rodríguez, G., 2010. A comparison of pointscoring procedures for species prioritization and allocation of seed collection resources in a mountain region. Biodivers. Conserv. 19, 3667–3684.

Jones, H.G., Archer, N., Rotenberg, E., casa, R., 2003. Radiation measurement for plant ecophysiology. J. Exp. Bot. 54, 879–889.

Jump, A.S., Huang, T.-J., Chou, C.-H., 2012. Rapid altitudinal migration of mountain plants in Taiwan and its implication for high altitude biodiversity. Ecography 35, 204–210.

Kelly, A.E., Goulden, M.L., 2008. Rapid shifts in plant distributions with recent climate change. Proc. Natl. Acad. Sci. U.S.A. 105, 11823–11826.

Kimball, S., Gremer, J.R., Angert, A.L., Huxman, T.E., Venable, D.L., 2012. Fitness and physiology in a variable environment. Oecologia 169, 319–329.

Kreyling, J., Wana, D., Beierkuhnlein, C., 2010. Potential consequences of climate warming for tropical plant species in high mountains of Southern Ethiopia. Divers. Distrib. 16, 593–605.

Laurence, W.F., Useche, D.C., Shoo, L.P., Herzog, S.K., Kessler, M., Escobar, F., et al., 2011. Global warming, elevational ranges and the vulnerability of tropical biota. Biol. Conserv. 144, 548–557.

Lenoir, J., Gégout, J.C., Marquet, P.A., De Ruffray, P., Brisse, H., 2008. A significant upward shift in plant species optimum elevation during the 20th century. Science 320, 1768–1771.

Liancourt, P., Boldgiv, B., Song, D.S., Spence, L.A., Helliker, B.R., Petraitis, P.S., et al., 2015. Leaf-trait plasticity and species vulnerability to climate change in a Mongolian steppe. Glob. Change Biol. 21, 3489–3498.

Malcom, J.R., Liu, C., Neilson, R.P., Hansen, L., Hannah, L., 2006. Global warming and extinctions of endemic species from biodiversity hotspots. Conserv. Biol. 20, 538–548.

McLachlan, J.S., Hellmann, J.J., Schwartz, M.W., 2007. A framework for debate of assisted migration in an era of climate change. Conserv. Biol. 21, 297–302.

Medeiros, J.S., Ward, J.K., 2013. Increasing atmospheric [CO_2] from glacial to future concentrations affects drought tolerance via impacts on leaves, xylem and their integrated function. New Phytol. 199, 738–748.

Merilä, J., 2012. Evolution in response to climate change: in pursuit of the missing evidence. BioEssays 34, 811–818.

Myers, N., Mittermeier, R.A., Mittermeier, C.G., Da Fonseca, G.A.B., Kent, J., 2000. Biodiversity hotspots for conservation priorities. Nature 403, 853–858.

Nogué, S., Rull, V., Vegas-Vilarrúbia, T., 2009. Modeling biodiversity loss by global warming in Pantepui, northern South America: projected upward migration and potential habitat loss. Clim. Change 94, 77–85.

Nogué, S., Rull, V., Vegas-Vilarrúbia, T., 2013. Elevational gradients in the neotropical table mountains: patterns of endemism and implications for conservation. Divers. Distrib. 17, 676–687.

Novo, I., Díaz, D., 2007. Informe Final de la Evaluación del Parque Nacional Canaima, Venezuela, como Sitio de Patrimonio Natural de la Humanidad. INPARQUES, Caracas.

Pärtel, M., Kalamees, R., Reier, Ü., Tuvi, E.-L., Roosaluste, E., Vellak, A., et al., 2005. Grouping and prioritization of vascular plant species for conservation: combining natural rarity and management need. Biol. Conserv. 123, 271–278.

Pauli, H., Gottfried, M., Hohenwallner, D., Reieter, K., Grabherr, G., 2005. "Ecological climate impact research in high mountain environments: GLORIA—its roots, its purpose and the long-term perspectives". In: Huber, U. M., Bugmann, H.K.M., Reasoner, M.A. (Eds.), Global Change and Mountain Regions. An Overview of Current Knowledge. Springer, Dordrecht, pp. 383–392.

Pauli, H., Gottfried, M., Reiter, K., Klettner, C., Grabherr, G., 2007. Signals of range expansions and contractions of vascular plants in the high Alps: observations (1994–2004) at the GLORIA master site Schrankogel, Tyrol, Austria. Glob. Change Biol. 13, 147–156.

Pauli, H., Gottfried, M., Dullinger, S., Abdaladze, O., Akhalkatsi, M., Benito Alonso, J.L., et al., 2012. Recent plant diversity changes on Europe's mountain summits. Science 336, 353–355.

Primack, R.B., 2002. Essentials of Conservation Biology. Sinauer, Sunderland.

Ricciardi, A., Simberloff, D., 2009. Assisted colonization is not a viable conservation strategy. Trends Ecol. Evol. 24, 248–253.

Rödder, D., Schlüter, A., Lötters, S., 2010. "Is the 'Lost World' lost? High endemism of amphibians and reptiles on South American tepuis in a changing climate". In: Habel, J.C., Assmann, T. (Eds.), Relict Species: Phylogeography and Conservation Biology. Springer, Berlin, pp. 401–416.

Rull, V., 2009. Microrefugia. J. Biogeogr. 36, 481–484.

Rull, V., Vegas-Vilarrúbia, T., 2006. Unexpected biodiversity loss under global warming in the neotropical Guayana Highlands: a preliminary appraisal. Glob. Change Biol. 12, 1–6.

Rull, V., Vegas-Vilarrúbia, T., 2008. Biopiracy rules hinder conservation efforts. Nature 453, 26.
Rull, V., Vegas-Vilarrúbia, T., 2017. Potential responses of vascular plants from the pristine 'Lost World' of the neotropical Guayana Highlands to global warming: review and perspectives. Front. Plant Sci. 8, 81.
Rull, V., Vegas-Vilarrúbia, T., Nogué, S., Montoya, E., 2008. Bureaucratic obstruction of conservation science in the Guayana Highlands. Conserv. Biol. 22, 508–509.
Rull, V., Vegas-Vilarrúbia, T., Nogué, S., Huber, O., 2009. Conservation of the unique neotropical vascular flora from the Guayana Highlands in the face of global warming. Conserv. Biol. 23, 1323–1327.
Rull, V., Vegas-Vilarrúbia, T., Safont, E., 2016. The lost World's pristinity at risk. Divers. Distrib. 22, 995–999.
Safont, E., Vegas-Vilarrúbia, T., Rull, V., 2012. Use of Environmental Impact Assessment (EIA) tools to set priorities and optimize strategies in biodiversity conservation. Biol. Conserv. 149, 113–121.
Safont, E., Rull, V., Vegas-Vilarrúbia, T., Holst, B.K., Huber, O., Nozawa, S., et al., 2014. Establishing a baseline of plant diversity and endemism on a neotropical mountain summit for future comparative studies assessing upward migration: an approach from biogeography and nature conservation. Syst. Biodiv. 12, 292–314.
Safont, E., Rull, V., Vegas-Vilarrúbia, T., Montoya, E., Huber, O., Holst, B.K., 2016. Late-Holocene vegetation and fire dynamics on the summits of the Guayana Highlands: the Uei-tepui palynological record. Palaeogeogr. Palaeoclimatol. Palaeoecol. 455, 33–43.
Salamí, E., Barrado, C., Pastor, E., 2014. UAV flight experiments applied to remote sensing in vegetated areas. Remote Sensing 6, 11051–11081.
Seddon, P.J., Armstrong, D.P., Soorae, P., Launay, F., Walker, S., Ruiz-Miranda, C.R., et al., 2009. The risks of assisted colonization. Conserv. Biol. 23, 788–789.
Sheldon, K.S., Yang, S., Tewkbury, J.J., 2011. Climate change and community disassembly: impacts of warming on tropical and temperate montane community structure. Ecol. Lett. 14, 1191–1200.
Solomon, S., Qin, D., Manning, M., Chen, Z., Marquis, M., Averyt, K.B., et al., 2007. Climate Change 2007: The Physical Science Basis. Cambridge University Press, Cambridge.
Stocker, T.F., Qin, D., Plattner, G.-K., Tignor, M.M.B., Allen, S.K., Boschung, J., et al., 2013. Climate Change 2013. The Physical Science Basis. Cambridge Univ. Press, Cambridge.
Thuiller, W., Lavorel, S., Araújo, M.B., Sykes, M.T., Colin Prentice, I., 2005. Climate change threats to plant diversity in Europe. Proc. Natl. Acad. Sci. U.S.A. 102, 8245–8250.
Vegas-Vilarrúbia, T., Nogué, S., Rull, V., 2012. Global warming, habitat shifts and potential refugia for biodiversity conservation in the neotropical Guayana Highlands. Biol. Conserv. 152, 159–168.
Vitt, P., Havens, K., Kramens, A.T., Sollenberger, D., Yates, E., 2010. Assisted migration of plants: changes in latitudes, changes in attitudes. Biol. Conserv. 143, 18–27.
Walther, G.-R., 2010. Community and ecosystem responses to recent climate change. Philos. Trans. R. Soc. B 365, 2019–2024.
Xu, Z., Zhou, C., 2006. Combined effects of water stress and high temperature on photosynthesis, nitrogen metabolism and lipid peroxidation of a perennial grass *Leymus chinensis*. Planta 224, 1080–1090.
Xu, Z., Shimizu, H., Ito, S., Yagasaki, Y., Zou, C., Zhou, G., et al., 2014. Effects of elevated CO_2, warming and precipitation change on plant growth, photosynthesis and peroxidation in dominant species from China grassland. Planta 239, 421–435.
Xu, Z., Jiang, Y., Jia, B., Zhou, G., 2016. Elevated CO_2 response of stomata and its dependence on environmental factors. Front. Plant Sci. 7, 657.

Taxonomic Index

Note: Page numbers followed by "*f*" and "*t*" refer to figures and tables, respectively.

A

B

C

D

F

G

H

Q

T

Subject Index

Note: Page numbers followed by "*f*" and "*t*" refer to figures and tables, respectively.

C

H

I

J

FSC
www.fsc.org
MIX
Papier | Fördert
gute Waldnutzung
FSC® C083411

Kolibri 360 GmbH
Ferdinand-Jühlke-Straße 7
99095 Erfurt, Deutschland
produktsicherheit@kolibri360.de